Bacterial Spores

Current Research and Applications

Edited by

Ernesto Abel-Santos

Department of Chemistry
University of Nevada
Las Vegas, NV
USA

Caister Academic Press

Caister Academic Press
Norfolk, UK

www.caister.com

British Library Cataloguing-in-Publication Data
A catalogue record for this book is available from the British Library

ISBN: 978-1-908230-00-3

Description or mention of instrumentation, software, or other products in this book does not imply endorsement by the author or publisher. The author and publisher do not assume responsibility for the validity of any products or procedures mentioned or described in this book or for the consequences of their use.

Cover design adapted from Figure 5.4

Printed and bound in Great Britain

Contents

Contributors v

1 Historical Notes and Introduction to Bacterial Spores 1

Mark Torred, Elias Benjelloun, Framelle Dibala, Ernesto Abel-Santos and Christian Ross

2 Gene Regulation of Sporulation in *Bacillus subtilis* 9

Eduardo A. Robleto, Holly A. Martin, Amber M. Pepper and Mario Pedraza-Reyes

3 Fruiting Body Development in *Myxococcus xanthus*: A Multicellular Developmental Programme that Leads to Sporulation 19

Krista M. Giglio and Anthony G. Garza

4 *Streptomyces* Sporulation 39

Julia P. Swiercz and Marie A. Elliot

5 The Structure and Composition of the Outer Layers of the Bacterial Spore 57

Arthur I. Aronson

6 Mechanisms of Bacterial Spore Survival 73

Mario Pedraza-Reyes, Norma Ramírez-Ramírez, Luz E. Vidales-Rodríguez and Eduardo A. Robleto

7 Initiation of Germination in *Bacillus* and *Clostridium* Spores 89

Graham Christie

8 Germination of *Bacillus anthracis* Spores 107

Jonathan D. Giebel, Katherine A. Carr and Philip C. Hanna

9 Degradation of Spore Peptidoglycan During Germination 121

David L. Popham, Jared D. Heffron and Emily A. Lambert

10 Water and Cations Flux During Sporulation and Germination 143

Daniela Bassi, Fabrizio Cappa and Pier Sandro Cocconcelli

11 The Spore as an Infectious Agent: Anthrax Disease as a Paradigm 169

James M. Vergis, Christy L. Ventura, Louise D. Teel and Alison D. O'Brien

12 Heterogeneity in Bacterial Spore Populations 199
Peter Setlow, Jintao Liu and James R. Faeder

13 Detection of Bacterial Spores: Prospects and Challenges 215
Sainath Rao Shilpakala, Chintamani D. Atreya and Krishna Mohan V. Ketha

14 Properties and Detection Methods of *Bacilli* Spores in Food and in Medical Settings 237
Olga Tarasenko, Pierre Alusta, Sergey Kazakov and Kalle Levon

15 Expression of Recombinant Proteins using *Bacillus subtilis* Spores 263
Luis Carlos de Souza Ferreira and Wolfgang Schumann

Index 273

Colour plate A1

Contributors

Ernesto Abel-Santos
Department of Chemistry
University of Nevada
Las Vegas, NV
USA

ernesto.abelsantos@unlv.edu

Pierre Alusta
Bioinformatics Program
University of Arkansas at Little Rock and University of Arkansas for Medical Sciences
Little Rock, AR
USA

pxalusta@ualr.edu

Arthur I. Aronson
Department of Biological Sciences
Purdue University
Lafayette, IN
USA

aronson@purdue.edu

Chintamani D. Atreya
Section of Cell Biology
Laboratory of Cellular Hematology
Center for Biologics Evaluation and Research, Food and Drug Administration
Bethesda, MD
USA

chintamani.atreya@fda.hhs.gov

Daniela Bassi
Microbiology Institute
C.R.B. Centre for Biotechnological Research
Università Cattolica del Sacro Cuore
Piacenza, Cremona
Italy

daniela.bassi@unicat.it

Elias Benjelloun
Department of Chemistry
University of Nevada
Las Vegas, NV
USA

benjello@unlv.nevada.edu

Fabrizio Cappa
Microbiology Institute
C.R.B. Centre for Biotechnological Research
Università Cattolica del Sacro Cuore
Piacenza, Cremona
Italy

fabrizio.cappa@unicat.it

Katherine A. Carr
United States Army Medical Research Institute of Infectious Diseases
Bacteriology Division
Frederick, MD
USA

carr.katie.a@gmail.com

Graham Christie
Department of Chemical Engineering and Biotechnology
Institute of Biotechnology
University of Cambridge
Cambridge
UK

gc301@cam.ac.uk

Pier Sandro Cocconcelli
Microbiology Institute
C.R.B. Centre for Biotechnological Research
Università Cattolica del Sacro Cuore
Piacenza, Cremona
Italy

pier.cocconcelli@unicat.it

Eramelle Dibala
Department of Chemistry
University of Nevada
Las Vegas, NV
USA

dibalaez@unlv.nevada.edu

Marie A. Elliot
Department of Biology and Institute for Infectious Disease Research
McMaster University
Hamilton, ON
Canada

melliot@mcmaster.ca

James R. Faeder
Department of Computational and Systems Biology
University of Pittsburgh
Pittsburgh, PA
USA

faeder@pitt.edu

Luis Carlos de Souza Ferreira
Department of Microbiology
Institute of Biomedical Sciences
University of São Paulo
Brazil

lcsf@usp.br

Anthony G. Garza
Department of Biology
Syracuse University
Syracuse, NY
USA

agarza@syr.edu

Jonathan D. Giebel
Bayer CropScience
Raleigh, NC
USA

giebel.jonathan@gmail.com

Krista M. Giglio
Department of Biology
Syracuse University
Syracuse, NY
USA

kgiglio@syr.edu

Philip C. Hanna
Department of Microbiology and Immunology
University of Michigan Medical School
Ann Arbor, MI
USA

pchanna@umich.edu

Jared D. Heffron
United States Army Medical Research Institute of Infectious Diseases
Bacteriology Division
Frederick, MD
USA

jared.d.heffron.ctr@us.army.mil

Sergey Kazakov
Department of Chemistry & Physical Sciences
Pace University
Pleasantville, NY
USA

skazakov@pace.edu

Emily A. Lambert
Department of Biology
Bluefield College
Bluefield, VA
USA

elambert@bluefield.edu

Kalle Levon
Department of Chemical and Biological Sciences and Engineering
Polytechnic Institute
New York University
Brooklyn, NY
USA

klevov@poly.edu

Jintao Liu
Departments of Physics and Astronomy and Computational and Systems Biology
University of Pittsburgh
Pittsburgh, PA
USA

physfan.liv@gmail.com

Holly A. Martin
School of Life Sciences
University of Nevada
Las Vegas, NV
USA

mart1015@unlv.nevada.edu

Krishna Mohan V. Ketha
Section of Cell Biology
Laboratory of Cellular Hematology
Center for Biologics Evaluation and Research, Food and Drug Administration
Bethesda, MD
USA

krishna.ketha@fda.hhs.gov

Alison D. O'Brien
Department of Microbiology and Immunology
Uniformed Services University of the Health Sciences
Bethesda, MD
USA

aobrien@usuhs.mil

Mario Pedraza-Reyes
Department of Biology
University of Guanajuato
Guanajuato
Mexico

pedrama@ugto.mx

Amber M. Pepper
School of Life Sciences
University of Nevada
Las Vegas, NV
USA

stiley@unlv.nevada.edu

David L. Popham
Department of Biological Sciences
Virginia Tech
Blacksburg, VA
USA

dpopham@vt.edu

Norma Ramírez-Ramírez
Department of Biology
University of Guanajuato
Guanajuato
Mexico

noramram@hotmail.com

Eduardo A. Robleto
School of Life Sciences
University of Nevada
Las Vegas, NV
USA

eduardo.robleto@unlv.nevada.edu

Christian Ross
School of Life Sciences
University of Nevada
Las Vegas, NV
USA

rossc6@unlv.nevada.edu

Wolfgang Schumann
Institute of Genetics
University of Bayreuth
Bayreuth
Germany

wschumann@uni-bayreuth.de

Peter Setlow
Department of Molecular, Microbial and Structural Biology
University of Connecticut Health Center
Farmington, CT
USA

setlow@nso2.uchc.edu

Sainath Rao Shilpakala
Section of Cell Biology
Laboratory of Cellular Hematology
Center for Biologics Evaluation and Research, Food and Drug Administration
Bethesda, MD
USA

shilpakala.sainathrao@fda.hhs.gov

Julia P. Swiercz
Department of Biology and Institute for Infectious Disease Research
McMaster University
Hamilton, ON
Canada

swiercj@mcmaster.ca

Olga Tarasenko
Department of Biology
University of Arkansas at Little Rock
Little Rock, AR
USA

omtarasenko@ualr.edu

Louise D. Teel
Department of Microbiology and Immunology
Uniformed Services University of the Health Sciences
Bethesda, MD
USA

lteel@usuhs.mil

Mark Torred
Department of Chemistry
University of Nevada
Las Vegas, NV
USA

torredm@unlv.nevada.edu

Christy L. Ventura
Department of Microbiology and Immunology
Uniformed Services University of the Health Sciences
Bethesda, MD
USA

christy.ventura.ctr@usuhs.mil

James M. Vergis
Department of Microbiology and Immunology
Uniformed Services University of the Health Sciences
Bethesda, MD
USA

james.vergis.ctr@usuhs.mil

Luz Elena Vidales-Rodríguez
Department of Biology
University of Guanajuato
Guanajuato
Mexico

luelviro@hotmail.com

Current books of interest

Real-Time PCR in Food Science: Current Technology and Applications	2013
Bacterial Gene Regulation and Transcriptional Networks	2013
Bioremediation of Mercury: Current Research and Industrial Applications	2012
Neurospora: Genomics and Molecular Biology	2013
Rhabdoviruses	2012
Horizontal Gene Transfer in Microorganisms	2012
Microbial Ecological Theory: Current Perspectives	2012
Two-Component Systems in Bacteria	2012
Foodborne and Waterborne Bacterial Pathogens	2012
Yersinia: Systems Biology and Control	2012
Stress Response in Microbiology	2012
Bacterial Regulatory Networks	2012
Systems Microbiology: Current Topics and Applications	2012
Quantitative Real-time PCR in Applied Microbiology	2012
Bacterial Spores: Current Research and Applications	2012
Small DNA Tumour Viruses	2012
Extremophiles: Microbiology and Biotechnology	2012
Bacillus: Cellular and Molecular Biology (Second edition)	2012
Microbial Biofilms: Current Research and Applications	2012
Bacterial Glycomics: Current Research, Technology and Applications	2012
Non-coding RNAs and Epigenetic Regulation of Gene Expression	2012
Brucella: Molecular Microbiology and Genomics	2012
Molecular Virology and Control of Flaviviruses	2012
Bacterial Pathogenesis: Molecular and Cellular Mechanisms	2012
Bunyaviridae: Molecular and Cellular Biology	2011
Emerging Trends in Antibacterial Discovery: Answering the Call to Arms	2011
Epigenetics: A Reference Manual	2011
Metagenomics: Current Innovations and Future Trends	2011
Nitrogen Cycling in Bacteria: Molecular Analysis	2011
The Biology of Paramyxoviruses	2011
Helicobacter pylori	2011
Microbial Bioremediation of Non-metals: Current Research	2011
Lactic Acid Bacteria and Bifidobacteria: Current Progress in Advanced Research	2011
Viruses and Interferon: Current Research	2011
Essentials of Veterinary Parasitology	2011

Full details at www.caister.com

Historical Notes and Introduction to Bacterial Spores

1

Mark Torred, Elias Benjelloun, Eramelle Dibala, Ernesto Abel-Santos and Christian Ross

Bacterial endospores are resistant and dormant structures produced by stressed vegetative cells (Driks, 2002). The resistance and dormancy of spores are remarkable. *Bacillus* spores trapped in amber or salt crystals are easily revived after perhaps millions of years of dormancy (Vreeland *et al.*, 2000). Interestingly, genomic examinations of prehistoric spores show amazing conservation with modern organisms. Owing to their intrinsic resistance, endospores are perfect vehicles for infections. Indeed, diseases caused by spore-forming bacteria are described in ancient literature. In Chapter 9 of the biblical book of Exodus, the fifth plague visited upon Egypt was the death of livestock, and the sixth was boils in humans (Sternbach, 2003). Some biblical scholars believe that the fifth plague of Egypt was animal anthrax and the sixth plague was cutaneous anthrax (Blane, 1890; Hort, 1958). The fifth plague is probably the first written record of a true epizootic outbreak.

In the third book of Georgics, probably published in 29 BC, Virgil described what is believed to be an anthrax outbreak:

> Ghastly Tisphone rages, and ... drives before her Disease and Dread... The rivers and thirsty banks and sloping hills echo to the bleating of flocks and incessant lowing of kine. And now in droves she deals out death, and in the very stalls piles up the bodies, rotting with putrid foulness, till men learn to cover them in earth and bury them in pits.
>
> Guillemin (2004)

Virgil also noted that anthrax spreads to humans by contact with infected animals or animal products:

> For neither might the hides be used, nor could one cleanse the flesh by water or master it by fire. They could not even shear the fleeces, eaten up with sores and filth, nor touch the rotten web. Nay, if any man donned the loathsome garb, feverish blisters and foul sweat would run along his fetid limbs, and not long had he to wait ere the accursed fire was feeding on his stricken limbs.
>
> Guillemin (2004)

This paragraph is even more remarkable since it also provides an early description of the persistence and resistance of anthrax. The Christian priest Eusebius is the first author to use the word 'anthrax' to describe an epidemic:

> The customary rains and showers of the winter season ceased to fall in their wonted abundance upon the earth and an unexpected famine made its appearance, and in addition to this a pestilence, and another severe disease consisting of an ulcer, which on account of its fiery appearance was appropriately called a carbuncle. This, spreading over the whole body, greatly endangered the lives of those who suffered from it; but as it chiefly attacked the eyes, it deprived multitudes of men, women, and children of their sight.

Eusebius' epidemic is generally accepted to have occurred in 312–313 AD. It was associated with drought and severe famine and caused great mortality in cities, although there were even more deaths in the countryside where entire communities were wiped out. Eusebius described the epidemic as a new disease characterized by specific skin ulcers that covered the entire body, and a tendency towards blindness (Relief, 2000). Some authors have concluded that this epidemic was indeed caused by *B. anthracis* since people living in areas where anthrax is endemic develop carbuncles around the eyes (Swiderski, 2004; Zinsser, 1935).

Anthrax epizootics were responsible for devastating domestic livestock losses in Europe from the seventh to the nineteenth centuries. Anthrax caused a European panzootic in 1712, and then settled down to frequent small-scale outbreaks, punctuated by further panzootics in the periods 1757 to 1763, 1774 to 1780, and 1786 to 1793. 'Black Bain' is believed to be the major outbreak of anthrax that in 1613 is reported to have killed over 60,000 people (Wallace, 2008).

Anthrax was clinically and epidemiologically confusing since it affects many animals as well as man, and it produces different clinical pictures in different hosts, as well as different diseases in the same host, depending on its mode of acquisition (Morens, 2003).

By 1769, Nicolas Fournier set up a system dividing human 'charbonous' diseases into two forms: 'spontaneous' and 'contagious.' The spontaneous form seemed to correspond to sporadic cutaneous anthrax, although Fournier speculated that the 'spontaneous' skin lesions were caused by work under the hot sun (Morens, 2003). The second form, contagious charbon, existed in two clinical varieties: (1) one associated with a single cutaneous lesion and a lesser tendency to become systemic and (2) an 'internal' variety that was rapidly fatal. Both of these clinical manifestations occurred only after touching or eating animal products. These two charbon varieties clearly describe human cutaneous anthrax. However, Fournier did not specifically describe inhalational anthrax (Morens, 2003).

In 1881, Louis Pasteur published the development of the first live attenuated anthrax vaccine (Pasteur, 1881). Previously, Pasteur had noticed that some cows developed the anthrax disease more severely than others. Pasteur inoculated cattle with weakened anthrax bacteria. This treatment gave the animal a mild form of anthrax, but made them immune to the most severe forms of anthrax.

Similar to anthrax, tetanus is another endospore bacterial infection known since antiquity. In approximately 380 BC, Hippocrates, the father of western medicine, described a case of tetanus:

> The master of a large ship mashed the index finger of his right hand with the anchor. Seven days later a somewhat foul discharge appeared; then trouble with his tongue – he complained he could not speak properly. The presence of tetanus was diagnosed, his jaws became pressed together, his teeth were locked, and then symptoms appeared on his neck; on the third day opisthotonos appeared with sweating. Six days after the diagnosis was made he died.
>
> Pearce (1996)

Aretaeus of Cappadocia (first century AD) more fully characterized the symptoms of tetanus:

> Tetanus in all its varieties, is a spasm of an exceedingly painful nature, very swift to prove fatal, but neither easy to be removed ... there is tension in a straight line of the whole body, which is unbent and inflexible.
>
> Pearce (1996)

Sternberg reports in the first edition of *Sternberg's Manual of Bacteriology* the production of tetanus disease in animals by the injection of 'gutter-water and earth' (Mcfarland and Ranck, 1899). In 1884, Carle and Rattone showed that tetanus disease in animals could be caused by the introduction of pus from the wounds of infected human patients (Sternberg, 1892). Nicolaier, in 1885, through the introduction of earth into subcutaneous pockets in the tails of mice, guinea pigs and rabbits, generated tetanic wounds and noted the invariable presence of long bacilli in the pus produced. Nicolaier was unable to secure the causative agent in pure culture because he was cultivating the cultures under aerobic conditions (Sternberg, 1892). In 1889, Koch's pupil Katasato was able to isolate the '*Bacillus* of Nicolaier' in pure culture. His isolation technique depended on *C. tetani*'s ability to grow under anaerobic conditions and the capability of its spores to survive prolonged incubation at 80°C (Mcfarland and Ranck, 1899).

Tetanus antitoxin was created in 1890 by von Behring and Kitasano. It did not come into widespread use until decades later, perhaps because of its failure to ameliorate established tetanus infection (Wright, 1958). In 1897, Nocard established tetanus antitoxin's utility as a prophylaxis when he published a description of its use to eliminate tetanus following veterinary surgery (Wright, 1958). It was not until after the increased incidence of tetanus on the battlefields of the First World War, that the use of antitetanic toxin became widespread in humans. Subsequently, morbidity and mortality from tetanus waned (Bruce, 1920).

Endospore-forming bacteria are uniquely adept at finding new niches for infection. Botulism is an often-fatal disease of the nervous system that has probably accompanied mankind only after we began to preserve and store food. Few historical sources document food poisoning in antiquity. The first recorded cases of botulism were observed in Europe in 1735 (Erbguth, 2004). These cases were suspected of being associated with a German sausage and, thus, botulism was named after the Latin word for sausage, '*botulus*'. At the end of the eighteenth century, the number of cases of fatal food poisoning throughout southwest Germany increased as a result of increased consumption of blood sausages. In 1811, Johann Heinrich Ferdinand Autenrieth described the symptom of 'sausage poisoning' (Erbguth, 2004). In 1820, Justinus Kerner summarized the case histories of 76 patients and gave a complete clinical description of what physicians now recognize as botulism. Kerner mentioned interesting clinical detail regarding bodily secretions:

> The tear fluid disappears, the gullet becomes a dead and motionless tube; in all mucous cavities of the human machine the secretion of the normal mucus stands still, from the biggest, the stomach, toward the tear canal and the excretory ducts of the lingual glands. No saliva is secreted. No drop of wetness is felt in the mouth, no tear is secreted anymore. No earwax appears in the auditory canal, also signs of drying out of the Eustachian tube become apparent. No mucus is secreted in the nose; no more sperm is secreted, the testicles decrease. Urination can only be performed by standing and with difficulty. Extreme drying out of the palms, the soles and the eyelids occurs.
>
> Häfner (2009)

On 14 December 1895, an extraordinary outbreak of botulism occurred in Ellezelles (Belgium). After a meal of smoked ham, diners started to feel muscle paralysis and eventually three people died. Based on his observation of this outbreak, Emile von Ermengem published a manuscript recognizing that botulism was an intoxication, that the toxin was heat-labile and produced in preserved food products, and that a specific anaerobic bacterium *Bacillus botulinus* (*Clostridium botulinum*) produces the toxin (Erbguth, 2004). By 1925, outbreaks of botulism from factory-canned foods essentially ceased (Tanner, 1935). This largely coincides with the widespread adoption of standards relating to the amount of time canned foods must be heated.

In contrast to anthrax, tetanus, and botulism, *Clostridium difficile* is a newcomer in the world of sporulating pathogens. *C. difficile* is uniquely associated with antibiotic exposure since, unlike any other enteric pathogen, antibiotic treatment is almost a prerequisite for disease. Hence, prior to the discovery of antibiotics, *C. difficile* infections were rare. The original report of pseudomembranous lesions of the intestinal tract was provided by Finney in 1893 (Bartlett, 1994). Relatively little was written on pseudomembranous colitis for the next 50 years. In the 1950s, with the widespread use of antibiotics, pseudomembranous colitis became a common complication of antibiotic usage. Although *C. difficile* was originally identified in 1935 by Hall and O'Toole, it was not until the 1970s that a causal relationship was established between *C. difficile* and pseudomembranous colitis (Bartlett, 1994).

Although infections by spore-forming bacteria can be traced for thousands of years, discovery of the spore as the infective agent did not occur until late nineteenth century. Perty was probably the first to observe and describe bacterial spores (Perty, 1852). A few years later, Pasteur, incidental to his work on diseases of the silk worm, confirmed the observations of Perty (Pasteur, 1870). In 1872, Ferdinand Cohn established the name *Bacillus* to include rod-shaped bacteria that grow in filaments (Wood, 1995).

In 1874, H. Charlton Bastian found that some bacteria could regrow in boiled cheese and turnip extracts (Bastian, 1872). Bastian believed that since boiling should have sterilized the initial solutions, the new bacteria must have been spontaneously generated by a process he dubbed 'archebiosis'. In contrast, Ferdinand Cohn theorized that there might be a special 'germ' that survived boiling. The bacteria that appeared after boiling were bacillus rods, which he called *Bacillus subtilis*. Cohn found that after growing, many of the bacilli would swell at one end and became filled with oval, strongly refractive little bodies. Cohn believed that these bodies represented a stage in the life cycle of the bacilli and suggested that they were 'real spores, from which new Bacilli may develop'. Since it was known that spores survived high temperature, he concluded that these must also be spores that survived the boiling and then germinated to form bacteria. Indeed, in 1876 both Cohn (1876) and Robert Koch (1876) independently reported that *Bacillus* species are able to form resting forms that are resistant to boiling.

In 1876, J. Tyndall was growing bacteria in either beef broths or hay infusions. In the beef broths, he was always able to achieve sterile solutions by boiling the broth for 5 minutes. However, the hay infusion did not sterilize even after 5 hours of boiling. Citing Ferdinand Cohn, he contended that all bacteria are killed by boiling but some have a spore form that can survive boiling (Tyndall, 1881).

Tyndall ultimately found a way to eradicate the spores though a process that came to be known as 'Tyndallization'. Tyndallization consists of heating the substance to a boil for 15 minutes for 3 days in a row. On the second day most of the spores that survived the first day boiling will have germinated into bacterial cells. These cells will be killed by the second day's heating. The third day kills bacterial cells from late-germinating spores (Mesquita and Teixeira, 1998).

Tyndallization was developed to remove late germinating spores. An unrecognized implication in the tyndallization procedure was that spore populations are not homogeneous. Indeed, spore germination kinetics were later shown to be non-exponential and dormancy may vary between individual spores. The heterogeneity of bacterial endospores will be discussed in Chapter 14.

Tyndallization was an early attempt in preservation of liquid foodstuff by removing spores. The continuous risk of food and environment contamination has triggered a large body of knowledge for detecting and controlling spores. This issue has been further highlighted by the need for detection of bioterrorism threats at early times of exposure to biological agents. Detection of spores is discussed in detail in Chapters 15 and 16.

Numerous theories were evolved concerning the causes and conditions, both cellular and environmental, leading to spore formation (Knaysi and Hillier, 1949). In 1890, Buchner was the first to implicate starvation as the primary factor for sporulation (Buchner, 1890). However, Lehmann and Migula proposed that sporulation was most probably due to accumulation of metabolic byproducts from dividing cells (Migula, 1904). This explanation was refuted by Itano and Neil when they showed that slow-growing cells can readily sporulate (Itano, 1919).

By 1904 microscopic analysis allowed Preisz to give the most complete account of the process of sporulation in *B. anthracis* and divided it in three phases:

> 1. A membrane develops which sets off a clear, fertile, nucleated segment or spore primordium at one pole equal in diameter but not more than one-fifth to one-third the length of the cell. 2. The contents of this primordium increase in density and differentiate to form a 'vorspore' consisting of a more or less elongated, dense, refractive, oval body, surrounded by a less dense layer of cytoplasm. 3. The vorspore migrates to the middle of the rod, the granular content of the parent cell

> disappears and the ripe spore becomes surrounded by a wall derived from the cytoplasmic layer.
>
> (Preisz, 1904)

Preisz considered sporulation as a definite stage in the development of bacilli, and held that the optimum conditions for the initiation of the process are identical with those favouring maximum vegetative development. Matzuschita (1902) recognized lack of nutrient material as a factor, but attached more significance to oxygen accessibility in sporulation. He stated that aerobic bacilli under hydrogen or vacuum never form spores. Hall (1922) found that sporulation of the anaerobic bacilli was most abundant in media characterized by a low content of fermentable carbohydrates. Henrici (1924) showed that spore formation is initiated at the end of the active growth stage of a culture (Morrison, 1930).

In the 1980s a series of seminal works highlighted the importance of sigma factor in sporulation initiation. The cascade of signals required for spore formation was elucidated over the next decades and have now been shown in exquisite details. The issue of the regulation of sporulation will be discussed in Chapter 2.

Koser and Mclelland in 1917 showed that spores from different species are able to resist host metabolism and can be transported from the site of inoculation to different organs (Koser, 1917). Since then, a large body of knowledge has been accumulated for host response to both *Bacilli* and *Clostridia* diseases. The interactions between spore and host will be discussed in greater depth in Chapter 13.

Preisz (1918) also contributed a critical discussion on the finer structure of bacterial spores and changes incident to germination. By means of a delicate, intravital staining technique, he was able to observe the cytological changes occurring in the process of germination, which he described and illustrated in detail. It was not until 1949, that Hills showed the importance of L-alanine and other amino acids as germinants of *Bacillus anthracis* spores (Hills, 1949). Hills also showed that D-alanine was a strong inhibitor of *B. anthracis* spore germination. Church later showed that L-alanine is also a germinant for a number of other *Bacillus* species (Church *et al.*, 1954). Church discovered that spores have alanine racemase activity that is not necessary for spore germination (Church *et al.*, 1954). The issues of spore germination and its relation to anthrax infection will be discussed in Chapters 9 and 10.

In 1952, Powell and Strange showed that germinating spores excrete large quantities of calcium. In this work, they suggested that the high calcium context contribute to the heat resistance of spores (Powell, 1953). Powell and Strange also showed that spores excrete calcium together with a nitrogenous ultraviolet-absorbing compound that was identified as dipicolinic acid. They preliminarily determined that dipicolinic acids was present in every *Bacillus* spore they tested. In Chapter 12, we will discuss our present understanding on cation and water flux in endospores.

Milovidov (1935) first recorded that spores had granules in the periphery, but remained silent on the configuration of the interior of the cell. In 1943, Lewis showed that some spores are enclosed by an exosporium, but did not detect other structures (Lewis, 1934). By 1951, Robinow found that spores have multiple structures including inner membranes, outer membrane, and core (Robinow, 1951). In 1987, Losick and colleagues reported the cloning of the first spore coat proteins (Donovan *et al.*, 1987). The bulk of proteomic research into spore-specific coat proteins has to date been largely conducted on *B. anthracis* and *B. subtilis.* The assembly and composition of spore structures will be reviewed in Chapter 6. The degradation of spore structures during germination will be reviewed in Chapter 11.

Although the majority of proteins within the spore core are also represented during the other phases of the *Bacillus* life cycle, the *Bacillus* spore is enriched in a set of specialized metabolic and protective proteins (Liu *et al.*, 2004). A number of studies in the 1970s showed that the cortex and especially the coat layers are necessary for spore resistance (Driks, 2002). Other resistance properties of the spore result from the cortex peptidoglycan which is responsible for maintaining dehydration (Popham, 2002). Further resistance is conferred by core dehydration, high concentrations of cations/DPA and the presence of DNA-binding proteins. This protein family (SASPs) was discovered in the 1980s. SASPs were determined to coat the spore genome to protect DNA from degradation. The issue of spore resistance mechanism will be discussed in Chapter 7.

Other bacterial genera have been shown to form resistant forms when placed under starvation stress. *Streptomyces* are non-motile Gram-positive bacteria. This genus was probably first described in 1891 by Rossi Doria (Pridham and Lyons, 1961). *Streptomyces* resemble fungi in that they produce branching filaments. The unique developmental programme required for filament formation also lead to the eventual release of resistance spores. The signalling pathways required for *Streptomyces* spore development will be covered in Chapter 5.

In contrast to Firmicutes and *Streptomyces*, *Myxococcus xanthus* is a Gram-negative soil bacterium. *M. xanthus* was first isolated by Roland Thaxter in 1892 (Roland, 1892). Under starvation, *M. xanthus* creates a unique structure (fruiting body) which can contain thousands of individual cells. Because most of the *M. xanthus* life cycle requires cell to cell contacts, this organism's behaviour can be considered as a precursor to multicellularity. The unique properties of *M. xanthus* sporulation will be covered in Chapter 4.

Although a number of endospore producing bacteria can cause diseases, other species have been used in technological applications. Nattō is a traditional Japanese food made from soybeans fermented with *Bacillus subtilis* (Sarkar, 2005). The materials and tools needed to produce nattō have been commonly available in Japan since ancient times; but sources differ about the earliest origin of nattō. one source puts the first use of nattō in the Jōmon period (10 000–300 BC). According to other sources the product may have originated in China during the Zhou Dynasty (1134–246 BC). A third source places the origin of nattō more recently, in the Edo period (1603–1867). One popular theory traces its origin to ancient samurai who used to carry boiled soybeans wrapped in straw as rations during battles (Sarkar, 2005).

Bacterial endospores have been also used as biocides. *B. thuringiensis* was first discovered in 1901 by Shigetane Ishiwatari (Sanahuja *et al.*, 2011). It was not until 1911, that Ernst Berliner isolated *B. thuringiensis* as a disease agent in moth caterpillars (Sanahuja *et al.*, 2011). *B. thuringiensis* spores are sold commercially worldwide as a 'green' pesticide. Tragically, the spores of closely related species, *B. anthracis* have been developed as a bioweapon (Gursky *et al.*, 2003). More recently, the resistant properties of endospores have been used as platforms for developing a series of biotechnology tools. The biotechnological application of spores will be discussed in Chapter 17.

It is clear that endospore-forming bacteria have existed quietly with human-kind for our entire recorded history, often as predator, sometimes as partner. Indeed, they have had a significant influence on all life on earth – the remarkable properties of the endospore allow it to be scattered ubiquitously throughout the environment, where it can bide its time for an appropriate opportunity to germinate and become biologically active. In the contemporary era, through the techniques of the modern biological sciences, the spore-forming bacteria are slowly revealing the biochemical and biophysical tricks that they have accumulated over their long evolutionary history. This book attempts to distil the current state of our knowledge about endospores and how one of our wild bacterial nemeses is being tamed.

References

Bartlett, J.G. (1994). *Clostridium difficile*: History of its role as an enteric pathogen and the current state of knowledge about the organism. Clin. Infect. Dis. J. *4*, 65–72.

Bastian, H.C. (1872). The Beginnings of Life: Being Some Account of Nature, Mode of Origin, and Transformation of Lower Organisms (MacMillan & Co., London).

Blane, H.W. (1890). Anthrax: The disease of the Egyptian plagues. New Orleans Med. Surg. J. *12*, 233–236.

Bruce, D. (1920). Tetanus: analysis of 1458 cases, which occurred in Home Military Hospitals during the years 1914–1918. J. Hygeine *19*, 1–32.

Buchner, H. (1890). Ueber die Ursache der Sporenbildung beim Milzbrandbacillus. Zentr. Bakt. Parasitenk *8*, 1–6.

Church, B.D., Halvorson, H., and Halvorson, H.O. (1954). Studies on spore germination: Its independence from alanine racemace activity. J. Bacteriol. *68*, 393–399.

Donovan, W., Zheng, L.B., Sandman, K., and Losick, R. (1987). Genes enconding spore coat polypeptides from *Bacillus subtilis*. J. Mol. Biol. *196*, 1–10.

Driks, A. (2002). Maximum shields: the assembly and function of the bacterial spore coat. Trends Microbiol. *6*, 251–254.

Cohn, F. (1876). Untersuchungen Uber Bakterien. Beitr. Biol. Pflanz. *2*, 249–276.

Erbguth, F.J. (2004). Historical notes on botulism *Clostridium botulinum*, botulinum toxin, and the idea of the therapeautic use of the toxin. Mov. Disord. *19*, S2–S6.

Guillemin, J. (2004). Anthrax: the investigation of a deadly outbreak. Isis *95*, 330–331.

Gursky, E., Inglesby, T.V., and O'Toole, T. (2003). Anthrax 2001: observations on the medical and public health response. Biosecur. Bioterror. *1*, 97–110.

Häfner, S. (2009). Justinus Kerner's heroic experiments with botulinum toxin. Aktuelle Neurologie *36*, 412–417.

Hills, G.M. (1949). Chemical factors in the germination of spore-bearing aerobes; the effect of yeast extract on the germination of *Bacillus anthracis* and its replacement by adenosine. Biochem. J. *45*, 353–362.

Hort, G. (1958). The plagues of Egypt. Zeitschrift fur die Altesttamentliche Wissenschaft *70*, 48–59.

Itano, A.N.J. (1919). Influence of temperature and hydrogen ion concentration upon the spore cycle of *Bacillus subtilis*. J. Gen. Physiol. *4*, 421–428.

Knaysi, G., and Hillier, J. (1949). Preliminary observations on the germination of the endospore in *Bacillus megatherium* and the structure of the spore coat. J. Bacteriol. *57*, 23–29.

Koch, R. (1876). Untersuchungen ueber Bakterien V. Die Aetiologie der Milzbrand-Krankheit, begruendent auf die Entwicklungsgeschichte des *Bacillus* Anthracis. Beitr. z. Biol. D. Pflanzen *2*, 277–310.

Koser, S.A. (1917). The fate of bacterial spores in the animal body. J. Med. Res. *32*, 259–268.

Lewis, J.M. (1934). Cell inclusions and endospores formation in *Bacillus mycoides*. J. Bacteriol. *28*, 133–144.

Liu, H., Bergman, N.H., Thomason, B., Shallom, S., Hazen, A., Crossno, J., Rasko, D.A., Ravel, D.A., Read, T.D., Peterson, S.N., *et al.* (2004). Formation and composition of the *Bacillus anthracis* endospore. J. Bacteriol. *186*, 164–178.

Mcfarland, J., and Ranck, E.M. (1899). Tetanus toxin and antitoxin. J. Comp. Med. Surg. *20*, 694–699.

Mesquita, J.A.M., and Teixeira, M.A. (1998). Tyndallization of goats' milk in glass bottles. J. Animal Sci./J. Dairy Sci. *76* (Suppl. 1); *81* (Suppl. 1), 21.

Migula, W. (1904). Lafar's Handb. techn. Mykol. *1*, 129–149.

Morens, D.M. (2003). Characterizing a 'new' disease: Epizootic and epidemic anthrax, 1769–1780. Am. J. Pub. Health *93*, 886–893.

Morrison, E.W. (1930). Bacteroa; spores II. A study of bacterial spore germination in relation to environment. J. Bacteriol. *5*, 313–342.

Pasteur, L. (1870). Etudes sur la maladie des vers a soie. (Masson et Cie, Paris, France).

Pasteur, L. (1881). De l'attenuation des virus et de leur retore a la virulence. Comptes Rendus des Sean *90*, 430–435.

Pearce, J.M. (1996). Notes on tetanus (lockjaw). J. Neurol. Neurosurg. Psychiatry *60*, 332.

Perty, M. (1852). Kenntnis Kleinster Lebensformen (Jent & Reinert., Bern)

Popham, D.L. (2002). Specialized peptidoglycan of the bacterial endospore: the inner wall of the lockbox. Cell Mol. Life Sci. *59*, 426–433.

Powell, J.F. (1953). Biochemical changes occurring during the germination ofbacterial spores. Biochem. J. *54*, 205–209.

Preisz, H. (1904). Untersuchugen iiber die feinere structur und die enchwickelung der sporen beim milzrand-Bacillus. Centralbl. f. Bakt 35, 280.

Pridham, T.G., and Lyons, A.J., Jr. (1961). *Streptomyces albus* (rossi-doria) waksman et henrici: taxonomic study of strains labeled *Streptomyces albus*. J. Bacteriol. *81*, 431–441.

Relief, L.C. (2000). Epidemics of the Roman Empire, 27 BC–AD 176. South African Med. J. *90*, 267–272.

Robinow, C.F. (1951). Observations on the structure of *Bacillus* spores. J. Gen. Microbiol. *3*, 439–457.

Roland, T. (1892). On the Myxobacteriaceae, a new order of Schizomyetes. Bot. Gaz. J. Bacteriol. *2*, 394.

Sanahuja, G., Banakar, R., Twyman, R.M., Capell, T., and Christou, P. (2011). *Bacillus thuringiensis*: a century of research, development and commercial applications. Plant Biotechnol. J. *9*, 283–300.

Sarkar, P.K. (2005). Natto. In Microbial Diversity: Current Perspectives and Potential Applications, Sternbach, G., ed. (I.K. International Pvt Ltd, New Delhi).

Sternbach, G. (2003). The history of anthrax. J. Emerg. Med. *24*, 463–467.

Sternberg, G. (1892). A Manual of Bacteriology (William Wood & Co., New York).

Swiderski, R.M. (2004). Anthrax: A History (McFarland & Co., Jefferson, NC).

Tanner, F.W. (1935). Home canning and public health. Am. J. Publ. Health *25*, 301–313.

Tyndall, J. (1881). Essays on the Floating-matter of the Air in Relation to Putrefaction and Infection (Johnson Reprint Corp., New York).

Vreeland, R.H., Rosenzweig, W.D., and Powers, D.W. (2000). Isolation of a 250 million-year-old halotolerant bacterium from a primary salt crystal. Nature *407*, 897–900.

Wallace, R.B. (2008). Anthrax. Wallace/Maxcy-Rosenau-Last Public Health & Preventative Medicine (McGraw-Hill Medical, New York).

Wood, J.B. (1995). The Lactic Acid Bacteria: The Genera of Lactic Acid Bacteria (Chapman & Hall, London).

Wright, G.P. (1958). Discussion on tetanus. Proc. Royal Soc. Med. *51*, 997–1007.

Zinsser, H. (1935). Rats, Lice, and History (Atlantic Monthly Press: Little, Brown, & Co, Boston, MA).

Gene Regulation of Sporulation in *Bacillus subtilis*

2

Eduardo A. Robleto, Holly A. Martin, Amber M. Pepper and Mario Pedraza-Reyes

Abstract

Endospore formation in the bacilli has been used extensively as a proxy for cell growth and differentiation. Research that spans many years has revealed that endospore formation is a complex and elegant process that exquisitely coordinates the development of a dormant spore. The spore structure allows survival when nutrients and moisture are scarce. Once growth-conducive conditions are restored, the process of germination gives rise to vegetative cells. The cell differentiation programme of sporulation is activated upon cessation of growth. Comparison of gene regulons activated during sporulation and stress survival suggest that the mode of regulation of endospore formation is well conserved between bacilli and clostridia.

Introduction

Bacillus and *Clostridium* spp. are used as bacterial models to study cell growth and differentiation in the context of endospore formation. Spores are depressed metabolically and resist environmental insults indefinitely. In growth-conducive environments, germinant receptors in the coat respond to species specific molecules, mediate activation of metabolic functions and the formation of a vegetative cell.

The well-characterized process of endospore development is initiated by environmental conditions that limit growth. The sporulation cycle leads to the formation of two cell compartments: the forespore and the mother cell. The final stage of sporulation results in the lysis of the mother cell and the release of the mature spore. This developmental programme is completed in 8–10 hours in *B. subtilis*.

The formation of the forespore and mother cell is divided into seven developmental states and is temporally controlled by five distinct sigma factors (σ) (Piggot and Coote, 1976; de Hoon *et al.*, 2010). In undifferentiated vegetative cells with two genome molecules (stage 0–I, Fig. 2.1), the transcription factor Spo0A and the sigma factor σ^H initiate the sporulation process by promoting an asymmetric cell septum that parts the cell into the forespore and the mother cell (stage II). The formation of the asymmetric septum also results in incomplete entrapment of a genome molecule into the forespore compartment. A forespore specific sigma factor (σ^F) completes the translocation of the septum-trapped genome molecule into the forespore. The σ^F factor also directs transcription of genes that mediate the engulfment of the forespore into the mother cell (stage III). After engulfment, gene expression required for cortex formation is directed by σ^E from the compartment of the early mother cell. σ^E also triggers the activation of σ^G in the maturing spore (stage IV–V). σ^G coordinates gene expression for DNA protection and repair and late gene expression in the mother cell via activation of σ^K. Transcriptional activity mediated by σ^K directs the formation of spore coat proteins, mother cell lysis, and spore release (stage V–VI) (VII) (Fig. 2.1).

The development of sporulation not only depends on transcriptional control by different sigma factors, but also involves several transcription factors, protein–protein interactions,

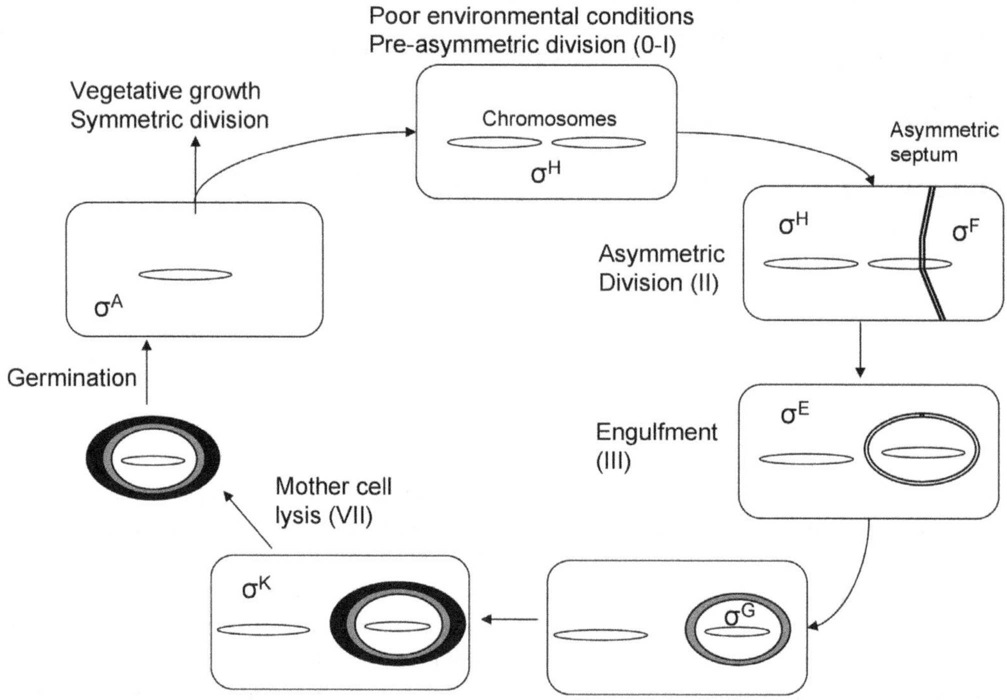

Figure 2.1 Overview of the sporulation programme in *Bacillus subtilis*. Gene regulation is temporally, compartment-specific and mediated by sigma factors (adapted from Hoon *et al.*, 2010).

communication between the mother cell/forespore compartments, and forward feedback loops (FFL) that activate and repress genes in each of the sigma regulons (Fig. 2.2 and Table 2.1). Transcription factors within each of the five sigma factor regulons further refine gene expression by coherent and incoherent FFLs. A coherent FFL is a mechanisms for lengthened or late gene expression within a regulon, whereas an incoherent FFL results in burst or pulse expression pattern. Here, we provide a brief overview of how the process of sporulation is coordinated and describe gene expression in the different cell stages.

The predivisional cell (σ^H)

The sporulation differentiation programme is initiated in vegetative cells that have activated the Spo0A response regulator. In *Bacillus subtilis,* different sensor kinases (*kin* genes) respond to environmental cues and phosphorylate Spo0A through Spo0F and Spo0B (Hoch and Varughese, 2001). The sensor kinases are antagonized by phosphatases that target Spo0A-P or Spo0F-P (Perego *et al.*, 1996; Bongiorni *et al.*, 2007). Active SpoA-P binds DNA (Hoch, 1993) and directly affects transcription of 121 genes. SpoA-P also regulates the transcription of the *sigH* gene (Molle *et al.*, 2003), which in turn affects expression of 87 genes (Britton *et al.*, 2002). The transcriptional activation of *sigH* is mediated by the Spo0A-P repression of AbrB expression, a repressor of *sigH* (Weir *et al.*, 1991).

Characterization of DNA regions affected by Spo0A was based on chromatin precipitation, transcription profiling and electrophoretic mobility shift assays (Molle *et al.*, 2003). To study Spo0A regulation, IPTG was used to induce the production of an active form of Spo0A during growing conditions. The results indicated that Spo0A directly activate transcription or binding DNA regions coding for sporulation and its delay (e.g. *spo0A, spo0F, kin, sdp*), post-exponential phase (*sinR*), antibiotic production and resistance (*yfml*), chromosome remodelling (*racA*), production of inactive mother cell (σ^E) and forepore (σ^F) sigma factors, production of killing factors that target non-sporulating cells (*skf* genes).

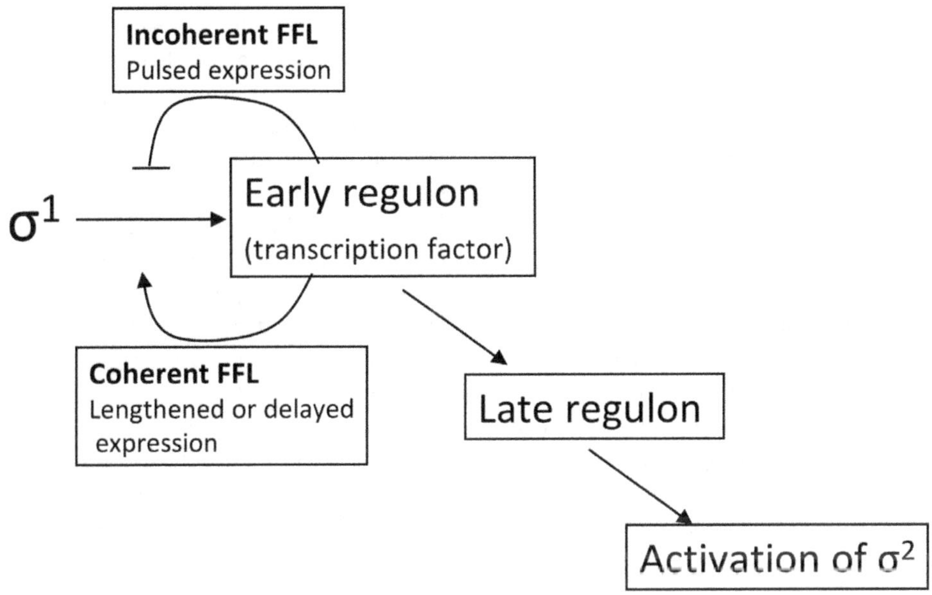

Figure 2.2 Diagram of gene expression during sporulation. A sigma factor regulon is composed of an early regulon, which includes one or more transcription factors. Transcription factors in turn repress, activate or further stimulate the late regulon, which results in activation of a subsequent sigma factor. Repression results in an incoherent feed forward loop (FFL) and produces pulsed expression, whereas activation generates a coherent FFL and lengthened or delayed expression (adapted from Eichenberger *et al.*, 2004).

Table 2.1 Sporulation sigmas and transcriptional factors and the size of their regulons

Cell state	Sigma factor	Transcriptional factor	Number of genes	Reference
Pre septum formation	H	–	87	Britton *et al.* (2002)
		Spo0A	121	Molle *et al.* (2003)
		AbrB	57	Hamon *et al.* (2004)
Forepore early	F		48	Wang *et al.* (2006)
		RsfA	at least 2	Wang *et al.* (2006)
Mother cell early	E	–	262	Eichenberger *et al.* (2004)
		SpoIIID, GerR	136	Eichenberger *et al.* (2004)
Forespore late	G	–	95	Wang *et al.* (2006)
		SpoVT	47	Wang *et al.* (2006)
Mother cell late	K	–	144	Eichenberger *et al.* (2004)
		GerE	91	Eichenberger *et al.* (2004)

Spo0A-dependent direct negative regulation takes place in DNA regions that control DNA replication (*dnaA, holB* and *dnaG*), cell division (*divIVA, ftsX, ftsE*), competence (*com* genes), motility and chemotaxis (*flgB, flgC*). Based on the electrophoretic mobility shift assays a consensus sequence for Spo0A (TGTCGAA) binding (Spo0A box) was further refined in previous studies (Molle *et al.*, 2003).

Further studies indicated that the increase of Spo0A-P concentration in the cell is gradual, depends on a σ^H-feed forward loop, and results in activation or repression of different genes at low and high levels of this factor. Studies using transcriptional fusions indicated that low levels of Spo0A-P activate genes that are not part of the sporulation process, delay sporulation or promote biofilm formation. In contrast, high levels of Spo0A-P promote the formation of an asymmetric septum and sigma factors that will be active in the mother cell and forespore (Hamon *et al.*, 2004; Fujita *et al.*, 2005). It is postulated that the activation of genes at different Spo0A concentrations conduces to a preparatory sequence of events to the formation of spores. Alternatively, these differences in gene activation may be seen as alternative stress responses. Cells commit to sporulation, an energetically consuming programme only when stress conditions (gauged by the level of Spo0A) are extreme. Under mild stress, sporulation is averted and the cell engages in sessile behaviour, antibiotic production, or competence (Fujita and Losick, 2005; Fujita *et al.*, 2005).

As previously mentioned, *abrB* is repressed at low levels of Spo0A which causes an increase in gene transcription of the post-exponential sigma factor *sigH* (Weir *et al.*, 1991; Chumsakul *et al.*, 2010). The σ^H regulon comprises more than 80 genes that control sporulation, competence and several other metabolic functions that alleviate stress (Britton *et al.*, 2002). Some of the genes directly affected by σ^H are also involved in the phosphorylation system that activates Spo0A (*kin* genes) and *spo0A* itself. The genes encoding pre-processed peptides (inactive) for the sigma factors for the early mother cell (*sigE*) and early forespore (*sigF*) are dependent on activation by σ^H. σ^H also activate the *rap* and *phr* genes, which encode for secreted peptide signals that inhibit Rap proteins (Britton *et al.*, 2002). Rap proteins are phosphatases that reduce the level of Spo0A-P and affect the activity of the transcription factor ComA required for the initiation of competence (Bongiorni *et al.*, 2005). Other genes activated by σ^H are involved in cytochrome biosynthesis (*ccdA*), nutrient transport (*gltP*) exopolysaccharide biosynthesis (*yve* genes), antibiotic resistance (*yoeA*) and cell wall metabolism (SpoVG, *yojLM*). Consensus sequences have been proposed for σ^H binding promoters, (A/G)NGGA(A/T)(A/T)(A/A) (–35 region) and (A/G)NNGAAT(A/T)(A/T) (–10 region) (Britton *et al.*, 2002).

A different set of transcriptional events mediate asymmetric septum formation and results in the formation of the mother cell and endospore compartments. While growing cells form the divisional septum at the midcell, the combined SpoA-P and σ^H-dependent transcriptional activation of *ftsZA*, *racA* and *spoIIE* mediates the formation of the septum at the poles of the cell (Ben-Yehuda and Losick, 2002). In vegetative cells, Min and Noc proteins additively control the location of the divisional septum by inhibiting FtsZ and forcing the divisome (proteins that form the septum) to assemble at midcell (Errington *et al.*, 2003). However, entry into post-exponential growth results in an over expression of and FtsZ and FtsA proteins, resulting in bypassing the inhibition by Min proteins. The protein RacA anchors the origins of the chromosomes to the poles resulting in the formation of two stretched nucleoids – the axial filament (Ben-Yehuda *et al.*, 2003). The formation of the axial filament in turn inhibits the assembly of the divisome at the midcell point. The phosphatase SpoIIE locates to the membrane where it interacts with FtsZ and also activates the early endospore sigma factor σ^F (Arigoni *et al.*, 1996). Concomitantly or sequentially, SpoIIE–FtsZ interaction recruits the divisome to the poles (Barak and Youngman, 1996; Feucht *et al.*, 1996).

Early forespore (σ^F)

The asymmetric compartment splits the originating cell into a small and a large compartment. The small and large compartments engage in differential activation of genes because of compartment-specific σ factors. In the forming forespore, σ^F activates transcription of 48 genes (Wang *et al.*, 2006). The consensus sequence $(N)_4$GTATAAA$(N)_{12}$TGGNAATAATA, from promoter mapping experiments, has been proposed for σ^F (Wang *et al.*, 2006).

An interesting feature of the formation of the forespore is only part of the chromosome that is proximal to the origin has been translocated into this compartment. Prior to septum formation, an anti-sigma factor, SpoIIAB, prevents σ^F from binding RNAP (Duncan *et al.*, 1996). The *sigF*, *spoIIAA* and *spoIIAB* genes are arranged in one transcriptional unit located near the terminus of chromosomal replication and its activation is dependent on Spo0A-P and σ^H (Wu *et al.*, 1989). The location of the *spoIIA* operon excludes it from the forepsore compartment and reduces new synthesis of SpoIIAB.

Active σ^F requires release by the anti-sigma factor which is mediated by unphosphorylated anti-anti-sigma factor SpoIIAA. Interestingly, SpoIIAB phosphorylates a SpoIIAA serine residue and prevents σ^F release (Min *et al.*, 1993). It has been proposed that the increase of the concentration of unphosphorylated SpoIIAA that leads to the build up of active σ^F in the forespore compartment is mediated by two mechanisms. The phosphatase SpoIIE (see above) perhaps acts in both the forespore and mother cell compartments. Because of the forespore small volume in relation to the mother cell, the concentration of unphosphorylated SpoIIA will be higher in the smaller compartment. This event may be occurring concomitantly with the degradation of SpoIIAB by the protease activity of ClpC (Pan *et al.*, 2001).

Once σ^F is active, transcription of genes that facilitate different functions takes place in the forespore compartment (Steil *et al.*, 2005; Wang *et al.*, 2006). Genes dependent on σ^F are involved in DNA binding (*rsfA*), synthesis of germination receptors (*gerA* operon), development of spore cortex components (*ssp*, *tlp*), detoxification (*ars–arsC*), DNA repair (*mutTA*) and communication with the mother cell compartment (*spoIIR*, *spoIIQ*, *spoIVB*). Interestingly, there is temporal distribution in gene expression correlated with the chromosomal position of the affected gene. For example, *spoIIR* is expressed early and is origin proximal (321°/360°), whereas *sigG* is expressed late (135°/360°).

Communication between the mother cell and the forespore is first mediated by the expression of *spoIIR* early in the forespore compartment (Karow *et al.*, 1995). SpoIIR localizes to the inner forespore membrane and facilitates proteolysis of the inactive form of σ^E, pro-σ^E (Rubio and Pogliano, 2004). Pro-σ^E is encoded by the gene *spoIIGB* and its expression is Spo0A-dependent (Kenney and Moran, 1987; Satola *et al.*, 1991). Encoded in the same operon as *spoIIGB*, SpoIIGA is membrane-bound and processes pro-σ^E into its active form switching on early mother cell gene expression (LaBell *et al.*, 1987; Stragier *et al.*, 1988). This processing does not occur unless the outside portion of the membrane-bound SpoIIGA interacts with SpoIIR that is produced in the forepore and secreted to the space between compartments. Expression of *spoIIR* is repressed by another σ^F activated gene, *rsfA* which limits the amount of SpoIIR made (Wang *et al.*, 2006). RsfA functions as a fine tuning transcriptional factor within the σ^F regulon and works as a repressor and activator of genes, which results in incoherent and coherent FFLs, respectively (Juan Wu and Errington, 2000; Wang *et al.*, 2006).

Early mother cell (σ^E)

A little more than 260 genes are transcriptionally affected by σ^E. A consensus sequence for σ^E promoters has been put forward [(T/G)N(A/C) TATTT$(N)_{12}$CATANANT]. This sigma factor works in conjunction with the DNA binding proteins, SpoIIID and GerR, and generates coherent and incoherent FFLs (Eichenberger *et al.*, 2004). The σ^E gene expression programme influences engulfment (the forespore is pinched off within the mother cell), coat and cortex formation, as well as mother cell metabolism (amino acid and extracellular proteases synthesis, as well as competence) (Errington, 1993). About half of the genes initially activated by σ^E are repressed by the action of SpoIIID and GerR and are therefore expressed in an incoherent FFL. The genes that are part of incoherent FFls are perhaps required early or in small amounts (Eichenberger *et al.*, 2004). An example of a SpoIIID-dependent incoherent FFL is the control of genes that mediate engulfment. Engulfment is mediated by proteins that are expressed in the predivisonal cell, the early forespore and the early mother cell. σ^E-controlled engulfment genes, *spoIID*, *spoIIM* and *spoIIP*, are essential for this process. These three gene products mediate the degradation of the peptidoglycan layer between the mother cell and the forespore (septal thinning). They are first found at the midpoint in the septum and subsequently move towards the edges of the septum (Frandsen and Stragier, 1995; Abanes-De Mello *et al.*, 2002). Other factors affecting engulfment are SpoIIB and SpoVG which are involved in the initiating steps of engulfment, as well as SpoIIIE which is important for membrane fusion at the end of engulfment. SpoIIIE has the added function of translocating the mother cell-trapped portion of the chromosome into the forepore (Wu and Errington, 1994; Sharp and Pogliano, 2002) and localizes to the mid septum point (Sharp and Pogliano, 1997). Genes of the *spoV* operon in addition to *cwlD* are important for spore cortex formation whereas several *cot* genes, *spoIVA* and *spoVID*, important for coat assembly, are also activated during this stage.

SpoIIID is another transcriptional activator during this stage. Promoter analysis of the genes affected by SpoIIID suggests TNGGACAAGC as its consensus sequence. SpoIIID increases transcription of ten genes, some of which are involved in the activation of σ^K, the late mother cell sigma factor. Communication with the forespore to activate the late forespore sigma factor σ^G is mediated by the transcriptional activation of the *spoIIIA* operon and *yqhV*, which are also repressed by SpoIIID. Inactivation of spoIIIA results in sporulation blocks after engulfment (Illing and Errington, 1991). An inactive form of σ^G (encoded by *spoIIIG*) is made in the forespore under the transcriptional control of σ^F (Sun *et al.*, 1989). The inactive form is processed by the interaction between SpoIIIAH (mother cell) and SpoIIQ (forepsore). This interaction has been proposed to serve as engulfment checkpoint and activate σ^G-dependent gene expression in the forespore (Blaylock *et al.*, 2004).

Late forespore (σ^G)

The last forespore sigma factor recognizes and transcriptionally activates G(C/T)ATAAAT(N)11(T/A)NG(N)$_2$CA(T/A)A(A/C)TA(A/T) sequences (Wang *et al.*, 2006). More than 90 genes are activated by σ^G, which include products for the transcriptional factor SpoVT. SpoVT acts as an activator and repressor for half of the genes in the σ^G regulon and forms coherent and incoherent FFLs. Genes activated in this regulon mediate continued morphogenesis of the spore cortex, the spores' resistance properties and germination functions (*ger* genes), as well as activation of the last sigma factor of the sporulation programme in the mother cell. Spore resistance characteristics are provided by protecting DNA from damage. This is accomplished by the synthesis of small acid-soluble proteins (SASPs), which remodel the forespore nucleoid into a ring (Pogliano *et al.*, 1995; Setlow, 1995). Proteins that repair double stranded breaks via non-homologous end-joining are expressed in this regulon and pack the spore with factors that prevent genotoxic events, particularly those induced by UV damaged (Slieman and Nicholson, 2000; Slieman *et al.*, 2000; Weller *et al.*, 2002).

Communication with the mother cell and activation of σ^K is mediated through the SpoVT-dependent activation of SpoIVB, a protease (Hoa *et al.*, 2002), which cleaves SpoIVFA. Activation of σ^K comes via proteolysis of the N-terminus of pro-σ^K and its premature activation results in abnormal spores (Cutting *et al.*, 1990). A protein complex that contains SpoIVFA, SpoIVFB and BofC is embedded in the outer forespore membrane (Rudner and Losick, 2002). Cleavage of SpoIVFA by SpoIVB dissociates the tripolypeptide complex and relieves inhibition of SpoIVFB-dependent proteolysis of pro-σ^K, thereby activating the last sigma factor in sporulation.

Late mother cell (σ^K)

A final step required for the activation of σ^K is the excision of *skin*, a DNA element that contains several genes (Stragier *et al.*, 1989). Such element excision is accomplished by the activity of SpoIVCA, a DNA recombinase, and translationally fuses the gene products of SpoIVCB and SpoIIIC. The *spoIVCA* gene is under the control of σ^E and SpoIIID, is located within the *skin* element which is subsequently lost from the mother cell chromosome but maintained in the forespore chromosome. The translational fusion is under the control of the previous mother cell transcription programme and generates the gene product for pro-σ^K. The σ^K factor [GCACA(N)$_{16}$ GCATATGNTA] and its cognate transcription element GerE affect gene expression of 144 genes (Eichenberger *et al.*, 2004). This regulon mediates the final steps in formation of the spore coat (*cot* genes and polysaccharide synthesis genes). The activation of the *spoVA* operon results in transfer of dipicolinic acid from the mother to the forespore. Defects on *spoVA* genes increases the water content of the core of the spore, which renders spores susceptible to heat and destabilizes the spore protoplast (Tovar-Rojo *et al.*, 2002). The last step in this stage is mother cell lysis which is controlled by σ^K and GerE as part of a coherent FFL that expresses *cwlC* and *cwlH* whose products are cell wall hydrolases (Eichenberger *et al.*, 2004).

Concluding remarks

Here we provide a brief overview of the main features of the programme for gene expression during the development of endospores in *Bacillus subtilis*. The formation of an endospore is accomplished by compartment-specific and temporal gene activation. The controlling elements for each of the developmental stages are the synthesis and activation of one predivisional-cell and four compartment-specific sigma factors. The regulons for each of the sporulating sigmas are fine-tuned by transcription factors that result in the development of forward feedback loops that effect the synthesis of gene products in a pulse or burst manner (incoherent loop), lengthened or delayed expression (coherent loops). A significant feature in this process is communication between compartments and the existence of check points that allow progression of the sporulation programme. This is accomplished by establishing protein–protein interactions between the mother and forespore compartment. Recent analysis of multiple endospore-forming bacterial species indicates high conservation of the regulatory network (sigma factors and FFLs) compared with conservation for the presence of sporulation genes. This suggests that the study of regulation of a cell differentiation pathway in a model system applies to closely related species (de Hoon *et al.*, 2010).

References

Abanes-De Mello, A., Sun, Y.L., Aung, S., and Pogliano, K. (2002). A cytoskeleton-like role for the bacterial cell wall during engulfment of the *Bacillus subtilis* forespore. Genes Dev. *16*, 3253–3264.

Arigoni, F., Duncan, L., Alper, S., Losick, R., and Stragier, P. (1996). SpoIIE governs the phosphorylation state of a protein regulating transcription factor sigma F during sporulation in *Bacillus subtilis*. Proc. Natl. Acad. Sci. U.S.A. *93*, 3238–3242.

Barak, I., and Youngman, P. (1996). SpoIIE mutants of *Bacillus subtilis* comprise two distinct phenotypic classes consistent with a dual functional role for the SpoIIE protein. J. Bacteriol. *178*, 4984–4989.

Ben-Yehuda, S., and Losick, R. (2002). Asymmetric cell division in *B. subtilis* involves a spiral-like intermediate of the cytokinetic protein FtsZ. Cell *109*, 257–266.

Ben-Yehuda, S., Rudner, D.Z., and Losick, R. (2003). RacA, a bacterial protein that anchors chromosomes to the cell poles. Science *299*, 532–536.

Blaylock, B., Jiang, X., Rubio, A., Moran, C.P., Jr., and Pogliano, K. (2004). Zipper-like interaction between proteins in adjacent daughter cells mediates protein localization. Genes Dev. *18*, 2916–2928.

Bongiorni, C., Ishikawa, S., Stephenson, S., Ogasawara, N., and Perego, M. (2005). Synergistic regulation of competence development in *Bacillus subtilis* by two Rap-Phr systems. J. Bacteriol. *187*, 4353–4361.

Bongiorni, C., Stoessel, R., and Perego, M. (2007). Negative regulation of *Bacillus anthracis* sporulation by the Spo0E family of phosphatases. J. Bacteriol. *189*, 2637–2645.

Britton, R.A., Eichenberger, P., Gonzalez-Pastor, J.E., Fawcett, P., Monson, R., Losick, R., and Grossman, A.D. (2002). Genome-wide analysis of the stationary-phase sigma factor (sigma-H) regulon of *Bacillus subtilis*. J. Bacteriol. *184*, 4881–4890.

Chumsakul, O., Takahashi, H., Oshima, T., Hishimoto, T., Kanaya, S., Ogasawara, N., and Ishikawa, S. (2010). Genome-wide binding profiles of the *Bacillus subtilis* transition state regulator AbrB and its homolog Abh reveals their interactive role in transcriptional regulation. Nucleic Acids Res. *39*, 414–428.

Cutting, S., Oke, V., Driks, A., Losick, R., Lu, S., and Kroos, L. (1990). A forespore checkpoint for mother cell gene expression during development in *B. subtilis*. Cell *62*, 239–250.

Duncan, L., Alper, S., and Losick, R. (1996). SpoIIAA governs the release of the cell-type specific transcription factor sigma F from its anti-sigma factor SpoIIAB. J. Mol. Biol. *260*, 147–164.

Eichenberger, P., Fujita, M., Jensen, S.T., Conlon, E.M., Rudner, D.Z., Wang, S.T., Ferguson, C., Haga, K., Sato, T., Liu, J.S., *et al.* (2004). The program of gene transcription for a single differentiating cell type during sporulation in *Bacillus subtilis*. PLoS Biol. 2, e328.

Errington, J. (1993). *Bacillus subtilis* sporulation: regulation of gene expression and control of morphogenesis. Microbiol. Rev. *57*, 1–33.

Errington, J., Daniel, R.A., and Scheffers, D.J. (2003). Cytokinesis in bacteria. Microbiol. Mol. Biol. Rev. *67*, 52–65, table of contents.

Feucht, A., Magnin, T., Yudkin, M.D., and Errington, J. (1996). Bifunctional protein required for asymmetric cell division and cell-specific transcription in *Bacillus subtilis*. Genes Dev. *10*, 794–803.

Frandsen, N., and Stragier, P. (1995). Identification and characterization of the *Bacillus subtilis spoIIP* locus. J. Bacteriol. *177*, 716–722.

Fujita, M., Gonzalez-Pastor, J.E., and Losick, R. (2005). High- and low-threshold genes in the Spo0A regulon of *Bacillus subtilis*. J. Bacteriol. *187*, 1357–1368.

Fujita, M., and Losick, R. (2002). An investigation into the compartmentalization of the sporulation transcription factor sigmaE in *Bacillus subtilis*. Mol. Microbiol *43*, 27–38.

Fujita, M., and Losick, R. (2005). Evidence that entry into sporulation in *Bacillus subtilis* is governed by a gradual increase in the level and activity of the master regulator Spo0A. Genes Dev. *19*, 2236–2244.

Hamon, M.A., Stanley, N.R., Britton, R.A., Grossman, A.D., and Lazazzera, B.A. (2004). Identification of AbrB-regulated genes involved in biofilm formation by *Bacillus subtilis*. Mol. Microbiol. *52*, 847–860.

Hoa, N.T., Brannigan, J.A., and Cutting, S.M. (2002). The *Bacillus subtilis* signaling protein SpoIVB defines a new family of serine peptidases. J. Bacteriol. *184*, 191–199.

Hoch, J.A. (1993). Regulation of the phosphorelay and the initiation of sporulation in *Bacillus subtilis*. Annu. Rev. Microbiol. *47*, 441–465.

Hoch, J.A., and Varughese, K.I. (2001). Keeping signals straight in phosphorelay signal transduction. J. Bacteriol. *183*, 4941–4949.

de Hoon, M.J., Eichenberger, P., and Vitkup, D. (2010). Hierarchical evolution of the bacterial sporulation network. Curr. Biol. *20*, R735–745.

Illing, N., and Errington, J. (1991). Genetic regulation of morphogenesis in *Bacillus subtilis*: roles of sigma E and sigma F in prespore engulfment. J. Bacteriol. *173*, 3159–3169.

Juan Wu, L., and Errington, J. (2000). Identification and characterization of a new prespore-specific regulatory gene, *rsfA*, of *Bacillus subtilis*. J. Bacteriol. *182*, 418–424.

Karow, M.L., Glaser, P., and Piggot, P.J. (1995). Identification of a gene, *spoIIR*, that links the activation of sigma E to the transcriptional activity of sigma F during sporulation in *Bacillus subtilis*. Proc. Natl. Acad. Sci. U.S.A. *92*, 2012–2016.

Kenney, T.J., and Moran, C.P., Jr. (1987). Organization and regulation of an operon that encodes a sporulation-essential sigma factor in *Bacillus subtilis*. J. Bacteriol. *169*, 3329–3339.

LaBell, T.L., Trempy, J.E., and Haldenwang, W.G. (1987). Sporulation-specific sigma factor sigma 29 of *Bacillus subtilis* is synthesized from a precursor protein, P31. Proc. Natl. Acad. Sci. U.S.A. *84*, 1784–1788.

Min, K.T., Hilditch, C.M., Diederich, B., Errington, J., and Yudkin, M.D. (1993). Sigma F, the first compartment-specific transcription factor of *B. subtilis*, is regulated by an anti-sigma factor that is also a protein kinase. *Cell 74*, 735–742.

Molle, V., Fujita, M., Jensen, S.T., Eichenberger, P., Gonzalez-Pastor, J.E., Liu, J.S., and Losick, R. (2003). The Spo0A regulon of *Bacillus subtilis*. Mol. Microbiol. *50*, 1683–1701.

Pan, Q., Garsin, D.A., and Losick, R. (2001). Self-reinforcing activation of a cell-specific transcription factor by proteolysis of an anti-sigma factor in *B. subtilis*. Mol. Cell *8*, 873–883.

Perego, M., Glaser, P., and Hoch, J.A. (1996). Aspartyl-phosphate phosphatases deactivate the response regulator components of the sporulation signal transduction system in *Bacillus subtilis*. Mol. Microbiol. *19*, 1151–1157.

Piggot, P.J., and Coote, J.G. (1976). Genetic aspects of bacterial endospore formation. Bacteriol. Rev. *40*, 908–962.

Pogliano, K., Harry, E., and Losick, R. (1995). Visualization of the subcellular location of sporulation proteins in *Bacillus subtilis* using immunofluorescence microscopy. Mol. Microbiol. *18*, 459–470.

Rubio, A., and Pogliano, K. (2004). Septal localization of forespore membrane proteins during engulfment in *Bacillus subtilis*. Embo. J. *23*, 1636–1646.

Rudner, D.Z., and Losick, R. (2002). A sporulation membrane protein tethers the pro-sigmaK processing enzyme to its inhibitor and dictates its subcellular localization. Genes Dev. *16*, 1007–1018.

Satola, S., Kirchman, P.A., and Moran, C.P., Jr. (1991). Spo0A binds to a promoter used by sigma A RNA polymerase during sporulation in *Bacillus subtilis*. Proc. Natl. Acad. Sci. U.S.A. *88*, 4533–4537.

Setlow, P. (1995). Mechanisms for the prevention of damage to DNA in spores of *Bacillus* species. Annu. Rev. Microbiol. *49*, 29–54.

Slieman, T.A., and Nicholson, W.L. (2000). Artificial and solar UV radiation induces strand breaks and cyclobutane pyrimidine dimers in *Bacillus subtilis* spore DNA. Appl. Environ. Microbiol. *66*, 199–205.

Slieman, T.A., Rebeil, R., and Nicholson, W.L. (2000). Spore photoproduct (SP) lyase from *Bacillus subtilis* specifically binds to and cleaves SP (5-thyminyl-5,6-dihydrothymine) but not cyclobutane pyrimidine dimers in UV-irradiated DNA. J. Bacteriol. *182*, 6412–6417.

Steil, L., Serrano, M., Henriques, A.O., and Volker, U. (2005). Genome-wide analysis of temporally regulated and compartment-specific gene expression in sporulating cells of *Bacillus subtilis*. Microbiology *151*, 399–420.

Stragier, P., Bonamy, C., and Karmazyn-Campelli, C. (1988). Processing of a sporulation sigma factor in *Bacillus subtilis*: how morphological structure could control gene expression. Cell *52*, 697–704.

Stragier, P., Kunkel, B., Kroos, L., and Losick, R. (1989). Chromosomal rearrangement generating a composite gene for a developmental transcription factor. Science *243*, 507–512.

Sun, D.X., Stragier, P., and Setlow, P. (1989). Identification of a new sigma-factor involved in compartmentalized gene expression during sporulation of *Bacillus subtilis*. Genes Dev. *3*, 141–149.

Tovar-Rojo, F., Chander, M., Setlow, B., and Setlow, P. (2002). The products of the spoVA operon are involved in dipicolinic acid uptake into developing spores of *Bacillus subtilis*. J. Bacteriol. *184*, 584–587.

Wang, S.T., Setlow, B., Conlon, E.M., Lyon, J.L., Imamura, D., Sato, T., Setlow, P., Losick, R., and Eichenberger, P. (2006). The forespore line of gene expression in *Bacillus subtilis*. J. Mol. Biol. *358*, 16–37.

Weir, J., Predich, M., Dubnau, E., Nair, G., and Smith, I. (1991). Regulation of *spo0H*, a gene coding for the *Bacillus subtilis* sigma H factor. J. Bacteriol. *173*, 521–529.

Weller, G.R., Kysela, B., Roy, R., Tonkin, L.M., Scanlan, E., Della, M., Devine, S.K., Day, J.P., Wilkinson, A., d'Adda di Fagagna, F., *et al.* (2002). Identification of a DNA nonhomologous end-joining complex in bacteria. Science 297, 1686–1689.

Wu, J.J., Howard, M.G., and Piggot, P.J. (1989). Regulation of transcription of the *Bacillus subtilis spoIIA* locus. J. Bacteriol. *171*, 692–698.

Fruiting Body Development in *Myxococcus xanthus*: A Multicellular Developmental Programme that Leads to Sporulation

3

Krista M. Giglio and Anthony G. Garza

Abstract

The myxobacteria are Gram-negative spore formers. Although the spores of myxobacteria and those of *Bacillus* species, which are Gram positive bacteria, have many properties in common, the developmental processes that lead to their formation are fundamentally different. In this chapter, we give a general overview of sporulation in the model myxobacterium *Myxoccocus xanthus*. We discuss the developmental process that yields *M. xanthus* spores, the structural and biochemical properties of the spores and the genes that are associated with sporulation in this organism, and we compare each to its counterpart in *Bacillus*.

Introduction

Multicellularity, once thought to be the domain of eukaryotes, has been uncovered in a variety of prokaryotic organisms, including the heterocyst-forming members of the cyanobacteria (reviewed by Adams, 2000) and the numerous bacterial species that form biofilms (reviewed by Lopez *et al.*, 2010). In these processes, cooperative behaviour is initiated among a population of cells in response to some environmental change. Sporulation in the bacterium *Myxococcus xanthus* is one such process. *M. xanthus* is a member of the myxobacteria, which is a group of organisms that was first studied in the late 1800s because they were believed to have an unusual (for prokaryotes) social ecology. First recognized by mycologists, myxobacteria were once thought to be fungi rather than bacteria due to the elaborate multicellular fruiting bodies they produce (Pfister, 1993). Myxobacteria are Gram-negative, rod-shaped bacteria. Many species of myxobacteria are social predators that dine on prey bacteria collectively as a swarm. When prey bacteria are scarce, myxobacteria launch a multicellular developmental programme that directs a series of coordinated cells movements. These orchestrated cell movements lead to the formation of aggregation centres and eventually to the formation of elaborate fruiting body structures that contain thousands of cells. Within these fruiting bodies, cells differentiate into metabolically dormant, stress-resistant myxospores (spores of the myxobacteria). Spore formation in myxobacteria differs from endospore formation in *Bacillus* sp., which is the best-characterized model of bacterial sporulation, in two fundamental ways: (1) myxobacterial sporulation is a group event that is dependent upon a multicellular developmental process, whereas *Bacillus* sporulation is an individual cell event; (2) myxobacterial cells are physically converted into spherical-shaped spores, whereas endospores are built from scratch inside the protective environment of a mother cell. Because it is one of the few myxobacteria to have its genome sequenced and a host of tools are available to analyse its cellular physiology and behaviour at a genetic level (Murphy and Garza, 2008), *M. xanthus* is an excellent model for studying myxobacterial sporulation. Indeed, *M. xanthus* is by far the best characterized system of myxobacterial sporulation and the focus of this chapter.

Starvation-induced sporulation is dependent upon the formation of multicellular fruiting bodies

The distinguishing feature of myxospore formation in natural settings is that it requires the efforts of a community of cells: *M. xanthus* cells do not initiate the sporulation process until they are inside a fruiting body (for a review, see Kaiser, 2003). Therefore, one can view the formation of fruiting bodies as a series of morphological steps that lead to sporulation. Fruiting body development is quite complex, requiring cells to carefully monitor the ambient nutrient levels, to detect and respond to intracellular and extracellular signals, and to regulate extensive changes in gene expression both temporally and spatially. Below is a summary of these important processes and their regulation, with a focus on late events that are more directly connected to sporulation.

Overview of fruiting body development

Fruiting body development proceeds through a series of stages: the vegetative growth to fruiting body development transition, pre-aggregation, aggregation, and sporulation. The transition from growth to development occurs when starvation for any amino acid, carbon, phosphorus or energy is detected, triggering the intracellular accumulation of ppGpp(p) (Manoil and Kaiser, 1980) and the initiation of a stringent response (see Cashel *et al.*, 1996, for a review of the stringent response). The pre-aggregation stage of development occurs after starvation is encountered and a stringent response has been launched, and it continues until the start of aggregation. The hallmark of the pre-aggregation stage of development is the accumulation of A-signal. A-signal is a diffusible quorum signal that allows *M. xanthus* cells to ascertain the local population density (Kuspa *et al.*, 1992a). A-signal accumulation is important for expression of many early developmental genes (Kuspa *et al.*, 1986) and for aggregation (Hagen *et al.*, 1978; Kuspa *et al.*, 1986; Kuspa and Kaiser, 1989; Cho and Zusman, 1999; Garza *et al.*, 2000). The aggregation and sporulation stages of development, and the changes in gene expression that are associated with these events are controlled by a second cell–cell signal called C-signal (Hagen *et al.*, 1978; Kroos *et al.*, 1986; Kroos and Kaiser, 1987). A diagram showing the different stages of the *M. xanthus* developmental cycle relative to the developmental timeline, and the intracellular and extracellular signals that control these developmental stages is presented in Fig. 3.1.

Early developmental events: ppGpp(p) and A-signal accumulation

When *M. xanthus* cells encounter low ambient nutrient levels, they begin to synthesize the intracellular starvation signal guanosine penta- and tetra-phosphate [ppGpp(p)] (Manoil and Kaiser, 1980). Harris *et al.* (1998) found that a strain carrying a mutation in the *relA* gene, which codes for the sole *M. xanthus* ppGpp(p) synthetase (Goldman *et al.*, 2006), is defective in ppGpp(p) production, aggregation and sporulation. Indeed, the accumulation of ppGpp(p) is necessary and sufficient to initiate the transition from vegetative growth to development (Singer and Kaiser, 1995; Harris *et al.*, 1998). Presumably, ppGpp(p) accumulation tells *M. xanthus* cells that the bout of starvation is severe enough and long enough to justify expending their remaining energy reserves on fruiting body development and sporulation.

As a consequence of ppGpp(p) accumulation, cells start to produce the quorum-indicating A-signal, which is hallmark of the pre-aggregation stage of development. It has been proposed that A-signal, which is a mixture of amino acids and peptides, is generated via proteolysis (Plamann *et al.*, 1992; Kuspa *et al.*, 1992b). A-signal is generated in proportion to the number of starving cells (Kuspa *et al.*, 1992b), indicating that it provides cell density information. Since A-signal accumulation occurs only after production of ppGpp(p), it is an assessment of how widespread starvation is among the local population. If there are enough starving cells present to build a fruiting body, then cells will begin the movements needed to form aggregation centres. Five genes are known to be important for A-signal production: *asgA*, *asgB*, *asgC*, *asgD*, and *asgE* (Kuspa and Kaiser, 1989; Mayo and Kaiser, 1989; Cho and Zusman, 1999; Garza *et al.*, 2000). Most of the *asg* genes code for regulatory proteins (Plamann *et al.*, 1994; Davis *et al.*, 1995; Cho and Zusman, 1999). Presumably,

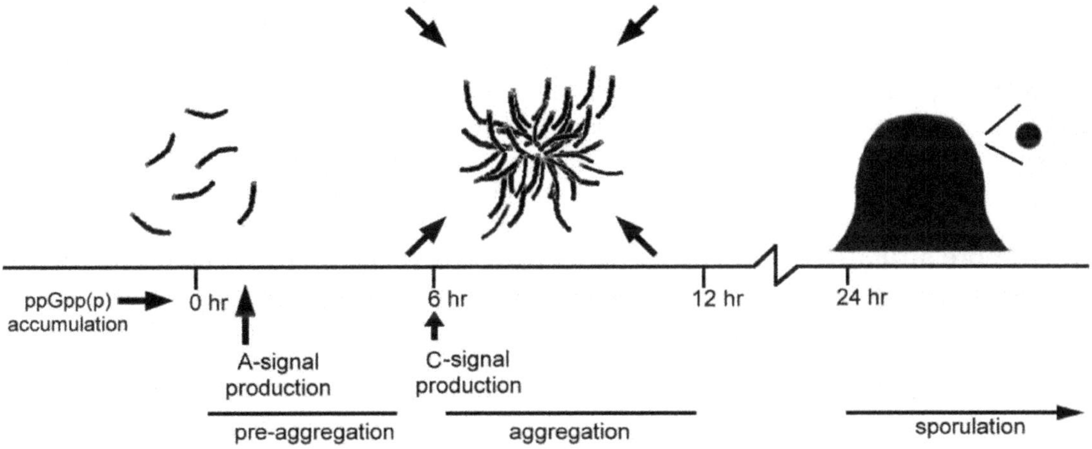

Figure 3.1 The *M. xanthus* developmental cycle. The morphological changes that occur during development are shown above the developmental timeline. The lines above pre-aggregation and aggregation indicate the approximate extent of these stages (1–5 hours and 6–12 hours, respectively) and the arrow above sporulation indicates that this stage begins at about 24 hours and continues for several days (24–120 hours). The arrows indicate that (*p*)ppGpp accumulation is required to make the transition from growth to development, that A-signal is required in the early part of pre-aggregation or that C-signal is required at the start of aggregation.

these proteins regulate genes that are directly involved in production of A-signal. A-signal accumulation is needed for full expression of important early developmental genes, including *fruA* (Kuspa *et al.*, 1986; Ogawa *et al.*, 1996; Ellehauge *et al.*, 1998), which is a component in the later-functioning C-signal pathway (Ellehauge *et al.*, 1998). Therefore, once A-signal accumulates, it passes control of development to the C-signalling system.

Late developmental events: aggregation and sporulation

The aggregation and sporulation phases of development are controlled by C-signal, a short-range cell–cell signal that is associated with the cell surface (Shimkets and Rafiee, 1990; Kim and Kaiser, 1990a). Transmission of C-signal requires cell–cell contact and cell alignment (Kim and Kaiser, 1990a–c). C-signal is the proteolytic cleavage product of the CsgA protein (Shimkets *et al.*, 1983; Kim and Kaiser, 1990a; Lobedanz and Søgaard-Andersen, 2003). CsgA acts in a concentration-dependent manner such that increasing levels of this protein bring about different morphological changes during fruiting body development: low levels of CsgA are important for expression of early C-signal dependent genes and for aggregation, and higher levels of CsgA activate expression of late C-signal dependent genes and initiate myxospore formation (Kim and Kaiser, 1991; Li *et al.*, 1992).

During *M. xanthus* development, cells inside fruiting bodies become spherical spores that are highly resistant to stresses such as heat, sonication, desiccation and UV radiation (Sudo and Dworkin, 1969), whereas cells outside the fruiting bodies (peripheral rods) remain rod-shaped and, presumably, less stress resistant (O'Connor and Zusman, 1991a). It has been proposed that C-signal is responsible for this spatial patterning of sporulation (Sager and Kaiser, 1993; Julien *et al.*, 2000). In particular, it has been suggested that as cells start to form fruiting bodies and their local density increases, they have a greater opportunity to make cell–cell contacts and, therefore, to transmit C-signal than peripheral rods. Cells inside fruiting bodies also synthesize more CsgA than peripheral rods. Consequently, the levels of CsgA in fruiting body cells are amplified to promote more C-signalling. Both of these events, and possibly cell alignment, lead to relatively high levels of C-signalling inside fruiting bodies. This in turn leads to the differential expression of sporulation genes: expression of several C-signal dependent genes that are important for sporulation is relatively high in cells within fruiting bodies. In this way, *M. xanthus* cells can spatially regulate the expression of sporulation genes, and thus spatially restrict sporulation to the cells within the fruiting bodies.

Myxospore structure

In order to form spores, *M. xanthus* cells must undergo a cell shape change: they must go from rod-shaped to spherical-shaped cells (Dworkin and Voelz, 1962; Dworkin and Gibson, 1964). Because the acquisition of optical refractility occurs soon after cells take on a spherical shape, it is often used as a marker for the progress of myxospore differentiation. Physically, spore differentiation is accomplished by a progressive shortening at the long axes of the cell, which makes it become ovoid, and then shortly thereafter, spherical-shaped (Voelz and Dworkin, 1962; Dworkin and Sadler, 1966a). Subsequently, the spherical-shaped cells mature into stress-resistant spores and, in nature, the mature spores remain together inside fruiting bodies until they encounter prey bacteria, which trigger spore germination and eventually group feeding. Spore development in *M. xanthus* and other myxobacteria is distinct from endospore development in Gram-positive bacteria such as *Bacillus* sp. In particular, endospores are built from scratch inside the protective environment of a mother cell and then released (Kroos, 2007). Although spore formation in *M. xanthus* is distinct from endospore formation, mature myxospores do share heat, ultraviolet radiation, and desiccation resistance properties with their endospore counterparts (Setlow, 1995, 2006; Sudo and Dworkin, 1969, 1973). However, myxospores are less resistant than endospores in all these areas (Sudo and Dworkin, 1969; Gould and Dring, 1975), which might result from specific differences in structure, or differences in protein content. The gross morphological structures found in myxospores are also similar to those found in Gram-positive endospores, and these structures are arranged in similar pattern in both spore types. From the inside out, myxospores contain a core, an inner membrane and outer membrane, a cortex and a coat (see Fig. 3.2). The myxospore core contains two copies of the *M. xanthus* chromosome (Tzeng and Singer, 2005). Presumably, having two chromosomes ensures that one copy of each essential gene remains intact when spores are exposed to DNA-damaging agents such as UV light. During the early stages of myxospore formation, large vacuole-like structures can be seen in the core, but they are not present in mature myxospores (Voelz and Dworkin, 1962). Voelz and Dworkin (1962) speculated that these structures might be invaginations of the cell membrane. The work of Bode and colleagues (Ring *et al.*, 2006; Hoiczyk *et al.*, 2009) suggest that the vacuole-like structures might be lipid bodies. Lipid bodies were found in cells that would eventually become spores and they got smaller as the spores matured. By the time the spores had fully matured, lipid bodies were almost non-existent. It was proposed that lipid bodies provide fuel for important developmental processes such as cell motility, which is needed to construct fruiting bodies, and the synthesis of new proteins, which is needed for spore formation (Hoiczyk *et al.*, 2009). Lipid bodies were also found in germinating spores, but not in peripheral rods, which do not differentiate into spores. Thus,

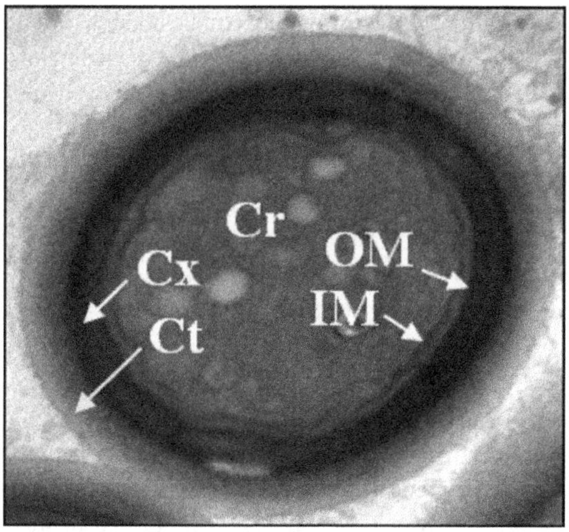

Figure 3.2 Myxospore structure. *M. xanthus* spores were isolated from fruiting bodies after 5 days of development on TPM starvation agar and analysed using transmission electron microscopy. A representative spore is shown with the core (Cr), inner membrane (IM), outer membrane (OM), cortex (Cx) and coat (Ct). Bar = 500 nm (taken from Dahl *et al.*, 2007, with permission).

it seems that lipid bodies are good landmarks for cell fate during fruiting body development (Hoiczyk *et al.*, 2009).

The myxospore core is surrounded by both the inner and outer membranes, followed by the cortex. The cortex of Gram-positive endospores is composed of peptidoglycan (Warth and Strominger, 1969) and the cortex of myxospores might be composed of such material as well (White *et al.*, 1968). Surrounding the cortex is the outer most layer of the myxospore, which is a thick coat containing polysaccharide (approximately 75% of the dry weight), protein (14% of the dry weight), glycine (8% of the dry weight) and organic phosphorus (< 1% of the dry weight) (Kottel *et al.*, 1975). The myxospore cortex and coat are discussed in more detail in the 'Myxospore formation' section, below.

Myxospore formation

The majority of work on myxospore morphogenesis and myxospore biochemistry has come from studying the process of chemically induced sporulation, which occurs in the absence of starvation and fruiting body formation. In 1964, Dowrkin and Gibson found that you can induce sporulation in *M. xanthus* by adding certain chemicals to cells growing in nutrient-rich broth. Glycerol (0.5 M) is by far the most frequently used chemical inducer, and the spores that result from glycerol induction share many properties with starvation-induced fruiting body spores. Namely, glycerol-induced spores are resistant to heat, sonication and UV radiation (Dworkin and Gibson, 1964). However, glycerol-induced spores lack some of the structural features of fruiting body spores. Specifically, these spores have a very thin coat and cortex compared with fruiting body spores (Inouye *et al.*, 1979a). Furthermore, they do not contain protein C, nor do they contain protein S, which are associated with the spore coat (Inouye *et al.*, 1979a; McCleary *et al.*, 1990), and they are unable to support self assembly of exogenous protein S on their coats (Inouye *et al.*, 1979a). One important advantage of studying *M. xanthus* sporulation via glycerol induction instead of in the context of fruiting body development is that it takes hours instead of days. After glycerol is added, rod-shaped vegetative cells convert into spherical-shaped cells in a little over an hour, and the spherical-shaped cells slowly gain optical refractility over the next 50–60 minutes (Dworkin and Sadler, 1966a); during fruiting body development, this process can take days. Another important advantage of using glycerol-induction is that sporulation is synchronized; during fruiting body development, sporulation is asynchronous and only a subset of cells in the population (those in fruiting bodies) becomes spores. It is for these reasons and the fact that glycerol-induced spores and fruiting body spores have shared morphologies, biochemical properties and stress-resistance properties that glycerol induction has been the preferred method to study many aspects of spore morphogenesis and spore biochemistry in *M. xanthus*. Several such studies are discussed below.

The peptidoglycan layer of the cell wall plays a crucial role in maintaining the shape of a bacterial cell (Vollmer *et al.*, 2008). Hence, *M. xanthus* must alter its peptidoglycan layer in order to make the transition from a rod-shaped cell to a spherical-shaped myxospore. At about the same time, the spore coat is synthesized. As mentioned below, the coordination of these two processes might help cells maintain the integrity of their envelopes and maintain osmotic stability while the cell is in a vulnerable state. Such problems do not have to be taken into account in endospore-forming bacteria because the spore is synthesized inside the protective environment of the mother cell.

The structural changes that peptidoglycan undergoes during myxospore formation have been examined in only a few studies. In one such study, White *et al.* (1968) found that glycerol-induced myxospores in the dark sphere stage, which are not fully mature (they have undergone morphological shape change, but have not yet become optically refractile), contain around 0.6% peptidoglycan, which is similar to the amount found in vegetatively growing cells. Dark sphere stage myxospores were used in the study because mature, optically refractile myxospores are insensitive to the detergents that are normally used to lyse cells. When the authors attempted to circumvent this problem by disrupting the myxospores with glass beads and extracting with phenol and detergents, the peptidoglycan was contaminated with significant amounts of amino acids. Glycine was particularly abundant, existing in quantities 10–15× that of diaminopimelic acid (DAP), which is a major constituent of peptidoglycan. Additionally, in the course of their study, White *et al.* (1968) were unable to isolate an intact peptidoglycan sac from vegetatively growing *M. xanthus* cells; when incubated with trypsin and SDS, the peptidoglycan was completely disaggregated. It was suggested that the peptidoglycan in the cell envelope of vegetative cells might be discontinuous or laid down in patches.

In the same study, White *et al.* (1968) compared the composition of the peptidoglycan of vegetative cells and dark sphere-stage myxospores. They found that muramic acid makes up around 0.2% of dry weight of both vegetatively growing cells and the dark sphere stage myxospores. They also found that the composition and molar ratios of the major components of the peptidoglycan of vegetatively growing cells and the dark sphere stage myxospores are similar: they contain glutamic acid, DAP, alanine, *N*-acetyl-glucosamine, and *N*-acetyl-muramic acid in ratios of 1:1:1.7:0.75:0.75, respectively. Muramidase digestions of the peptidoglycan revealed that the peptidoglycan of vegetative cells contains subunits that are mostly dimers, where as the main subunits in dark sphere stage myxospores were considerably larger (White *et al.*, 1968). The authors inferred that the larger subunits indicate that there is an increase in cross-linkages during myxospore formation, that new components were added to the peptidoglycan, or both.

In additional experiments, White and colleagues (1968) examined the effects of the antibiotics penicillin G and D-cycloserine on cells that were growing vegetatively and cells that were differentiating into myxospores. Penicillin G inhibits the formation of peptidoglycan cross-links, whereas D-cycloserine inhibits the synthesis of peptidoglycan components. Vegetatively growing cells were found to be sensitive to both antibiotics. In particular, in the presence of either inhibitor, cells displayed aberrant shapes. These findings demonstrate the importance of peptidoglycan synthesis and integrity to *M. xanthus* vegetative cell morphology. When cells were making the change from rods to spheres, a progressive loss in sensitivity to penicillin G was observed. In contrast, cells retained their sensitivity to D-cycloserine until the later stages of the cell shape change. One explanation for these results is that myxospore morphogenesis is accompanied by a transient decrease in peptidoglycan cross-linking and the synthesis of new peptidoglycan components. Of course, there are other explanations for these data. For example, penicillin G and D-cycloserine might lose access to the cell interior at different stages of myxospore morphogenesis. White and colleagues (1968) did, however, detect an increase in the amount of free amino group of DAP immediately following this period of time, which is consistent with a decrease in peptidoglycan cross-linking around this time.

A more recent study by Bui *et al.* (2009) analysed the peptidoglycan of *M. xanthus* vegetative cells and glycerol-induced myxospores that were between 2.5 and 3 hours post induction and, presumably, optically refractile. The peptidoglycan sacculi of vegetative cells were labelled with anti-peptidoglycan antibody and then with a protein A–gold conjugate to allow them to be visualized using TEM. Vegetative cells contained peptidoglycan sacculi that were uniformly layered with gold conjugate, indicating that the peptidoglycan in vegetative cells is a continuous layer. The authors suggested that perhaps the discontinuous vegetative peptidoglycan layer isolated in the earlier study resulted from degradation by endogenous peptidoglycan hydrolases, which created technical problems that were noted in this study and in the earlier study (White *et al.*, 1968). The components of the vegetative cell peptidoglycan were digested with cellosyl, which is a muramidase, and then analysed using HPLC and mass spectrometry. These studies revealed that the peptidoglycan of *M. xanthus* vegetative cells had a unique structural feature: about 30% of the DAP was in the form of LL-diaminopimelic acid instead of the typical meso-diaminopimelic acid, which is found in the peptidoglycan of other Gram-negative bacteria. The LL-DAP form is also seen in the peptidoglycan of certain Gram-positive species.

In the Bui *et al.* (2009) study, the sacculi from myxospores were not labelled with gold conjugate and they did not release muropeptides when incubated with the muramidase cellosyl. Samples from cells transitioning from the vegetative state to myxospores contained sacculi that were not labelled with gold conjugate and cellular material of various sizes and shapes that was gold conjugate-labelled. In some cases, these labelled pieces were attached to the spores, suggesting that the material might have originated from the particular cell to which they were attached. Based on their findings, the authors speculated that glycerol-induced myxospores do not contain peptidoglycan, or that they contain peptidoglycan that could not be hydrolysed by muramidases or labelled using anti-peptidoglycan antibody. If the second scenario is true, then the peptidoglycan of glycerol-induced myxospores must have been altered in some way or become inaccessible. To further examine the myxospore sacculus, Bui *et al.* (2009) hydrolysed the sacculus material and analysed the composition. They found that the composition of this material differed from peptidoglycan in that it contained variable amounts of all the amino acids found in proteins, with glycine being the most abundant. This material also contained glucosamine, and small amounts of DAP and muramic acid. Since the later two are specific to peptidoglycan, the sacculus material probably contained at least some peptidoglycan. However, the authors speculated that the sacculi that they isolated from myxospores were coats. In summary, Bui *et al.* (2009) made three major conclusions in their analysis:

1 The peptidoglycan of vegetative cells forms a continuous sac.
2 The peptidoglycan of vegetative *M. xanthus* cells contains unique structural features.
3 Much, if not most, of the peptidoglycan is degraded when glycerol induces the transition from vegetative cells to myxospores.

Although the conclusions of the two studies on *M. xanthus* peptidoglycan differ in many respects, they do have some common ground. Both studies identified a peptidoglycan layer in the cell envelopes of vegetatively growing *M. xanthus* cells, although they do differ in their conclusion regarding its structure. Both studies concluded that the peptidoglycan undergoes structural changes during the formation of glycerol-induced myxospores and these changes, at some point in the process of morphogenesis, probably include a decrease in cross linking. Finally, both studies found that the peptidoglycan material that they isolated from myxospores was associated with new components. These new components included amino acids, with glycine being particularly abundant.

Restructuring of the peptidoglycan during spore development is not unique to myxospore formation. For example, in *Bacillus subtilis*, cross-linking per muramic acid decreases from about 33% in vegetatively growing cells to about 3% in endospores (Atrih *et al.*, 1996) and 50% of the muramic acid has been replaced with muramic δ-lactam subunits in endospores (Warth and Strominger, 1969; Atrih *et al.*, 1996). Both the White *et al.* (1968) and the Bui *et al.* (2009) studies noted structural changes in the peptidoglycan of myxospores as compared with that of vegetative cells. Perhaps the latter study was unable to identify myxospore peptidoglycan because they worked with myxospores that were at a different stage of development than those in the former study. It is important to note that the peptidoglycan structure in *B. subtilis* spores and in the spores of related species is dependent upon the conditions in which the spores are produced (Atrih and Foster, 2001; Vollmer *et al.*, 2008). This seems to be true in *M. xanthus* as well because, based on electron microscopy, mature glycerol-induced spores have little or no cortex, which is the spore structure that is thought to contain a peptidoglycan layer, and fruiting body spores have a relatively large cortex (Inouye *et al.*, 1979a). Furthermore, glycerol-induced spores are less resistant to heat than fruiting body spores (Sudo and Dworkin, 1969), and heat resistance is a property known to be associated with the spore cortex (Ellar, 1978). Therefore, it is unclear whether the changes in peptidoglycan structure observed in glycerol-induced spores are similar to those in fruiting body spores. To date, no studies have been done to investigate the peptidoglycan structure of the fruiting body spore cortex because it is technically difficult to isolate a homogeneous population of spores from a group of developing *M. xanthus* cells. Tengra *et al.* (2006) did, however, provide genetic data to support the idea that the cortex of fruiting body spores is important for heat resistance. In this study, the *M. xanthus cbgA* gene, which codes for a homologue of *Bacillus subtilis* SpoVR, was characterized via mutational analysis. SpoVR is involved in endospore cortex formation, and a mutation in *spoVR* results in spores with little or no cortex as visualized using electron microscopy (Beall and Moran, 1994). Similarly, mutations in the *M. xanthus cbgA* gene produce myxospores that either lack a cortex altogether, or have a very thin cortex, and the *cbgA* mutant myxospores are less resistant to heat than wild-type myxospores. Interestingly, *cbgA* mutant spores appear to have intact spore coats (Tengra *et al.*, 2006), which suggest that the myxospore coat synthesis is independent of the cortex. White *et al.* (1968) and Bui *et al.* (2009) have proposed that the spore coat is synthesized during spore morphogenesis to help maintain osmotic stability and cell integrity. The fact that *cbgA* mutant spores, which often lack a cortex layer, maintain their spherical shape and remain viable, suggest that the coat might indeed be capable of doing so.

The myxospore coat begins to accumulate around 1 hour after induction with glycerol and continues to accumulate up to 5 hours (Kottel *et al.*, 1975). However, it is noteworthy, that glycerol spores lack protein S in their coats, and protein S is found to be a major component in the coats of fruiting body spores (Inouye *et al.*, 1979a,b). Nevertheless, it appears that the spore coat, even in protein S-lacking glycerol spores, contributes widely to the resistance properties of the myxospore. Isolated spore coats from glycerol spores were found to be resistant to a variety of treatments, including enzymatic digestion, sonication, and chemical agents such as SDS or dithiothreitol (Kottel *et al.*, 1975). Therefore, the spore coat might have two important roles in the *M. xanthus* sporulation process: maintaining cell integrity during development and providing enhanced resistance.

Is chromosome replication regulated during myxospore formation?

Rosenberg *et al.* (1967) addressed the question of chromosome replication during myxospore formation by examining DNA synthesis rates during this process and developing models from the data they collected. To monitor DNA synthesis, radiolabelled thymidine was added to vegetatively growing *M. xanthus* cells, glycerol-spore formation was induced, and the rate of radiolabelled thymidine incorporation into DNA was measured. They found that the culture continued to synthesize DNA after glycerol induction, but at a decreasing rate for 140 minutes, resulting in a 19% increase in DNA content. After this time, no DNA synthesis or turnover was detected. Based on these data and some additional experiments, the authors developed a model with a key point: after glycerol induces sporulation, cells do no initiate replication of new chromosomes. This model suggests that chromosome replication is regulated in sporulating cells.

Decades later, Tzeng and Singer (2005) approached the subject of DNA replication control during *M. xanthus* sporulation. In this study, DNA replication in *M. xanthus* was monitored during vegetative exponential growth, stationary phase, and the development of fruiting bodies. This study uncovered several important pieces of information. First, *M. xanthus* cells that are growing exponentially contain one to two copies of their chromosome. Second, upon entering stationary phase, the chromosome copy number drops to one per cell. Third, fruiting body-derived myxospores contain two copies of their chromosome, indicating that the chromosomal copy number is tightly regulated during *M. xanthus* development. Fourth, peripheral rods contain one chromosomal copy per cell, which is similar to that of stationary phase cells. This finding supports previous data suggesting that the peripheral rods exist in a state that is distinct from exponentially growing vegetative cells and myxospores (O'Connor and Zusman, 1991a–c). Fifth, the chromosomal copy number of glycerol-induced spores varied: they had one, two and sometimes even more chromosomal copies. Hence, as stated previously, glycerol-induced spores share many properties with fruiting body spores but there are some clear differences, including the regulation of chromosomal copy number. Why do myxospores have two copies of their chromosomal DNA?

Tzeng and Singer (2005) speculated that having an extra copy of essential genes enhances spore survival when they are exposed to DNA damaging agents such as UV light. It is improbable that both copies of an essential gene would get damaged when the spore is exposed to such agents and when the spore starts to germinate the undamaged copy could serve as a template for repair of the damaged copy prior to cell division. It has been shown in other species, that diploid (or higher) cells are more resistant to radiation and better able to repair DNA damage (Mortimer, 1958; Woese, 1958; Krasin and Hutchinson, 1977). In a second study, Tzeng *at al.* (2006) found that DNA replication inhibitors arrested development during the aggregation phase, blocking the formation of myxospores. These findings suggest that *M. xanthus* has a checkpoint during fruiting body development that monitors whether there are two chromosomal copies available for spore formation.

In another study, Sadler and Dworkin (1966b) examined whether transcription and protein synthesis are important for spore formation. To do this, chloramphenicol (a protein synthesis inhibitor) or actinomycin D (a transcription inhibitor) was added to glycerol-induced cultures and the progress of spore formation was monitored. Sadler and Dworkin (1966b) found that both inhibitors arrested myxospore formation at the stage at which they were added, indicating transcription and protein synthesis are crucial throughout the process of glycerol-induced sporulation. In an additional study, Ring *et al.* (2006) showed that specific iso-fatty acid-derived lipids accumulate in fruiting body spores and mutants having reduced levels of iso-FAs are defective for fruiting body development and for sporulation. Other biochemical changes that were detected during glycerol-induced sporulation include an increase in glyoxylate cycle enzyme activity when cells make the transition from rods to spheres (Orlowski *et al.*, 1972), a decrease in acetate metabolism after spores become optically refractile (Watson and Dworkin, 1968), and a 5-fold increase in isocitrate dehydrogenase activity after spores become optically refractile (Watson and Dworkin, 1968).

Genes associated with sporulation

Expression studies suggest that more than 500 genes may be involved in spore development in the model organism *Bacillus subtilis* (Eichenbeger *et al.*, 2004; Wang *et al.*, 2006). Expression studies in *M. xanthus* suggests that a large number of genes are probably involved in myxospore development as well: global transcriptional analyses showed that about 1500 genes are differentially regulated during glycerol-induced sporulation and fruiting body development in *M. xanthus* (Müller *et al.*, 2010; Giglio *et al.*, 2011). Surprisingly, though, only a small number of genes have been directly linked to *M. xanthus* sporulation. These genes were implicated in sporulation using one of the following methods:

1. Expression of their protein products was localized to the myxospore.
2. Mutations in the genes cause defects in both fruiting body sporulation and glycerol-induced sporulation, which bypasses the need for starvation and aggregation.
3. Mutations in the genes cause defects in fruiting body sporulation but not in aggregation. Table 3.1 lists these sporulation genes according to the methods used to identify them and their functions or putative functions.

The first group listed in Table 3.1 contains genes whose protein products were isolated from myxospores or whose protein products are differentially expressed in myxospores. The proteins encoded by *tps*, *ops*, *pru*, *prc* (*prc* is the name that we will use for the gene encoding protein C) and *prw* can be found in myxospores. Protein S is the product of the *tps* gene (Inouye *et al.*, 1983). Protein S was not detected in glycerol spores (Inouye *et al.*, 1979a,b); however, in fruiting body spores it associates with the spore surface and it makes up about 15% of total protein synthesis during the late stages of fruiting body development (Inouye *et al.*, 1979a,b). Using electron microscopy, Inouye *et al.* (1979a) showed that protein S-deficient fruiting body spores have no observable spore coat layer and that protein S self-assembles on the surfaces of fruiting body spores. The results of several

Table 3.1 *Myxococcus xanthus* genes directly involved in sporulation

Gene	MXAN #	Function/putative function	Reference
Genes whose expression is localized to the myxospore			
tps	5430	Protein S	Inouye *et al.* (1983)
ops	5430	Protein S1	Inouye *et al.* (1983)
pru	3885	Spore surface protein	Gollop *et al.* (1991)
prw	2491	Structural spore protein	Otani *et al.* (1998)
prc	unknown	Spore surface protein	McCleary *et al.* (1991)
mspA	2269	Hypothetical protein	Dahl *et al.* (2007)
mspB	2432	Hypothetical protein	Dahl *et al.* (2007)
mspC	6969	Hypothetical protein	Dahl *et al.* (2007)
Genes that are important for both fruiting body sporulation and glycerol-induced sporulation			
exo	3227	Exopolysaccharide transport	Licking *et al.* (2000)
nla4	2516	Enhancer binding protein	Caberoy *et al.* (2003)
nla6	4042	Enhancer binding protein	Caberoy *et al.* (2003)
nla18	3692	Enhancer binding protein	Caberoy *et al.* (2003)
nla24	7440	Enhancer binding protein	Caberoy *et al.* (2003)
nla28	1167	Enhancer binding protein	Caberoy *et al.* (2003)
nfsA	3371	Hypothetical protein	Müller *et al.* (2010)
nfsB	3372	Hypothetical protein	Müller *et al.* (2010)
nfsC	3373	Hypothetical protein	Müller *et al.* (2010)
nfsD	3374	TPR domain repeat protein	Müller *et al.* (2010)
nfsE	3375	Hypothetical protein	Müller *et al.* (2010)
nfsF	3376	Hypothetical protein	Müller *et al.* (2010)
nfsG	3377	Phosphopeptide binding protein	Müller *et al.* (2010)
nfsH	3378	Hypothetical protein	Müller *et al.* (2010)
Genes that are important for fruiting body sporulation but not aggregation			
sigB	3357	Regulatory; RNAP sigma factor	Apelian and Inouye (1990)
aglU	3008	Gliding motility protein	White and Hartzell (2000)
mrpA	5123	Histidine kinase	Sun and Shi (2001)
cbgA	5828	SpoVR-like protein	Tengra *et al.* (2006)

studies indicate that Protein S is not required for spore stress resistance: Protein S is not expressed in glycerol spores, yet glycerol spores still show resistance to heat and sonication (Dworkin and Gibson, 1964; Sudo and Dworkin, 1969; Downard and Zusman, 1985) and mutations in the *tps* gene do not significantly affect production of stress-resistant fruiting body spores (Downard *et al.*, 1984; Komano *et al.*, 1984). Interestingly, Inouye *et al.* (1983) found a second gene (*ops*) in the chromosome that is homologous to *tps*. Like protein S, protein S1, which is the product of the *ops* gene, is capable of assembling on the surface of *M. xanthus* spores when it is added exogenously (Teintze *et al.*, 1985). During development, however, protein S1 remains inside spores rather than reaching the spore's outer surface. Since neither protein S nor protein S1 has an obvious signal peptide (Inouye *et al.*, 1983), it has been suggested that the differences in protein S's and protein S1's location in spores is due to the timing of their expression (Teintze *et al.*, 1985). In particular, the *tps* and ops genes have different patterns of developmental regulation; *tps* is expressed early in development, around the time of aggregation, and *ops* is expressed late in development, around the time that sporulation starts. When the *ops* gene was fused to the promoter of the *tps* gene, protein S1 was deposited on the spore surface. The authors proposed that protein S is produced early in cells that eventually lyse. Cell lysis releases protein S and it self-assembles on the surface of developing spores. In contrast, protein S1 is produced later in development and only in cells that differentiate into spores (Teintze *et al.*, 1985). Consequently, protein S1 remains inside the myxospore (Teintze *et al.*, 1985). At this time, the function of protein S1 is unclear.

The products of three additional sporulation genes, *pru*, *prc*, and *prw*, have been isolated from myxospores. The *pru* gene, which codes for protein U, is expressed late in development (Inouye *et al.*, 1979b). Based on sequence data, protein U has a signal peptide for translocation across the membrane (Gollop *et al.*, 1991). Indeed, protein U assembles on the surface of myxospores. The *prc* gene codes for protein C. Protein C is not found in vegetative cells nor in glycerol spores, but it starts to be detectable in starving cells around the time of aggregation and it continues to accumulate during sporulation (McCleary *et al.*, 1991). Protein C is capable of binding to the surface of myxospores in which the coats had been removed with NaOH (McCleary *et al.*, 1991), suggesting that protein C is a spore-surface-associated protein. The functions of Protein U and Protein C have not yet been determined. Protein W, which is the product of the *prw* gene, is a spore specific protein that, like protein S1, is found to accumulate in the spore core (Otani *et al.*, 1998). Protein W forms dense particles inside myxospores. Structurally, protein W has many charged residues and the particles that it forms in the spore interior contain high levels of phosphate, which led Otani *et al.* (1998) to propose that protein W might be used as an energy store for germination.

In more recent work, three genes (*mspA*, *mspB* and *mspC*) encoding proteins that are differentially expressed in myxospores were identified using a global proteomics approach (Dahl *et al.*, 2007). MspA, B and C (major sporulation protein A, B and C) are expressed at much higher (> 20-fold) levels in fruiting body spores than in vegetatively growing cells (Dahl *et al.*, 2007). As a proof of concept, the same study confirmed sporulation specific markers, such as proteins S and S1 are present at relatively high levels in myxospores. Strains carrying a mutation in any of the three *msp* genes are capable of aggregating. However, the *mspA* or *mspB* mutants produced fruiting bodies with abnormal shapes and all three of the *msp* mutants produced defects in sporulation. Myxospores produced by the *msp* mutants had altered cortex layers, which were thinner in all cases compared with that of wild-type myxospores, and they were more sensitive to both heat and SDS than the myxospores of wild-type cells (Dahl *et al.*, 2007). The authors conclude that all three genes play a role in spore maturation and the development of spore resistance properties.

Mutations in the genes listed in group 2 in Table 3.1 produce defects in both fruiting body sporulation and glycerol spore formation. Many mutants that have fruiting body sporulation defects are still capable of producing glycerol-induced myxospores. Presumably, this is because glycerol-induced sporulation bypasses many of the early developmental events that are necessary for fruiting body spore formation, such as the starvation-induced stringent response, A-signal production and aggregation. In doing so, glycerol-induced sporulation proceeds by directly activating at least some of the *M. xanthus* sporulation programme. If a mutation in a particular chromosomal locus produces defects in both glycerol-induced sporulation and fruiting body sporulation, then it is presumed that locus contains one or more genes that are directly involved in the sporulation process. Seven such loci have been identified in *M. xanthus*: *exo*, *nla4*, *nla6*, *nla18*, *nla24*, *nal28*, and *nfsA-H*.

In 2000, Licking *et al.*, found that a strain carrying the Ω7536 Tn*5lac* insertion in the *exo* locus was blocked at the developmental stage when rod-shaped cells are morphologically restructured into spherical-shaped spores. Cells carrying Ω7536 were no longer capable of making the rod to sphere shape conversion during either glycerol spore formation or fruiting body development. Since the Ω7536 Tn*5lac* insertion created a promoter fusion to *lacZ* reporter gene, *exo* expression was studied by monitoring β-galactosidase activity during both glycerol spore and fruiting body spore formation. The authors found that the timing of *exo* activation coincided with the timing of the cell shape change in each process: one hour after inducing sporulation with glycerol, and at around 18 hours post-starvation during fruiting body development (Licking *et al.*, 2000). Based on DNA sequence analysis, the *exo* operon contains 5 genes that function in exopolysaccharide biosynthesis and export (Goldman *et al.*, 2006), which suggests that the extracellular deposition of polysaccharide might be important for the cell shape change that occurs during glycerol-induced sporulation and fruiting body sporulation.

The sequences of the Nla (NtrC-like activator) proteins, which are the products of the *nla* genes, are similar to that of *E. coli* NtrC. NtrC is an enhancer binding protein (EBP) (for a review see Wedel and Kustu, 1995) and as such, it functions as a transcriptional activator at σ^{54} promoter elements. σ^{54} promoters contain a σ^{54}-RNA-polymerase binding site located −12 and −24 bp upstream from the start of transcription and an enhancer element, which is an EBP binding site that is typically located 75–200 bp upstream of the σ^{54}-RNA polymerase binding site (for review of the σ^{54} system see Xu and Hoover, 2001). EBPs are absolutely required for transcriptional activation at σ^{54} promoters. The *nla4, 6, 18, 24,* and *28* genes, which code for NtrC-like EBPs, were initially characterized in a study by Caberoy *et al.* (2003). Inactivation of any of these *nla* genes leads to defects in both glycerol-induced sporulation and fruiting body sporulation. With the exception of *nla24,* the *nla* genes will be discussed further in the section below on transcriptional regulation during fruiting body formation because they have been found to function in a regulatory network that controls gene expression during fruiting body development and sporulation (Giglio *et al.*, 2011). Because all of these genes code for EBPs, which are transcriptional activators, it seems probable that they regulate genes that have sporulation-specific functions. Indeed, the developmental gene targets of the Nla28 EBP have been identified, and seven of these Nla28 target genes have been linked to both glycerol-induced sporulation and fruiting body sporulation (Caberoy *et al.*, in preparation). The Nla28 target genes that are important for sporulation and the putative functions (Goldman *et al.*, 2006) of these genes are shown in Table 3.2.

The remaining eight genes in group 2 (Table 3.1) appear to be part of a single operon, the *nfs* operon. These genes were identified in a recent global transcriptional study by Müller *et al.* (2010) in which DNA microarrays were used to analyse the gene expression patters of glycerol-induced spores at specific time points over the course of their development. The genes in the *nfs* operon were classified as sporulation genes because they were greater than 10-fold upregulated during glycerol-induced sporulation, an in-frame deletion of the *nfs* operon creates a mutant strain that is defective for both glycerol-induced and fruiting body sporulation, and the operon is differentially expressed in sporulating cells during fruiting body development (Müller *et al.*, 2010). Functional characterization of the individual Nfs proteins has not been done as of yet; however, six out of eight of the Nfs proteins are predicted to localize to the cell envelope. Furthermore, NfsD and NfsE are predicted (via bioninformatic analysis) to contain

Table 3.2 Additional sporulation genes: developmental targets of Nla28

Gene	Annotation
MXAN0881	Lipoprotein
MXAN0989	Cation efflux protein
MXAN2511	General secretion pathway protein
MXAN5040	Aldehyde dehydrogenase family protein
MXAN6732	Aldolase
MXAN7147	ABC transporter permease
MXAN7280	CRISPR-associated protein

tetra-tricopeptide repeats (TPRs), which function in facilitating protein–protein interactions (Smith, 2004), and one protein, NfsG, appears to have a forkhead-associated (FHA) domain (Muller *et al.*, 2010), which suggest that it might be involved in a signal transduction system (for review of the FHA domain see Hammet *et al.*, 2003). Taken collectively, these data led the authors to speculate that the Nsf proteins form a complex that is associated with the cell envelope, that the Nsf complex functions during sporulation, and that the Nsf complex is important for spore viability.

The last group of genes that was linked to sporulation in *M. xanthus* (see Table 3.1) has a common property: when the genes are mutated, the mutations cause defects in sporulation, but they do not affect aggregation or other early developmental events. It should be noted that these genes are not important for glycerol spore formation or they have not been tested for a role in glycerol sporulation. These genes are *sigB*, *aglU*, *mrpA*, and *cgbA*.

The product of the *sigB* gene has similarity to bacterial RNA polymerase sigma factors (Apelian and Inouye, 1990), which are proteins that provide RNA polymerase with promoter specificity. Expression studies using a strain that carries a *lacZ* reporter fusion to the *sigB* promoter revealed that the *sigB* gene is differentially expressed in fruiting body spores (Apelian and Inouye, 1990). Furthermore, a *sigB* mutant strain is defective for spore development and for production of spore-associated protein S1. Since it does not seem to be involved in spore germination, it was suggested that SigB might serve as a regulator of genes involved in myxospore development. The *aglU* gene codes for a WD repeat motif protein that is predicted to attach to the outer membrane (White and Hartzell, 2000). An *aglU* mutant is capable of aggregating and forming fruiting bodies, and cells in the *aglU* fruiting bodies initiate sporulation, as indicated by their ovoid shape. However, *aglU* spores fail to become refractile spheres and they show a dramatic increase in heat sensitivity compared with wild-type spores. Based on these results, White and Hartzell (2000) proposed that AglU might be anchor for proteins involved in the assembly of the outer layers of myxospores. The *mrpA* gene appears to be part of a two gene operon that contains a gene (*mrpA*) for a histidine kinase (HK) sensor and a gene (*mrpB*) for an EBP-type response regulator (RR) (Sun and Shi, 2001). HK and RR proteins form two-component signal transduction systems (TCSs). In TCSs, the HK sensor responds to environmental stimuli by autophosphorylating. The phosphoyl group is then transferred to the cognate RR, which modulates its activity (for review see Bekker *et al.*, 2006). Typically, RRs regulate changes in gene expression in response to HK-mediated phosphorylation. Based on mutational analyses, only the *mrpA* gene product has a function that is specific to sporulation: a mutation in *mrpA* causes defects in sporulation but not in aggregation (Sun and Shi, 2001).

Table 3.3 Phenotype of strains carrying a mutation in an EBP gene.

Strain	MXAN gene #	Phenotypea	Reference
actB-	MXAN 3214	agg– spo– –	Gorski *et al.* (1998)
nla4-	MXAN 2516	ag– spo– –	Caberoy *et al.* (2003)
nla6-	MXAN 4042	agg+/– spo– –	Caberoy *et al.* (2003)
nla18-	MXAN 3692	agg+/– spo– –	Caberoy *et al.* (2003)
nla28-	MXAN 1167	agg+/– spo–	Caberoy *et al.* (2003)
nla36-	MXAN 4899	agg+/– spo– –	Jelsbak *et al.* (2005)

[a]Aggregation and sporulation phenotypes of mutants on starvation agar (TPM). Mutant phenotypes were compared with that of a wild-type strain. Sporulation efficiencies were determined by counting viable spores after treatment with heat and sonication. agg +/–, strains that show an aggregation delay; agg–, strains that show incomplete aggregation; spo–, strains that have about a 10-fold reduction in viable spores; spo– –, strains that have a 100-fold or greater reduction in viable spores.

The last gene in this group, *cbgA*, was tagged as a potential *M. xanthus* sporulation gene based on the sequence similarity of CbgA to the *B. subtilis* SpoVR protein (Tengra *et al.*, 2006), which has been shown to be involved in endospore cortex formation (Beall *et al.*, 1994). As mentioned above, cells carrying an insertion in *cbgA* are capable of aggregating, but *cbgA* fruiting bodies have altered morphologies. Furthermore, fruiting body spores produced but the *cbgA* mutant are less resistant to heat and SDS treatment. Subsequent structural analyses showed that *cbgA* mutant spores lack a cortex all together or had a very thin cortex layer compared with the spores of wild-type cells. These findings suggest that the cortex of fruiting body spores is important for heat resistance. Cortex defects have also been linked to loss of heat resistance in *B. subtilis* (Ellar, 1978).

Transcriptional regulation during fruiting body development

During the complex process of fruiting body development, which leads to sporulation, *M. xanthus* cells must regulate expression of many genes both temporally and spatially. Microarray data show that hundreds or perhaps thousands of genes are differentially expressed during fruiting body development (Jakobsen *et al.*, 2004; Giglio *et al.*, 2011). Many of these genes are activated in the late stages of fruiting body development, when the process of sporulation has commenced. In another microarray study, the gene expression patterns of cells that were inducted to sporulate via the glycerol method were examined; about 1500 genes were differentially regulated during this process (Müller *et al.*, 2010). Clearly, changes in gene transcription are essential for the morphological changes taking place during fruiting body development and sporulation.

Some well-known multicellular processes that require a complex system of signal transduction networks are fruiting body development in the eukaryotic organism *Dictyostelium disciodeum* and biofilm formation in the bacterium *Pseudomonas aeruginosa* (Williams, 2006; Parsek and Tolker-Nielson, 2008). In *M. xanthus*, signal transduction networks are needed to coordinate changes in morphology and gene expression. These signal transduction systems appear to be complex, with multiple networks regulating the transition into and the progression through development, including the process of sporulation. Below is a brief description of a few of these networks.

Recent work has uncovered a key transcriptional regulator that functions in the pathway leading to cell death, and in the C-signalling pathway that controls late developmental events. During fruiting body development, a large percentage of *M. xanthus* cells lyse instead of becoming spores (Wireman and Dworkin, 1977; Nariya and Inouye, 2008). The MrpC protein functions in a toxin/anti-toxin system with the MazF protein to regulate cell lysis during development (Nariya and Inouye, 2008). MrpC is also a regulator of developmental gene expression. During vegetative growth, phosphorylation of MrpC inhibits its transcriptional activator function (Nariya and Inouye, 2005). During fruiting body development, MrpC is processed to MrpC2, which presumably lacks the site(s) of phosphorylation, allowing it to function as a transcriptional activator of important developmental genes (Nariya and Inouye, 2006). MrpC2 works with FruA, another transcriptional regulator, to co-regulate genes in the C-signal transduction pathway such as *fmgA, fmgBC, and fmgD*, which are activated in the later stages of fruiting body development (Mittal and Kroos, 2009a,b; Lee *et al.*, 2011). Interestingly, MrpC2 and FruA bind cooperatively to the *fmg* promoters, which suggest that they link information from different developmental cues to transcription of late developmental genes (Mittal and Kroos, 2009a,b; Lee *et al.*, 2011). Hence, the MrpC signal transduction pathway regulates two very different developmental events: cell death and expression of late developmental genes. Are these two events connected? The answer to this question is unclear because the function of cell lysis is unknown. However, it is possible that MrpC-mediated cell lysis plays an important role in late developmental events. For example, lysed cells might provide nutrients to the remaining developing population so that they are not in danger of running out of energy during fruiting body development and sporulation. Regardless of the answer to the question above, it is clear that the regulation of late gene expression is important for fruiting body development and sporulation. It is also clear that MrpC regulates such genes in an unusual way. Namely, MrpC co-regulates its late developmental gene targets in conjunction with FruA. This regulatory mechanism, which integrates input from two signal transduction systems, is reminiscent of the signal responsive networks found in eukaryotes. Perhaps it will turn out to be a common mechanism for gene regulation in multicellular bacterial systems as well.

Signal transduction networks have also been found to temporally regulate the events of the developmental cycle. Higgs *et al.* (2005) and Jagadeesan *et al.* (2009) found a phosphorelay system consisting of four TCS proteins that regulates the timing of key developmental events. Two of the proteins are RRs, one protein is HK sensor, and one is an HK-like protein that cannot autophosphorylate *in vitro*. The HK-like protein lacks some of the conserved residues of typical HK proteins, and interestingly, accepts a phosphate from one of the RR proteins in the system (Jagadeesan *et al.*, 2009). There are, however, precedents for *M. xanthus* TCS proteins that lack some of the conserved residues or functional regions that are typically found in TCS proteins in other bacteria (Ellehauge *et al.*, 1998; Z. Sarwar and A.G. Garza, unpublished data).

Recently, Giglio *et al.* (2011) found that *M. xanthus* uses a cascade of EBPs to regulate entry into development and to propagate the developmental programme leading to sporulation (Giglio *et al.*, 2012). Many lines of evidence indicate that EBPs and the σ^{54} system are important for *M. xanthus* fruiting body development. First, the completed sequence of the *M. xanthus* genome revealed an abundance of genes (53) that code for EBPs (Goldman *et al.*, 2006). Second, mutations in 24 of the 53 EBP genes produce mutant strains with defects in aggregation and/or sporulation (Kaufman and Nixon, 1996; Wu and Kaiser, 1997; Gorski and Kaiser, 1998; Guo *et al.*, 2000; Hager *et al.*, 2001; Sun and Shi, 2001; Caberoy *et al.*, 2003; Kirby and Zusman, 2003; Jakobsen *et al.*, 2004; Jelsbak *et al.*, 2005; Giglio *et al.*, 2010). Third, many genes that play a role in development are known to be under transcriptional control of EBPs and σ^{54} promoters (Romeo and Zusman, 1991; Keseler and Kaiser, 1995; Wu and Kaiser, 1997; Garza *et al.*, 1998, 2000; Sun and Shi, 2001; Gronewold and Kaiser, 2007; Giglio *et al.*, 2011)

We found that six of the developmental EBPs form a regulatory network that guides early and late events in fruiting body development (Giglio *et al.*, 2011). The six EBPs are ActB, Nla4, Nla6, Nla18, Nla28, and MXAN4899 (see Table 3.3 for a summary of the developmental phenotypes of the EBP mutant strains). A variety of data indicates that these EBPs begin functioning at specific stages in the developmental process: Nla4 and Nla18 function during the transition from vegetative growth to fruiting body development; Nla6 and Nla28 begin functioning during pre-aggregation; and ActB and MXAN4899 begin functioning during aggregation (Gorski *et al.*, 1998; Caberoy *et al.*, 2003; Jelsbak *et al.*, 2005; Giglio *et al.*, 2011). Giglio *et al.* (2011) found that these six EBPs form a transcriptional cascade that is propagated when EBPs that function at one stage of development directly regulate transcription of an EBP gene that is important for the next developmental stage. It also seems that the EBPs in the cascade are part of a much larger regulatory network that is important for aggregation and sporulation: using bioinformatics, putative σ^{54} promoters were found upstream of hundreds of developmentally regulated operons (Giglio *et al.*, 2011). Recently, we found that several of these operons are directly regulated by the Nla28 EBP (Caberoy *et al.*, in preparation). Furthermore, we found that many of the Nla28 target genes are important for spore maturation during fruiting body development and glycerol-induced sporulation. Thus, one of the functions of Nla28 is to activate genes that are important for the later part of the sporulation process.

Germination

Germination has been investigated in both glycerol-induced spores and fruiting body spores. During germination, spores make a sphere to rod shape change. Germination of glycerol-induced spores occurs synchronously when the spores are placed in nutrient-rich media (Shimkets and Seale, 1975). Several properties of glycerol spore germination were seen in the TEM preparations of Shimkets and Seale (1975):

1. Immediately after placement into nutrient-rich media, spores start to appear ovoid and elongate.
2. A vegetative cell emerges from one end of the myxospore.
3. The spore coat appeared to fuse with the cell wall of the emerging cell.
4. Interestingly, newly emerged cells appeared to have a distorted surface and they did not regain a smooth shape (as is consistent with exponentially growing vegetative cells) until around 7 hours after placement into nutrient-rich media.

Fruiting body spores take several times longer to germinate than glycerol-induced spores, and the germination process has several distinct differences (Shimkets and Seale, 1975). For example, fruiting body spores germinate asynchronously. Furthermore, the spore coat is shed from fruiting body spores upon germination and the cells that emerge are structurally and morphologically similar to exponentially growing cells. Shimkets and Seale (1975) speculated that the differences in glycerol-induced spore germination and fruiting body spore germination might be due to the available nutrient pools during sporulation: glycerol-induced sporulation occurs in rich media, whereas fruiting body sporulation occurs in nutrient-depleted media.

A study by Elías and Murillo (1991) revealed that fruiting body spore germination proceeds in an ordered series of stages. In the first stage, spores lose their resistance properties, in the second stage they lose their light refractility (as monitored under a light microscope), and in the last stage they elongate. This study also found that germination of fruiting body spores was asynchronous, which is different than the observed synchrony of glycerol spore germination (Shimkets and Seale, 1975). Additionally, Elías and Murillo (1991) investigated the nutritional requirements for germination and found that a mixture of amino acids was effective at inducing germination, whereas several different sugars that were tested were ineffective. There were several mixes of amino acids that were effective; however, no mix was as effective as complex media such as CTT or casamino acids. In all cases tested, calcium made the process of spore germination more rapid and synchronous. A 2-hour pulse of the correct nutrient mix was sufficient for a large percentage of spores to complete the physiological and morphological changes associated with germination. However, these morphological changes can be severely impaired or blocked by the addition of certain protease inhibitors or by the addition of ethylene glycol tetraacetic acid (EGTA), indicating that myxospore germination requires one or more proteases and Ca^{2+} (Otani *et al.*, 1995).

Summary and future work

In nature, *M. xanthus* sporulation is a multicellular process. This is in contrast to sporulation in endospore-forming bacteria, which proceeds through the efforts of individual cells. Because it is multicellular, spore development in *M. xanthus* requires genes and layers of regulation that are not necessary for the development of an endospore. Another distinction between endospore formation and myxospore formation is that endospores are constructed from scratch within the mother cell, while myxospores are generated from a morphological transformation of rod-shaped vegetative cells to spherical-shaped spores. Despite these fundamental differences, the end products of both developmental processes are similar: a spore that contains a cortex and a protective coat, and is highly resistant to environmental stress. Because of their morphological similarities, one might predict some overlap in the gene sets required for endospore formation and myxospore formation. However, with a few exceptions such as *cbgA*, very little overlap has been uncovered. On the surface, it might seem surprising that relatively few sporulation-specific genes have been identified in *M. xanthus* after decades of genetic analyses. However, one must keep in mind that traditional genetic screens have been the routine way to identify genes that are important for *M. xanthus* development and it is technically difficult to screen for mutations that affect sporulation, but not aggregation. Consequently, the vast majority of *M. xanthus* genes that have sporulation-specific functions are probably still out there waiting to be discovered. If the number of genes that are involved in endospore formation is any indication, then the number of undiscovered genes is probably in the hundreds. Among the genes that have been discovered are those involved in myxospore coat and cortex formation, and those involved transcriptional regulation. Our recent approach for the discovery of new sporulation genes was to find the targets of transcriptional regulators of sporulation. This worked quite well in the case of the Nla28 EBP and we are currently using other late-functioning EBPs as probes for additional sporulation genes. Another approach would be to take advantage of recent genome-wide expression data (Muller *et al.*, 2010; Giglio *et al.*, 2011). Namely, characterize genes that are activated late in fruiting body development and during glycerol-spore formation. Regardless of the method that is used, the challenge for the future will be to identify the molecular functions of new sporulation genes and to relate these functions to the morphological events and stress-resistance properties of myxospores.

References

Adams, D.G. (2000). Heterocyst formation in cyanobacteria. Curr. Opin. Microbiol. *6*, 618–624.

Apelian, D., and Inouye, S. (1990). Development-specific sigma-factor essential for late-stage differentiation of *Myxococcus xanthus*. Genes Dev. *4*, 1396–1403.

Atrih, A., and Foster, S.J. (2001). Analysis of the role of bacterial endospore cortex structure in resistance properties and demonstration of its conservation amongst species. J. Appl. Microbiol. *91*, 364–372.

Atrih, A., Zöllner, P., Allmaier, G., and foster, S.J. (1996). Strucutral analysis of *Bacillus subtilis* 168 endospore peptidoglycan and its role during differentiation. J. Bacteriol. *21*, 6173–6183.

Beall, B., and Moran, C.P., Jr. (1994). Cloning and characterization of *spoVR*, a gene from *Bacillus subtilis* involved in spore cortex formation. J. Bacteriol. *7*, 2003–2012.

Bekker, M., Teixeria de Mattos, M.J., and Hellingwerf, K.J. (2006). The role of two-component regulation systems in the physiology of the bacterial cell. Sci. Prog. *89*, 213–242.

Bui, N.K., Gray, J., Schwarz, H., Schumann, P., Blanot, D., and Vollmer, W. (2009). The peptidoglycan sacculus of *Myxococcus xanthus* has unusual structural features and is degraded during glycerol-induced myxospore development. J. Bacteriol. *191*, 494–505.

Caberoy, N.B., Li, T., Giglio, K.M., Murphy, K.A., Suen, G., Welch, R.D., and Garza, A.G. An enhancer binding protein recognizes tandem repeat sequences in early developmental promoters of *Myxococcus xanthus*. (In preparation.)

Caberoy, N.B., Welch, R.D., Jakobsen, J.S., Slater, S.C., and Garza, .A.G. (2003). Global mutational analysis of NtrC-like activators in *Myxococcus xanthus:* identifying activator mutants defective for motility and fruiting body development. J. Bacteriol. *185*, 6083–6094.

Cashel, M., Gentry, D.R., Hernandez, J.V., and Vinella, D. (1996). The stringent response. In *Escherichia coli* and *Salmonella*: Cellular and Molecular Biology, Neidhardt, F.C., ed. (American Society for Microbiology, Washington D.C.), pp. 1485–1496.

Cho, K., and Zusman, D.R. (1999). AsgD, a new two-component regulator required for A-signalling and nutrient sensing during early development of *Myxococcus xanthus*. Mol. Microbiol. *34*, 268–281.

Dahl, J.L., Tengra, F.K., Dutton, D., Yan, J., Andacht, T.M., Coyne, L., Windell, V., and Garza, A.G. (2007). Identification of major sporulation proteins of *Myxococcus xanthus* using a proteomic approach. J. Bacteriol. *8*, 3187–3197.

Davis, J.M., Mayor, J., and Plamann, L. (1995). A missense mutation in *ropD* results in an A-signalling defect in *Myxococcus xanthus*. Mol. Microbiol. *18*, 943–952.

Downard, J.S., Kupfer, D., and Zusman, D.R. (1984). Gene expression during development of *Myxococcus xanthus*. Analysis of the genes for protein S. J. Mol. Biol. *175*, 469–492.

Downard, J.S., and Zusman, D.R. (1985). Differential expression of protein S genes during *Myxococcus xanthus* development. J. Bacteriol. *161*, 1146–1155.

Dworkin, M., and Gibson, S.M. (1964). A system for studying microbial morphogenesis: rapid formation of microcysts in *Myxococcus xanthus*. Science *146*, 243–244.

Dworkin, M., and Voelz, H. (1962). The formation and germination of microcysts in *Myxococcus xanthus*. J. Gen. Microbiol. *28*, 81–85.

Eichenberger, P., Fujita, M., Jensen, S.T., Conlon, E.M., Rudner, D.Z., Wang, S.T., Ferguson, C., Haga, K., Sato, T., Liu, J.S., *et al.* (2004). The program of gene transcription for a single differentiating cell type during sporulation in *Bacillus subtilis*. PLoS Biol. 2, e328.

Elías, M., and Murillo, F.J. (1991). Induction of germination in *Myxococcus xanthus* fruiting body spores. J. Gen. Microbiol. *137*, 381–388.

Ellar, D.J. (1978). Spore specific structures and their functions. Symposium Soc. Gen. Microbiol. *28*, 295–334.

Ellehauge, E., Nørregaard-Madsen, M., and Søgaard-Andersen, L. (1998). The FruA signal transduction protein provides a checkpoint for the temporal co-ordination of intercellular signals in *Myxococcus xanthus*. Mol. Microbiol. *30*, 807–817.

Garza, A.G., Harris, B.Z., Greenberg, B.M., and Singer, M. (2000). Control of *asgE* expression during growth and development of *Myxococcus xanthus*. J. Bacteriol. *182*, 6622–6629.

Garza, A.G., Pollack, J.S., Harris, B.Z., Lee, A., Keseler, I.M., Licking, E.F., and Singer, M. (1998). SdeK is required for early fruiting body development in *Myxococcus xanthus*. J. Bacteriol. *180*, 4628–4637.

Giglio, K.M., Eisenstatt, J., and Garza, A.G. (2010). Identification of enhancer binding proteins important for *Myxococcus xanthus* development. J. Bacteriol. *19*, 2360–2364.

Giglio, K.M., Caberoy, N.B., Suen, G., Kaiser, D., and Garza, A.G. (2011). A cascade of co-regulating enhancer binding proteins initiates and propagates a multicellular developmental program. Proc. Natl. Acad. Sci. U.S.A. *108*, E431–E439.

Giglio, K.M., Kummer, S., and Garza, A.G. The enhncer binding protein NIa6 regulates developmental genes that are important for quorum signal production and sporulation in *Myxococcus xanthus*. (Submitted.)

Goldman, B.S., Nierman, W.C., Daiser, D., Slater, S.C., Durkin, A.S., Eisen, J.A., Ronning, C.M., Barbazuk, W.B., Blanchard, M., Field, C., *et al.* (2006). Evolution of sensory complexity recorded in a myxobacterial genome. Proc. Natl. Acad. Sci. U.S.A. *103*, 15200–15205.

Gollop, R., Inouye, M., and Inouye, S. (1991). Protein U, a late-developmental spore coat protein of *Myxococcus xanthus*, is a secretory protein. J. Bacteriol. *11*, 3597–3600.

Gorski, L., and Kaiser, D. (1998). Targeted mutagenesis of σ^{54} activator proteins in *Myxococcus xanthus*. J. Bacteriol. *180*, 5896–5905.

Gould, G.W., and Dring, G.J. (1975). Heat resistance of bacterial endospores and concept of an expanded osmoregulatory cortex. Nature *258*, 402–405.

Gronewold, T.M., and Kaiser, D. (2007). Mutations of the *act* promoter in *Myxococcus xanthus*. J. Bacteriol. *185*, 1836–1844.

Guo, D., Wu, Y., and Kaplan, H.B. (2000). Identification and characterization of genes required for early *Myxococcus xanthus* gene expression. J. Bacteriol. *182*, 4564–4571.

Hagen, D.C., Bretscher, A.P., and Kaiser, D. (1978). Synergism between morphogenetic mutants of *Myxococcus xanthus*. Dev. Biol. *64*, 284–296.

Hager, E., Tse, H., and Gill, R.E. (2001). Identification and characterization of *spdR* mutations that bypass the BsgA protease-dependent regulation of developmental gene expression in *Myxococcus xanthus*. Mol. Microbiol. *39*, 765–780.

Hammet, A., Pike, B.L., McNees, C.J., Conlan, L.A., Tenis, N., and Heierhorst, J. (2003). FHA domains as phosphor-threonine binding modules in cell signaling. IUBMB Life *55*, 23–27.

Harris, B.Z., Kaiser, D., and Singer, M. (1998). The guanosine nucleotide (*p*)ppGpp initiates development and A-factor production in *Myxococcus xanthus*. Genes Dev. *12*, 1022–1035.

Higgs, P.I., Cho, K., Whitworth, D.E., Evans, L.S., and Zusman, D.R. (2005). Four unusual two-component signal transduction homologs, RedC to RedF, are necessary for timely development in *Myxococcus xanthus*. J. Bacteriol. *187*, 8191–8195.

Hoiczyk, E., Ring, M.W., McHugh, C.A., Schwar, G., and Bode, E. (2009). Lipid body formatin plays a central role in cell fate determination during developmental differentiation of *Myxococcus xanthus*. Mol. Microbiol. *74*, 497–517.

Inouye, M., Inouye, S., and Zusman, D.R. (1979a). Biosynthesis and self-assembly of protein S, a development-specific protein of *Myxococcus xanthus*. Proc. Natl. Acad. Sci. U.S.A. *76*, 209–213.

Inouye, M., Inouye, S., and Zusman, D.R. (1979b). Gene expression during development of *Myxococcus xanthus*: Pattern of protein synthesis. Dev. Biol. *68*, 579–591.

Inouye, S., Ike, Y., and Inouye, M. (1983). Tandem repeat of the genes for protein S, a development-specific protein of *Myxococcus xanthus*. J. Biol. Chem. *258*, 38–40.

Jagadeesan, S., Mann, P., Schink, C.W., and Higgs, P.I. (2009). A novel 'four-component' two-component signal transduction mechanism regulates developmental progression in *Myxococcus xanthus*. J. Biol. Chem. *284*, 4416–4426.

Jakobsen, J.S., Jelsbak, L., Welch, R.D., Cummings, C., Goldman, B., Stark, E., Stalter, S., and Kaiser, D. (2004). Sigma54 enhancer binding proteins and *Myxococcus xanthus* fruiting body development. J. Bacteriol. *186*, 4361–4368.

Jelsbak, L., Givskov, M., and Kaiser, D. (2005). Enhancer-binding proteins with a forkhead-associated domain and the σ^{54} regulon in *Myxococcus xanthus* fruiting body development. Proc. Natl. Acad. Sci. U.S.A. *102*, 3010–3015.

Julien, B., Kaiser, A.D., and Garza, A. (2000). Spatial control of cell differentiation in *Myxococcus xanthus*. Proc. Natl. Acad. Sci. U.S.A. *97*, 9098–9103.

Kaiser, D. (2003). Coupling cell movement to multicellular development in myxobacteria. Nat. Rev. Microbiol. *1*, 45–54.

Kaufman, R.I., and Nixon, B.T. (1996). Use of PCR to isolate genes encoding σ^{54}-dependent activators from diverse bacteria. J. Bacteriol. *178*, 3967–3970.

Keseler, I.M., and Kaiser, D. (1995). An early A-signal-dependent gene in *Myxococcus xanthus* has a sigma 54-like promoter. J. Bacteriol. *177*, 4638–4644.

Kim, S.K., and Kaiser, D. (1990a). C-factor: a cell–cell signaling protein required for fruiting body morphogenesis of *M. xanthus*. Cell *61*, 19–26.

Kim, S.K., and Kaiser, D. (1990b). Cell alignment required in differentiation of *Myxococcus xanthus*. Science *249*, 926–928.

Kim, S.K., and Kaiser, D. (1990c). Cell motility is required for the transmission of C-factor, an intercellular signal that coordinates fruiting body morphogenesis of *Myxococcus xanthus*. Genes Dev. *4*, 896–904.

Kim, S.K., and Kaiser, D. (1991). C-factor has distinct aggregation and sporulation thresholds during *Myxococcus* development. J. Bacteriol. *173*, 1722–1728.

Kirby, J.R., and Zusman, D.R. (2003). Chemosensory regulation of developmental gene expression in *Myxococcus xanthus*. Proc. Natl. Acad. Sci. U.S.A. *100*, 2008–2013.

Komano, T., Furuichi, T., Teintze, M., Inouye, M., and Inouye, S. (1984). Effects of deletion of the gene for the development-specific protein S on differentiation in *Myxococcus xanthus*. J. Bacteriol. *158*, 1195–1197.

Kottel, R.H., Bacon, K., Clutter, D., and White, D. (1975). Coats from *Myxoccus xanthus*: characterization and synthesis during myxospore differentiation. J. Bacteriol. *124*, 550–557.

Krasin, F., and Hutchinson, F. (1977). Repair of DNA double-strand breaks in *Escherichia coli*, which requires *recA* function and the presence of a duplicate genome. J. Mol. Biol. *116*, 81–98.

Kroos, L. (2007). The *Bacillus* and *Myxococcus* developmental networks and their transcriptional regulators. Annu. Rev. Genet. *41*, 13–39.

Kroos, L., and Kaiser, D. (1987). Expression of many developmentally regulated genes in *Myxococcus* depends on a sequence of cell interactions. Genes Dev. *8*, 840–854.

Kroos, L., Kuspa, A., and Kaiser, D. (1986). A global analysis of developmentally regulated genes in *Myxococcus xanthus*. Dev. Biol. *117*, 252–266.

Kuspa, A., and Kaiser, D. (1989). Genes required for developmental signaling in *Myxococcus xanthus*: three *asg* loci. J. Bacteriol. *171*, 2762–2772.

Kuspa, A., Kroos, L., and Kaiser, D. (1986). Intercellular signalling is required for developmental gene expression in *Myxococcus xanthus*. Dev. Biol. *117*, 267–276.

Kuspa, A., Plamann, L., and Kaiser, D. (1992a) A-signalling and the cell density requirement for *Myxococcus xanthus* development. J. Bacteriol. *174*, 7360–7369.

Kuspa, A., Plamann, L., and Kaiser, D. (1992b) Identification of heat-stable A-factor from *Myxococcus xanthus*. J. Bacteriol. *174*, 3319–3326.

Lee, J.S., Son, B., Viswanathan, P., Lethy, P.M., and Kroos, L. (2011). Combinatorial regulation of *fmgD* by MrpC2 and FruA during *Myxococcus xanthus* development. J. Bacteriol. 7, 1681–1689.

Li, S.F., Lee, B.U., and Shimkets, L.J. (1992). *csgA* expression entrains *Myxococcus xanthus* development. Genes Dev. *6*, 401–410.

Licking, E., Gorski, L., and Kaiser, D. (2000). A common step for changing cell shape in fruiting body and starvation-independent sporulation of *Myxococcus xanthus*. J. Bacteriol. *182*, 3553–3558.

Lobedanz, S., and Søgaard-Andersen, L. (2003). Identification of the C-signal, a contact-dependent morphogen coordinating multiple developmental responses in *Myxococcus xanthus*. Genes Dev. *17*, 2151–2161.

López, D., Vlamakis, H., and Kolter, R. (2010). Biofilms. Cold Spring Harb. Prespect. Biol. 7, a000398.

Manoil, C., and Kaiser, D. (1980). Accumulation of guanosine tetraphosphate and guanosine pentaphosphate in *Myxococcus xanthus* during starvation and myxospore formation. J. Bacteriol. *141*, 297–304.

Mayo, K.A., and Kaiser, D. (1989). *asgB*, a gene required early for developmental signaling, aggregation and sporulation of *Myxococcus xanthus*. Mol. Gen. Genet. *218*, 409–418.

McCleary, W.R., Esmon, B., and Zusman, D.R. (1991). *Myxococcus xanthus* protein C is a major spore surface protein. J. Bacteriol. *173*, 2141–2145.

Mittal, S., and Kroos, L. (2009a). A combination of unusual transcription factors binds cooperatively to control *Myxococcus xanthus* developmental gene expression. Proc. Natl. Acad. Sci. U.S.A. *106*, 1965–1970.

Mittal, S., and Kroos, L. (2009b). Combinatorial regulation by a novel arrangement of FruA and MrpC2 transcription factors during *Myxococcus xanthus* development. J. Bacteriol. *191*, 2753–2763.

Mortimer, R.K. (1958). Radiobiological and genetic studies on a polyploidy series (haploid to hexaploid) of *Saccharomyces cerevisiae*. Radiat. Res. *9*, 312–326.

Müller, F.D., Treuner-Lange, A., Heider, J., Huntley, S.M., and Higgs, P.I. (2010). Global transcriptome analysis of spore formation in *Myxococcus xanthus* reveals a locus necessary for cell differentiation. BMC Genomics *11*, 264.

Murphy, K.A., and Garza, A.G. (2008). Genetic tools for studying *Myxococcus xanthus* biology. In Myxobacteria: Multicellularity and Differentiation, Whitworth, D.E., ed. (ASM Press, Washington, DC), pp. 491–502.

Nariya, H., and Inouye, M. (2008). MazF, an mRNA interferase, mediates programmed cell death during multicellular *Myxococcus development*. Cell *132*, 55–66.

Nariya, H., and Inouye, S. (2005). Identification of a protein Ser/Thr kinase cascade that regulates essential transcriptional activators in *Myxococcus xanthus* development. Mol. Microbiol. *58*, 367–379.

Nariya, H., and Inouye, S. (2006). A protein Ser/Thr kinase cascade negatively regulates the DNA-binding activity of MrpC, a smaller form of which may be necessary for the *Myxococcus xanthus* development. Mol. Microbiol. *60*, 125–127.

O'Connor, K.A., and Zusman, D.R. (1991a). Development in *Myxococcus xanthus* involves differentiation into two cell types, peripheral rods and spores. J. Bacteriol. *11*, 3318–3333.

O'Connor, K.A., and Zusman, D.R. (1991b). Analysis of *Myxococcus xanthus* cell types by two-dimensional polyacrylamide gel electrophoresis. J. Bacteriol. *11*, 3334–3341.

O'Connor, K.A., and Zusman, D.R. (1991c). Behavior of peripheral rods and their role in the life cycle of *Myxococcus xanthus*. J. Bacteriol. *11*, 3342–3355.

Ogawa, M., Fujitani, S., Mao, X., Inouye, S., and Komano, T. (1996). FruA, a putative transcription factor essential for the development of *Myxococcus xanthus*. Mol. Microbiol. *22*, 757–767.

Orlowski, M., Martin, P., White, D., and Wong, M.C. (1972). Changes in activity of glyoxylate cycle enzymes during myxospore development in *Myxococcus xanthus*. J. Bacteriol. *3*, 784–790.

Otani, M., Inouye, M., and Inouye, S. (1995). Germination of myxospores from the fruiting bodies of *Myxococcus xanthus*. J. Bacteriol. *177*, 4261–4265.

Otani, M., Kozuka, S., Xu, C., Umezawa, C., Sano, K., and Inouye, S. (1998). Protein W, a spore-specific protein in *Myxococcus xanthus*, formation of a large electron-dense particle in a spore. Mol. Microbiol. *30*, 57–66.

Parsek, M.R., and Tolker-Nielson, T. (2008). Pattern formation in *Pseudomonas aeruginosa* biofilms Curr. Opin. Microbiol. *11*, 560–566.

Pfister, D.H. (1993). Roland Thaxter and the myxobacteria. In Myxobacteria II, Dworkin, M., and Kaiser, D., eds. (ASM Press, Washington D.C.), pp. 1–12.

Plamann, L., Davis, J.M., Cantwell, B., and Mayor, J. (1994). Evidence that *asgB* encodes a DNA-binding protein essential for growth and development of *Myxococcus xanthus*. J. Bacteriol. *176*, 2013–2020.

Plamann, L., Kuspa, A., and Kaiser, D. (1992). Proteins that rescue A-signal defective mutants of *Myxococcus xanthus*. J. Bacteriol. *174*, 3311–3318.

Ring, M.W., Schwär, G., Thiel, V., Dickschat, J.S., Kroppenstedt, R.M., Schulz, S., and Bode, H.B. (2006). Novel iso-branched ether lipids as specific markers of developmental sporulation in the myxobacterium *Myxococcus xanthus*. J. Biol. Chem. *48*, 36691–36700.

Romeo, J.M., and Zusman, D.R. (1991). Transcription of the myxobacterial hemagglutinin gene is mediated by a sigma 54-like promoter and a cis-acting upstream regulatory region of DNA. J. Bacteriol. *173*, 2969–2976.

Rosenberg, E., Katarski, M., and Gottlieb, P. (1967). Deoxyribonucleic acid synthesis during exponential growth and microcyst formation in *Myxococcus xanthus*. J. Bacteriol. *93*, 1402–1408.

Sadler, W., and Dworkin, M. (1966a). Induction of ç in *Myxococcus xanthus* I. General description. J. Bacteriol. *91*, 1516–1519.

Sadler, W., and Dworkin, M. (1966b). Induction of cellular morphogenesis in *Myxococcus xanthus*. II. Macromolecular synthesis and mechanism of inducer action. J. Bacteriol. *91*, 1520–1525.

Sager, B., and Kaiser, D., (1993). Spatial restriction of cellular differentiation. Genes Dev. *7*, 1645–1653.

Setlow, P. (1995). Mechamisms for the prevention of damage to DNA in spores of *Bacillus* species. Annu. Rev. Microbiol. *49*, 29–54.

Setlow, P. (2006). Spores of *Bacillus subtilis*: their resistance to and killing by radiation, heat and chemicals. J. Appl. Microbiol. *101*, 514–525.

Shimkets, L.J., Gill, R.E., and Kaiser, D. (1983). Developmental cell interactions in *Myxococcus xanthus* and the *spoC* locus. Proc. Natl. Acad. Sci. U.S.A. *80*, 1406–1410.

Shimkets, L.J., and Rafiee, H. (1990). CsgA, an extracellular protein essential for *Myxococcus xanthus* development. J. Bacteriol. *172*, 5299–5306.

Shimkets, L., and Seale, T.W. (1975). Fruiting-body formation and myxospore differentiation and germination in *Myxococcus xanthus* viewed by scanning electron microscopy. J. Bacteriol. *2*, 711–720.

Singer, M., and Kaiser, D., (1995). Ectopic production of guanosine penta- and tetraphosphate can initiate early developmental gene expression in *Myxococcus xanthus*. Genes Dev. *9*, 1633–1644.

Smith, D.F. (2004). Tetratricopeptide repeat cochaperones in steroid receptor complexes. Cell Stress Chaperones *9*, 109–121.

Sudo, S., and Dworkin, M. (1969). Resistance of vegetative cells and microcysts of *Myxococcus xanthus*. J. Bacteriol. *98*, 883–887.

Sudo, S., and Dworkin, M. (1973). Comparative biology of prokaryotic resting cells. Adv. Microb. Phys. *9*, 153–224.

Sun, H., and Shi, W. (2001). Genetic studies of *mrp*, a locus essential for cellular aggregation and sporulation of *Myxococcus xanthus*. *183*, 4786–4795.

Teintze, M., Thomas, R., Furuichi, T., Inouye, M., and Inouye, S. (1985). Two homologous genes coding for spore-specific proteins are expressed at different times during development of *Myxococcus xanthus*. J. Bacteriol. *163*, 121–125.

Tengra, F.K., Dahl, J.L., Dutton, S., Caberoy, N.B., Coyne, L., and Garza, A.G. (2006). CbgA, a protein involved in cortex formation and stress resistance in *Myxococcus xanthus* spores. J. Bacteriol. 23, 8299–8302.

Tzeng, L., Ellis, T.N., and Singer, M. (2006). DNA replication during aggregation phase is essential for *Myxococcus xanthus* development. J. Bacteriol. *188*, 2774–2779.

Tzeng, L., and Singer, M. (2005). DNA replication during sporulation in *Myxococcus xanthus* fruiting bodies. Proc. Natl. Acad. Sci. U.S.A. *102*, 14428–14433.

Voelz, H., and Dworkin, M. (1962). Fine Structure of *Myxococcus xanthus* during morphogenesis. J. Bacteriol. *84*, 943–952.

Vollmer, W., Blanot, D., and de Pedro, M.A. (2008.) Peptidoglycan structure and architecture. FEMS Microbiol. Rev. *32*, 149–167.

Wang, S.T., Setlow, B., Conlon, E.M., Lyon, J.L., Imamura, D., Sato, T., Setlow, P., Losick, R., and Eichenberger, P. (2006). The forespore line of gene expression in *Bacillus subtilis*. J. Mol. Biol. *358*, 16–37.

Warth, A.D., and Strominger, J.L. (1969). Structure of the peptidoglycan of bacterial spores: occurrence of the lactam of muramic acid. Proc. Natl. Acad. Sci. U.S.A. *64*, 528–535.

Watson, B.F., and Dworkin, M. (1968). Comparative intermediary metabolism of vegetative cells and microcysts of *Myxococcus xanthus*. J. Bacteriol. *5*, 1465–1473.

Wedel, A., and Kustu, S. (1995). The bacterial enhancer-binding protein NTRC is a molecular machine: ATP hydrolysis is couple to transcriptional activation. Genes Dev. *9*, 2042–2052.

White, D., Dworkin, M., and Tipper, D.J. (1968). Peptidoglycan of *Myxococcus xanthus*: Structure and relation to morphogenesis. J. Bacteriol. *95*, 2186–2197.

White, D.J., and Hartzell, P.L. (2000). AglU, a protein required for gliding motility and spore maturation of *Myxococcus xanthus*, is related to WD-repeat proteins. Mol. Microbiol. *36*, 662–678.

Williams, J.G. (2006). Transcriptional regulation of *Dictyostelium* pattern formation. EMBO Rep. *7*, 694–698.

Wireman, J.W., and Dworkin, M. (1977). Developmentally induced autolysis during fruiting body formation by *Myxococcus xanthus*. J. Bacteriol. *129*, 798–802.

Woese, C.R. (1958). Comparison of the x-ray sensitivity of bacterial spores. J. Bacteriol. *75*, 5–8.

Wu, S.S., and Kaiser, D. (1997). Regulation of the *pilA* gene in *Myxococcus xanthus*. J. Bacteriol. *179*, 7748–7758.

Xu, H., and Hoover, T.R. (2001). Transcriptional regulation at a distance in bacteria. Curr. Opin. Microbiol. *4*, 138–144.

Streptomyces Sporulation

4

Julia P. Swiercz and Marie A. Elliot

Abstract

Streptomyces are soil-dwelling Gram-positive bacteria with a complex, multicellular life cycle. The latter stages of their life cycle are defined by the metamorphosis of multigenomic aerial hyphae into chains of unigenomic exospores. Here, we discuss the classical studies that established a solid genetic understanding of aerial development and sporulation, and highlight important new advances in the areas of cell division and spore septum placement, chromosome segregation and condensation during sporulation, and spore maturation.

Introduction to *Streptomyces* development and sporulation

Streptomyces are soil saprophytes and are sporulating members of the high G+C, Gram-positive Actinobacteria group. They are renowned for both their morphological complexity and their ability to produce a wide variety of secondary metabolites having broad medical and veterinary application, including antibiotics, antifungal compounds, and anti-tumour agents (Hopwood, 2007). *Streptomyces coelicolor* is the best-characterized member of the streptomycetes, and was the first of this group to have its genome sequenced (Bentley *et al.*, 2002).

The life cycle of *S. coelicolor* is characterized by the progression through several distinct differentiated states. Growth initiates with the germination of a free spore. Apical extension of the initial germ tubes and subsequent hyphal branching leads to the formation of a dense network of filamentous vegetative hyphae known as the vegetative mycelium. Hyphal extension during this growth phase is directed by the coiled-coil protein DivIVA (Flärdh, 2003). DivIVA localizes to the growing cell poles and to sites of future branch emergence, where it is thought to recruit the cell wall biosynthetic machinery (Hempel *et al.*, 2008). DivIVA is probably targeted to the hyphal tips by geometric cues (negative curvature of the cell) as has been observed in *Bacillus* (Lenarcic *et al.*, 2009; Ramamurthi and Losick, 2009); however, it is not known what promotes its assembly at the future branch sites. Unlike most other bacteria, cell growth in the vegetative mycelium of the streptomycetes is not coupled with cell division. Instead, occasional crosswalls are laid down at irregular intervals within the vegetative hyphae, and consequently, these cells are multigenomic.

Vegetative growth is typically thought to continue until some sort of stress is encountered (*e.g.* nutrient deprivation), at which point reproductive aerial development and secondary metabolism ensue. This developmental and metabolic transition is governed by a complex regulatory cascade that has yet to be fully elucidated, and is marked by the emergence of aerial hyphae, which extend away from the surface of the developing colony, and by the onset of secondary metabolite production, notably the synthesis of two pigmented antibiotics in *S. coelicolor*, undecylprodigiosin (red) and actinorhodin (blue) (Fig. 4.1).

Aerial hyphae, like vegetative hyphae, grow by apical tip extension (Flärdh, 2003), but unlike the vegetative cells, they do not branch. Elongation of the aerial hyphae is terminated in response

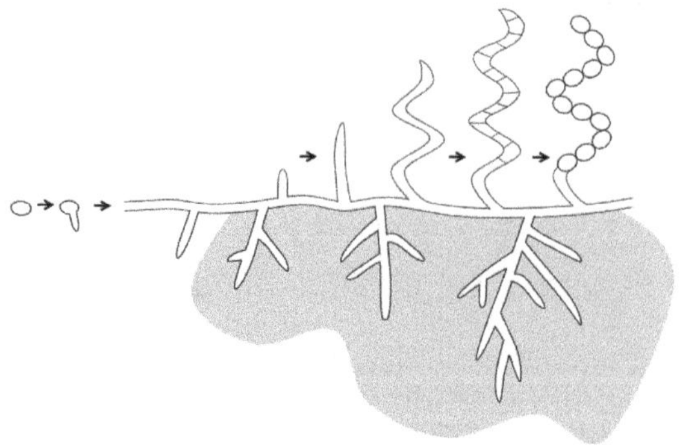

Figure 4.1 The *Streptomyces coelicolor* life cycle. The life cycle of *S. coelicolor* initiates with spore germination. Tip extension and hyphal branching lead to the formation of a vegetative mycelium. As the life cycle progresses, aerial hyphae emerge from the surface of the developing colony, extending into the air. The initiation of aerial hyphae formation coincides with secondary metabolite production (indicated in grey). The aerial hyphae undergo a synchronous round of cell division, and the resulting compartments mature, forming chains of reproductive spores.

to an as yet unknown signal, at which point the metamorphosis of these filaments into spore chains initiates. Sporulation involves a carefully orchestrated programme that begins with cell septation, followed by chromosome segregation and chromosome condensation, and ends with spore maturation (Chater, 2001). Prespore compartments are formed during a synchronous round of cell division where sporulation septa are laid down at regularly spaced intervals within the aerial hyphae, effectively subdividing each aerial cell into ~50 compartments. Coinciding with septation is a mass chromosome partitioning event that ensures each prespore compartment receives a single copy of the chromosome. The chromosome is then condensed and the prespore compartments undergo a maturation process that culminates in the deposition of a grey polyketide-derived pigment into the thick cell wall of a rounded spore, and release of individual spores from the confines of the spore chain. These mature spores have low metabolic activity, and are resistant to environmental and chemical insults including heat, desiccation, detergent exposure and antibiotic treatment.

The morphologically distinct developmental stages of the *Streptomyces* life cycle provide a unique opportunity to investigate cell differentiation in a bacterial context. The different 'tissue types' are defined by developmental checkpoints, and by temporally and spatially constrained gene expression, as outlined below.

Classical regulators of sporulation

Early genetic studies of *S. coelicolor* morphogenesis led to the identification of a number of genes that are critical for cellular differentiation and development (Chater, 1972; Merrick, 1976). These genes, termed the *bld* and *whi* genes, regulate aerial development and sporulation, respectively (as reviewed by Flärdh and Buttner, 2009). These name designations refer to the colony morphology of each mutant type: *bld* (or 'bald') mutants fail to raise aerial hyphae and lack the 'fuzzy' appearance conferred by aerial hyphae and spores, while the *whi* (or 'white') mutants fail to produce the grey spore pigment associated with mature spores, leaving colonies with a distinctive white appearance.

The majority of *bld* genes encode DNA binding regulatory proteins, and their functions and targets have been characterized to varying degrees. Of these, BldD is probably the best understood, functioning as a 'master regulator' during vegetative growth. Among other genes, it controls the expression of multiple sigma factors, numerous developmental genes including *bldA, bldC, bldH* and *bldN,* as well as its own expression (Elliot and Leskiw, 1999; Elliot *et al.*, 2001; den Hengst *et al.*, 2010). Its regulon has been fully fleshed out using a combination of 'ChIP on chip' (chromatin immunoprecipitation combined with microarray analysis) and transcription profiling experiments (den Hengst *et al.*, 2010). Somewhat surprisingly, BldD appears to act primarily as a repressor, and may function to ensure the appropriate temporal progression of development. While *bldD* mutants are defective in raising aerial hyphae, they also fail to produce the full complement of antibiotics. This characteristic is shared by a number of different *bld* mutants, including three BldD targets: *bldC, bldA*, and *bldH*. *bldC,* like *bldD*, encodes a DNA binding transcription factor, and based on sequence similarity, is predicted to belong to the MerR family of regulators (Hunt *et al.*, 2005). Unlike BldD, however, it is not yet known which genes are under BldC control.

Another BldD target, *bldA*, encodes an unusual regulatory molecule: a leucyl tRNA specific for the UUA codon (Lawlor *et al.*, 1987). Given the very G+C-rich nature of the *S. coelicolor* genome (72%), this 'TTA' sequence occurs rarely, with only 145 genes out of >7800 possessing this codon (Takano *et al.*, 2003). Deletion of *bldA* does not lead to a loss of viability, but does prevent aerial hyphae formation and antibiotic production. These phenotypic characteristics can be directly attributed to an inability to effectively translate specific mRNAs. In the case of antibiotic production, TTA codons are found in the genes of multiple 'pathway-specific regulators' that each govern the expression of a single antibiotic. For aerial development, a TTA codon is conserved within the *bldH* gene, and the aerial defects of a *bldA* mutant can be completely abrogated by changing the *bldH* TTA codon to another leucine codon (Nyugen *et al.*, 2003; Takano *et al.*, 2003). *bldH*, which is also known as *adpA*, encodes yet another DNA binding protein whose function has been studied most extensively in *Streptomyces griseus* (Ohnishi *et al.*, 2005). Its regulatory targets in *S. coelicolor* have yet to be fully elucidated, but it appears to impact the production of multiple proteases, as well as RamR (Nyugen *et al.*, 2003; Xu *et al.*, 2010), a regulator of the *ram* genes, which are discussed in more detail below. The importance of *bldH/adpA* in developmental regulation is underscored by the fact that it is subject to multilevel regulation: transcriptional regulation by BldD; translational regulation by the BldA tRNA, as well as post-transcriptional regulation by RNase III, a ribonuclease that specifically cleaves double-stranded RNA (Xu *et al.*, 2010).

The final *bld* gene under the control of BldD is *bldN*, which encodes an extra cytoplasmic function (ECF) sigma factor (Bibb *et al.*, 2000). *S. coelicolor* encodes > 60 sigma factors, a fact that effectively conveys the regulatory complexity, and flexibility, possessed by this bacterium (Bentley *et al.*, 2002). The *bldN* regulon is currently under investigation, but includes yet another *bld* gene, *bldM* (Bibb *et al.*, 2000), which encodes an 'orphaned' response regulator having no known targets (Molle and Buttner, 2000).

There remain three characterized *bld* gene products that are not under BldD control: *bldB*, *bldG* and *bldK*. *bldK* encompasses a multigene locus that collectively encodes an oligopeptide transporter proposed to be involved in the transport of a still elusive extracellular signalling molecule (Nodwell *et al.*, 1996). Like the *bldD* mutant, *bldK*, as well as *bldB* and *bldG* mutants, are defective in both aerial hyphae formation and antibiotic production. *bldB* mutants are further defective in the control of carbon utilization (Pope *et al.*, 1996). BldB encodes a small, actinomycete-specific protein with no obvious functional motifs (Pope *et al.*, 1998). It has been proposed to act as a transcription factor; however, there is as yet no evidence to support this function. In contrast, *bldG* has been shown to encode an anti-anti-sigma factor, and is adjacent to a gene specifying its cognate anti-anti sigma factor (Bignell *et al.*, 2000; Parashar *et al.*, 2009); however, there is no associated sigma factor encoded in the nearby vicinity. Recent work has suggested that BldG may control the activity of σ^H, whose gene is under BldD control, and that it may interact with a second, σ^H-specific, anti-sigma factor (PrsH) (Sevciková *et al.*, 2010). Interestingly, loss of σ^H-activity does not impact aerial hyphae formation (Sevciková *et al.*, 2001; Viollier *et al.*, 2003), but loss of both *sigH* and *prsH* results in a conditionally *bld* mutant phenotype (Viollier *et al.*, 2003).

While the majority of the *bld* genes encode regulatory proteins, and the downstream targets of many of their gene products have yet to be fully elucidated, there are two groups of proteins whose expression depends on the activity of most *bld* genes: SapB and 'the chaplins'. These are secreted proteins that associate with the cell surface and promote the growth of aerial hyphae. It is, however, worth noting that both are not expressed under all growth conditions: SapB is only present during growth on rich media, while the chaplins are expressed during growth on either nutrient-deficient or nutrient-replete media (Capstick *et al.*, 2007). SapB is a lantibiotic-like peptide produced by the products of the *ramCSAB* operon (a gene cluster under the control of the BldH/AdpA-dependent RamR regulator) (Kodani *et al.*,

2004). *ramS* encodes the SapB precursor, which is modified by the RamC lantibiotic synthase. RamAB encode subunits of an ABC-transporter, and may be involved in the secretion of the mature SapB peptide. Conversely, the chaplins comprise a group of eight proteins encoded from disparate sites in the genome and include three 'long' chaplins (ChpA, B and C), and five 'short' chaplins (ChpD–H) (Claessen *et al.*, 2003; Elliot *et al.*, 2003). All chaplins share a hydrophobic region termed the 'chaplin domain', with the long chaplins having two of these domains upstream of an extended C-terminus containing a motif that probably targets them for covalent attachment to the cell surface. SapB and the chaplins both have notable 'surfactant properties', which probably facilitate the raising of aerial hyphae into the air (Willey *et al.*, 2006). The chaplins remain associated with the aerial surfaces, conferring hydrophobic properties to the aerial structures, and contributing to the 'rodlet' ultrastructure that decorates the surface of mature spores.

Following the successful assembly/formation of aerial hyphae, these structures then undergo a maturation process that is controlled by the *whi* gene products. Mutations in the *whi* genes all confer a distinctive 'white' colony phenotype, due to the inability of these mutants to produce the grey spore pigmentation characteristic of mature spores. Despite the common phenotype shared by all *whi* mutant strains, however, *whi* mutations can be subdivided into those that affect spore septation (early *whi* mutations) and those that impact spore maturation (late *whi* mutations). The early-acting Whi proteins include classical representatives discovered in the 1970s (Chater, 1972): WhiG, WhiH, WhiI, WhiA, WhiB, as well as the more recently identified SsgA and SsgB (van Wezel *et al.*, 2000; Keijser *et al.*, 2003).

WhiG is typically situated at the top of the sporulation regulatory cascade (Méndez and Chater, 1987; Chater *et al.*, 1989), with *whiG* mutations being epistatic to all other classical *whi* mutations (Chater, 1975). *whiG* encodes a sporulation-specific sigma factor with exquisite specificity, having only two known target genes: *whiH*, which encodes a GntR family transcription factor (Ryding *et al.*, 1998), and *whiI*, which encodes an orphaned response regulator (Aínsa *et al.*, 1999). No direct targets have been identified for either of these regulators. *whiG* mutants form straight aerial hyphae devoid of obvious septation or chromosome condensation, while the aerial hyphae of *whiH* and *whiI* mutants appear more flexible, exhibiting a slight coiling (Flärdh *et al.*, 1999); chromosome condensation is also occasionally observed in a *whiH* mutant.

The cellular function of WhiA (a conserved protein of unknown function) (Aínsa *et al.*, 2000) and WhiB (a small actinomycete-specific protein) (Davis and Chater, 1992) remain to be determined. In stark contrast to the aerial hyphae of *whiG* mutants, *whiA* and *whiB* mutants develop aerial hyphae that are very tightly coiled and are significantly longer than those of the wild-type or any other *whi* mutant (Flärdh *et al.*, 1999). This suggests that *whiA* and *whiB* mutants may have lost the ability to sense the signal(s) that arrest aerial hyphal growth prior to initiating sporulation septation.

One of the most recently identified members of the early acting sporulation gene products is SsgA. SsgA is the founding member of a family of sporulation-associated proteins known as the SALPs, for SsgA-like proteins, of which there are seven members in *S. coelicolor* (SsgA–G) (Noens *et al.*, 2005). SsgA and SsgB are involved in cell division during sporulation and their function is discussed in more detail in the cell division section (see below).

Unlike the early *whi* mutants, the late *whi* mutant strains can form spores, but these are often misshapen and fail to acquire the dormant characteristics associated with mature spores. For example, *whiD* (a *whiB* homologue) mutant spores are thin walled and of varying sizes; they are also extremely heat sensitive (Molle *et al.*, 2000). How WhiD acts to promote spore maturation is not known. It is a small (~100 aa) iron–sulphur-containing protein (Jakimowicz *et al.*, 2005a, and is a member of an actinomycete-specific protein family having debatable function. Evidence from mycobacterial studies supports for a role for WhiB/WhiD-like proteins as transcription factors (Rybniker *et al.*, 2010), whilst other work suggests they may act as protein disulphide reductases (Alam *et al.*, 2007). Similar to *whiD* mutants, mutations in the gene encoding the late sporulation-specific sigma factor *sigF* also result in spores with thin, irregular walls (Potúková *et al.*, 1995); however, unlike *whiD*, *sigF* mutant spores are resistant to heat stress (Molle *et al.*, 2000). This suggests that σ^F and WhiD probably act independently during spore maturation.

The final stage in sporulation is the synthesis and deposition of a grey polyketide pigment into the spore coat (Davis and Chater, 1990). Spore pigmentation is a characteristic shared by most *Streptomyces* species (Yu *et al.*, 1998), but the function of these pigments is a complete mystery. Mutations in the *whiE* locus lead to defects in grey pigment production; however, apart from their coloration, *whiE* mutant spores are phenotypically indistinguishable from wild-type spores (Flärdh *et al.*, 1999). The *whiE* locus includes eight genes whose products constitute the biosynthetic machinery for the grey spore pigment (Bentley *et al.*, 2002). These genes are organized into two operons, with the second operon under the control of the late sporulation sigma factor σ^F (Kelemen *et al.*, 1998). Expression of both gene clusters, however, requires the activity of all 'early' *whi* gene products (Flärdh *et al.*, 1999).

Cell division

For approximately 30 years, investigations into *Streptomyces* development have focused on the transcriptional regulators governing this process (the *bld* and *whi* gene products). Over the last 10 years, however, sporulation studies in *S. coelicolor* have shifted away from the transcription factor control of sporulation, to focus more on proteins involved specifically in chromosome condensation, chromosome segregation and cell division. Cell division in the streptomycetes is only required during sporulation, not for vegetative growth. The fact that this is not an essential process makes these bacteria an attractive model system for probing division and division-associated chromosome dynamics. Cell division requires the participation of many different proteins. Some cell division-specific proteins are widely conserved across bacteria, while others seem to be more species specific (Adams and Errington, 2009). A universal participant in cell division is FtsZ, a tubulin-like protein that polymerizes into a ring-like structure ('Z-ring') at sites of future cell division (Margolin, 2005). In most bacteria, *ftsZ* is an essential gene. However, given the filamentous nature of the vegetative cells in *S. coelicolor*, FtsZ is dispensable for vegetative growth (McCormick *et al.*, 1994). FtsZ is, however, essential for sporulation in *S. coelicolor*, and this requirement is reflected in its expression profile. *ftsZ* is differentially expressed from at least three promoters: transcription from the first promoter provides low-level, constitutive expression; the third promoter is expressed at its highest level during vegetative growth; while the second promoter appears to be 'sporulation specific', with its expression being dramatically upregulated during aerial hyphae formation (Flärdh *et al.*, 2000). The enhanced transcription observed from the second promoter is spatially constrained to the aerial hyphae, and this is thought to provide sufficient levels of FtsZ to enable assembly of the Z-ring 'ladders' that initiate the sporulation process (Flärdh *et al.*, 2000). The striking increase in *ftsZ* transcription in the aerial hyphae is not observed in any of the early *whi* mutant strains (Flärdh *et al.*, 2000), suggesting that this may be a contributing factor in their inability to form sporulation septa.

FtsZ assembly in the aerial hyphae is a dynamic process. Helical filaments form first, and are later resolved into regularly spaced Z-rings that delineate the prespore compartments (Schwedock *et al.*, 1997; Grantcharova *et al.*, 2005) (Fig. 4.2). What directs the positioning and assembly of the Z-rings has been an important and unresolved question in *Streptomyces* biology. Several recent studies have implicated two different protein complexes in the assembly of Z-rings. Investigations into the cell division genes *ftsW* and *ftsI* revealed that mutations in these genes had no impact on the formation of the FtsZ spiral, but led to a failure to form Z-rings (Mistry *et al.*, 2008; Bennett *et al.*, 2009). *ftsW* and *ftsI* are located directly upstream of *ftsZ* on the chromosome. *ftsW* encodes a membrane protein associated with shape, elongation, division and sporulation (SEDS) (Henriques *et al.*, 1998), while *ftsI* is a membrane-associated penicillin binding protein believed to be involved in peptidoglycan remodelling (Botta and Park, 1981). While no direct interaction between either of these proteins and FtsZ has been demonstrated in *S. coelicolor*, *in vitro* studies of FtsW from *Mycobacterium tuberculosis* (a related actinomycete),

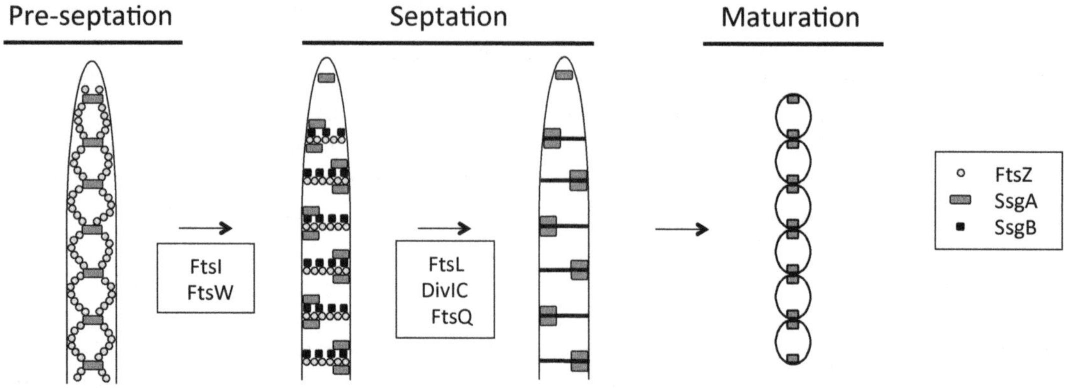

Figure 4.2 Localization of proteins involved in cell division. FtsZ initially forms a helix along the length of the aerial hypha. SsgA then localizes to the future site of cell division, and recruits SsgB, which in turn promotes the formation of Z-rings through interaction with FtsZ. Following the completion of septation, SsgA remains at the poles of prespore and mature spore compartments, ultimately marking the future sites of spore germination.

revealed it to interact with the C-terminus of FtsZ (Datta *et al.*, 2002), as well as with FtsI (Datta *et al.*, 2006). It seems reasonable to predict that similar interactions may occur in the streptomycetes.

More recently, SsgA and SsgB – the products of two early '*whi*' genes – have also been shown to have a critical role in Z-ring formation. Using combined cell biology, genetic and biochemical approaches, Willemse *et al.* (2011) showed that SsgA localizes to future cell division sites, where it then recruits SsgB. SsgB in turn promotes FtsZ ring assembly through its interaction with the N-terminus of FtsZ (Willemse *et al.*, 2011) (Fig. 4.2). An obvious question for future investigation is how SsgA localizes to the sites of future division to initiate this interaction cascade, particularly given that neither SsgA nor SsgB are membrane proteins. It will be interesting to see whether the membrane-bound FtsI or FtsW have a role in SsgA recruitment.

In addition to its role in organizing cell division, SsgA also appears to play a role in the downstream events following spore maturation: it remains associated with the sporulation septa following cell division, and serves to mark the poles of the spore from which germ tubes will later emerge (Noens *et al.*, 2007) (Fig. 4.2). SsgA activity is also not absolutely required for sporulation, as the *ssgA* mutant undergoes sporulation (albeit at reduced levels) during growth on mannitol-containing media (van Wezel *et al.*, 2000). This suggests that other proteins may be able to compensate for the lack of SsgA under these growth conditions, but whether these are other SALPs (SsgA-like proteins) or other cell division proteins, remains to be seen. Mutagenic analysis of the other five SALPs in *S. coelicolor* have hinted at disparate roles in sporulation (Noens *et al.*, 2005). SsgG seems to have a role in Z-ring formation, as loss of this protein results in 'missing' Z-rings but perfectly segregated chromosomes, while SsgC is proposed to act as an antagonist of SsgA activity, given that the *ssgC* mutant phenotype is reminiscent of an *ssgA* overexpression strain. SsgE and SsgF appear to have roles in spore maturation, and are discussed in more detail below.

While Z-ring assembly is a critical step in the sporulation process, ring formation is not observed prior to the cessation of aerial hyphae extension. This implies that there may be factors acting to prevent premature Z-ring formation. A possible candidate for this is CrgA (for coordination of reproductive growth). CrgA is an integral membrane protein that is expressed at maximal levels during the initiation of aerial hyphal growth (Del Sol *et al.*, 2003). Overexpressing CrgA prevents Z-ring formation, while deleting *crgA* leads to accelerated aerial hyphae formation and enhanced septation (Del Sol *et al.*, 2003). Interestingly, however, the *crgA* mutant phenotype is only evident during growth on rich media, suggesting that like for SsgA, there is another system controlling Z-ring formation during growth on mannitol-containing media.

Following the successful assembly of FtsZ rings within the aerial hyphae, membrane invagination and peptidoglycan synthesis are required to define the prespore compartments. Septal ingrowth appears to be coordinated by a number of proteins, including FtsL, DivIC and FtsQ (Fig. 4.2). These three membrane proteins associate to form a complex in both *Escherichia coli* and *Bacillus subtilis* (Buddelmeijer and Beckwith, 2002; Errington *et al.*, 2003). Two of the three are essential proteins in both *E. coli* and *B. subtilis* (FtsL and DivIC), but none are required for viability in *S. coelicolor* (McCormick and Losick, 1996; Bennett *et al.*, 2007; Mistry *et al.*, 2008). FtsL and DivIC are small proteins predicted to form a heterodimer in the membrane (Sievers and Errington, 2000). While their precise role in cell division has yet to be determined, it is proposed that the cytoplasmic domain of FtsL serves to recruit other members of the cell division machinery to the division site (Gonzalez *et al.*, 2010). Deletion of *ftsL* and *divIC* in *S. coelicolor*, both individually and together, results in a failure to complete septation in the aerial hyphae (Bennett *et al.*, 2007). In these mutant strains, occasional spore chains could be detected, but the majority of aerial hyphae contained incomplete – often asymmetric – constrictions, positioned at regular intervals along the length of the hyphae. This suggests that normal Z-ring formation is occurring in these strains, but that they are blocked in septum synthesis. These defects, like those of the *ssgA* and *crgA* mutants, were most pronounced under specific growth conditions, in this case high osmolarity rich media (Bennett *et al.*, 2007). Recent work in *Bacillus* has suggested that DivIC may protect FtsL from intra-membrane proteolysis by the protease RasP (Bramkamp *et al.*, 2006; Wadenpohl and Bramkamp, 2010). *S. coelicolor* encodes a RasP homologue (SCO5695), but whether it targets FtsL for degradation in an analogous manner to the *Bacillus* protein is currently unknown. FtsQ is the third member of this complex in *B. subtilis* and *E. coli*. The observation that an *ftsQ* mutant phenotype is similar to that of the *ftsL* and *divIC* mutants in *S. coelicolor* (Mistry *et al.*, 2008) suggests that an analogous tripartite complex may also be conserved in the streptomycetes.

Chromosome segregation

Cell division is inextricably linked with chromosome segregation, thereby ensuring that each new cell inherits a complete copy of the genomic material. Many bacteria employ a mitosis-like system for chromosome segregation comprising three major components: the ATPase protein ParA, the DNA binding protein ParB, and the centromere-like DNA sequences of *parS* that flank the origin of replication (Fogel and Waldor, 2006; Ptacin *et al.*, 2010). ParB binds to the *parS* sequences, and forms a nucleoprotein complex that interacts with ParA. ParA then either polymerizes or oscillates between independent chromosomes, promoting their segregation (Leonard *et al.*, 2005). *S. coelicolor* encodes a typical *parAB* operon, and has 24 *parS* sites situated around the origin region. Disruption of the *par* operon is not lethal and does not affect vegetative or aerial growth, but it does lead to a 10-fold increase in the number of anucleate spores that are formed, compared with wild-type strains (Kim *et al.*, 2000). This suggests that active chromosome segregation is most critical during sporulation. Intriguingly, an in-frame deletion of *parA* alone affects not only chromosome partitioning (with more than a quarter of all spore compartments lacking DNA), but also has adverse effects on septation, with significant variation in septum placement leading to an increase in the number of both elongated and mini-cell spore compartments (Jakimowicz *et al.*, 2007).

Like *ftsZ*, *parAB* is under the control of several promoters, including one which exhibits low level constitutive activity, and a second which is strongly induced prior to sporulation septation (Kim *et al.*, 2000); upregulation of transcription from this second promoter requires the early *whi* gene products (Jakimowicz *et al.*, 2006). During vegetative growth, ParB complexes are transient, small and irregularly spaced, although they are most prominent at the growing hyphal tips (Jakimowicz *et al.*, 2005). Presumably, increased *parAB* expression in the aerial hyphae serves a similar purpose as that of *ftsZ* upregulation: to increase the local concentration of each protein during a time when significant chromosome segregation is required (Jakimowicz *et al.*, 2006). ParA initially aggregates at the tip of the growing aerial hyphae, and as the hyphae extend, ParA forms a double helical filament that stretches from the tip down towards the base of the hyphae (Fig. 4.3). This helix formation precedes the appearance of FtsZ spirals, and disappears by the time Z-rings have formed. Unexpectedly, it appears that the ATPase activity of ParA is dispensable for helix formation, but is necessary for effective DNA partitioning and for interaction with ParB (Jakimowicz *et al.*, 2007).

ParB does not form helical filaments. Instead, it appears to be diffusely localized in the aerial hyphae prior to the appearance of the ParA helix, at which point it assembles into regularly spaced foci. These foci are positioned between Z-rings, in the centre of what will become the pre-spore compartments, and remain in place until septation is complete (Fig. 4.3). In the absence of ParA, ParB fails to assemble regular foci in the aerial hyphae, and is also absent from the tips of many growing vegetative hyphae. It is possible that either ParB requires the structural scaffold provided by ParA to form effective foci, or that ParA somehow promotes the interaction of ParB with its *parS* DNA target. While there is no *in vivo* data to support either possibility, *in vitro* work by Jakimowicz *et al.* (2007) has shown that DNA binding by ParB is enhanced in the presence of ParA, supporting the second possibility.

A key question is what stimulates the polymerization – and depolymerization – of ParA

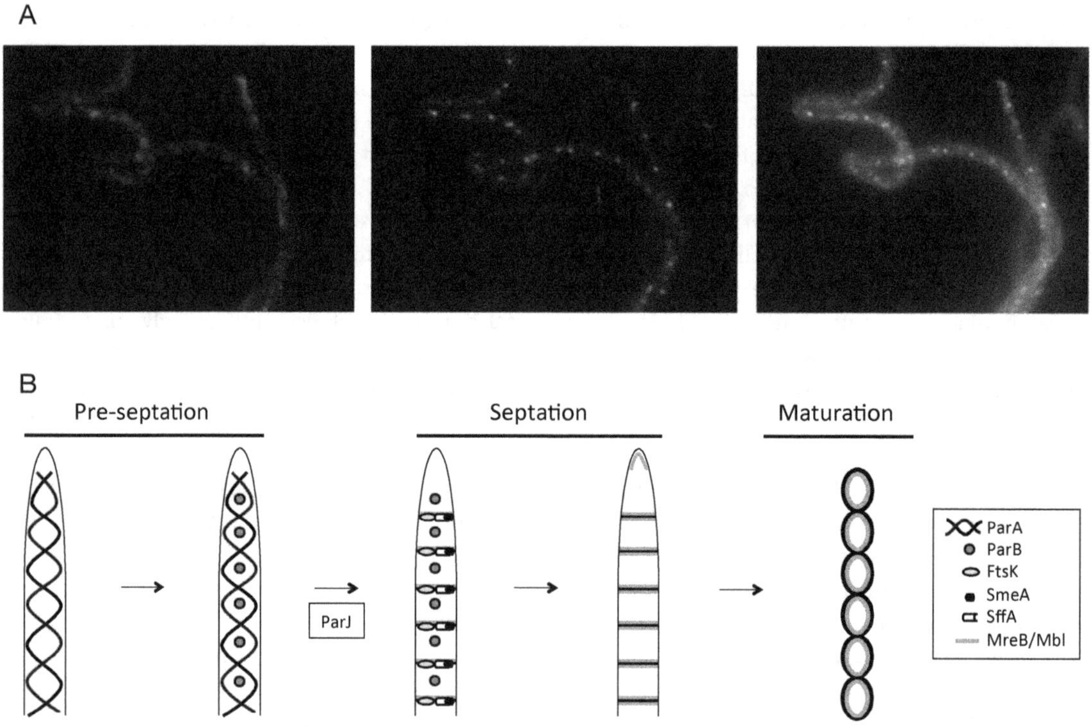

Figure 4.3 (A) Localization of ParA and ParB within the developing aerial hyphae. The ParA helix (left) is visualized using immunofluorescence; ParB (centre) is translationally fused to GFP, and in the merged image (right), the cell wall is stained with wheatgerm agglutin (WGA) Alexa Fluor 350 and false coloured grey (images courtesy of Dr Dagmara Jakimowicz). A colour version of this figure is available in the plate section at the back of the book. (B) Localization of proteins involved in chromosome segregation and spore maturation. After aerial hyphae are fully extended, ParA forms a double helix along the length of each cell. During this time, ParB assembles into regularly spaced foci in the centre of future prespore compartments. ParA filaments are proposed to be disassembled by ParJ prior to the initiation of septation. FtsK, SmeA and SffA, along with MreB/Mbl, then localize to the sites where cell division will occur. Following completion of septation, MreB envelopes the entire spore wall, probably promoting spore maturation.

during sporulation? Recent work has identified a potential regulator of ParA filament assembly. ParJ encodes an unusual ATPase (it lacks any obvious ATP binding motifs) that *in vitro*, can promote the disassembly of ParA filaments; interaction between the two proteins has been further supported by bacterial two hybrid analyses (Ditkowski *et al.*, 2010). The *parJ* mutant exhibits many of the phenotypic characteristics of a *parA* mutant in that it has increased anucleate spores and mini-compartments when compared with a wild-type strain (although not as many as observed for a *parA* mutant), suggesting modest defects in both chromosome segregation and septum placement. Localization studies have shown that ParJ assembles into foci along the length of the aerial hyphae, much like ParB, only in greater numbers; however, unlike ParB, its assembly does not require ParA. Further investigation is needed to determine whether the *in vitro* activity of ParJ in promoting the disassembly of ParA filaments reflects its *in vivo* role during sporulation.

While there is strong experimental evidence supporting a role for ParAB and ParJ in DNA segregation, the fact that > 75% of spores in *par* mutants appear to contain a complete chromosome complement suggests that there must be other systems contributing to chromosomal movement and segregation during sporulation. Packaging of single chromosomes into spore compartments requires not only the precise positioning of uncondensed chromosomes at uniform intervals along the length of the aerial hyphae, but also a mechanism to prevent guillotining of the chromosome during septation. Other bacteria (e.g. *E. coli* and *B. subtilis*) invoke a 'nucleoid occlusion' mechanism to prevent septum formation over the nucleoid (Wu and Errington, 2004; Bernhardt and de Boer, 2005); however, in *Streptomyces*, sporulation septation frequently initiates over uncondensed chromosomes, suggesting that they have evolved different means of ensuring chromosome clearance from the division site. Most bacteria also encode a membrane-bound 'DNA pump' (FtsK) (Iyer *et al.*, 2004) that is associated with the leading edge of the invaginating septum, and this pump is proposed to clear the septal region of residual chromosomal DNA by translocating it into its appropriate daughter cell (Pease *et al.*, 2005). FtsK has additional roles in chromosome decatenation (Espeli *et al.*, 2003) and in assembling the 'divisome' itself (Wang and Lutkenhaus, 1998), and thus plays a central role in coordinating cell division and chromosome segregation. In *S. coelicolor*, loss of FtsK does not impact sporulation per se, and does not appear to affect cell division. Furthermore, anucleate spores do not occur more frequently in a *ftsK* mutant than in wild-type cells (Wang *et al.*, 2007; Dedrick *et al.*, 2009), suggesting that FtsK does not play a role in the wholesale partitioning of chromosomes. Instead, deletion of *ftsK* appears to lead to large deletions at one or both of the chromosome ends (*Streptomyces* have linear chromosomes) in 15–20% of spores (Wang *et al.*, 2007). Colonies that develop from spores containing the truncated chromosomes inevitably display abnormal phenotypic characteristics. The fact that mutant cells containing truncated chromosomes are viable is probably due to the organization of genes in the linear *Streptomyces* chromosome. Essential genes tend to be clustered in a central 'core' region of the linear chromosome, while auxiliary genes are more frequently found in the chromosome 'arms' (Bentley *et al.*, 2002). FtsK localization is most pronounced during sporulation, where it is associated with the developing septa during prespore compartmentalization, but it disappears during spore maturation (Wang *et al.*, 2007; Ausmees *et al.*, 2007; Dedrick *et al.*, 2009) (Fig. 4.3B). Taken together, this suggests that FtsK functions much like FtsK/SpoIIIE in *Bacillus* (Sharp and Pogliano, 2002), pumping chromosomal DNA that may be 'trapped' by the ingrowing septum into the appropriate adjacent spore compartment.

The fact that ~80% of spores produced by an *ftsK* mutant appear to contain an intact chromosome suggests that like the *par* system, there may be additional proteins that can function in its place. *S. coelicolor* encodes multiple proteins belonging to the FtsK/SpoIIIE family, but apart from FtsK, only one of these has been characterized. SffA (for <u>S</u>poIIIE/<u>F</u>tsK <u>f</u>amily) is expressed together with SmeA, a small membrane protein that is frequently associated with FtsK-like proteins in *Streptomyces* genomes (Ausmees *et al.*, 2007). Deletion of the *smeA–sffA* operon results in a strain with spore defects that include reduced grey spore pigmentation, increased anucleate spores, thinner spore walls, increased heat sensitivity of spores, and heterogeneous spore size. Like FtsK, SffA localizes to the sporulation septa; however, this localization requires SmeA (Ausmees *et al.*, 2007) (Fig. 4.3B). Despite a common localization pattern, the differing phenotypes of the *ftsK* and *smeA/sffA* mutants suggest that their respective functions are not redundant, raising the possibility that additional FtsK/SpoIIIE family members (of which there are at least eight) may contribute to chromosome segregation during sporulation.

DNA condensation

Once chromosome segregation and septation are complete, spore maturation initiates. One of the first steps in this process is chromosome condensation. In *Streptomyces*, there are at least three classes of proteins that participate in this process: the SMC proteins, HU-like proteins, and Dps-like proteins.

SMC (structural maintenance of chromosomes) proteins are ubiquitous, found in eukaryotes, archaea and bacteria, where they play a key role in chromosome organization and higher order chromosome structure. SMC proteins typically act in conjunction with other proteins, and two predicted protein partners have been identified in *S. coelicolor*: ScpA and ScpB (segregation and condensation proteins A and B) (Dedrick *et al.*, 2009). As in other bacteria (Graumann, 2000), deletion of the *smc* locus in *S. coelicolor* results in poor nucleoid compaction (Dedrick *et al.*, 2009; Kois *et al.*, 2009). However, SMC also impacts chromosome segregation, with increasing numbers of anucleate spores being detected in the mutant strain. Loss of *scpA* and *scpB* confers a similar mutant phenotype, although the chromosome segregation defect is less severe, and the chromosome itself adopts a distinctive 'bilobed' appearance that is not seen in the *smc* mutant (Dedrick *et al.*, 2009). Surprisingly, immunofluorescence and fluorescence microscopy investigations into SMC localization during sporulation revealed a lack of regularly spaced foci. Instead, SMC foci were distributed at uneven intervals along the length of the aerial hyphae, with focal abundance increasing as the aerial hyphae reached full length, and then disappearing following completion of sporulation septation (Kois *et al.*, 2009). The chromosome segregation defects of the *smc* mutant raise the question of whether there are connections between this system and the *par* system. A double *smc parB* mutant has a greater segregation defect than would be expected based on simply additive effects, suggesting that the two systems may act cooperatively (Dedrick *et al.*, 2009). Supporting this observation is the work of Kois *et al.* (2009), who demonstrated that ParB foci are less intense when SMC is absent. It will be interesting to see how these systems act together in the years to come.

In addition to the SMC proteins, a second class of nucleoid-associated protein has recently been shown to impact chromosome condensation and DNA compaction during sporulation in *S. coelicolor*. The HU (heat unstable)-like proteins preferentially associate with DNA having a curved structure as opposed to a specific sequence (Kamashev *et al.*, 1999). DNA binding by HU, in conjunction with other nucleoid-associated proteins, is thought to promote supercoiling and compaction of the chromosome. *S. coelicolor* encodes two HU-like proteins, HupA and HupS, which are differentially expressed: *hupA* is expressed predominantly in the vegetative mycelium, while *hupS* is upregulated during aerial development and sporulation (Salerno *et al.*, 2009). Increased *hupS* expression is not seen in the early *whi* mutants (including *whiA*, *whiI* and *whiG*), suggesting that upregulation of *hupS* is contingent on the same sporulation-specific checkpoint as *ftsZ* and *parAB*. Localization studies and mutational analyses both support a role for HupS in chromosome compaction during sporulation. Fluorescence microscopy has revealed HupS to be nucleoid-associated, while loss of *hupS* leads to decreased chromosome compaction compared with its wild-type parent (Salerno *et al.*, 2009). Furthermore, *hupS* mutants are defective in spore maturation, generating spores that have reduced spore pigmentation and are less heat resistant than wild-type spores (Salerno *et al.*, 2009). This suggests that chromosome condensation may be an important prerequisite for either initiating or completing the maturation process.

A final group of proteins that contribute to

chromosome condensation during sporulation in *S. coelicolor* are the Dps (DNA-binding protein from starved cells)-like proteins. These proteins are widely conserved in bacteria, and protect the DNA from damage due to environmental stress. Like the HU proteins, Dps proteins typically bind DNA in a sequence-independent manner (Almiron *et al.*, 1992). Previous work has shown that DNA binding by the Dps proteins results in 'hyper compaction' of the chromosome which, along with the well-documented ferroxidase activity associated with Dps proteins, probably confers effective protection of the chromosomal DNA against oxidative damage (Nair and Finkel, 2004). *S. coelicolor* has three genetically distinct *dps* loci: *dpsA*, *dpsB* and *dpsC*. Throughout development, *dpsA* is expressed at a higher level than either *dpsB* or *dpsC*, and its expression is further increased during sporulation (Facey *et al.*, 2009). Mutational analyses have indicated that the three *dps* gene products have different roles during sporulation in *S. coelicolor*. A *dpsA* mutant forms aberrantly sized spore compartments, frequently containing more than one nucleoid, suggesting a role in both spore septum placement and chromosome segregation. In contrast, a *dpsB* mutant exhibited enhanced nucleoid condensation, while a *dpsC* mutant appeared more phenotypically wild-type. Unexpectedly, mutations in *dpsB* and *dpsC* appeared to be epistatic to a *dpsA* mutation, suggesting an intriguing and as yet unexplored relationship between these three proteins (Facey *et al.*, 2009).

It is clear that the processes of spore septation, chromosome segregation and chromosome condensation are necessarily intertwined, and additional study over the next few years will undoubtedly shed light on what is probably an intricate system controlling these processes in *Streptomyces*.

Spore maturation

Following the establishment of spore compartments containing a single condensed chromosome, the process of imparting dormant characteristics to the resulting spores ensues. This includes generating a thick spore coat, rounding up of spores within the context of spore chains, and ultimately liberating individual, free spores. The nature of the *Streptomyces* spore coat is not well defined. It is expected to comprise spore-specific proteins, few of which have yet been identified, along with a thick layer of peptidoglycan that may have a different architecture than the vegetative or aerial hyphae cell walls, based on analogies to the *Bacillus* spore wall (Atrih and Foster, 1999). Given this alteration of aerial hypha, cell walls would require the concerted effort of both peptidoglycan hydrolytic and biosynthetic enzymes. Recent work has implicated the activity of at least four different peptidoglycan hydrolases in spore wall remodelling: RpfA, SwlA, SwlB and SwlC (Haiser *et al.*, 2009). RpfA (resuscitation promoting factor A) is a member of an actinomycete-specific family of proteins that, in *Micrococcus* and *Mycobacterium*, is required for re-establishing the active growth of dormant cells through the controlled cleavage of peptidoglycan (Keep *et al.*, 2006), while SwlA, B and C (for *Streptomyces* cell wall lytic enzyme) have diverse peptidoglycan hydrolytic abilities (SwlA and C are predicted peptidases, while SwlB is a transglycosylase). Deletion of genes encoding each of these hydrolase enzymes results in strains producing spores of variable size (in the case of *swlA*), and having thinner cell walls and increased heat sensitivity (for all mutant strains). Three of the hydrolase mutants (*rpfA*, *swlB* and *swlC*) also exhibited defects in spore separation, suggesting a further role for these enzymes in the release of free spores (Haiser *et al.*, 2009). Localization studies will help to clarify the roles for each of these enzymes in spore maturation.

In rod-shaped bacteria, peptidoglycan growth and remodelling has been closely linked to the activity of the actin-like cytoskeletal element MreB (Margolin, 2009). Within the actinomycetes, MreB is only encoded by those species that develop spore chains within aerial hyphae, including all streptomycetes. *mreB* is not an essential gene in *S. coelicolor*, and like many of the cell division mutants, an *mreB* mutant is unaffected in vegetative growth (Mazza *et al.*, 2006). These mutants do, however, exhibit defects in aerial growth: they produce aerial hyphae that are prone to lysis, develop aberrant spores that lack dormant characteristics (are susceptible to heat and detergent) and germinate prematurely (when still associated with the spore chain) (Mazza *et al.*, 2006). MreB exhibits a dynamic localization during sporulation within the aerial hyphae. Initially, it is associated with the sporulation septa and remains associated with the septa until septation is complete, before it is apparently mobilized to envelop the entire spore wall (Mazza *et al.*, 2006) (Fig. 4.3B). Deletion of other genes within the '*mre* cluster', including *mreC* and *mreD*, which are thought to be co-transcribed with *mreB*, as well as *pbp2* and *sfr*, which are located immediately downstream of the *mreBCD* cluster but are transcribed independently, resulted in strains with nearly identical characteristics to an *mreB* mutant (Kleinschnitz *et al.*, 2011). This suggests that these gene products may act in the same pathway. *mreC*, *mreD* and *sfr* all encode integral membrane proteins, while *pbp2* encodes an extra-cytoplasmic protein involved in peptidoglycan biosynthesis. Investigations into potential interactions between these different proteins, and other proteins encoded by *S. coelicolor*, using a bacterial two hybrid system revealed that these proteins appear to act as part of a larger complex. MreB was seen to interact with RodZ, which is a multidomain integral membrane protein (Gerdes, 2009). RodZ in turn, interacted with as many as 12 other proteins, including MreC, MreD, PBP2, Sfr, as well as the cell division/septation-associated protein FtsI, and another predicted penicillin binding protein encoded by *SCO3901* (Kleinschnitz *et al.*, 2011). Interestingly, MreB localization did not depend on any of the other 'complex' members, suggesting that it may be required to initiate complex formation.

There are two additional MreB-like proteins encoded by *S. coelicolor*: Mbl and SCO6166. SCO6166 does not appear to influence sporulation (or vegetative growth), as deletion of its corresponding gene had no detectable phenotypic consequence (Heichlinger *et al.*, 2011). In contrast, Mbl, like MreB, appears to be important during sporulation, as knockout of the *mbl* gene led to the formation of defective spores with characteristics similar to those of an *mreB* mutant. The localization profile of Mbl also closely mirrored that of MreB (Fig. 4.3B), and this localization was abolished in an *mreB* mutant (Heichlinger *et al.*, 2011), suggesting that a major role for MreB may be in ensuring the proper functioning of Mbl. It appears that this MreB–Mbl-associated complex has a critical role in the engineering of a dormant spore wall, and additional study will yield important insight into the spore maturation process as a whole.

The final step in spore wall remodelling comes with the detachment of spores from each other, and the release of free spores. While autolytic enzymes, like the RpfA and Swl proteins appear to have a role in this process, two of the SALPs, SsgE and SsgF, also appear to function in spore separation. *ssgE* mutants exhibit premature spore separation, while *ssgF* mutants fail to effectively release spores (Noens *et al.*, 2005). How these proteins function in spore separation remains to be seen. An attractive possibility is that they are involved in localizing autolytic enzymes to the spore septum, analogous to the FtsZ positioning by SsgA and SsgB; however, this remains to be tested.

Conclusions and future directions

Many new players in the sporulation process have been identified in the last 5–10 years, and there has been significant progress made in understanding the interactions between the different processes that collectively lead to spore formation and maturation. Advances in post-genomic technologies are now providing ways of identifying the complete regulons for transcription factors of interest, and as such, the targets of the *bld* and *whi* gene products are within reach. We thus have an opportunity to establish a comprehensive and cohesive understanding of the regulatory networks governing aerial hyphae formation and sporulation. Beyond the regulation of development, there remain fundamental questions about the processes driving aerial morphogenesis and sporulation. What signals the cessation of aerial hyphal extension? What promotes SsgA localization, and thus controls spore size? How do the factors that impact Z-ring formation (SsgA, SsgB, FtsZ) interact with those proteins that promote septum formation (FtsW, FtsI)? What other factors influence chromosome segregation and compaction, and how do these work together (e.g. the *par* system and SMC proteins) to ensure that each spore inherits a complete chromosome? What drives the reorganization of MreB, and how does it, and its associated proteins, promote spore maturation? How are the spores primed to ensure efficient germination under appropriate growth conditions? We have established an outstanding foundation from which to proceed with investigations into *Streptomyces* sporulation, and the next few years will be very exciting, as connections between these different systems are further explored, and additional contributing factors are uncovered.

References

Adams, D.W., and Errington, J. (2009). Bacterial cell division: assembly, maintenance and disassembly of the Z ring. Nat. Rev. Micro. 7, 642–653.

Aínsa, J.A., Parry, H.D., and Chater, K.F. (1999). A response regulator-like protein that functions at an intermediate stage of sporulation in *Streptomyces coelicolor* A3(2). Mol. Microbiol. *34*, 607–619.

Aínsa, J.A., Ryding, N.J., Hartley, N., Findlay, K.C., Bruton, C.J., and Chater, K.F. (2000). WhiA, a protein of unknown function conserved among Gram-positive bacteria, is essential for sporulation in *Streptomyces coelicolor* A3(2). J. Bacteriol. *182*, 5470–5478.

Alam, M., Garg, S.K., and Agrawal, P. (2007). Molecular function of WhiB4/Rv3681c of *Mycobacterium tuberculosis* H37Rv: a [4Fe–4S] cluster co-ordinating protein disulphide reductase. Mol. Microbiol. *63*, 1414–1431.

Almiron, M., Link, A.J., Furlong, D., and Kolter, R. (1992). A novel DNA-binding protein with regulatory and protective roles in starved *Escherichia coli*. Genes Dev. *6*, 2646–2654.

Atrih, A., and Foster, S.J. (1999). The role of peptidoglycan structure and structural dynamics during endospore dormancy and germination. Antonie van Leeuwenhoek *75*, 299–307.

Ausmees, N., Wahlstedt, H., Bagchi, S., Elliot, M.A., Buttner, M.J., and Flärdh, K. (2007). SmeA, a small membrane protein with multiple functions in *Streptomyces* sporulation including targeting of a SpoIIIE/FtsK-like protein to cell division septa. Mol. Microbiol. *65*, 1458–1473.

Bennett, J.A., Aimino, R.M., and McCormick, J.R. (2007). *Streptomyces coelicolor* genes *ftsL* and *divIC* play a role in cell division but are dispensable for colony formation. J. Bacteriol. *189*, 8982–8992.

Bennett, J.A., Yarnall, J., Cadwallader, A.B., Kuennen, R., Bidey, P., Stadelmaier, B., and McCormick, J.R. (2009). Medium-dependent phenotypes of *Streptomyces coelicolor* with mutations in *ftsI* or *ftsW*. J. Bacteriol. *191*, 661–664.

Bentley, S.D., Chater, K.F., Cerdeno-Tarraga, A.M., Challis, G.L., Thomson, N.R., James, K.D., *et al.* (2002). Complete genome sequence of the model actinomycete *Streptomyces coelicolor* A3(2). Nature *417*, 141–147.

Bernhardt, T.G., and de Boer, P.A.J. (2005). SlmA, a nucleoid-associated FtsZ binding protein required for blocking septal ring assembly over chromosomes in *E. coli*. Mol. Cell *18*, 555–564.

Bibb, M.J., Molle, V., and Buttner, M.J. (2000). sigma(BldN), an extracytoplasmic function, RNA polymerase sigma factor required for aerial mycelium formation in *Streptomyces coelicolor*, A3(2). J. Bacteriol. *182*, 4606–4616.

Bignell, D.R., Warawa, J.L., Strap, J.L., Chater, K.F., and Leskiw, B.K. (2000). Study of the *bldG* locus suggests that an anti-anti-sigma factor and an anti-sigma factor may be involved in *Streptomyces coelicolor* antibiotic production and sporulation. Microbiology *146*, 2161–2173.

Botta, G.A., and Park, J.T. (1981). Evidence for involvement of penicillin-binding protein 3 in murein synthesis during septation but not during cell elongation. J. Bacteriol. *145*, 333–340.

Bramkamp, M., Weston, L., Daniel, R.A., and Errington, J. (2006). Regulated intramembrane proteolysis of FtsL protein and the control of cell division in *Bacillus subtilis*. Mol. Microbiol. *62*, 580–591.

Buddelmeijer, N., and Beckwith, J. (2002). Assembly of cell division proteins at the *E. coli* cell center. Curr. Opin. Microbiol. *5*, 553–557.

Capstick, D.S., Willey, J.M., Buttner, M.J., and Elliot, M.A. (2007). SapB and the chaplins: connections between morphogenetic proteins in *Streptomyces coelicolor*. Mol. Microbiol. *64*, 602–613.

Chater, K.F. (1972). A morphological and genetic mapping study of white colony mutants of *Streptomyces coelicolor*. J. Gen. Microbiol. *72*, 9–28.

Chater, K.F. (1975). Construction and phenotypes of double sporulation deficient mutants in *Streptomyces coelicolor* A3(2). J. Gen. Microbiol. *87*, 312–325.

Chater, K.F. (2001). Regulation of sporulation in *Streptomyces coelicolor* A3(2): a checkpoint multiplex? Curr. Opin. Microbiol. *4*, 667–673.

Chater, K.F., Bruton, C.J., Plaskitt, K.A., Buttner, M.J., Méndez, C., and Helmann, J.D. (1989). The developmental fate of *S. coelicolor* hyphae depends upon a gene product homologous with the motility [sigma] factor of *B. subtilis*. Cell 59, 133–143.

Claessen, D., Rink, R., de Jong, W., Siebring, J., de Vreugd, P., Boersma, F.G., *et al.* (2003). A novel class of secreted hydrophobic proteins is involved in aerial hyphae formation in *Streptomyces coelicolor* by forming amyloid-like fibrils. Genes Dev. *17*, 1714–1726.

Datta, P., Dasgupta, A., Bhakta, S., and Basu, J. (2002). Interaction between FtsZ and FtsW of *Mycobacterium tuberculosis*. J. Biol. Chem. *277*, 24983–24987.

Datta, P., Dasgupta, A., Singh, A.K., Mukherjee, P., Kundu, M., and Basu, J. (2006). Interaction between FtsW and penicillin-binding protein 3 (PBP3) directs PBP3 to mid-cell, controls cell septation and mediates the formation of a trimeric complex involving FtsZ, FtsW and PBP3 in mycobacteria. Mol. Microbiol. *62*, 1655–1673.

Davis, N.K., and Chater, K.F. (1990). Spore colour in *Streptomyces coelicolor* A3(2) involves the developmentally regulated synthesis of a compound biosynthetically related to polyketide antibiotics. Mol. Microbiol. *4*, 1679–1691.

Davis, N.K., and Chater, K.F. (1992). The *Streptomyces coelicolor whiB* gene encodes a small transcription factor-like protein dispensable for growth but essential for sporulation. Mol. Gen. Genet. *232*, 351–358.

Dedrick, R.M., Wildschutte, H., and McCormick, J.R. (2009). Genetic interactions of *smc*, *ftsK*, and *parB* genes in *Streptomyces coelicolor* and their developmental genome segregation phenotypes. J. Bacteriol. *191*, 320–332.

Del Sol, R., Pitman, A., Herron, P., and Dyson, P. (2003). The product of a developmental gene, *crgA*, that coordinates reproductive growth in *Streptomyces* belongs to a novel family of small actinomycete-specific proteins. J. Bacteriol. *185*, 6678–6685.

Ditkowski, B., Troć, P., Ginda, K., Donczew, M., Chater, K.F., Zakrzewska-Czerwińska, J., and Jakimowicz, D. (2010). The actinobacterial signature protein ParJ (SCO1662) regulates ParA polymerization and affects chromosome segregation and cell division during *Streptomyces* sporulation. Mol. Microbiol. *78*, 1403–1415.

Elliot, M.A., Bibb, M.J., Buttner, M.J., and Leskiw, B.K. (2001). BldD is a direct regulator of key developmental genes in *Streptomyces coelicolor* A3(2). Mol. Microbiol. *40*, 257–269.

Elliot, M.A., Karoonuthaisiri, N., Huang, J., Bibb, M.J., Cohen, S.N., Kao, C.M., *et al.* (2003). The chaplins: a family of hydrophobic cell-surface proteins involved in aerial mycelium formation in *Streptomyces coelicolor*. Genes Dev. *17*, 1727–1740.

Elliot, M.A., and Leskiw, B.K. (1999). The BldD protein from *Streptomyces coelicolor* is a DNA-binding protein. J. Bacteriol. *181*, 6832–6835.

Errington, J., Daniel, R.A., and Scheffers, D.J. (2003). Cytokinesis in bacteria. Microbiol. Mol. Biol. Rev. *67*, 52–65.

Espeli, O., Lee, C., and Marians, K.J. (2003). A physical and functional interaction between *Escherichia coli* FtsK and topoisomerase IV. J. Biol. Chem. *278*, 44639–44644.

Facey, P.D., Hitchings, M.D., Saavedra-Garcia, P., Fernandez-Martinez, L., Dyson, P.J., and Del Sol, R. (2009). *Streptomyces coelicolor* Dps-like proteins: differential dual roles in response to stress during vegetative growth and in nucleoid condensation during reproductive cell division. Mol. Microbiol. *73*, 1186–1202.

Flärdh, K. (2003). Essential role of DivIVA in polar growth and morphogenesis in *Streptomyces coelicolor* A3(2). Mol. Microbiol. *49*, 1523–1536.

Flärdh, K., and Buttner, M.J. (2009). *Streptomyces* morphogenetics: dissecting differentiation in a filamentous bacterium. Nat. Rev. Microbiol. *7*, 36–49.

Flärdh, K., Findlay, K.C., and Chater, K.F. (1999). Association of early sporulation genes with suggested developmental decision points in *Streptomyces coelicolor* A3(2). Microbiology *145*, 2229–2243.

Flärdh, K., Leibovitz, E., Buttner, M.J., and Chater, K.F. (2000). Generation of a non-sporulating strain of *Streptomyces coelicolor* A3(2) by the manipulation of a developmentally controlled *ftsZ* promoter. Mol. Microbiol. *38*, 737–749.

Fogel, M.A., and Waldor, M.K. (2006). A dynamic, mitotic-like mechanism for bacterial chromosome segregation. Genes Dev. *20*, 3269–3282.

Gerdes, K. (2009). RodZ, a new player in bacterial cell morphogenesis. EMBO J. *28*, 171–172.

Gonzalez, M.D., Akbay, E.A., Boyd, D., and Beckwith, J. (2010). Multiple interaction domains in FtsL, a protein component of the widely conserved bacterial FtsLBQ cell division complex. J. Bacteriol. *192*, 2757–2768.

Grantcharova, N., Lustig, U., and Flärdh, K. (2005). Dynamics of FtsZ assembly during sporulation in *Streptomyces coelicolor* A3(2). J. Bacteriol. *187*, 3227–3237.

Graumann, P.L. (2000). *Bacillus subtilis* SMC is required for proper arrangement of the chromosome and for efficient segregation of replication termini but not for bipolar movement of newly duplicated origin regions. J. Bacteriol. *182*, 6463–6471.

Haiser, H.J., Yousef, M.R., and Elliot, M.A. (2009). Cell wall hydrolases affect germination, vegetative growth and sporulation in *Streptomyces coelicolor*. J. Bacteriol. *191*, 6501–6512.

Heichlinger, A., Ammelburg, M., Kleinschnitz, E.M., Latus, A., Maldener, I., Flärdh, K., *et al.* (2011). The MreB-like protein Mbl of *S. coelicolor* A3(2) depends on MreB for proper localization and contributes to spore wall synthesis. J. Bacteriol. *193*, 1533–1542.

Hempel, A., Wang, S.B., Letek, M., Gil, J.A., and Flärdh, K. (2008). Assemblies of DivIVA mark sites for hyphal branching and can establish new zones of cell wall growth in *Streptomyces coelicolor*. J. Bacteriol. *190*, 7579–7583.

den Hengst, C.D., Tran, N.T., Bibb, M.J., Chandra, G., Leskiw, B.K., and Buttner, M.J. (2010). Genes essential for morphological development and antibiotic production in *Streptomyces coelicolor* are targets of BldD during vegetative growth. Mol. Microbiol. *78*, 361–379.

Henriques, A.O., Glaser, P., Piggot, P.J., and Moran, C.P., Jr. (1998). Control of cell shape and elongation by the *rodA* gene in *Bacillus subtilis*. Mol. Microbiol. *28*, 235–247.

Hopwood, D.A. (2007). *Streptomyces* in Nature and Medicine: The Antibiotic Makers. Oxford University Press, U.S.A.

Hunt, A.C., Servín-González, L., Kelemen, G.H., and Buttner, M.J. (2005). The *bldC* developmental locus of *Streptomyces coelicolor* encodes a member of a family of small DNA-binding proteins related to the DNA-binding domains of the MerR family. J. Bacteriol. *187*, 716–728.

Iyer, L.M., Makarova, K.S., Koonin, E.V., and Aravind, L. (2004). Comparative genomics of the FtsK-HerA superfamily of pumping ATPases: implications for the origins of chromosome segregation, cell division and viral capsid packaging. Nucl. Acids Res. 32, 5260–5279.

Jakimowicz, D., Cheesman, M.R., Bishai, W.R., Chater, K.F., Thompson, A.J., and Buttner, M.J. (2005a). Evidence that the *Streptomyces* developmental protein WhiD, a member of the WhiB family, binds a [4Fe-4S] cluster. J. Biol. Chem. *280*, 8309–8315.

Jakimowicz, D., Gust, B., Zakrzewska-Czerwińska, J., and Chater, K.F. (2005b). Developmental-stage-specific assembly of ParB complexes in *Streptomyces coelicolor* hyphae. J. Bacteriol. *187*, 3572–3580.

Jakimowicz, D., Mouz, S., Zakrzewska-Czerwińska, J., and Chater, K.F. (2006). Developmental control of a *parAB* promoter leads to formation of sporulation-associated ParB complexes in *Streptomyces coelicolor*. J. Bacteriol. *188*, 1710–1720.

Jakimowicz, D., Zydek, P., Kois, A., Zakrzewska-Czerwińska, J., and Chater, K.F. (2007). Alignment of multiple chromosomes along helical ParA scaffolding in sporulating *Streptomyces* hyphae. Mol. Microbiol. *65*, 625–641.

Kamashev, D., Balandina, A., and Rouviere-Yaniv, J. (1999). The binding motif recognized by HU on both nicked and cruciform DNA. EMBO J. *18*, 5434–5444.

Keep, N.H., Ward, J.M., Cohen-Gonsaud, M., and Henderson, B. (2006). Wake up! Peptidoglycan lysis and bacterial non-growth states. Trends Microbiol. *14*, 271–276.

Keijser, B.J.F., Noens, E.E.E., Kraal, B., Koerten, H.K., and Van Wezel, G.P. (2003). The *Streptomyces coelicolor ssgB* gene is required for early stages of sporulation. FEMS Microbiol. Lett. *225*, 59–67.

Kelemen, G.H., Brian, P., Flärdh, K., Chamberlin, L., Chater, K.F., and Buttner, M.J. (1998). Developmental regulation of transcription of *whiE*, a locus specifying the polyketide spore pigment in *Streptomyces coelicolor* A3(2). J. Bacteriol. *180*, 2515–2521.

Kim, H.J., Calcutt, M.J., Schmidt, F.J., and Chater, K.F. (2000). Partitioning of the linear chromosome during sporulation of *Streptomyces coelicolor* A3(2) involves an *oriC*-linked *parAB* locus. J. Bacteriol. *182*, 1313–1320.

Kleinschnitz, E.M., Heichlinger, A., Schirner, K., Winkler, J., Latus, A., Maldener, I., *et al.* (2011). Proteins encoded by the *mre* gene cluster in *Streptomyces coelicolor* A3(2) cooperate in spore wall synthesis. Mol. Microbiol. *79*, 1367–1379.

Kodani, S., Hudson, M.E., Durrant, M.C., Buttner, M.J., Nodwell, J.R., and Willey, J.M. (2004). The SapB morphogen is a lantibiotic-like peptide derived from the product of the developmental gene *ramS* in *Streptomyces coelicolor*. Proc. Natl. Acad. Sci. U.S.A. *101*, 11448–11453.

Kois, A., Swiatek, M., Jakimowicz, D., and Zakrzewska-Czerwińska, J. (2009). SMC protein-dependent chromosome condensation during aerial hyphal development in *Streptomyces*. J. Bacteriol. *191*, 310–319.

Lawlor, E.J., Baylis, H.A., and Chater, K.F. (1987). Pleiotropic morphological and antibiotic deficiencies result from mutations in a gene encoding a tRNA-like product in *Streptomyces coelicolor* A3(2). Genes Dev. *1*, 1305–1310.

Lenarcic, R., Halbedel, S., Visser, L., Shaw, M., Wu, L.J., Errington, J., *et al.* (2009). Localisation of DivIVA by targeting to negatively curved membranes. EMBO J. *28*, 2272–2282.

Leonard, T.A., Müller-Jensen, J., and Löwe, J. (2005). Towards understanding the molecular basis of bacterial DNA segregation. Phil. Trans. Royal Soc. B: Biol. Sci. *360*, 523–535.

Margolin, W. (2005). FtsZ and the division of prokaryotic cells and organelles. Nat. Rev. Mol. Cell Biol. *6*, 862–871.

Margolin, W. (2009). Sculpting the bacterial cell. Curr. Biol. *19* R812–R822.

Mazza, P., Noens, E.E., Schirner, K., Grantcharova, N., Mommaas, A.M., Koerten, H.K., *et al.* (2006). MreB of *Streptomyces coelicolor* is not essential for vegetative growth but is required for the integrity of aerial hyphae and spores. Mol. Microbiol. *60*, 838–852.

McCormick, J.R., and Losick, R. (1996). Cell division gene *ftsQ* is required for efficient sporulation but not growth and viability in *Streptomyces coelicolor* A3(2). J. Bacteriol. *178*, 5295–5301.

McCormick, J.R., Su, E.P., Driks, A., and Losick, R. (1994). Growth and viability of *Streptomyces coelicolor* mutant for the cell division gene *ftsZ*. Mol. Microbiol. *14*, 243–254.

Méndez, C., and Chater, K.F. (1987). Cloning of *whiG*, a gene critical for sporulation of *Streptomyces coelicolor* A3(2). J. Bacteriol. *169*, 5715–5720.

Merrick, M.J. (1976). A morphological and genetic mapping study of bald colony mutants of *Streptomyces coelicolor*. J. Gen. Microbiol. *96*, 299–315.

Mistry, B.V., Del Sol, R., Wright, C., Findlay, K., and Dyson, P. (2008). FtsW is a dispensable cell division protein required for Z-ring stabilization during sporulation septation in *Streptomyces coelicolor*. J. Bacteriol. *190*, 5555–5566.

Molle, V., and Buttner, M.J. (2000). Different alleles of the response regulator gene *bldM* arrest *Streptomyces coelicolor* development at distinct stages. Mol. Microbiol. *36*, 1265–1278.

Molle, V., Palframan, W.J., Findlay, K.C., and Buttner, M.J. (2000). WhiD and WhiB, homologous proteins required for different stages of sporulation in *Streptomyces coelicolor* A3(2). J. Bacteriol. *182*, 1286–1295.

Nair, S., and Finkel, S.E. (2004). Dps protects cells against multiple stresses during stationary phase. J. Bacteriol. *186*, 4192–4198.

Nguyen, K.T., Tenor, J., Stettler, H., Nguyen, L.T., Nguyen, L.D., and Thompson, C.J. (2003). Colonial differentiation in *Streptomyces coelicolor* depends on translation of a specific codon within the *adpA* gene. J. Bacteriol. *185*, 7291–7296.

Nodwell, J.R., McGovern, K., and Losick, R. (1996). An oligopeptide permease responsible for the import of an extracellular signal governing aerial mycelium formation in *Streptomyces coelicolor*. Mol. Microbiol. *22*, 881–893.

Noens, E.E., Mersinias, V., Willemse, J., Traag, B.A., Laing, E., Chater, K.F. *et al.* (2007). Loss of the controlled localization of growth stage-specific cell-wall synthesis pleiotropically affects developmental gene expression in an *ssgA* mutant of *Streptomyces coelicolor*. Mol. Microbiol. *64*, 1244–1259.

Noens, E.E.E., Mersinias, V., Traag, B.A., Smith, C.P., Koerten, H.K., and Van Wezel, G.P. (2005). SsgA-like proteins determine the fate of peptidoglycan during sporulation of *Streptomyces coelicolor*. Mol. Microbiol. *58*, 929–944.

Ohnishi, Y., Yamazaki, H., Kato, J.Y., Tomono, A., and Horinouchi, S. (2005). AdpA, a central transcriptional regulator in the A-factor regulatory cascade that leads to morphological development and secondary metabolism in *Streptomyces griseus*. Biosci. Biotechnol. Biochem. *69*, 431–439.

Parashar, A., Colvin, K.R., Bignell, D.R., and Leskiw, B.K. (2009). BldG and SCO3548 interact antagonistically to control key developmental processes in *Streptomyces coelicolor*. J. Bacteriol. *191*, 2541–2550.

Pease, P.J., Levy, O., Cost, G.J., Gore, J., Ptacin, J.L., Sherratt, D., *et al.* (2005). Sequence-directed DNA translocation by purified FtsK. Science *307*, 586–590.

Pope, M.K., Green, B.D., and Westpheling, J. (1996). The *bld* mutants of *Streptomyces coelicolor* are defective in the regulation of carbon utilization, morphogenesis and cell–cell signalling. Mol. Microbiol. *19*, 747–756.

Pope, M.K., Green, B., and Westpheling, J. (1998). The *bldB* gene encodes a small protein required for morphogenesis, antibiotic production, and catabolite control in *Streptomyces coelicolor*. J. Bacteriol. *180*, 1556–1562.

Potúkov, L., Kelemen, G.H., Findlay, K.C., Lonetto, M.A., Buttner, M.J., and Kormanec, J. (1995). A new RNA polymerase sigma factor σ is required for the late stages of morphological differentiation in *Streptomyces* spp. Mol. Microbiol. *17*, 37–48.

Ptacin, J.L., Lee, S.F., Garner, E.C., Toro, E., Eckart, M., Comolli, L.R. *et al.* (2010). A spindle-like apparatus guides bacterial chromosome segregation. Nat. Cell. Biol. *12*, 791–798.

Ramamurthi, K.S., and Losick, R. (2009). Negative curvature as a cue for subcellular localization of a bacterial protein. Proc. Natl. Acad. Sci. U.S.A. *106*, 13541–13545.

Rybniker, J., Nowag, A., Van Gumpel, E., Nissen, N., Robinson, N., Plum, G., and Hartmann, P. (2010). Insights into the function of the WhiB-like protein of mycobacteriophage TM4 – a transcriptional inhibitor of WhiB2. Mol. Microbiol. *77*, 642–657.

Ryding, N.J., Kelemen, G.H., Whatling, C.A., Flärdh, K., Buttner, M.J., and Chater, K.F. (1998). A developmentally regulated gene encoding a repressor-like protein is essential for sporulation in *Streptomyces coelicolor* A3(2). Mol. Microbiol. *29*, 343–357.

Salerno, P., Larsson, J., Bucca, G., Laing, E., Smith, C.P., and Flärdh, K. (2009). One of the two genes encoding nucleoid-associated HU proteins in *Streptomyces coelicolor* is developmentally regulated and specifically involved in spore maturation. J. Bacteriol. *191*, 6489–6500.

Schwedock, J., McCormick, J.R., Angert, E.R., Nodwell, J.R., and Losick, R. (1997). Assembly of the cell division protein FtsZ into ladder-like structures in the aerial hyphae of *Streptomyces coelicolor*. Mol. Microbiol. *25*, 847–858.

Sevciková, B., Benada, O., Kofronova, O., and Kormanec, J. (2001). Stress-response sigma factor sigma(H) is essential for morphological differentiation of *Streptomyces coelicolor* A3(2). Arch. Microbiol. *177*, 98–106.

Sevciková, B., Rezuchova, B., Homerova, D., and Kormanec, J. (2010). The anti-anti-sigma factor BldG is involved in activation of the stress response sigma factor σ(H) in *Streptomyces coelicolor* A3(2). J. Bacteriol. *192*, 5674–5681.

Sharp, M.D., and Pogliano, K. (2002). Role of cell-specific SpoIIIE assembly in polarity of DNA transfer. Science *295*, 137–139.

Sievers, J., and Errington, J. (2000). The *Bacillus subtilis* cell division protein FtsL localizes to sites of septation and interacts with DivIC. Mol. Microbiol. *36*, 846–855.

Takano, E., Tao, M., Long, F., Bibb, M.J., Wang, L., Li, W., *et al.* (2003). A rare leucine codon in *adpA* is implicated in the morphological defect of *bldA* mutants of *Streptomyces coelicolor*. Mol. Microbiol. *50*, 475–486.

Viollier, P.H., Kelemen, G.H., Dale, G.E., Nguyen, K.T., Buttner, M.J., and Thompson, C.J. (2003). Specialized osmotic stress response systems involve multiple SigB-like sigma factors in *Streptomyces coelicolor*. Mol. Microbiol. *47*, 699–714.

Wadenpohl, I., and Bramkamp, M. (2010). DivIC stabilizes FtsL against RasP cleavage. J. Bacteriol. *192*, 5260–5263.

Wang, L., Yu, Y., He, X., Zhou, X., Deng, Z., Chater, K.F., and Tao, M. (2007). Role of an FtsK-like protein in genetic stability in *Streptomyces coelicolor* A3(2). J. Bacteriol. *189*, 2310–2318.

Wang, L., and Lutkenhaus, J. (1998). FtsK is an essential cell division protein that is localized to the septum and induced as part of the SOS response. Mol. Microbiol. *29*, 731–740.

van Wezel, G.P., van der Meulen, J., Kawamoto, S., Luiten, R.G.M., Koerten, H.K., and Kraal, B. (2000). *ssgA* is essential for sporulation of *Streptomyces coelicolor* A3(2) and affects hyphal development by stimulating septum formation. J. Bacteriol. *182*, 5653–5662.

Willemse, J., Borst, J.W., de Waal, E., Bisseling, T., and Van Wezel, G.P. (2011). Positive control of cell division: FtsZ is recruited by SsgB during sporulation of *Streptomyces*. Genes Dev. *25*, 89–99.

Willey, J.M., Willems, A., Kodani, S., and Nodwell, J.R. (2006). Morphogenetic surfactants and their role in the formation of aerial hyphae in *Streptomyces coelicolor*. Mol. Microbiol. *59*, 731–742.

Wu, L.J., and Errington, J. (2004). Coordination of cell division and chromosome segregation by a nucleoid occlusion protein in *Bacillus subtilis*. Cell *117*, 915–925.

Xu, W., Huang, J., Lin, R., Shi, J., and Cohen, S.N. (2010). Regulation of morphological differentiation in, *S. coelicolor* by RNase III (AbsB) cleavage of mRNA encoding the AdpA transcription factor. Mol. Microbiol. *75*, 781–791.

Yu, T.W., Shen, Y., McDaniel, R., Floss, H.G., Khosla, C., Hopwood, D.A., and Moore, B.S. (1998). Engineered biosynthesis of novel polyketides from *Streptomyces* spore pigment polyketide synthases. J. Am. Chem. Soc. *120*, 7749–7759.

The Structure and Composition of the Outer Layers of the Bacterial Spore

5

Arthur I. Aronson

Abstract

This review summarizes the basic structural features, composition and protein interactions of the bacterial spore. Aspects of the possible interactions between the external layers of the bacterial spore, the coat and exosporium and the functions of each is discussed in detail. An important theme is that spores are not simply dormant resistant structures for long term survival under conditions unsuitable for germination. Rather they are adaptable to physical and nutritional conditions in order to enhance their dormancy (resistance), infectivity and dispersal with the spore coat and exosporium having significant roles. Multiple functions beyond protection of the spore cytoplasm, a role in germination and attachment to substrata can begin to account for the complexity of each of these structures.

Morphology

The topic of the external layers of the bacterial spore, the coat and exosporium, has been periodically well reviewed (Driks, 1999; Henriques and Moran, 2007). The surface morphology of bacterial spores varies considerably among sporulating species when examined by freeze etching (Holt and Leadbetter, 1969; Aronson and Fitz-James, 1976) or atomic force microscopy (AFM; Chada *et al.*, 2003; Plomp *et al.*, 2005). Spores of the *Bacillus cereus* group (also includes *B. thuringiensis* and *B. anthracis* plus a few others) have a cross-hatched appearance in their outer coat layers beneath the exosporium (peeled back in Fig. 5.1) in contrast to *Bacillus subtilis* spores (and presumably related species such as *B. lichenformis*, *B. amyloliquefans*, *B pumilis*, *B. brevis*) with a striated surface beneath a thin surface layer (SL in Fig. 31 in Aronson and Fitz-James, 1976). *Bacillus megaterium* KM spores provide a third less prominent external morphology (Aronson and Fitz-James, 1976) and there are other species variations (Holt and Leadbetter, 1969). Beneath the cross-hatched surface of *B. cereus* spores is a distinct pitted layer (Fig. 5.2) clearly indicating at least two morphological components to the coat. Such an undercoat was not evident in freeze etches of *B. subtilis* spores although etching down to the cortex did show that the coat was much thicker.

The AFM surface pictures revealed a convoluted or wrinkled surface appearance which may be species-specific in its details (Chada *et al.*, 2003). Upon germination or hydration when the spore core hydrates, this convolution disappears indicating expansion of the coat in preparation for outgrowth (Plomp *et al.*, 2005) demonstrating flexibility built into the coat structure.

The morphological differences are reinforced by the appearance of the surface layers in thin sections of dormant spores (Fig. 5.1a). *B. cereus* spores have a thin multilayered coat (CT) with a prominent exosporium (EX) whereas *B. subtilis* spores have a much thicker multilayered coat with no exosporium in evidence (Fig. 28 in Aronson and Fitz-James, 1976) although as mentioned above, there is a thin surface layer (SL in Fig. 31 in Aronson and Fitz-James, 1976) adhering closely to the spore surface. This layer stains with ruthenium red indicating that it is glycosylated as are some of the exosporial proteins in the *B.*

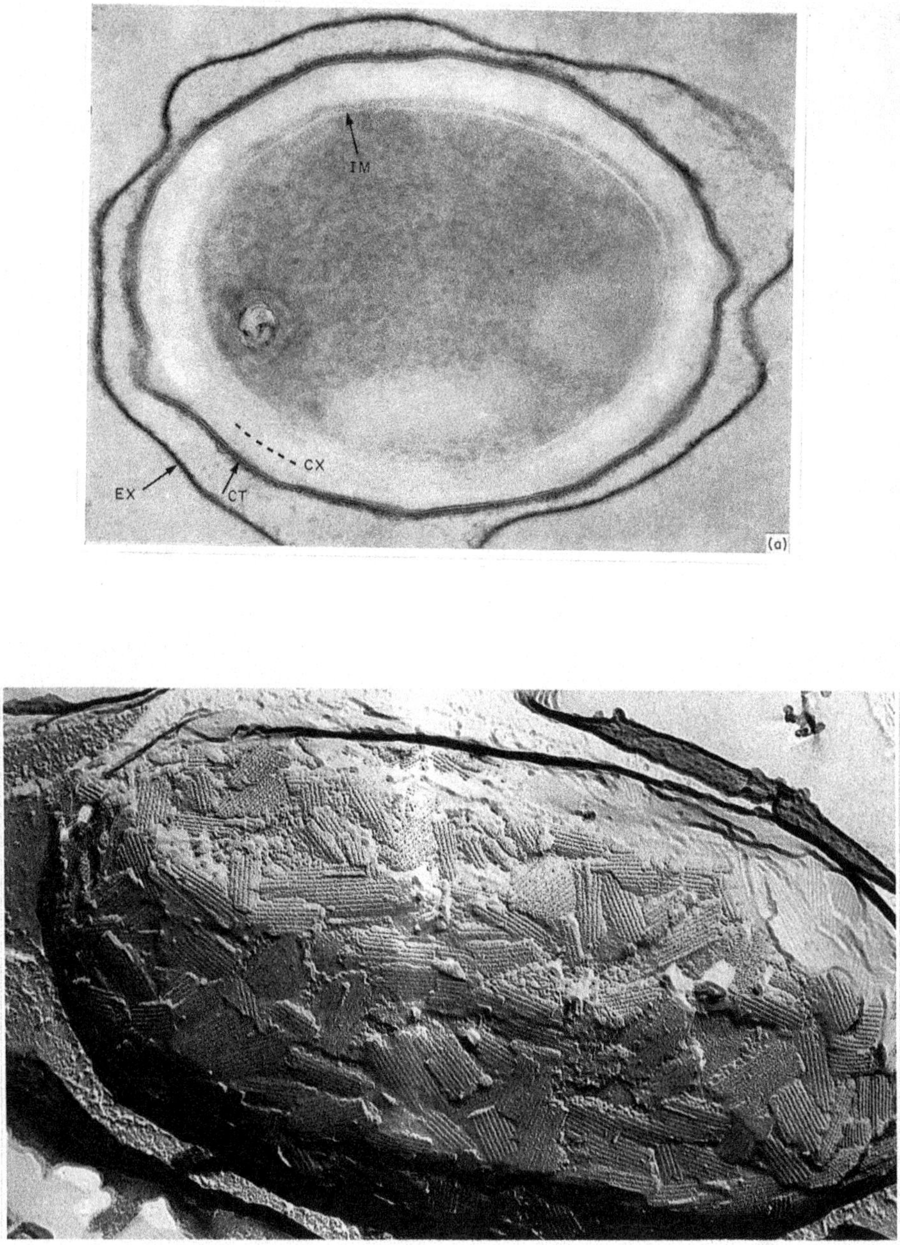

Figure 5.1 Section (top) and freeze-etch (bottom) of *Bacillus cereus* spores grown on enriched medium agar plates. In the top panel, the inner spore membrane (IM), coat layers (CT) and exosporium (EX) are indicated. The presumptive demarcation between the cortex (CX) and inner coat layers is marked with a dashed line. The lower panel shows the peeled back exosporium revealing the cross-patched outer coat layer. The individual rods in this layer are about 13 nm wide by 105 nm long. ×78,000.

cereus group. The lack of a distinct exosporium in spores of *B. subtilis* and related species constitutes a major difference in appearance of these spores versus those in the *B. cereus* and *B. megaterium* groups. As anticipated from the lack of a distinct exosporium in *B. subtilis,* there is an absence in that species of a number of *B. cereus* exosporial proteins (Table 2 in Henriques and Moran, 2007). The functional significance of this difference will be discussed below.

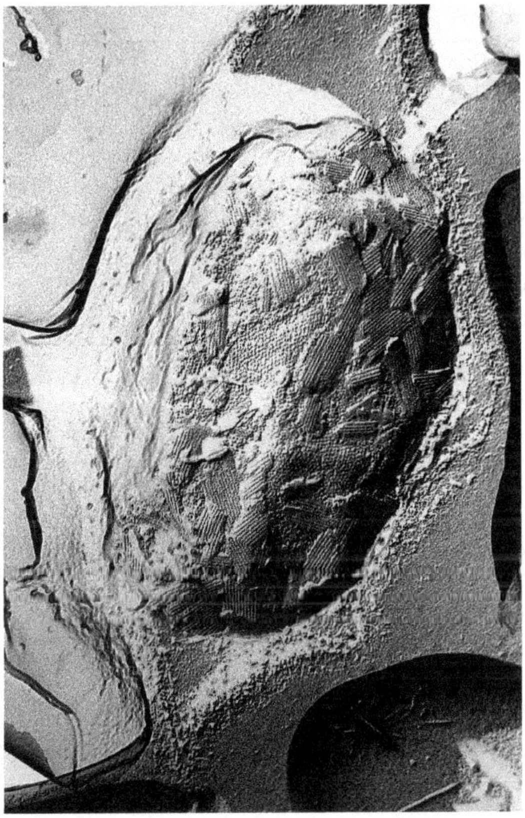

Figure 5.2 A deeper etching of a *Bacillus cereus* spore (as in Fig. 5.1) revealing the inner pitted spore coat layer beneath the outer cross-patched layer. ×60,000.

On the basis of the thin sections, coat layers have been designated as just inner and outer or perhaps as three layers, inner, outer and an amorphous under coat layer (Aronson and Fitz-James, 1976; Fig. 1 in Henriques and Moran, 2007). An undercoat may comprise the proteins involved in the initial assembly steps. In the absence of definitive evidence for the assignment of specific proteins to such an 'under coat' layer, the issue of two or three layers has not been resolved and may be semantic rather than structural.

Other less well-studied spore formers such as *Clostridia* and *Bacillus sphaericus* (*Lysinobacillus*) among others, contain their own unique surface morphology in freeze etchings (Holt and Leadbetter, 1969; Aronson and Fitz-James, 1976) but usually features similar to those of *B. cereus* and *B. subtilis* in thin sections although some *B. megaterium* strains, among others, have very thick outer layers (Aronson and Fitz-James, 1976). These may be part of the coat or an ancillary layer that is a form of a surface appendage. Spores of some species, such as certain *Clostridia* and *Geobacillus* species, contain a distinct and prominent surface appendage (Krasil'nikov *et al.*, 1963; Rode *et al.*, 1967) with a composition and possible function discussed later.

Number and function of spore coat proteins

There is a large number of presumptive spore coat proteins in the best studied species, *B. subtilis* (Table 5.1), and some are present in other species (Table 5.2). Initially, an assignment of these proteins to the coat was based on reverse genetics, i.e. partial sequencing of major bands extracted from one or two-dimensional SDS-PAGE profiles of an extract from purified spores and analysis by direct sequencing (Donovan *et al.*, 1987; Zhang *et al.*, 1993) or later by mass spectrometry (Kim *et al.*, 2006). The method of spore purification is of course critical as some of the minor bands may be contaminants or components released from lysed spores. It is also possible that the purification removed some loosely associated coat proteins. In some cases, the assignments were confirmed by isolation of mutants deficient in the particular protein with some alteration in phenotype such as germination, sensitivity to lysozyme or perhaps the appearance in electron micrograph sections (see, for example, Zhang *et al.*, 1993; Kim *et al.*, 2004).

Another major criterion for assigning proteins to the coat is the localization of a GFP fusion. Fluorescent labelling of spores, either via labelled antibodies to a purified presumptive coat protein or a GFP fusion to such a protein was useful for determining the sites of deposition around the forespore membrane and whether the particular protein stayed localized or spread around the spore surface (vanOoij *et al.*, 2004; Kim *et al.*, 2006). A further refinement in the technique permitted distinguishing between coat protein localization to the inner or outer spore coat (Imamura *et al.*, 2010).

Dependence for deposition upon a functional CotE protein or one of the other morphogenetic proteins (Table 5.1) is also an important criterion

Table 5.1 Classification of *Bacillus subtilis* spore coat proteins

1	Mutations affect structure/composition: CotE, SpoIVA, SpoVID[1], YrbA[1]
2	Dependent on CotE, CotH, GerE or SpoVID for incorporation: GerQ YaaH, YlbD, YtxO, CotA, CotB, CotG[2], CotI, CotR, CotSA
3	Mutations affect function (germination, lysozyme sensitivity): CotT[2], CotO[3], CotH[2,3], SpoVIF[1], SafA[3]
4	Coat localization by fluorescence (Gfp fusion or labelled antibody)[4]: YaaH[3], YabG[3], YeeK[3], YabP, YbaN[2], YtrI[2], YhaX, YhbB, YheD, YhjR, YirY, YknT, YmaG, YodL, YppG, YsnD, YxE, YutH, YuzC, YybI, CotG, CotW[3], GerQ, CotQ, CotU CotJA, CotJB, CotJC, YusA[3]
5	Defined by reverse genetics: CotV, CotW, CotX[3], CotY, CotZ[2], CotD, CotN, YrbB, CotM, CotS[3], CotC[3]
6	Enzymatic activity: Tgl, CwlJ[3], SodA[2], OxdD[3], laccase, lipC (YcsK), (YabG)

[1]Part of coat and/or morphogenetic.
[2]Non-assembly mutations result in a detectable defect in coat structure.
[3]Also CotE dependent.
[4]Patterns of localization differ.

Table 5.2 Distribution of spore coat proteins among sporulating species

A	Only in ***B. subtilis*** and related species (*B. lichenformis*, *B. amyliquafaciens*, *B. pumilis*, *B. atropheus*, *B. pseudofirmus*, *B. couhuilensis*): YrbB[1,2], CotC[1], OxdD[1], YuzC, CotG[1], CotU[1], YtrI, YheD, YknT, YmaG[1], YodL, YrbA[1], YsnD, YxeE, YdhD, YtxO[1], YhbB, CotR, CotT[1], CotQ[3], CotB[4], CotO, CotV, SpoVIF, YeeK[1], YabP, YbaN, YuzC, YybI[1], YppG, lipC, YodI[4], YopQ, YhaX[1]
B	***B. subtilis*** plus one or more of: *B. cereus*, *B. thuringiensis* subsp. *israelensis* (Bti), *B. megaterium* (*B. meg*), *B. sphaericus*, *Geobacillus* (Geo), *B. brevis*, *B. coagulans*: CotX, CotY, CotN(Bti, Oceanobacillus, *B. halouridans*), CotM, YhjR (*Lysinobacillus. sphaericus*), CotB (Geo), CotG (Geo), YsxE(B. meg, *B. coagulans*, *B. cereus*), CotD, SpoVID, CotH, CotW, CotZ[5], CotQ, CotA, CotP, CotJA, CotJB, YutH, YirY, YaaH
C	***B. subtilis*** plus Clostridia: YybI (also in *B. sphaericus*, *Lysinobacillus*), CotI
D	All major species (includes *B. subtilis* and related; *B. cereus* group; *B. megaterium*, *Clostridia*, *Geobacillus*): CotS, CwlJ, CotF, GerQ, CotJC, SafA, YabG[6], SpoIVA, YisY, YsxE, YusA., CotSA, SodA, CotE, Tgl

[1]Found only in *B. subtilis* spores.
[2]Also the exosporial protein, ExsA in *B. cereus.*
[3]Probably lipoprotein so exact role in coat structure not clear.
[4]Also present in *Geobacillus*.
[5]Also in *Lysinobacillus* and the *B. cereus* exosporium.
[6]Probably protease or peptidase.

for assignment. CotE is a major morphogenetic protein, originally characterized in *B. subtilis*, regulating deposition of coat proteins to either the inner or outer coat layers (Zheng *et al.*, 1988). Many but not all *B. subtilis* coat proteins require a functional CotE for their deposition so a screening by SDS-PAGE of coat extracts from the wild type versus a *cotE* mutant is a valuable method for assigning coat proteins (Kim *et al.*, 2006). There are other such 'regulatory' proteins such as CotH (CotG) (Zilhao *et al.*, 2004), CotO, SpoIVA and SpoVID controlling coat deposition of a smaller subset of coat proteins (Table 5.3). CotH is present in the coats of some but not all spore formers (Table 5.2).

SpoIVA is deposited early around or attached to the outer forespore membrane (Roels *et al.*, 1992) via the SpoVM peptide (Ramamurthi *et al.*, 2006) and is a key coat morphogenetic protein in most sporulating species. SpoVID which, is also widely distributed among spore formers, interacts with SpoIVA and is necessary for spore encasement (Wang *et al.*, 2009). It also binds directly with another coat protein, SafA (or YrbA), (Ozin *et al.*, 2001) and helps to initiate the addition of CotJA and CotJC (Seyler *et al.*, 1997) and thus the

Table 5.3 *Bacillus subtilis* spore coat proteins with well defined functions in coat assembly

Protein	Location	Function
SpoVM	Anchored to inner spore membrane	Interacts with SpoIVA to initiate pre-coat formation
SpoIVA	Early in precoat as a shell	Interacts with SpoVM, SpoIVD and SafA
SpoVID	Early precoat	Interacts with SafA and SpoJA, SpoJC
SafA	Part of early precoat and then under coat	Also influences presence of CotG in outer coat
SpoJA, JC	Components of the inner coat via dependence on SafA and CotE	–
CotE	Forms a ring between future inner and outer coat layers	Responsible for the deposition of several outer coat proteins and for the completion of the inner coat
CotH	–	Dependence on CotE and in turn regulates the deposition of CotS and SA to the inner coat and CotB to the surface of the outer coat. Also partially involved in deposition of other coat proteins: CotG, Q, U, B, C, G and Z
Tgl	–	Transglutaminase with a role in polymerization of GerQ and perhaps other coat insoluble proteins such as CotV, W, X in the outer coat. Cross-linking is a late event just prior to spore release from the mother cell and the extent may vary with sporulation conditions[1]
CotG	Also cross-linked	Regulates deposition of CotU and CotQ in the outer coat
GerQ	Cross-linked as multimers in the outer coat.	Role in the deposition of the cortex lytic enzyme, CwlJ
SodA	Superoxide dismutase in outer coat	May have a role in the cross-linking via dityrosines of some outer coat proteins such as CotV W,X
CotO	–	Regulates the deposition of a subset of outer coat proteins (CotB C G, U, W, Z)

[1]There is also disulfide bond cross-linking of some outer coat proteins such as Cot Y and Z. This may occur just prior to spore release and may be dependent upon suitable oxidative conditions in the environment (rather than a specific enzyme).

formation of the inner or under coat. Deposition of CotE as a ring then directs the further addition of some coat proteins to either the inner or outer coat layers. The subset of proteins regulated by CotH including CotC, CotG and CotB aid in the completion of the coat layers (Isticato *et al.*, 2004; Zilhao *et al.*, 2004; Table 5.3). Another subset appears to be controlled by YabG (Takamatsu *et al.*, 2000). Processes such as multimerization (probably both homo and hetero) and protease modifications occur during or shortly after deposition. Proteolytic cleavage of *B. subtilis* proteins CotT (Bourne *et al.*, 1991) and CotF (Cutting *et al.*, 1991), are necessary for deposition, in particular for the formation of filaments by CotT (Aronson *et al.*, 1992). There are proteases associated with the spore (such as YabG) but the exact role of each is not known. Processing of some such as of GerQ (not to be confused with the *gerQ* operon in *B. cereus*), may be essential for polymerization (Ragkousi and Setlow, 2004).

The final stages probably involve the cross-linking of certain coat proteins by dityrosine due to oxidation of tyrosine residues by an enzyme such as SodA (Henriques *et al.*, 1998), γ-glutamyl-lysine cross-links catalysed by enzymes such as the Tgl protein (Kobayashi *et al.*, 1996; Ragkousi and Setlow, 2004) and disulphide cross links. About 20–30% of the coat proteins in *B. subtilis* including the CotV W,X Y,Z proteins, are found in the insoluble fraction (Zhang *et al.*, 1993). The genes encoding these five protein are closely linked and are present in two operons encoding Cot V W,X (and a monocistronic cotX promoter) and a separate *cotY Z* operon, all transcribed by sigmaK RNA polymerase (Zhang *et al.*, 1994). Cot Y and

Z are rich in cysteine residues and probably cross-linked by disulfide bonds. The other three are probably cross-linked by γ-glutamyl-lysine residues. Disruption of the *cotx, y* and *z* genes resulted in spores with a deficient outer coat (Zhang *et al.*, 1993) implying that these cross-linked proteins are located there. Another well-characterized cross-linked protein is GerQ, which is a multimeric coat protein, probably in the outer layer. It functions to make the cortex lytic enzyme, CwlJ available (Ragkousi and Setlow, 2004). The assembly process of a *B. subtilis* spore coat from precoat to completed inner and outer coat layers including the extensive network of dependency for assembly is summarized in Table 5.3 and in figures in Kim *et al.* (2006) and in Henriques and Moran (2007).

The extent of cross linking in the *B. cereus* group and perhaps in other spore formers appears to be considerably less as a larger fraction of the coat protein can be solubilized by a solution containing SDS and urea with a pH greater than 10.0 (in the case of *B. cereus*, >90%, compared with about 70% in *B. subtilis*). The Tgl protein is widespread but the nature of the cross-linked proteins in species other than *B. subtilis* is not known as homologues of CotX Y, and Z are not found very broadly although GerQ is (Table 5.2).

All of the crosslinks appear to be late events in spore assembly even some at the time of release of the endospore from the mother cell (Ragkousi and Setlow, 2004). The formation and extent of a particular cross link could thus depend upon the sporulation conditions implying the ability of the spore to alter surface properties in response to environmental conditions.

The absence of a phenotypic effect due to the deletion of a presumptive coat protein gene may be due to redundant functions, lack of a suitable selection or inadequate screening procedures so there is a strong reliance on the criteria discussed above for assigning proteins to the coat. Considering all of these factors, the number of well defined *B. subtilis* coat proteins to date is between 65 and 70, somewhat fewer than reported by Henriques and Moran (2007) because some in their Table 5.1 did not meet the above criteria. It is not yet completely known how many of the assigned coat proteins are for assembly, for protection (i.e. resistance to lysozyme and perhaps engulfment by protozoa), to enhance germination or to contribute to spore flexibility by responding to nutritional and physical conditions. This latter function could aid spore survival, the capacity to germinate or perhaps in the case of pathogens, enhance their invasive properties.

On the basis of an NCBI Blast search, a number of the *B. subtilis* spore coat proteins are either limited to this species or present only in closely related species (Table 5.2). Other spore formers may contain an equally extensive array and some may be unique with functions homologous to those in *B. subtilis*. Some may be present only in a particular species contributing to differences in appearance (Figs. 5.1 and 5.2), ecology, germination conditions, etc. For example, spores of the *B. cereus* group are already known to contain at least three novel coat proteins (designated Cotα, Cotβ and Cotγ) (Kim *et al.*, 2004; Mallozzi *et al.*, 2008). In addition, some *B. subtilis* coat proteins such as CotE and CotY are present in extracts from purified exosporia in the *B. cereus* group (Todd *et al.*, 2003) implying an additional or different function in this group of spore formers.

The coat is obviously a complex structure with an extensive regulatory network for assembly as described in Kim *et al.* (2006). Somehow, this network must result in the deposition of two or three distinct coat layers each with its own profile of coat proteins. There are several subsets of deposition dependency as summarized in Kim *et al.* (2006) and Henriques and Moran (2007) as well as other studies showing direct interactions via immunological 'pull-down' assays such as for SafA and SpoVID (Ozin *et al.*, 2001). Such interactions would ensure the co-localization of two or more proteins within the coat and thus a rudimentary understanding of the assembly process.

The widespread presence of the key assembly proteins among most if not all spore formers surveyed to date (Table 5.2) indicates that the basic mechanisms for producing spore coat layers are probably similar to that in *B. subtilis*. The ecology of *B. subtilis* appears to be diverse and includes soil and probably other parts of the phylloplane. It is generally not a pathogen and thus not invasive and its life style (not completely understood) must account for the relatively thick coat with a large number of different proteins. Other species such as *B. anthracis* with a thinner coat and perhaps a less complex array of coat proteins beyond those required for assembly, survives in soil but is invasive. The thinner coat may provide flexibility for survival and a rapid germination response outside of a host. The coat/exosporium complex contributes to invasive properties and the ability to resist host defence mechanisms (see the later discussion of exosporia). There is obviously a long way to go in defining the array of coat proteins in various species and how this may reflect their special survival needs.

There has been a recent surge in the characterization of new species of spore formers resulting in large part from mass sequencing projects of DNA from various sources. These species include a variety of *Bacillus* species, including *Bacillus coahuilensis, Bacillus pseudofirmus, Bacillus atrophaeus, Bacillus cellulyticus, Bacillus* cytotoxicus, as well as *Turicibacter, Prevotella, Anoxybacillus, Syntrophomonas, Geobacillus, Oceanobacillus* and *Heliobacterium*. They all have one or more of the well characterized coat proteins from *B. subtilis* or from the *B. cereus* group as identified in a BLAST search. Each species almost certainly has its own unique array as well with either functions similar to those already defined (homologues) and very probably some confined to the particular species. Such diversity and uniqueness of spore proteins is already evident from a comparison of the array of germination proteins (and germination conditions) found in bacilli and clostridia (Paredes-Sabja *et al.*, 2010). These differences must reflect the very different ecology of these species. For example, the unique ability of *B. thuringiensis* subsp. *israelensis* spores to respond to alkaline conditions for germination is due to the presence of a novel three gene *ger* operon encoded on the major pathogenic plasmid (Abdoarrahem *et al.*, 2007). These spores can be taken up by Dipteran larvae where the midgut pH is very high. In such an environment, toxic inclusion proteins are solubilized resulting in larval death with the nutrient-rich haemolymph becoming available for bacterial growth. The ability of these spores to germinate in the alkaline midgut would shorten the lag period for subsequent growth once this nutrient-rich haemolymph became available. This diversity and uniqueness of spore germination properties among species is probably true of coat proteins as well.

Overall, the coat is not a rigid structure as shown by ATM with changes in the profile in response to the spores being fully hydrated, air dried or undergoing germination (Plomp *et al.*, 2005). Other factors such as temperature (Melly *et al.*, 2002), solid or liquid media (Rose *et al.*, 2007), nutrient variation and presence of solvents (Nicholson *et al.*, 2000) probably influence coat and/or coat/exosporium composition and morphology. This diversity and flexibility is not surprising given the need to be protective and to sustain dormancy. The coat must also expand during hydration usually in response to germinants and for the subsequent opening of this structure for outgrowth. Other factors such as the extent of spore hydrophobicity, ability to respond rapidly to a particular set of germinants, and invasive properties could influence the appearance and composition of the spore outer layers.

Several enzymes are present in the coat (Tables 5.1 and 5.2) some of which are involved in coat structure such as Tgl for cross-linking GerQ (Ragkousi and Setlow, 2004) and perhaps CotV W and X. As discussed above, the *B. subtilis* GerQ is involved in germination via activation of the cortex lytic enzyme, CwlJ. It appears to be a coat protein rather than one of the Ger proteins found in the inner spore membrane (usually as three gene operons). The CwlJ protein embedded in the coat is involved in cortex degradation during the early stages of germination. Each of these proteins must be positioned to interact with their substrates, a further indication of considerable order in coat assembly. Other enzymes (SodA, OxdD, laccase) may be located on or near the coat surface and provide protection to the dormant spore. A role for superoxide dismutase (SodA) would be beneficial in any environment where toxic oxidative compounds are present or can be generated.. It is more difficult to appreciate the need for oxalate decarboxylase or laccase which presumably function for a select group of *B. subtilis* growing in special environments. If these enzymes were critical for spore survival, germination, etc. then they should be located so as to access substrates efficiently. A surface location is a distinct possibility, although the permeability of the coat to small molecules, which has not yet been assessed, may permit other locations. It is conceivable that some coat proteins with no known functions serve to make these enzymes accessible to their substrates.

Alternatively, these 'enzymes' may only function in the spore structure. If they were enzymatic and thus reflecting the environment with a need for protection or perhaps to generate germinants, then each spore former would be expected to have its own array. In fact, a different set of enzymes is found in exosporial extracts of *B. cereus* group spores as will be discussed later.

As noted above, the spores of many species (*B. cereus* group, *B. megaterium*, *Clostridia*) contain thinner coats than *B. subtilis* but with a similar inner and outer coat arrangement (and perhaps a third under coat layer) (Fig. 5.2). While the spores of these species do not appear to contain close homologues of many of the *B. subtilis* coat proteins (Table 5.2), the SDS-PAGE profiles of coat extracts are relatively complex and do differ (Driks, 1999) in all cases examined to date indicating that there are probably numerous coat proteins unique to each. These may be functional analogues of those in *B. subtilis* or may reflect the special environmental conditions, i.e. germinants, protection, retaining dormancy, etc., for each species. For example, at least three novel proteins have been found in coat extracts of *B. anthracis* spores (Kim *et al.*, 2004; Mallozzi *et al.*, 2008). One of ca. 13 kDa, Cotα, is present exclusively in members of the *B. cereus* group and is a major component of the outer coat (Kim *et al.*, 2004). Another protein found in the under coat of *B. anthracis* spores, Cotγ, has a significant function in altering the association between the exosporium and the coat as will be discussed later. Cotβ was also isolated from *B. anthracis* spore coat extracts (Mallozzi *et al.*, 2008) but has no well defined function. It is anticipated that other such species-specific coat proteins will be found. Learning about their functions and how they contribute to spore survival will be important for understanding the breadth of spore coat functions as well as unique structural aspects.

An example of a specialty spore surface structure is the major appendage(s) attached to either the spore coat or exosporium of some anaerobes, especially clostridial species, as first described by Krasil'nikov *et al.* (1963) and then Rode *et al.* (1967). These appendages vary in shape, size and number but are most prevalent among anaerobes. Four proteins were isolated from such an appendage from *Clostridium taeniosporum* including a glycoprotein and three are unique (Walker *et al.*, 2007). To date, the most distinctive feature of this appendage is the ability to adhere to various substrates. It could thus serve as an anchor perhaps to ensure that the spores remain in an anaerobic environment for germination and subsequent growth. There is much to learn about the composition, assembly and diversity of these unique structures.

Another novel property of the spores of certain *B. cereus* strains is the presence of silica in the spore coat (Hirota *et al.*, 2010). One particular strain took up silica quite rapidly and it appeared to be localized to the surface of the outer coat layer. Spores of this strain were more resistant to acid treatment than the same strain grown without silica. This particular isolate from a paddy field in Japan may confront lethal acidic environments and thus a selective advantage for the presence of silica. It will be interesting to see if there is a special spore coat protein needed to bind the silica or perhaps an overall conformational difference in the spore coat involving numerous proteins for silica binding. This strain and the presence of appendages in anaerobes underscore the potential variety among spore formers of coat proteins and structures unique to particular species in order to enhance spore survival in special environments. As mentioned above, some coat proteins may have equivalent functions perhaps in protecting the spore core. Others could be involved in the response to unique environmental conditions for germination, survival in special environmental conditions, dispersal, invasiveness, etc.

A distinct exosporium

As originally defined by Gerhardt and Ribi (1964) for *B. cereus*, the exosporium consists of two principal layers, a paracrystalline basal structure and a hair-like nap [see Ball *et al.* (2008) for a more recent structural analysis]. As noted above, spores of the *B. subtilis* group do not have a distinct exosporium although a thin outer layer is visible in freeze etchings (Aronson and Fitz-James, 1973). This layer stains with ruthenium red indicating that it is a glycosylated. As discussed below, glycosylation is a feature of some exosporial proteins in the *B. cereus* group. The function of this thin external layer in *B. subtilis* is not known.

While a distinct exosporium is present in many species other than *B. subtilis* and related species, it has been studied most extensively in the *B. cereus* group, especially *B. anthracis*. A summary of the proteins identified in *B. cereus*-type exosporia is in the review by Henriques and Moran (2007) and is updated with possible functions in Table 5.4. Interestingly, most of these proteins are confined to the *B. cereus* group indicating that other species with clear morphological evidence for an exosporium must have sets of unique exosporial proteins. How extensively they overlap in structure and function with those in the *B. cereus* group will be of great interest for defining exosporial function.

In the *B. cereus* group, there are about 20 different proteins in the basal layer and the nap consists primarily of the glycosylated BclA protein (Sylvestre *et al.*, 2002) and a second less prevalent glycosylated protein, BclB (Thompson *et al.*, 2007). ExsFA (BxpB) is required for proper assembly of BclA in the nap (Sylvestre *et al.*, 2005; Giorno *et al.*, 2009; Thompson *et al.*, 2011). The principal sugars in a tetrascaccharide in BclA are rhamnopyranoside and dideoxyglucopyranoide (anthrose), with the latter unique to *B. anthracis* exosporia (Daubenspeck *et al.*, 2004). ExsB is a highly phosphorylated protein (McPherson *et al.*, 2010), but there is controversy whether it is primarily an exosporial protein or would be better described as the inner coat protein, Cotγ (A. Aronson, unpublished results). Immunogold

Table 5.4 Exosporial proteins from *Bacillus cereus/Bacillus anthracis* and their functions

Protein	Location	Function
Alr	–	Alanine racemase
Sod	–	Superoxide dismutase; protection
InH (IunH)	–	Inosine hydrolase
BclA	Outer nap	Cell attachment; glycosylated
BclB	High MW complex In outer nap	Glycosylated
ExsY	Basal layer	Assembly (CotZ orthologue)
CotY/Z	Basal layer	Assembly (ExsY homologue)
ExsA	Coat?	Assembly onto spore (SafA orthologue); repeats
CotB	–	–
ExsB[1]	High MW complex	Cot G analogue
ExsC	Basal layer	–
ExsD	Basal layer	–
ExsE	Basal layer	–
ExsFB	Basal layer	For nap (homologue of ExsFA)
ExsG	Basal layer	–
ExsJ	–	Glycosylated
ExsK[2]	High MW complex Outer basal; space	–
ExsM	–	Germination; exo structure
ExsFA (BxpB)	Basal layer	Needed for nap (BclA incorporation)

[1]Also called Cotγ; see text for discussion.
[2]On spore surface and/or in exosporium/coat space?
Note: GroEL has been found in exosporial extracts; ExsH found in *B. thuringiensis*.

labelling with specific antibody indicates that this protein is present in the under coat or perhaps even in the cortex (Fig 5.3). If so, its proper designation is Cotγ and more about its function will be presented later.

As anticipated, there is interaction and interdependence for deposition among these basal layer proteins. Overall, deposition proceeds from one pole of the spore, the so-called bottle cap model (Steichen *et al.*, 2007). It is through this site that the germinating vegetative cell emerges. CotE, the same protein responsible for directing coat protein to the inner and outer coat layers in *B. subtilis*, is a key regulator of exosporium deposition in *B. anthracis* (Giorno *et al.*, 2007). It appears to have an additional or different role in *B. anthracis* than in *B. subtilis* but in both cases, it directs or traffics proteins to their proper location. Another *B. subtilis* coat protein, CotY and its homologue, ExsY, have been either extracted from *B. cereus* exosporia or identified in a transposon screening and each affects exosporial deposition or structure (Todd *et al.*, 2003; Johnson *et al.*, 2006). The double mutant lacks exosporia and has a very thin coat permeable to lysozyme indicating dual roles for these proteins.

The assembled exosporium appears in most electron micrograph sections to be loosely associated with the spore and well separated from the coat (Fig. 5.1, upper panel). This arrangement can vary depending upon the medium used for sporulation. In a glucose-containing medium (the sugar is usually absent in most sporulation media), the exosporium is more closely juxtaposed with the coat (Fig. 5.4). This configuration may be due to an altered composition of the outer coat layer or of the exosporium that could lead to some modification in the interaction of the two layers. This altered position of the exosporium relative to the coat probably contributes to the lower hydrophobicity of the spore and could thus be a factor in spore dispersal. This medium dependency is due to the regulated synthesis of the Cotγ protein mentioned above. The transcription of *cotγ* is decreased in a glucose medium and this is correlated with the change in the position of the exosporium. The same result is found in a mutant with a deletion of the *cotγ* gene and the wild-type phenotype is restored by introducing a plasmid containing this gene (A. Aronson, unpublished results). It appears that the medium-dependent transcriptional regulation of *cotγ* encoding an

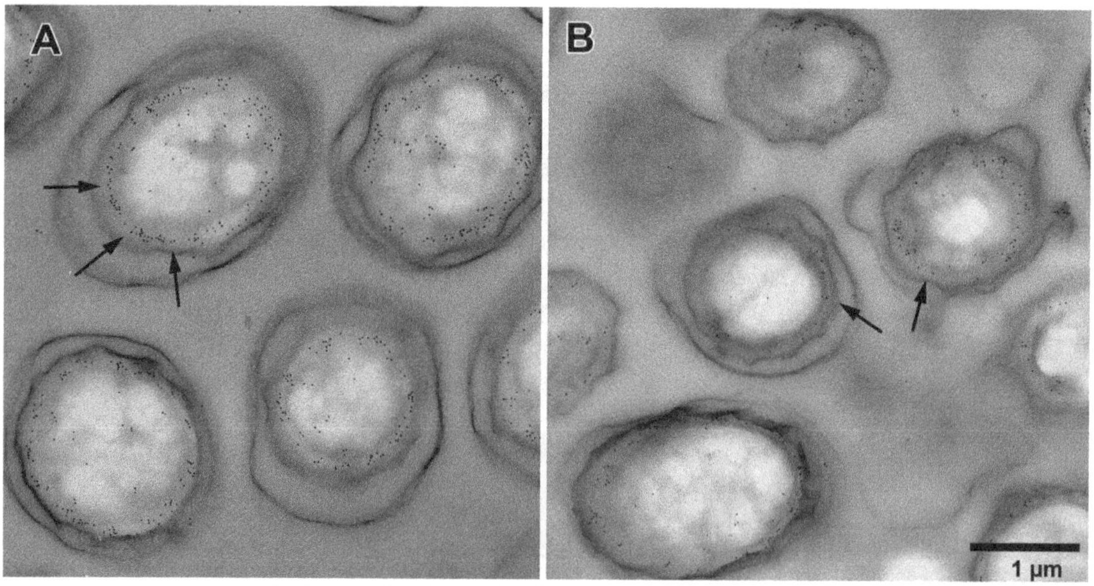

Figure 5.3 Localization of Cotγ in spore sections by immunogold labelling with anti-Cotγ antibody. A: spores produced on NSM plates at 30°C; B: spores produced on G plates at 30°C. Two arrows on left in (A) point to grains localized to inner coat/cortex region; the other arrow in (A) and the two in (B) point to stained outer spore coat. The average difference in grain number per section (50 sections) was 2-fold.

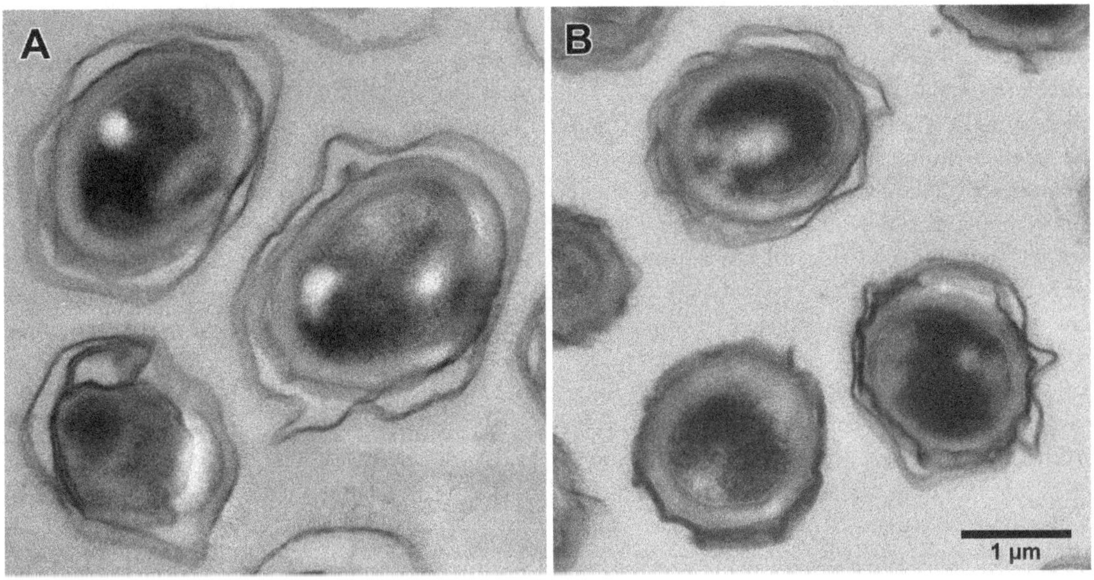

Figure 5.4 Appearance of the spore coat and exosporium in electron micrograph sections of spores. (A) NSM liquid at 30°C and (B) G liquid at 30°C. In each case, about 100 spores were examined and these are representative.

inner spore coat protein affects the juxtaposition of the coat and exosporium and thus the hydrophobicity of the spore.

Since the *cotγ* gene appears to be transcribed by a sigmaK promoter when much of the spore structure has been completed, the sporulating cell is able to respond, even at such a late stage, to environmental conditions in order to alter spore properties. Presumably, the lower hydrophobicity, would decrease dispersal perhaps to keep spores in a nutrient-rich environment for germination and subsequent growth. Another 'inner coat' protein in *B. cereus*, ExsA, a homologue of the *B. subtilis* SafA protein, has a cortex binding motif and is involved in both coat and exosporium assembly (Bailey-Smith *et al.*, 2005). A disruption of the *exsA* gene resulted in lysozyme sensitivity of the spores and a delay in the late stages of germination. The ExsFA protein mentioned above may also have a dual role in assembly of the exosporial nap and the outer spore coat (Giorno *et al.*, 2009). The roles of these 'exosporial' proteins as well as that of Cotγ in both coat and exosporium structure/assembly clearly indicate an assembly and probably functional interactions between the two layers. This could involve the positioning of the exosporium relative to the coat which could affect several critical spore properties. These would include hydrophobicity, attachment (and thus infectivity; see below), protection due to an altered distribution of exosporial (and perhaps spore coat) enzymes and response to germinants. It is probable that the coat and exosporium function coordinately in order to enhance or modify certain spore properties.

The space between the exosporium and coat has been dubbed the interspace and it has been hypothesized that certain exosporial proteins, such as ExsK, ExsFB and IunH are present in this region as well as in the exosporium (Giorno *et al.*, 2009; Severson *et al.*, 2009). In addition to the exosporial structural proteins, several enzyme proteins (perhaps as many as twenty; Redmond *et al.*, 2004) have been identified in exosporial fractions. These include alanine racemase (Alr), nucleoside hydrolase (lunH or InH), superoxide dismutase (Sod) and in some isolates, arginase. There is also a unique manganese oxidizing activity found in the exosporium from a marine *Bacillus* sp. (strain SG-1) (Francis *et al.*, 2002). This is probably one of many special exosporial enzymes designed to aid the survival of spores from species in certain environments. The specific location of these enzymes within the exosporium is not known nor is their enzymatic activity confirmed in all cases. Since the exosporium is readily permeable to small molecules (Ball *et al.*, 2008), they could be in either the nap or basal layer and be accessible to substrates in either of these locations. If active, the hydrolase and racemase could serve to prevent premature germination in an environment unfavourable for spore outgrowth or in the case of *B. anthracis* spores, in the hostile environment within a macrophage. Ungerminated spores survive within macrophages but are rapidly killed upon germination (Hu *et al.*, 2006). The dismutase could prevent spore killing by superoxide radicals as could the coat-associated SodA in *B. subtilis* spores. Arginase appears to remove arginine which is a source of nitric oxide for spore killing within macrophages (Weaver *et al.*, 2007).

The glycosylated BclA protein in the hair-like nap contains collagen-like repeats of variable number among different isolates (Sylvestre *et al.*, 2002). The probable interaction of monomers to form trimers would be important in the overall structure of the nap. This protein in *B. anthracis* is critical for the binding of spores to cells since spores require it to enter professional phagocytes but only those containing a Mac-1 receptor (Oliva *et al.*, 2008). Similarly, spore entry into epithelial cells is dependent upon the presence of the BclA protein and an integrin α2β1 as receptor (Xue *et al.*, 2011). Such binding and subsequent entry are critical for the successful movement of spores (or germinated cells) to lymph nodes for dissemination into the blood stream with the subsequent synthesis of the *B. anthracis* toxins by germinated cells.

Spores lacking or deficient in an exosporium were able to enter some cell types but not macrophages and were infective in a mouse model (Bozue *et al.*, 2007; Giorno *et al.*, 2007). The specificity of spore binding to certain cell types attributable to BclA, however, does imply that the exosporium has an important role in the disease process for *B. anthracis* by controlling the types of cells that the spores can enter and eventually initiate a successful infection. Interrupting such exosporial-dependent binding, especially in *B. anthracis*, could be a valuable approach for disease abatement.

BclA or a related surface glycoprotein such as ExsJ may serve a similar function for related Bacilli such as *B. cereus* food-borne pathogens. It may also function for specific attachment in non-pathogenic *B. cereus* and in *B. thuringiensis*. The latter species produces toxic insecticidal proteins as inclusions as well as on the spore surface. A surface glycoprotein such as BclA may help in spore attachment to larval midgut cells and thus enhance the toxin available for larval killing as well as the retention of spores for subsequent germination and growth.

Conclusions

1. The foundations of spore coat and exosporium assembly including the major players are known and may be the same across all species (Tables 5.2 and 5.3).
2. In addition, there are a number of other proteins which appear to be involved in basic coat structure but their number and species-specificity vary considerably. It is not clear, for example, why the *B. subtilis* coat is relatively thick and contains a large number of unique proteins. Some such as CotT and the cross-linked proteins including GerQ and CotV, W, X, Y, and Z may be involved in the basic structure providing protection but the roles of so many others are not known.
3. There are potential enzyme functions in the *B. subtilis* coat and in the exosporia of the *B. cereus* group which could aid in protection and germination. The array of such enzymes would be expected to be species specific, reflecting germination conditions and the environment or habitat of a particular spore former. It is possible that several or even many spore coat proteins with no well-defined functions serve to position these enzymes to ensure optimal activity.
4. To date, coat function is considered to be primarily protective as reflected by resistance to lysozyme and other potentially harmful chemicals. There also appears to be a secondary role in germination (such as for GerP, Behravan *et al.*, 2000) that may include trafficking germinants to the Ger proteins in the inner spore membrane as well as the localization of cortex lytic enzymes. The ability

of the coat to rapidly expand in response to hydration and to be opened for germination are also important properties of the coat involving unknown arrays of coat proteins.

5 Spores are known to vary in size and coat composition in response to the conditions of sporulation (nutrients, temperature, etc.) perhaps reflecting the roles of a special subset of regulated coat proteins. For example, the Cotγ protein influences the interaction between the coat and exosporium and thus the hydrophobicity of the spores. There are probably other more subtle coat/exosporium interactions that affect spore properties and thus the adaptability of various spore formers to their environments.

6 The prominent surface appendages in clostridia, the binding of silica by a particular isolate of *B. cereus* and the oxidation of manganese by a spore former isolated from a marine *Bacillus* are examples of species-specific adaptability in coat/surface composition.

7 The characterization of exosporia has to date been largely confined to the *B. cereus* group so it is not known if this structure will be more invariant than the coat but with some species-specific characteristics (at least in terms of enzymatic functions). The assembly of the coat and exosporium is probably co-regulated implying cooperation or interaction of some of the functions of these two spore components.

8 There is still much to learn about spore coat/exosporial structure and function especially the variable and species-specific properties and composition required for the subtle interactions of spores with their environments. These would include special protective properties, ability to respond to certain germinants, adherence, dispersal and infectivity, survival in the dormant state in special environments and other properties contributing to spores as dynamic entities.

References

Abdoarrahem, M.M., Gammon, K., Dancer, B.N., and Berry, C. (2009). Genetic basis for alkaline activation of germination in *Bacillus thuringiensis* subsp. *israelensis*. Appl. Environ. Microbiol. 75, 6410–6413.

Aronson, A.I., Ekanayake, L., and Fitz-James, P.C. (1992). Protein filaments may initiate the assembly of the *Bacillus subtilis* spore coat. Biochimie 74, 661–667.

Aronson, A.I., and Fitz-James, P. (1976). Structure and morphogenesis of the bacterial spore coat. Bacteriol. Rev. *40*, 360–402.

Bailey-Smith, K., Todd, S.J., Southworth, T.W., Proctor, J., and Moir, A.. (2005). the ExsA protein of *Bacillus cereus* is required for assembly of coat and exosporium onto the spore surface. J. Bacteriol. *187*, 3800–3806.

Ball, D.A., Taylor, R., Todd, S.J., Redmund, C., Couture-Tosi, E., Sylvestre, P., Moir, A., and Bullough, P.A. (2008). Structure of the exosporium and sublayers of spores of the *Bacillus cereus* family revealed by electron crystallography. Mol. Microbiol. *68*, 947–958.

Behravan, J., Chirakkal, H., Masson, A., and Moir, A. (2000). Mutations in the *gerP* locus of *Bacillus subtilis* and *Bacillus cereus* affect access of germinants to their targets in spores. J. Bacteriol. *182*, 1987–1994.

Bourne, N., Fitz-James, P.C., and Aronson, A.I. (1991). Structural and germination defects of *Bacillus subtilis* spores with altered contents of a spore coat protein. J. Bacteriol. *173*, 6618–6625.

Bozue, J., Moody, K.L., Cote, C.K., Stiles, B.G., Friedlander, A.M., Welkos, S.L., and Hale, M.L. (2007). *Bacillus anthracis* spores of the *bclA* mutant exhibit increased adherence to epithelial cells, fibroblasts and endothelial cells but not to macrophages. Infect. Immun. *75*, 4498–4505.

Chada, V.G.R., Sanstad, E.A., Wang, R., and Driks, A. (2003). Morphogenesis of *Bacillus* spore surfaces. J. Bacteriol. *185*, 6255–6261.

Cutting, S., Zheng, L., and Losick, R. (1991). Gene encoding two alkali-soluble components of the spore coat from *Bacillus subtilis*. J. Bacteriol. *173*, 2915–2919.

Daubenspeck, J.M., Zeng, H., Chen, P., Dong, S., Steichen, C.T., Krishna, N.R., Pritchard, D.G., and Turnbough, C.L., Jr. (2004). Novel oligosaccharide sidechains of the collagen-like region of BclA, the major glycoprotein of the *Bacillus anthracis* exosporium. J. Biol. Chem. *279*, 30945–30953.

Donovan, W.L., Zheng, B., Sandman, K., and Losick, R. (1987). Genes encoding spore coat polypeptides from *Bacillus subtilis*. J. Mol. Biol. *196*, 1–10.

Driks, A. (1999). *Bacillus subtilis* spore coat. Microbiol. Mol. Biol. Rev. *63*, 1–20.

Francis, C.A., Casciotti, K.L., and Tebo, B.M. (2002). Localization of Mn(II)-oxidizing activity and the putative multicopper oxidase, MnxG, to the exosporium of the marine *Bacillus* sp. strain SG-1. Arch. Microbiol. *178*, 450–456.

Gerhardt, P., and Ribi, E. (1964). Ultrastructure of the exosporium enveloping spores of *Bacillus cereus*. J. Bacteriol. *88*, 1774–1789.

Giorno, R., Bozue, J., Cote, T., Wenzel, T., Moody, K.S., Mallozzi, M., Ryan, M., Wang, R., Zielke, R., Maddock, J.R., *et al.* (2007). Morphogenesis of the *Bacillus anthracis* spore. J. Bacteriol. *189*, 691–705.

Giorno, R., Mallozzi, M., Bozue, J., Moody, K.S., Slack, A., Qiu, D., Wang, R., Friedlander, A., Welkas, S., and Driks, A. (2009). Localization and assembly of proteins comprising the outer structures of the *Bacillus anthracis* spore. Microbiol. *155*, 1133–1145.

Henriques, A.O., Melsen, L.R., and Moran, C.P., Jr. (1998). Involvement of superoxide dismutase in spore coat assembly in *Bacillus subtilis*. J. Bacteriol. *180*, 2285–2291.

Henriques, A.O., and Moran, C.P., Jr. (2007). Structure, assembly, and function of the spore surface layers. Annu. Rev. Microbiol. *61*, 555–588.

Hirota, R., Hata, Y., Ikeda, T., Ishida, T., and Kuroda, A. (2010). The silicon layer supports acid resistance of *Bacillus cereus* spores. J. Bacteriol. *192*, 111–116.

Holt, S.C., and Leadbetter, E.R. (1969). Comparative ultrastructure of selected aerobic spore-forming bacteria: a freeze etching study. Bacteriol. Rev. *33*, 346–378.

Hu, H., Sa, Q., Koehler, T.M., Aronson, A.I., and Zhou, D. (2006). Inactivation of *Bacillus anthracis* spores in murine primary macrophages. Cell Microbiol. *8*, 1634–1642.

Imamura, D., Kuwana, R., Takamatsu, H., and Watabe, K. (2010). Localization of proteins to different layers and regions of *Bacillus subtilis* spore coats. J. Bacteriol. *192*, 518–524.

Isticato, R., Esposito, G., Zilhao, R., Nolasco, S., Cangiano, G., De Felice, M., Henriques, A.O., and Ricca, E. (2004). Assembly of multiple CotC forms into the *Bacillus subtilis* spore coat. J. Bacteriol. *186*, 1129–1135.

Johnson, M.J., Todd, S.J., Ball, D.A., Shepherd, A.M., Sylvestre, P., and Moir, A. (2006). ExsY and CotY are required for the correct assembly of the exosporium and spore coat of *Bacillus cereus*. J. Bacteriol. *188*, 7905–7913.

Kim, H.-S., Sherman, D., Johnson, F., and Aronson, A.I. (2004). Characterization of a major *Bacillus anthracis* spore coat protein and its role in spore inactivation. J. Bacteriol. *186*, 2413–2417.

Kim, H., Hahn, M., Grabowski, P., McPherson, D.C., Otte, M.M., Wang, R., Ferguson, C.C., Eichenberger, P., and Driks, A. (2006). The *Bacillus subtilis* spore coat interaction network. Mol. Microbial. *59*, 487–502.

Kobayashi, K., Kumazawa, Y., Miwa, K., and Yamanaka, S. (1996). ε-(γ-Glutaminyl)lysine cross-links of spore coat proteins and transglutaminase activity in *Bacillus subtilis*. FEMS Microbiol. Lett. *144*, 157–160.

Krasil'nikov, N.A., Duda, V.I., and Sokolov, A.A. (1963). Outward growth on spores of anaerobic bacteria of the genus *Clostridium*. Proc. Acad. Sci. U.S.S.R. Microbiol. Sect. *152*, 735–736.

McPherson, S.A., Li, M., Kearney, J.F., and Turnbough, C.L., Jr. (2010). ExsB, an unusually highly phosphorylated protein required for the stable attachment of the exosporium of *Bacillus anthracis*. Mol. Microbiol. *76*, 1527–1538.

Mallozzi, M., Bozue, J., Giorno, R., Moody, K.S., Slack, A., Cote, C., Qiu, D., Wang, R., McKenney, P., Lai, E.M., *et al.* (2008). Characterization of a *Bacillus anthracis* spore coat-surface protein that influences coat-surface morphology. FEMS Microbial. Lett. *289*, 110–117.

Melly, E., Genest, P.C., Gilmore, M.E., Little, S., Popham, D.L., Driks, A., and Setlow, P. (2002). Analysis of the properties of spores of *Bacillus subtilis* prepared at different temperatures. J. Appl. Microbiol. *92*, 1105–1116.

Nicholson, W.L., Munakata, N., Horneck, G., Melosh, H.J., and Setlow, P. (2000). Resistance of *Bacillus* endospores to extreme terrestrial and extraterrestrial environments. Microbiol. Mol. Biol. Rev. *64*, 548572.

Oliva, C.R., Swiecki, M.K., Griguer, C.E., Lisnaby, M.W., Bullard, D.C., Turnbough, C.L., Jr., and Kearney, J.F. (2008). The integrin Mac-1 (CR3) mediates internalization and directs *Bacillus anthracis* spores into professional phagocytes. Proc. Natl. Acad. Sci. U.S.A. *105*, 1261–1268.

Ozin, A.J., Samford, C.S., Henriques, A.O., and Moran, C.P., Jr. (2001). SpoVID guides SafA to the spore coat in *Bacillus subtilis*. J. Bacteriol. *183*, 3041–3049.

Paredes-Sabja, D., Setlow, P., and Sarker, M.R. (2011). Germination of spores of Bacillales and Clostridiales species: mechanisms and proteins involved. Trends Microbiol. *19*, 85–94.

Plomp, M., Leighton, T.J., Wheeler, K.E., and Malkin, A.J. (2005). The high-resolution architecture and structural dynamics of *Bacillus* spores. Biophys. J. *88*, 603–608.

Ragkousi, K., and Setlow, P. (2004). Transglutaminase–mediated cross-linking of GerQ in the coats of *Bacillus subtilis* spores. J. Bacteriol. *186*, 5567–5575.

Ramamurthi, K.S., Clapham, K.R., and Losick, R. (2006). Peptide anchoring spore coat assembly to the outer forespore membrane in *Bacillus subtilis*. Mol. Microbiol. *62*, 1547–1557.

Redmond, C., Baillie, L.W.J., Hibbs, S., Moir, A.J.G., and Moir, A. (2004). Identification of proteins in the exosporium of *Bacillus anthracis*. Microbiol. *150*, 355–363.

Rode, L.J., Crawford, M.A., and Williams, M.G. (1967). *Clostridium* spores with ribbon-like appendages. J. Bacteriol. *93*, 1160–1173.

Roels, S., Driks, A., and Losick, R. (1992). Characterization of *spoIVA*, a sporulation gene involved in coat morphogenesis in *Bacillus subtilis*. J. Bacteriol. *174*, 575–585.

Rose, R., Setlow, B., Monroe, A., Mallozzi, M., Driks, A., and Setlow, P. (2007). Comparison of the properties of *Bacillus subtilis* spores made in liquid or on agar plates. J. Appl. Microbiol, *103*, 691–699.

Severson, K.M., Mallozzi, M., Bozue, J., Welkos, S.L., Cote, C.K., Knight, K.L., and Driks, A. (2009). Roles of the *Bacillus anthracis* spore protein ExsK in exosporium maturation and germination. J. Bacteriol. *191*, 7587–7596.

Seyler, R.W., Jr., Henriques, A.O., Ozin, A.J., and Moran, C.P., Jr. (1997). Assembly and interactions of the cotJ-encoded proteins constituents of the inner layers of the *Bacillus subtilis* spore coat. Mol. Microbiol. *25*, 955–966.

Steichen, C.T., Kearney, J.F., and Turnbough, C.L., Jr. (2007). Non-uniform assembly of *the Bacillus anthracis* exosporium and a bottle cap model for spore germination and outgrowth. Mol. Microbiol. *64*, 359–367.

Sylvestre, P., Couture-Tosi, E., and Mock, M. (2002). A collagen-like surface glycoprotein is a structural component of the *Bacillus anthracis* exosporium. Mol. Microbiol. *45*, 169–182.

Sylvestre, P., Couture-Tosi, E., and Mock, M. (2005). Contribution of ExsFA and ExsFB proteins to the localizatioin of BclA on the spore surface and to the stability of the *Bacillus anthracis* exosporium. J. Bacteriol. *187*, 5122–5128.

Takamatsu, H., Kodama, T., Imamura, A., Asai, K., Kobayashi, K., Nakayama, T., Ogasawara, N., and Watanabe, K. (2000). The *Bacillus subtilis yabG* gene is transcribed by SigK RNA polymerase during sporulation and *yabG* mutant spores have altered coat protein composition. J. Bacteriol. *182*, 1883–1888.

Thompson, B.M., Waller, L.N., Fox, K.F., Fox, A., and Stewart, G.C. (2007). The BclB glycoprotein of *Bacillus anthracis* is involved in exosporium integrity. J. Bacteriol. *189*, 6704–6713.

Thompson, B.M., Hsieh, H.-Y., Spreng, K.A., and Stewart, G.C. (2011). The co-dependence of BxpB/ExsFA and BclA for proper incorporation into the exosporium of *Bacillus anthracis*. Mol. Microbiol. *79*, 799–813.

Todd, S.J., Moir, A.J.G., Johnson, M.J., and Moir, A. (2003). Genes of *Bacillus cereus* and *Bacillus anthracis* encoding proteins of the exosporium. J. Bacteriol. *185*, 3373–3378.

Van Ooij, C., Eichenberger, P., and Losick, R. (2004). Dynamic patterns of subcellular protein localization during spore coat morphogenesis in *Bacillus subtilis*. J. Bacteriol. *186*, 4441–4448.

Walker, J.R., Gnaman, A.J., Blinkova, A.L., Hermandson, M.J., Karymov, M.A., Lyubchenko, Y.L., Graves, P.R., Haystead, T.A., and Linse, K.D. (2007). *Clostridium taeniosporum* spore ribbon-like appendage structure, composition and genes. Mol. Microbiol. *63*, 629–643.

Wang, K.H., Isidro, A.L., Domingues, L., Eskandarian, H.A., McKenney, P.T., Drew, K., Grabowski, P., Chua, M.H., Barry, S.N., Guan, M., *et al.* (2009). The coat morphogenetic protein SpoVID is necessary for spore encasement in *Bacillus subtilis*. Mol. Microbiol. *74*, 634–649.

Weaver, J., Kang, T.J., Raines, K.W., Cao, G.-L., Hibbs, S., Tsai, P., Baillie, L., Rosen, G.M., and Cross, A.S. (2007). Protective role of *Bacillus anthracis* exosporium in macrophage-mediated killing by nitric oxide. Infect. Immune. *75*, 3894–3901.

Xue, Q., Gu, C., Rivera, J., Höök, M., Chen, X., Pozzi, A., and Xu, Y. (2011). Entry of *Bacillus anthracis* spores into epithelial cells is mediated by the spore surface protein BclA, integrin α2β1, and complement component C1q. Cell. Microbiol. *13*, 620–634.

Zhang, J., Fitz-James, P.C., and Aronson, A.I. (1993). Cloning and characterization of a cluster of genes encoding polypeptides present in the insoluble fraction of the spore coat of *Bacillus subtilis*. J. Bacteriol. *175*, 3757–3766.

Zhang, J., Ichikawa, H., Halberg, R., Kroos, L., and Aronson, A.I. (1994). Regulation of the transcription of a cluster of *Bacillus subtilis* spore coat genes. J. Mol. Biol. *240*, 405–415.

Zheng, L., Donovan, W.P., Fitz-James, P.C., and Losick, R. (1988). Gene encoding a morphogenetic protein required in the assembly of the outer coat of the *Bacillus subtilis* endospore. Genes Dev. *2*, 1047–1054.

Zilhao, R., Serrano, R., Isticato, R., Ricca, E., Moran, C.P., Jr., and Henriques, A.O. (2004). Interactions among CotB, CotG, and CotH during assembly of the *Bacillus subtilis* spore coat. J. Bacteriol. *186*, 1110–1119.

Mechanisms of Bacterial Spore Survival

6

Mario Pedraza-Reyes, Norma Ramírez-Ramírez, Luz E. Vidales-Rodríguez and Eduardo A. Robleto

Abstract

Sporulation is the process by which a spore is formed and is generally induced when microorganisms of the genera *Bacillus* and *Clostridia* are challenged by nutritional stress. Spores of *Bacillus* species contaminate food and other products of human consumption, compromising health. thus, a deeper understanding of the mechanisms involved in long-term survival is important to develop rational strategies for spore killing. Spores can survive for extended periods of time and possess the ability to return to life if nutrients become available. For decades bacterial spores have served as extraordinary paradigms to understand the consequences of long-term exposure to environmental factors that potentially compromise cell viability. This special type of cells can withstand a wide range of assaults that would otherwise destroy its vegetative cell counterparts. Spores of the genus *Bacillus* are several times more resistant to heat, desiccation, UV-C light, oxidizing chemicals and other genotoxic drugs. The particular cell architecture of spores and multiple mechanisms to protect its genome from cytotoxic and genotoxic factors explain in great part such resistance. The present chapter will describe the processes underlying the capability of the genus *Bacillus* spores to survive under extreme conditions.

Introduction

The endospore-forming bacteria belong to two classes of the phylum Firmicutes that include the Bacilli (aerobic Firmicutes) and the Clostridia (anaerobic Firmicutes) clusters. Both classes can be divided into a number of orders, some with sporulating genera and some with non-sporulating genera. It is generally assumed that the common ancestor of all Bacilli and Clostridia was an endospore-forming organism, even though several related genera (such as *Listeria, Staphylococcus, Streptococcus* and *Lactobacillus*) that presumably evolved from that common ancestor have lost the ability to sporulate. Consistent with this interpretation, loss of sporulation appears to be advantageous for cells living in a fairly constant environment (Maughan *et al.*, 2009). Owing to the large number of genes that are essential to endospore formation, it is improbable that the ability to sporulate could have been gained by other phyla as a result of horizontal gene transfer (reviewed in de Hoon *et al.*, 2010).

The exposure of actively growing Bacilli and Clostridia cells to limited nutrients results in the entrance to stationary phase (Errington, 1993; Driks, 2002a; de Hoon *et al.*, 2010). During the transition to stationary phase, bacterial cells initiate some or all of several adaptive processes such as competence development, motility, secondary metabolites production, stationary-phase mutagenesis and spore formation, among others. These stationary phase associated processes allow adaptation of cells to adverse environment conditions (Yasbin and Pedraza-Reyes, 2004).

Bacillus subtilis, the best characterized Gram-positive microorganism, has been extremely useful as a paradigm to study the network of cellular development and differentiation that generates an endospore. The sporulation process is genetically controlled by a transcriptional regulatory network (Errington, 1993; Hilbert and Piggot, 2004). The main role in this network is played by sigma factors, whose activity is restricted to specific compartments (Losick and Stragier, 1992; Hilbert and Piggot, 2004). In *Bacillus subtilis* the sporulation regulatory network is organized spatially and temporally into modules, with the sporulation-specific sigma factors σ^F, σ^E, σ^G, and σ^K directing gene expression in the early forespore, early mother cell, late forespore, and late mother cell, respectively (Losick and Stragier, 1992; Hilbert and Piggot, 2004). The phylogenetic analysis of the sporulation network in most of the endospore-formers revealed that the sporulation process follows a hierarchical and functional organization (de Hoon *et al.*, 2010). This analysis revealed that sporulation sigma factors and the master regulator of sporulation, Spo0A, are highly conserved in all spore formers. Moreover, the sequential activation of these global regulators is also strongly conserved (de Hoon *et al.*, 2010). In addition, the regulatory pathways (feed-forward loops) which are probably used to fine-tune waves of gene expression within regulatory modules show an intermediate level of conservation (de Hoon *et al.*, 2010). However, even though these loops are less conserved than sigma factors, they are significantly more conserved than structural sporulation genes, which form the lowest level in the functional and evolutionary hierarchy of the sporulation network (de Hoon *et al.*, 2010).

Factors contributing to bacterial spore resistance

Spore architecture

For many years, the structure and components of bacterial spores has been widely studied in order to understand the high resistance of spores to chemical, physical agents and different environmental stressful conditions. The extraordinary spore's resistance properties are due, in great part, to a well-structured multilayer morphology. Starting from the outside, the spore layers include the exosporium, coats, outer membrane, cortex, germ cell wall, inner membrane and the central core. The most external layer of bacterial spores is the exosporium, a large loose-fitting structure found in some organisms of the *Bacillus cereus* group (Driks, 2002b; Lai *et al.*, 2003; Redmond *et al.*, 2004; Waller *et al.*, 2004). The exosporium is a multilayered shell that surrounds the spore but is not attached to the spore. Spores of many species, including *B. subtilis*, either do not contain an exosporium or if they do it is greatly reduced in size and tightly associated with the outer coat (Sousa *et al.*, 1978a,b). It has been proposed that the exosporium may be an expanded version of the outermost coat layer since homologues of coat proteins in *B. subtilis* are found in the exosporium of spores of *Bacillus anthracis* (Lai *et al.*, 2003; Todd *et al.*, 2003; Redmond *et al.*, 2004). Moreover, the exosporium is composed of specific glycoproteins found only in spores that contain this layer (Lai *et al.*, 2003; Todd *et al.*, 2003; Daubenspeck *et al.*, 2004; Redmond *et al.*, 2004).

Proceeding inwardly, spores contain a multilayered proteinaceous structure called the coat. The coat is mostly composed of tyrosine and cysteine-rich proteins and small amounts of carbohydrate and lipids (Warth *et al.*, 1963; Murrel, 1969). In *B. subtilis* there are two major coat layers easily distinguishable by electron microscopy: an inner coat and an outer coat (Driks, 1999). The coat contains > 50 different proteins, most of which are spore-specific gene products (Driks, 1999, 2002b; Lai *et al.*, 2003). Although the functions of most coat proteins are not known, some of coat proteins are morphogenic proteins involved in the assembly of the spore coat or exosporium (Bailey-Smith *et al.*, 2005).

The spore coat constitutes the first line of defence of the spore, due to its multiple protein layers that acts as a chemical filter. Indeed, the coat plays a key role in spore resistance to some chemicals, to peptidoglycan lytic enzymes, and to predation by protozoa (Driks, 1999). Even more, the coat is believed to have a role in spore germination by facilitating permeation of nutrient germinant molecules into the inner membrane of spore, where specific germination receptors are located (Moir and Smith, 1990; Moir *et al.*, 1994; Setlow, 2003; Pelczar and Setlow, 2008). However, the coat is not considered as a permeability barrier for small molecules (Setlow, 2006).

The next structure, the outer membrane, is an essential structure for spore formation (Popham,

2002). However, in mature spores this membrane may not retain its integrity (Popham, 2002). It has been proposed that the outer membrane, as the coat, do not constitute a significant permeability barrier against small molecules accessing the interior of the spore (Piggot and Hilbert, 2004).

Just beneath the spore outer membrane is located a thicker structure called the cortex (Driks, 2002b). The cortex has a similar structure to the peptidoglycan layer (PG) of the vegetative cells but contains several spore-specific modifications (Popham, 2002). The major differences of the cortex peptidoglycan compared with its vegetative wall counterpart is the replacement of around 50% of peptide side chain muramic acid residues by muramic acid lactam resulting in a more flexible and loosely cross-linked peptidoglycan (Warth and Strominger, 1969, 1972). Thus, the cortex presumably serves as a retaining structure to withstand turgor pressures generated by the high concentration of solutes in the core (Popham, 2002). In addition to the cortex, the spore peptidoglycan contains an inner layer called the germ cell wall that appears to have a structure similar to that of vegetative cell wall peptidoglycan (Tipper and Linnet, 1976) and serves as the initial cell wall during spore outgrowth.

The inner membrane and the core are the final spore layers and are the analogue of the growing cell cytoplasmic membrane and protoplast, respectively (Cortezzo and Setlow, 2005). The inner membrane constitutes a strong permeability barrier that plays a key role in the resistance of spores to many DNA-damaging chemicals (Nicholson *et al.*, 2000; Setlow, 2000; Cortezzo and Setlow, 2005).

The spore core have several unique characteristics that contributes to increased spore resistance, including, low water content, high levels of minerals, decreased spore core permeability and saturation of the spore chromosome with small, acid-soluble proteins (SASP) (Beaman and Gerhardt, 1986; Mason and Setlow, 1986; Mohr *et al.*, 1991; Nakashio and Gerhardt, 1995; Huster *et al.*, 1997; Setlow, 1998; Mathai *et al.*, 2008; Sunde *et al.*, 2009).

Permeability of the inner membrane is a determinant factor in maintenance of the low water content in spores and also contributes in the resistance to DNA-damaging agents. However, the mechanism by which the spore achieves dehydration remains unclear (Popham *et al.*, 1995; Popham, 2002). In dormant spores, the inner membrane is a very compressed structure and its volume increases approximately 2-fold during the first minutes of spore germination in the absence of ATP production (Cowan *et al.*, 2004). The composition of lipid molecules of this membrane is very similar to that of the plasma membrane of growing cells (Cortezzo *et al.*, 2004; Cortezzo and Setlow, 2005). However, in dormant spores the lipid molecules remains largely immobile but become fully mobile upon spore germination (Cowan *et al.*, 2004). Consistent with the immobilization state of the membrane lipids, water permeability of the inner membrane is about 3–4 orders of magnitude lower than prototypical lipid bilayer membranes (Huster *et al.*, 1997; Mathai *et al.*, 2008; Sunde *et al.*, 2009).

Spore water content

Bacterial spores contain a highly reduced level of water as compared with its vegetative cell counterparts. In protoplast of growing cells the water content is about 75–80% of wet weight whereas in spores the core water constitutes only about 27–55% of wet weight (Gerhardt and Marquis, 1989). Furthermore, the spore core takes up water rapidly in the first minutes of spore germination and outgrowth when macromolecular motion and enzyme activity in the core are restored (Cowan *et al.*, 2003; Setlow, 2003).

The low water content of the core is the major factor contributing to the spore enzymatic dormancy and heat resistance (Gerhardt and Marquis, 1989; Cowan *et al.*, 2003). Indeed, the ability of spores to survive moist heat treatment (wet heat resistance) correlates with low water spore content, since in *B. subtilis* spores heat resistance decreases exponentially with increasing core water content (Beaman, 1986). This behaviour was explained by the finding that most spore proteins are rotationally immobilized preventing the irreversible aggregation of heat-denatured proteins in bacterial spores (Sunde *et al.*, 2009).

Although the mechanism for the reduction in spore core water content during sporulation is not known, the maintenance of low water content has been attributable to the low permeability of the inner membrane (IM) and to the accumulation of DPA (pyridine-2,6-dicarboxylic acid or dipicolinic acid) in the spore core (Setlow, 2006). Several years ago it was proposed that the spore core existed in a glassy-state (Gould, 1986; Sapru and Labuza, 1993; Ablett *et al.*, 1999). This glassy state explained the enzymatic dormancy simply by the extreme retardation of diffusive molecular motions in the highly viscous glass state. However, spin relaxation rates in H_2O/D_2O exchange in *B. subtilis* spores, indicate that water have a high mobility throughout the spore and as a consequence, the spore dormancy is probably linked to the dehydration-induced conformational changes of enzymes instead to a retarded diffusion of substrates (Sunde *et al.*, 2009). Moreover, it was established that core water molecules are bound to the hydration layers of the proteins but still remain mobile (Sunde *et al.*, 2009).

Spore mineral content

The mineral content of the core plays a key role in spore resistance of *Bacillus* and *Clostridium* species (Nicholson, 2000). Spore mineralization is determined by the presence of pyridine-2,6-dicarboxylic acid (dipicolinic acid, DPA). DPA is located in the spore's core as a 1:1 chelate with divalent cations, predominantly Ca^{2+} (Setlow, 2006). DPA is a small molecule that constitutes 5–15% of the dry weight of spores (Gerhardt and Marquis, 1989). DPA is synthesized only in the mother cell compartment of the sporulating cell and taken up into the forespore (Tovar-Rojo *et al.*, 2002; Huang *et al.*, 2007). The inner membrane proteins encoded by the *spoVA* operon and transcribed exclusively in the developing forespore were proposed to be involved in DPA uptake into the forespore during sporulation (Errington, 1993; Tovar-Rojo *et al.*, 2002; Huang *et al.*, 2007) and in Ca-DPA release during spore germination, probably as part of a channel in the spore's inner membrane (Vepachedu and Setlow, 2007). However, the mechanism by which Ca-DPA is released during germination remains unclear. Although the Ca-DPA levels differ significantly among individual spores in a population, the Ca-DPA concentration in the spore core is well above Ca-DPA solubility (> 800 mM to 1 M) indicating that the majority of Ca-DPA in the spore core is in an insoluble form (Huang *et al.*, 2007).

The accumulation of Ca-DPA in the spore core contributes to reduction in core water content during the spore formation. Indeed, genetically DPA-less dormant spores have more core water than wild-type dormant spores (Paidhungat *et al.*, 2000). Thus, DPA plays a role in spore wet heat resistance (Gerhardt and Marquis, 1989). In addition, DPA plays a major role in the UV photochemistry of spore DNA (Douki *et al.*, 2005a) affecting both DNA photoreactivity and the nature of photoproducts produced (Setlow and Setlow, 1996; Douki *et al.*, 2005b).It has been proposed that the ability of DPA to favour the production of 5-thyminyl-5,6-dihidrothymine [spore photoproduct (SP)] over cyclobutane dimers (CPDs) obeys to a mechanism of energy transfer from an excited state of DPA to DNA indicating that DPA must be in close proximity to spore DNA (Douki *et al.*, 2005b). Moreover, the DPA-DNA proximity is also consistent with spore protection against DNA damage promoted by moist-less heating (dry heat) (Setlow and Setlow, 1995).

α/β-type SASPs

The third type of core molecule that plays an important role in spore resistance is a group of small, acid-soluble spore proteins (SASP) of the α/β-type (Setlow 1992, 1995; Driks, 2002b). The α/β-type SASP are synthesized only in the developing forespore in the later stages of sporulation, slightly prior to DPA uptake. These small (60–75 amino acid) proteins are extremely abundant in spores and constitutes 3–6% of total spore protein. The α/β-type SASP are the products of a multigene family varying from four to seven members in *Bacillus* species which are scattered around the genome (Setlow, 2007). All members of this family appear to be expressed in parallel, generally with two proteins expressed at high levels and others at much lower levels (Setlow, 1988). The amino acid sequences of the α/β-type SASP are very highly conserved both within and across species but do not have homologues in non-spore formers (Setlow, 1988; Fairhead *et al.*, 1993).

The α/β-type SASP bind and saturate the spore DNA (Driks and Setlow, 1999; Setlow,

2000, 2006, 2007; Lee *et al.*, 2008). The binding of α/β-type SASP to DNA causes structural changes in DNA (A-B type conformation) and alters DNA photochemistry (Mohr *et al.*, 1991; Lee *et al.*, 2008). In consequence, the α/β-type SASP play significant roles in spore resistance to heat and many chemicals, and a major factor in spore resistance to UV radiation (Mason and Setlow, 1986; Nicholson *et al.*, 2000; Setlow, 2006, 2007; Raju *et al.*, 2006). Indeed, During germination, the α/β-type SASP are degraded early in spore outgrowth, and if not, their continued presence interferes significantly with subsequent development, presumably by blocking DNA transcription (Sanchez-Salas *et al.*, 1992; Hayes and Setlow, 2001).

Mutants in genes encoding α- and β-type SASPs lack approximately 80% of the α/β-type SASP pool (Setlow *et al.*, 2002) and present a reduced resistance to a variety of treatments, including UV radiation, heat, peroxides, ionizing radiation and other sporicidal treatments (Setlow *et al.*, 2000; Tennen *et al.*, 2000; Genest *et al.*, 2002; Young and Setlow, 2003, 2004a,b; Shapiro *et al.*, 2004; Moeller *et al.*, 2008, 2009). Furthermore, *Clostridium perfringens* SASP genes expressed in SASP-deficient *B. subtilis* spores restored the resistance of these mutant spores to UV, nitrous acid, dry heat, UV radiation and moist heat. This indicates the interchangeability of α/β-type SASP in DNA protection in spores (Leyva-Illades *et al.*, 2007).

Mechanism of spore resistance to physical and chemical factors

Wet heat resistance

Moist heat is used routinely for inactivation of spores, generally at temperatures of ≥ 100°C for short to moderate periods of time. This method is routinely used for inactivation of spores in a food product (Setlow, 2000).

In general, spores are resistant to moist-heat temperatures that are ~45°C higher than those that inactivate growing cells of the same organism (Warth, 1978). A number of factors are responsible for spore moist-heat resistance, including: (i) the optimum growth temperature of the bacterial strain and the sporulation temperature, (ii) the spore core's level of dipicolinic acid (DPA) and its associated divalent cations (Ca-DPA), (iii) the type of divalent cations associated with DPA, (iv) the protection of spore DNA by its saturation with a group of α/β-type small, acid-soluble spore proteins, and (v) the low water content in the spore core, which may contain as little as 25% of its wet weight as water in the most resistant spores (Gerhardt and Marquis, 1989; Nicholson *et al.*, 2000; Setlow, 2006).

Despite the existence of a number of factors important in spore wet-heat resistance, the mechanism for wet-heat killing of spores is not fully understood. It is well known that wet heat does not kill spores by DNA damage or oxidative damage (Setlow and Setlow, 1998, 2006). Instead, spore killing by wet heat is associated with protein denaturation and enzyme inactivation (Warth, 1980; Belliveau, 1992; Coleman *et al.*, 2007), although specific damaged proteins that result in spore death have not been identified. Wet-heat treatment also often results in the release of the spore core's large depot of Ca-DPA. The mechanism for this Ca-DPA release is not known but is presumably associated with the rupture of the spore's inner membrane (Coleman *et al.*, 2007). In addition, the relationship between protein denaturation and Ca-DPA release is not clear, although recent work suggests that significant protein denaturation can occur prior to Ca-DPA release (Coleman *et al.*, 2007). Ca-DPA release was also shown to take place more slowly than killing of *Bacillus cereus* and *Bacillus megaterium* spores after moist heat treatment, while inactivation of some spore core enzymes was actually faster than loss in spore viability (Warth, 1980; Belliveau *et al.*, 1992).

Dry heat resistance

Most factors potentially responsible for spore resistance to dry heat have not been investigated thoroughly (Alderton and Snell, 1969), although the α/β-type SASP seems to play a major role in *B. subtilis* spores (Setlow and Setlow, 1995; Setlow, 2006). Indeed, SASP protect against dry heat damage to DNA both *in vivo* and *in vitro* (Setlow and Setlow, 1995; Nicholson, 2000; Setlow, 2006).

In contrast to spore killing by wet heat, killing of wild-type spores by dry heat is accompanied by accumulation of DNA damage including strand breaks and probably AP sites (Setlow and Setlow, 1995; Setlow, 2000, 2007; del Huesca-Espitia *et al.*, 2002; Urtiz-Estrada, *et al.*, 2003). As expected since DNA damage accompanies spore killing by dry heat, DNA repair in spore outgrowth is also important in spore dry heat resistance. RecA-deficient *B. subtilis* spores are much more sensitive to dry heat than are wild-type spores, suggesting that RecA-dependent DNA repair systems are involved in the repair of dry heat damage to spore DNA. Indeed, the genes for a number of DNA repair enzymes, including *exoA*, *nfo* and the products of the *ykoUVW* operon are transcribed in the developing spore under the control of the forespore specific sigma factor, σ^G, for RNA polymerase, and have been shown to be involved in the repair of dry heat damage to DNA in wild-type spores (Weller *et al.*, 2002; Urtiz-Estrada *et al.*, 2003; Salas-Pacheco *et al.* 2005; Barraza-Salas *et al.*, 2010). In particular, ExoA and Nfo seem to be important in protecting spores against dry-heat-induced DNA damage (Salas Pacheco *et al.*, 2005). On the other hand, it has been reported that NHEJ- and α/β-type SASP-deficient (and the respective double-mutant) spores were significantly more sensitive to dry heat than wild-type spores (Wang *et al.*, 2006; Moeller *et al.*, 2007).

Spores of *B. subtilis* contain a single chromosome arranged in a toroidal shape (Ragkousi *et al.*, 2000; Frenkiel-Krispin *et al.*, 2004); therefore, the homologous recombination pathway, which requires at least two homologous chromosomes, cannot operate on DSB during spore germination (Wang *et al.*, 2006). An alternative repair pathway for DSB induced in spore DNA, is non-homologous-end joining (NHEJ) (Bowater *et al.*, 2001). The first step in the NHEJ DNA repair mechanism involves the binding of the Ku complex to the two DNA ends. The next proteins to be recruited to the complex are those that lead to resection of the ends of a DSB. The final steps in NHEJ join the two ends by a specific DNA ligase, thus restoring the integrity of the DNA. The ligation of DNA strands is an energy-dependent process, and ATP is required for the formation of a phosphodiester bond at the site of an SSB in a duplex DNA (Wilson *et al.*, 2003). Recently, a Ku homologue (encoded by the *ykoV* gene) was identified in *B. subtilis* (Weller *et al.*, 2002), which has the biochemical characteristics of the eukaryotic Ku heterodimer. The bacterial Ku specifically recruits a DNA ligase (encoded by *ykoU*) to DNA ends and thereby stimulates DNA ligation. It is worthwhile to note that dormant spores do not contain detectable amounts of endogenous ATP; rather, high-energy phosphate is stored in the spore as a large depot of 3-phosphoglyceric acid. In the first minutes of spore germination, 3-phosphoglyceric acid rapidly generates ATP by the lower part of glycolysis (Setlow and Kornberg, 1970; Setlow *et al.*, 2001). *B. subtilis* spores defective in the non-homologous end joining (NHEJ) DSB repair system were significantly more sensitive to DNA SSB and DSB generated by treatments such as ultra-high vacuum, mono- and polychromatic UV, and X-rays, thus establishing a major role for NHEJ in spore resistance to these extreme environmental parameters (Pitcher *et al.*, 2005). Recent data indicated that YkoU and YkoV are functional in the ionizing radiation stress response of *B. subtilis*, *Mycobacterium tuberculosis*, and *Mycobacterium smegmatis* (Aravind and Koonin, 2001; Weller *et al.*, 2002; Wilson *et al.*, 2003; Della *et al.*, 2004; Pitcher *et al.*, 2005).

In conclusion, α/β-type SASP, NHEJ and Nfo/ExoA appear to be important factors in the prevention and repair of ionizing-radiation-induced DNA damage while homologous recombination and/or the SOS response mediated by *recA* and *polA*-encoded repair DNA polymerase I play lesser but still significant roles in repair of X-ray induced damage. Finally, SP lyase encoded by *splB* plays only a minor role, if any, in repair of X-ray-induced DNA damage in spores (Moller *et al.*, 2007b, 2008, 2011)

Desiccation resistance

Intrinsic to spore resistance to dry heat is the spore resistance to desiccation. Indeed, wild-type *B. subtilis* spores are resistant to multiple cycles of desiccation and rehydration (reviewed in Setlow, 2002, 2007). Spores are also more resistant than their growing counterparts to multiple cycles of freeze-drying (Roberts and Hitchins, 1969;

Gould, 1983; Setlow, 2000). It has been reported that extreme desiccation (10^{-6} Pa at 77 K for 24 h) resulted in complete killing of *Escherichia coli* and *Halobacterium halobium* cells, but spores of *Clostridium mangenoti* and *B. subtilis* survived to 55 and 75%, respectively (Koike *et al.*, 1992). To date the only factor that has been shown to contribute to spore desiccation resistance is the protection of spore DNA by the α/β-type SASP and killing by desiccation is accompanied by mutagenesis and DNA damage, e.g. single-strand breaks and DNA–protein cross-links (Dose *et al.*, 1992; Setlow, 1995).

Although wild-type spores are quite resistant to extreme desiccation, DNA repair-deficient spores showed significant sensitivity to extended dryness (Reviewed in Nicholson *et al.*, 2000). Although significant killing of wild-type spores was not observed, the survivors exhibited increased mutagenesis (Munakata *et al.*, 1997), indicating that DNA was targeted. Indeed, spores carrying *recA*, *ykoU*, *ykoV* or *ykoU ykoV* mutations were dramatically more sensitive to high vacuum. The extreme sensitivity of *ykoU ykoV*-deficient spores to high vacuum-induced desiccation is entirely consistent with the observation that exposure to high vacuum induces SSB and DSB formation in DNA (Dose *et al.*, 1992, 1995; Nicholson *et al.*, 2000). Thus, NHEJ is an extremely important determinant of spore survival of the desiccation and the ultrahigh vacuum prevailing in space (Moeller *et al.*, 2007b).

Chemical resistance

Spores are extremely resistant to a variety of chemicals, including acids, bases, oxidizing agents, alkylating agents, aldehydes and organic solvents (Nicholson, 2000; Setlow, 2006). As a consequence, spores are often the most resistant cell type that chemical decontaminants are designed to deal with (Setlow, 2006). In a few cases (e.g. formaldehyde, nitrous acid and alkylating agents) the mechanism of spore killing is via DNA damage, as the survivors accumulate mutations and a *recA* mutation sensitizes spores to these agents (Setlow *et al.*, 1998; Loshon *et al.*, 1999; Tennen *et al.*, 2000). However, this is only true for a few genotoxic chemicals, and even oxidizing agents such as hydrogen peroxide that mutagenize growing cells do not do so to spores (Setlow, 2006).

The coat is an important determinant in the resistance of spores to many, albeit not all, reactive chemicals (Setlow, 2006). Coat-deficient *B. subtilis* spores were sensitive to killing by sodium hypochlorite (Gosh *et al.*, 2008). Indeed, coat-mutant spores were at least as sensitive to this agent as were growing cells. The precise coat component responsible for spore resistance to hypochlorite and other reactive chemicals is not known, but coat proteins might react with and detoxify such chemicals before they can gain access to and damage more sensitive targets further in the spore's interior (Gosh *et al.*, 2008). The spore coat also appears to contain enzymes that may assist in detoxification of some reactive toxic chemicals (Driks, 1999; Henriques and Moran Jr, 2007).

Most oxidizing agents appear to kill spores by causing damage to spores' external layers, principally the spore's inner membrane (Cortezzo *et al.*, 2004; Setlow, 2006). In these cases, treated spores germinate, the damaged membrane ruptures, and spore death ensues (Loshon *et al.*, 2001; Genest *et al.*, 2002; Melly *et al.*, 2002; Young and Setlow, 2003, 2004a,b; Shapiro *et al.*, 2004; Paul, 2006). Interestingly, sublethal treatment of spores with oxidizing agents also sensitizes survivors to other treatments (e.g. wet heat) to which an undamaged inner membrane may be required for full spore resistance (Cortezzo *et al.*, 2004).

The precise nature of the inner membrane damage caused by oxidizing agents is not known, although it is not the oxidation of unsaturated fatty acids. Changing the lipid composition of the *B. subtilis* inner membrane did not affect sporulation, although modest effects on spore germination and wet heat and oxidizing agent sensitivity were observed, especially when multiple lipids were absent (Griffiths and Setlow, 2010).

In contrast to other oxidizing agents, H_2O_2 is thought to act in the core and thus would need to traverse the inner membrane in order to cause spore killing (Melly *et al.*, 2002). Mutants of the major cardiolipin (CL) synthase resulted in decreased resistance to H_2O_2. This could be because CL plays some direct or indirect role in reducing the ability of H_2O_2 to cause oxidative damage to the core (Griffiths and Setlow, 2010). Recent work has also identified polycyclic isoprenoids called sporulenes within spores, which also appear to be important for resistance to oxidizing agents (Bosak *et al.*, 2008). However, these isoprenoids have not yet been localized within the spore, and may well not be in the inner membrane. The mechanisms by which sporulenes and CL contribute to resistance to H_2O_2 are not known and require further study.

The third factor important in spore resistance to some chemicals is the saturation of spore DNA with α/β-type SASP (Setlow, 1988, 1992, 2007). The α/β-type SASP also protect DNA *in vitro* against damage by hydrogen peroxide (Loshon *et al.*, 1999; Tennen *et al.*, 2000). However, the α/β-type SASP do not protect spore DNA against all chemicals, as alkylating agents such as ethyl methanesulphonate kill wild-type and $\alpha^- \beta^-$ spores at the same rate and the α/β-type SASP do not block DNA alkylation by this type of agent *in vitro* (Setlow *et al.*, 1998). Concerning DNA repair mechanisms, it has been shown that although the AP endonucleases Nfo and ExoA do not contribute in protecting wild-type and $\alpha^-\beta^-$ spores from H_2O_2 (Salas-Pacheco *et al.*, 2005), these enzymes in conjunction with RecA and NER play important roles in eliminating H_2O_2-induced DNA-damage during spore germination/outgrowth (Ibarra *et al.*, 2008).

Ionizing radiation resistance

The cellular response to ionizing radiation is complex due to the variety of targets in a cell (Kiefer, 1990; Qiu *et al.*, 2006). Ionizing radiation can damage cellular components though direct deposition of radiation energy into biomolecules and also indirectly by generating reactive oxygen species (ROS). Hydrogen peroxide (H_2O_2) and hydroxyl radicals (HO·) are major oxidizing species produced by the radiolysis of water, and superoxide ions (O_2) are formed in the presence of dissolved oxygen (Nicholson *et al.*, 2000). Generally, the cytotoxic and mutagenic effects induced by ionizing radiation are thought to be the result of DNA damage caused during the course of irradiation, which includes single-strand breaks (SSB), double-strand breaks (DSB), base modification, abasic sites, and sugar modification (Hutchinson, 1985; Goodhead, 1994; Micke, *et al.*, 1992, 1994). Although in spores, it is thought that DNA DSB are the most critical damage caused by ionizing radiation (Weller *et al.*, 2002). Indeed, the absence of the AP-endonucleases Nfo and ExoA, that process AP sites and single strand brakes in *B. subtilis* (Salas-Pacheco *et al.*, 2003, 2005; Ibarra *et al.*, 2008), generated spores that were more sensitive than wild-type spores to ionizing radiation (Moeller *et al.*, 2011).

The amounts of DNA damage caused by given doses of ionizing radiation for different bacteria are very similar, although the range of ionizing radiation resistance levels is wide (Nicholson *et al.*, 2000). It has been suggested that one factor which may result in increased spore ionizing-radiation resistance is the decreased level of spore core water, which may reduce the amount of hydroxyl radicals formed by exposure to ionizing radiation, a suggestion that has not yet been tested directly (Nicholson *et al.*, 2000). Several further factors have been investigated for their impact on spore resistance to ionizing radiation, in particular protection by sulphur-rich coat proteins (Hitchins *et al.*, 1966) and dipicolinic acid (Kamat and Pradhan, 1987). However, sulphur-rich spore coat proteins and dipicolinic acid seem to have only minor influence on spore resistance to ionizing radiation, mainly acting as ROS scavengers of radicals formed by water radiolysis (Hutchinson, 1985; Nicastro *et al.*, 2002).

UV radiation resistance

Depending on the species analysed, spores are 10–50 times more resistant than growing cells to UV radiation at 254 nm (Nicholson *et al.*, 2000, 2005; Setlow, 2001, 2006). At longer wavelengths the difference in spore and growing cell resistance is generally smaller, although spores are still more resistant (Nicholson *et al.*, 2000). It has been proposed that bacterial spore resistance to UV-A and UV-B obeys to the presence of melanin like pigments in the spore outer layer; however such pigments are not present in spores of all *Bacillus* species. There are two main reasons for the increased UV resistance of spores: a difference in the UV photochemistry of DNA in spores and the efficient and relatively error-free repair of the novel photoproduct formed by

UV light in spore DNA (Nicholson *et al.*, 2000; Setlow, 2001).

The major UV photoproduct formed in spores is the unique thymine adduct 5-thyminyl-5,6-dihydrothymine (Donnellan and Setlow, 1965; Varghese, 1970), which is called 'spore photoproduct' or 'SP'. After UV-C irradiation only very small amounts of cyclobutane pyrimidine dimers (CPDs), pyrimidine 6–4 primidone adducts (6–4 PPs), and single- and double-strand breaks are produced in spore DNA (Slieman and Nicholson, 2000; Douki *et al.*, 2005). Quantitatively SP formation is much higher on UV-C radiation: radiation at 254 nm is 10^3 times more efficient for SP induction than that at 313 nm (UV-B) and 10^6 times more efficient than that at 365 nm (UV-A) (Tyrrell, 1978). Maximum SP formation is reached near 260 nm (Lindberg and Horneck, 1992).

Two physical (temperature and pressure) and three chemical (α/β-type SASP, water, and Ca-DPA) factors have been clearly identified as being particularly critical for SP formation. Quantitatively, the optimum temperature for SP production is −80°C. At this temperature, SP formation is twice than that observed at 22°C and four times than that observed at −196°C for a UV-C dose of 0.02 J/cm^2. The combined effects of the humidity and temperature of the spore core on SP formation have not yet been explored (Desnous *et al.*, 2010). Under the Earth's atmospheric pressure (101.3 kPa), the UV-C irradiation of spores leads almost exclusively to the formation of SP. Under an ultrahigh vacuum of 2×10^{-6} Pa with a 254 nm UV dose of 0.5 J/cm^2, SP remains the main photoadduct (69%) but CPD are also formed (Lindberg and Horneck, 1991; Lindberg and Horneck, 1992; Nicholson *et al.*, 2002).

Binding of α/β-type SASP to spore DNA is necessary for spore UV-C resistance in the laboratory (Mason and Setlow, 1986) and in the solar UV environment (Nicholson *et al.*, 2000; Setlow, 2006). α/β-SASP binding promotes SP formation and suppresses CPD and 6–4PP formation, probably because of conformational change in DNA upon α/β-type SASP binding (Setlow, 1992, 1995; Frenkiel-Krispin *et al.*, 2004). However, the precise change in DNA structure in the α/β-type SASP–DNA complex is not clear. Because SP is formed in α/β-type SASP-less mutants, α/β-type SASP is clearly not the only factor governing SP formation in spores.

Among other factors influencing SP formation is dipicolinic acid, which is present in the spore as a Ca^{2+} chelate. The participation of Ca-DPA in SP formation may involve a selective triplet-state energy transfer from the UV-C-excited Ca-DPA to the thymine bases (Douki *et al.*, 2005b). Such selective energy transfer could explain the exclusive involvement of dithymine in SP formation and consequently the absence of spore photoproducts involving cytosine residues (Douki *et al.*, 2005a,b). A full understanding of the mechanism of the involvement of Ca-DPA remains elusive. As noted above, Ca-DPA also induces a decrease in the core dehydration state (Paidunghat *et al.*, 2000). This has led to the suggestion that Ca-DPA may also act indirectly and participate with α/β-type SASP in the highly dehydrated state of the spore core. In addition, Ca-DPA may bind through intercalation to DNA. This could provide an additional hydration reduction mechanism in the immediate vicinity of the spore DNA (Lindsay and Murrell, 1986). The involvement of Ca-DPA in the conformational modification of core DNA has also been raised. However, at present this hypothesis is not strongly supported and the role of Ca-DPA is considered to be chiefly as a photosensitizer (Douki *et al.*, 2005b)

Low DNA hydration levels have been implicated in causing SP formation in DNA *in vitro* (Rahn and Hosszu, 1969; Patrick and Grey, 1976), suggesting that the low core water content may also play a role in spore DNA photochemistry. Recent reports revealed that spore core dehydration is an important factor in the resistance of spores to environmentally relevant solar UV radiation, particularly the longer UV-A wavelengths, and that core dehydration specifically protects spore DNA by virtue of reducing the production and/or migration of ROS within the spore core (Moeller *et al.*, 2009).

Since dormant spores are metabolically inactive, repair of UV-C-induced DNA damage occurs during the process of germination, when spores reactivate and prepare to return to vegetative growth. Studies with mutants have identified two major DNA-repair systems involved at this stage:

(i) the general nucleotide excision repair (NER) pathway, which excises SP and fills in the resulting single-stranded gap, and (ii) an SP-specific repair enzyme called SP lyase that cleaves SP back to the two original thymine residues in a monomerization reaction (Reviewed in Nicholson *et al.*, 2000; Setlow, 2001).

The NER system is broadly conserved among prokaryotes (Au *et al.*, 2005). Three subunits compose the Uvr excinuclease (uvrABC) of the *B. subtilis* NER mechanism (Yasbin *et al.*, 1993). The genes encoding the NER system are under control of the SOS response mediated by DinA/RecA and is induced by DNA damage during vegetative growth (Cheo *et al.*, 1991, 1992; Yasbin *et al.*, 1993) and germination. SP lyase is encoded by *splB*, the second cistron of the *splAB* operon (Fajardo-Cavazos *et al.*, 1993, 2000), and is widely conserved among endospore-forming bacteria (Nicholson *et al.*, 2000). SP lyase is produced upon transcription of *splB* by σ^G RNA polymerase at morphological stage III in the developing forespore, where the enzyme is packaged into the dormant spore and activated during germination (Pedraza-Reyes *et al.*, 1994, 1997). Transcription of *splB* is not observed during either vegetative growth or germination, nor is it induced by DNA damage during those processes (Pedraza-Reyes *et al.*, 1994). *splB* encodes an iron–sulphur protein which specifically splits accumulated SP photoproducts during spore germination in a reaction that requires S-adenosylmethionine (SAM) as a cofactor (Rebeil *et al.*, 1998; Rebeil and Nicholson, 2001).The RecA-mediated pathway is also involved in DNA repair, albeit to a lesser extent (Nicholson *et al.*, 2000; Moeller *et al.*, 2007b). RecA, as a multifunctional protein mainly involved in homologous recombination, ensures genome integrity. After UV radiation, RecA participates in the overall repair and closing of DNA strand breaks. It might also support homologous recombination during NER by gap filling after photoproduct excision.

After a germination time of 60 min, >75% of the SP is repaired. This efficient repair of DNA photoproducts was mainly performed by SP lyase (Moeller *et al.*, 2007a). The repair of all other bipyrimidine photoproducts was slower (at least 50–75%) than that of SP (Moeller *et al.*, 2007a). This observation is in agreement with the concept of an evolutionary advantage provided by the specific formation of SP in spores, a lesion that is more efficiently repaired than the other photoproducts formed in other cell types and in vegetative bacteria (Nicholson *et al.*, 2000; Desnous *et al.*, 2010). Combined NER and RecA deficiency yielded the lowest amount of DNA-photoproduct repair. These results show that NER and homologous recombination (RecA) can at least partly substitute for SP lyase and thereby prevent deleterious UV effects at least to a certain extent in spores lacking this enzyme (Moeller *et al.*, 2007a).

Future perspectives

The mechanisms employed by spores to withstand with different type of stresses is well known in the genus *Bacillus* and is evident that atop of these is the saturation of spore DNA with α/β-type SASPs. However, the absence of genes encoding these proteins in other spore-forming bacteria from the *Streptomyces* and *Myxobacteria* species suggest the existence of different strategies for spore survival. Therefore studies aimed to investigate different mechanism of bacterial spore survival are expected in the near future. On the other hand, although is generally accepted that repair of accumulated DNA damage during spore latency occurs during germination, little efforts to investigate the mechanistic details of these processes have been made (Moeller *et al.*, 2007a; Ibarra *et al.*, 2008).

A recent report revealed that the error-prone polymerases PolYI and PolY2 protect the sporulating cells of *B. subtilis* from DNA damage and are apparently required for an efficient sporulation process (Rivas-Castillo *et al.*, 2010). This report is important since, during sporulation no net replication of the chromosomes in the two compartments of the sporangia occurs; however, insulted DNA must be repaired otherwise the delicate transcriptional programme conducing to generate a spore can be compromised. The mechanisms and strategies that protect or eliminate DNA damage during this developing programme remains to be elucidated.

Ackowledgements

Work at MPR laboratory receives support from University of Guanajuato and from Consejo Nacional de Ciencia y Tecnología (CONACYT; grant 84482) of México.

References

Ablett, S., Darke, A.H., Lillford, P.J., and Martin, D.R. (1999). Glass formation and dormancy in bacterial spores. Int. J. Food Sci. Technol. *34*, 59–69.

Alderton, G., and Snell, N. (1969). Chemical states of bacterial spores: dry-heat resistance. Appl. Microbiol. *17*, 745–749.

Aravind, L., and Koonin, E.V. (2001). Prokaryotic homologues of eukaryotic DNA-end-binding protein Ku, novel domains in the Ku protein and prediction of a prokaryotic double-strand break repair system. Genome Res. *11*, 1365–1374.

Au, N., Kuester-Schoeck, E., Mandava, V., Bothwell, L.E., Canny, S.P., Chachu, K., Colavito, S.A., Fuller, S.N., Groban, E.S., Hensley, L.A., *et al.* (2005). Genetic composition of the *Bacillus subtilis* SOS system. J. Bacteriol. *187*, 7655–7666.

Barraza-Salas, M., Ibarra-Rodríguez, J.R., Mellado, S.J., Salas-Pacheco, J.M., Setlow, P., and Pedraza-Reyes, M. (2010). Effects of forespore-specific overexpression of apurinic/apyrimidinic endonuclease Nfo on the DNA-damage resistance properties of *Bacillus subtilis* spores. FEMS Microbiol. Lett. *302*, 159–165.

Beaman, T.C., and Gerhardt, P. (1986). Heat resistance of bacterial spores correlated with protoplast dehydration, mineralization, and thermal adaption. Appl. Environ. Microbiol. *52*, 1242–1246.

Belliveau, B.H., Beaman, T.C., Pankratz, S., and Gerhardt, P. (1992). Heat killing of bacterial spores analyzed by differential scanning calorimetry. J. Bacteriol. *174*, 4463–4474.

Bowater, R., and Doherty, A.J. (2006). Making ends meet: repairing breaks in bacterial DNA by non-homologous end-joining. PLOS Genet. *2*, e8–e15.

Cheo, D.L., Bayles, K.W., and Yasbin, R.E. (1991). Cloning and characterization of DNA damage-inducible promoter regions from *Bacillus subtilis*. J. Bacteriol. *173*, 1696–1703.

Cheo, D.L., Bayles, K.W., and Yasbin, R.E. (1992). Molecular characterization of regulatory elements controlling expression of the *Bacillus subtilis recA1* gene. Biochimie *74*, 755–762.

Coleman, W.H., Chen, D., Li, Y.-Q., Cowan, A.E., and Setlow, P. (2007). How moist heat kills spores of *Bacillus subtilis*. J. Bacteriol. *189*, 8458–8466.

Cortezzo, D.E., Koziol-Dube, K., Setlow, B., and Setlow, P. (2004). Treatment with oxidizing agents damages the inner membrane of spores of *Bacillus subtilis* and sensitizes the spores to subsequent stress. J. Appl. Microbiol. *97*, 838–852.

Cortezzo, D.E., and Setlow, P. (2005). Analysis of factors influencing the sensitivity of spores of *Bacillus subtilis* to DNA damaging chemicals. J. Appl. Microbiol. *98*, 606–617.

Cowan, A.E., Koppel, D.E., Setlow, B., and Setlow, P. (2003). A soluble protein is immobile in dormant spores of *Bacillus subtilis* but is mobile in germinated spores: implications for spore dormancy. Proc. Natl. Acad. Sci. U.S.A. *100*, 4209–4214.

Cowan, A.E., Olivastro, E.M., Koppel, D.E., Loshon, C.A., Setlow, B., and Setlow, P. (2004). Lipids in the inner membrane of dormant spores of *Bacillus* species are immobile. Proc. Natl. Acad. Sci. U.S.A. *101*, 7733–7738.

Daubenspeck, J.M., Zeng, H., Chen, P., Dong, S., Steichen, C.T., Krishna, N.R., Pritchard, D.G., and Turnbough, C.L. (2004). Novel oligosaccharide chains of the collagen-like region of BclA, the major glycoprotein of the *Bacillus anthracis* exosporium. J. Biol. Chem. *279*, 30945–30953.

De Hoon, M.J., Eichenberger, P., and Vitkup, D. (2010). Hierarchical evolution of the bacterial sporulation network. Curr. Biol. *20*, 735–745.

Della, M., Palmbos, P.L., Tseng, H.-M., Tonkin, L.M., Daley, J.M., *et al.* (2004). *Mycobacterial* Ku and ligase proteins constitute a two-component NHEJ repair machine. Science *306*, 683–685.

Desnous, C., Guillaume, D., and Clivio, P. (2010). Spore photoproduct: a key to bacterial eternal. Chem. Rev. *110*, 1213–1232.

Donnellan, J.E., Jr., and Setlow, R.B. (1965). Thymine photoproducts but not thymine dimers are found in ultraviolet irradiated bacterial spores. Science *149*, 308–310.

Dose, K., and Gill, M. (1995). DNA stability and survival of *Bacillus subtilis* spores in extreme dryness. Orig. Life Evol. Biosphere *25*, 277–293.

Dose, K., Bieger-Dose, A., Kerz, O., and Gill, M. (1991). DNA-strand breaks limit survival in extreme dryness. Orig. Life Evol. Biosphere *2*, 177–187.

Dose, K., Bieger-Dose, A., Labusch, M., and Gill, M. (1992). Survival in extreme dryness and DNA-single-strand breaks. Adv. Space Res. *12*, 221–229.

Douki, T., Setlow, B., and Setlow, P. (2005a). Effects of the binding of α/β-type small, acid-soluble spore proteins on the photochemistry of DNA in spores of *Bacillus subtilis* and *in vitro*. Photochem. Photobiol. *81*, 163–169.

Douki, T., Setlow, B., and Setlow, P. (2005b). Photosensitization of DNA by dipicolinic acid, a major component of spores of *Bacillus* species. Photochem. Photobiol. Sci. *4*, 591–597.

Driks, A. (1999). The *Bacillus subtilis* spore coat. Microbiol. Mol. Biol. Rev. *63*, 1–20.

Driks, A. (2002a). Overview: development in bacteria: spore formation in *Bacillus subtilis*. Cell Mol. Life. Sci. *59*, 389–391.

Driks, A. (2002b). Proteins of the spore core and coat. In *Bacillus subtilis* and its Closest Relatives: From Genes to Cells, Sonenshein, A.L., Hoch, J.A., and Losick, R., eds. (American Society for Microbiology, Washington, DC), pp. 527–535.

Errington, J. (1993). *Bacillus subtilis* sporulation: regulation of gene expression and control of morphogenesis. Microbiol. Rev. *57*, 1–33.

Fairhead, H., Setlow, B., and Setlow, P. (1993). Prevention of DNA damage in spores and *in vitro* by small, acid-soluble proteins from *Bacillus subtilis*. J. Bacteriol. *175*, 1367–1374.

Fajardo-Cavazos, P., and Nicholson, W.L. (2000). The TRAP-like SplA protein is a trans-acting negative regulator of spore photoproduct lyase synthesis during *Bacillus subtilis* sporulation. J. Bacteriol. *182*, 555–560.

Fajardo-Cavazos, P., Salazar, C., and Nicholson, W.L. (1993). Molecular cloning and characterization of the *Bacillus subtilis* spore photoproduct lyase (*spl*) gene, which is involved in repair of UV radiation-induced DNA damage during spore germination. J. Bacteriol. *175*, 1735–1744.

Frenkiel-Krispin, D., Sack, R., Englander, J., Shimoni, E., Eisenstein, M., Bullitt, E., Horowitz-Scherer, R., Hayes, C.S., Setlow, P., Minsky, A., *et al.* (2004). Structure of the DNA–SspC complex: implications for DNA packaging, protection, and repair in bacterial spores. J. Bacteriol. *186*, 3525–3530.

Genest, P.C., Setlow, B., Melly, E., and Setlow, P. (2002). Killing of spores of *Bacillus subtilis* by peroxynitrite appears to be caused by membrane damage. Microbiology *148*, 307–314.

Gerhardt, P., and Marquis, R.E. (1989). Spore thermoresistance mechanisms. In Regulation of Procaryotic Development, Smith, I., Slepecky, R.A., and Setlow, P., eds. (American Society for Microbiology, Washington, DC), pp. 43–63.

Ghosh, S., Setlow, B., Wahome, P.G., Cowan, A.E., Plomp, M., Malkin, A.J., and Setlow, P. (2008). Characterization of spores of *Bacillus subtilis* that lack most coat layers. J. Bacteriol. *190*, 6741–6748.

Goodhead, D.T. (1994). Initial events in the cellular effects of ionizing radiations: clustered damage in DNA. Int. J. Radiat. Biol. *65*, 7–17.

Gould, G.W. (1983). Mechanisms of resistance and dormancy. In The Bacterial Spore, Gould, G.W., and Hurst, A., eds. (Academic Press, London), vol. 2. pp. 397–444.

Gould, G.W. (1986). Water and the survival of bacterial spores. In Membranes, Metabolism and Dry Organisms, Leopold, A.C. ed. (Cornell University Press, Ithaca, NY), pp. 143–156.

Griffiths, K.K., and Setlow, P. (2010). Effects of modification of membrane lipid composition on *Bacillus subtilis* sporulation and spore properties. J. Appl. Microbiol. *106*, 2064–2078.

Hayes, C.S., and Setlow, P. (2001). An α/β-type small, acid-soluble spore protein which has a very high affinity for DNA prevents outgrowth of *Bacillus subtilis* spores. J. Bacteriol. *183*, 3982–3990.

Henriques, A.O., and Moran, C.P., Jr. (2007). Structure, assembly and function of the spore surface layers. Annu. Rev. Microbiol. *61*, 555–588.

Hitchins, A.D., King, W.L., and Gould, G.W. (1966). Role of disulphide bonds in the resistance of *Bacillus cereus* spores to gamma irradiation and heat. J. Appl. Bacteriol. *29*, 505–511.

Huang, S.S., Chen, D., Pelczar, P.L., Vepachedu, V.R., Setlow, P., and Li, Y.Q. (2007). Levels of Ca^{2+}-dipicolinic acid in individual *Bacillus* spores determined using microfluidic Raman tweezers. J. Bacteriol. *189*, 4681–4687.

del Huesca-Espita, L.C., Caley, C., Bagyan, I., and Setlow, P. (2002). Base-change mutations induced by various treatments of *Bacillus subtilis* spores with and without DNA protective small, acid-soluble spore proteins. Mutat. Res. *503*, 77–84.

Huster, D., Jin, A.J., Arnold, K., and Gawrisch, K. (1997). Water permeability of polyunsaturated lipid membranes measured by ^{17}O NMR. Biophys. J. *73*, 855–864.

Hutchinson, F. (1985). Chemical changes induced in DNA by ionizing radiation. Prog. Nucleic Acid Res. Mol. Biol. 32, 115–154.

Ibarra, J.R., Orozco, A.D., Rojas, J.A., López, K., Setlow, P., Yasbin, R.E. and Pedraza-Reyes, M. (2008). Role of the Nfo and ExoA apurinic/apyrimidinic endonucleases in repair of DNA damage during outgrowth of *Bacillus subtilis* spores. J. Bacteriol. *190*, 2031-2038.

Kamat, A.S., and Pradhan, D.S. (1987). Involvement of calcium and dipicolinic acid in the resistance of *Bacillus cereus* BIS-59 spores to UV and gamma radiations. Int. J. Radiat. Biol. Relat. Stud. Phys. Chem. Med. *51*, 7–18.

Koike, J., Oshima, T., Koike, K.A., Taguchi, H., Tanaka, R., Nishimura, K., and Miyaji, M. (1992). Survival rates of some terrestrial microorganisms under simulated space conditions. Adv. Space Res. *12*, 271–274.

Lai, E.M., Phadke, N.D., Kachman, M.T., Giorno, R., Vazquez, S., Vazquez, J.A., Maddock, J.R., and Driks, A. (2003). Proteomic analysis of the spore coats of *Bacillus subtilis* and *Bacillus anthracis*. J. Bacteriol. *185*, 1443–1454.

Lee, K.S., Bumbaca, D., Kosman, J., Setlow, P., and Jedrzejas, M.J. (2008.) Structure of a protein–DNA complex essential for DNA protection in spores of *Bacillus* species. Proc. Natl. Acad. Sci. U.S.A. *105*, 2806–2811.

Leyva-Illades, J.F., Setlow, B., Sarker, M.R., and Setlow, P. (2007). Effect of a small, acid-soluble spore protein from *Clostridium perfringens* on the resistance properties of *Bacillus subtilis* spores. J. Bacteriol. *189*, 7927–7931.

Lindberg, C., and Horneck, G.J. (1991). Action spectra for survival and spore photoproduct formation of *Bacillus subtilis* irradiated with short-wavelength (200–300 nm) UV at atmospheric pressure and in vacuo. J. Photochem. Photobiol. *11*, 69–80.

Lindberg, C., and Horneck, G.J. (1992). Thymine photoproduct formation and inactivation of intact spores of *Bacillus subtilis* irradiated with short wavelength UV (200–300nm) at atmospheric pressure and in vacuo. Adv. Space Res. *12*, 275–279.

Lindsay, J.A., and Murrell, W.G. (1986). Solution spectroscopy of dipicolinic acid interaction with nucleic acids: Role in spore heat resistance. Curr. Microbiol. *13*, 225–259.

Loshon, C.A., Melly, E., Setlow, B., and Setlow, P. (2001). Analysis of the killing of spores of *Bacillus subtilis* by a new disinfectant, Sterilox. J. Appl. Microbiol. *91*, 1051–1058.

Losick, R., and Stragier, P. (1992). Crisscross regulation of cell-type-specific gene expression during development in *B. subtilis*. Nature. *355*, 601–604.

Mathai, J.C., Tristram–Nagle, S., Nagle, J.F., and Zeidel, M.L. (2008). Structural determinants of water permeability through the lipid membrane. J. Gen. Physiol. *131*, 69–76.

Maughan, H., Birky, C.W., Jr., and Nicholson, W.L. (2009). Transcriptome divergence and the loss of plasticity in *Bacillus subtilis* after 6000 generations of evolution under relaxed selection for sporulation. J. Bacteriol. *191*, 428–433.

Melly, E., Cowan, A.E., and Setlow, P. (2002). Studies on the mechanism of killing of *Bacillus subtilis* spores by hydrogen peroxide. J. Appl. Microbiol. *93*, 316–325.

Micke, U., Schaefer, M., Anton, A., Horneck, G., and Buecker, H. (1992). Heavy ion induced double strand breaks in bacteria and bacteriophages. Adv. Space Res. *12*, 59–63.

Micke, U., Horneck, G., and Kozubek, S. (1994). Double strand breaks in the DNA of *Bacillus subtilis* cells irradiated by heavy ions. Adv. Space Res. *14*, 207–211.

Moeller, R., Douki, T., Cadet, J., Stackebrandt, E., Nicholson, W.L., Rettberg, P., Reitz, G., and Horneck, G. (2007a). UV-radiation-induced formation of DNA bipyrimidine photoproducts in *Bacillus subtilis* endospores and their repair during germination. Int. Microbiol. *10*, 39–46.

Moeller, R., Stackebrandt, E., Reitz, G., Berger, T., Rettberg, P., Doherty, A.J., Horneck, G., and Nicholson, W.L. (2007b). Role of DNA Repair by nonhomologous-end joining in *Bacillus subtilis* spore resistance to extreme dryness, mono- and polychromatic UV, and ionizing radiation. J. Bacteriol. *189*, 3306–3311.

Moeller, R., Setlow, P., Horneck, G., Berger, T., Reitz, G., Rettberg, P., Doherty, A.J., Okayasu, R., and Nicholson, W.L. (2008). Roles of the major, small, acid-soluble spore proteins and spore-specific and universal DNA repair mechansims in resistance of *Bacillus subtilis* spores to ionizing radiation from X rays and high-energy charged-particle bombardment. J. Bacteriol. *190*, 1134–1140.

Moeller, R., Setlow, P., Reitz, G., and Nicholson, W.L. (2009). Roles of small, acid-soluble spore proteins and core water content in survival of *Bacillus subtilis* spores exposed to environmental solar UV radiation. Appl. Environ. Microbiol. *16*, 5202–5208.

Moeller, R., Setlow, P., Pedraza-Reyes, M., Okayasu, R., Reitz, G., and Nicholson, W.L. (2011). Role of the Nfo and ExoA apurinic/apyrimidinic (AP) endonucleases in the radiation resistance and radiation-induced mutagenesis of *Bacillus subtilis* spores. J. Bacteriol. *193*, 2875–2879.

Mohr, S.C., Sokolov, N.V., He, C.M., and Setlow, P. (1991). Binding of small acid-soluble spore proteins from *Bacillus subtilis* changes the conformation of DNA from B to A. Proc. Natl. Acad. Sci. U.S.A. *88*, 77–81.

Moir, A., and Smith, D.A. (1990). The genetics of bacterial spore germination. Annu. Rev. Microbiol. *44*, 531–553.

Moir, A., Kemp, E.H., Robinson, C., and Corfe, B.M. (1994). The genetic analysis of bacterial spore germination. J. Appl. Bacteriol. *76*, 9S-16S.

Nakashio, S., and Gerhardt, P. (1985). Protoplast dehydration correlated with heat resistance of bacterial spores. J. Bacteriol. *162*, 571–578.

Nicastro, A.J., Vreeland, R.H., and Rosenzweig, W.D. (2002). Limits imposed by ionizing radiation on the long-term survival of trapped bacterial spores: beta radiation. Int. J. Radiat. Biol. *78*, 891–901.

Nicholson, W.L., Munakata, N., Horneck, G., Melosh, H.J., and Setlow, P. (2000). Resistance of *Bacillus* endopsores to extreme terrestrial and extraterrestrial environments. Microbiol. Mol. Biol. Rev. *64*, 548–572.

Paidhungat, M., Setlow, B., Driks, A., and Setlow, P. (2000). Characterization of spores of *Bacillus subtilis* which lack dipicolinic acid. J. Bacteriol. *182*, 5505–5512.

Patrick, M.H., and Gray, D.M. (1976). Independence of photoproduct formation on DNA conformation. Photochem. Photobiol. *24*, 507–513.

Paul, M., Atluri, S., Setlow, B., and Setlow, P. (2006). Mechanisms of killing of spores of *Bacillus subtilis* by dimethyldioxirane. J. Appl. Microbiol. *101*, 1161–1168.

Pedraza-Reyes, M., Gutierrez-Corona, F., and Nicholson, W.L. (1994). Temporal regulation and forespore-specific expression of the spore photoproduct lyase gene by sigma-G RNA polymerase during *Bacillus subtilis* sporulation. J. Bacteriol. *176*, 3983–3991.

Pedraza-Reyes, M., Gutierrez-Corona, F., and Nicholson, W.L. (1997). Spore photoproduct lyase operon (splAB) regulation during *Bacillus subtilis* sporulation: modulation of *splB–lacZ* fusion expression by P1 promoter mutations and by an *in-frame* deletion of *splA*. Curr. Microbiol. *34*, 133–137.

Pelczar, P.L., and Setlow, P. (2008). Localization of the germination protein GerD to the Inner membrane in *Bacillus subtilis* spores. J. Bacteriol. *190*, 5635–5641.

Piggot, P.J., and Hilbert, D.W. (2004). Sporulation of *Bacillus subtilis*. Curr. Opin. Microbiol. *7*, 579–586.

Pitcher, R.S., Wilson, T.E., and Doherty, A.J. (2005). New insights into NHEJ repair processes in prokaryotes. Cell Cycle *4*, 675–680.

Popham, D.L. (2002). Specialized peptidogycan of the bacterial endospore: the inner wall of the lockbox. Cell. Mol. Life Sci. *59*, 426–433.

Popham, D.L., Illades-Aguiar, B., and Setlow, P. (1995). The *Bacillus subtilis dacB* gene, encoding penicillin-binding protein 5*, is part of a three-gene operon required for proper spore cortex synthesis and spore core dehydration. J. Bacteriol. *17*, 4721–4729.

Qiu, X., Daly, M.J., Vasilenko, A., Omelchenko, M.V., Gaidamakova, E.K., Wu, L., Zhou, J., Sundin, G.W., and Tiedje, J.M. (2006). Transcriptome analysis applied to survival of *Shewanella oneidensis* MR-1 exposed to ionizing radiation. J. Bacteriol. *188*, 1199–1204.

Ragkousi, K., Cowan, A.E., Ross, M.A., and Setlow, P. (2000). Analysis of nucleoid morphology during germination and outgrowth of spores of *Bacillus* species. J. Bacteriol. *182*, 5556–5562.

Rahn, R.O., and Hosszu, J.L. (1969). Influence of relative humidity on the photochemistry of DNA films. Biochim. Biophys. Acta *190*, 126131.

Raju, D., Waters, M., Setlow, P., and Sarker, M.R. (2006). Investigating the role of small, acid-soluble spore proteins (SASPs) in the resistance of *Clostridium perfringens* spores to heat. BMC Microbiol. *6*, 50–63.

Rebeil, R., and Nicholson, W.L. (2001). The subunit structure and catalytic mechanism of the *Bacillus subtilis* DNA repair enzyme spore photoproduct lyase. Proc. Natl. Acad. Sci. U.S.A. *98*, 9038–9043.

Rebeil, R., Sun, Y., Chooback, L., Pedraza-Reyes, M., Kinsland, C., Begley, T.P., and Nicholson, W.L. (1998). Spore photoproduct lyase from *Bacillus subtilis* spores is a novel iron–sulfur DNA repair enzyme which shares features with proteins such as class III anaerobic ribonucleotide reductases and pyruvate-formate lyases. J. Bacteriol. *180*, 4879–4885.

Redmond, C., Baillie, L.W., Hibbs, S., Moir, A.J., and Moir, A. (2004). Identification of proteins in the exosporium of *Bacillus anthracis*. Microbiology *150*, 355–363.

Rivas-Castillo, A.M., Yasbin, R.E., Robleto, E., Nicholson, W.L., and Pedraza-Reyes, M. (2010). Role of the Y-family DNA polymerases YqjH and YqjW in protecting sporulating *Bacillus subtilis* cells from DNA damage. Curr. Microbiol. *60*, 263–267.

Roberts, T.A., and Hitchins, A.D. (1969). Resistance of spores. In The Bacterial Spore, Gould, G.W., and Hurst, A. , eds. (Academic Press, New York), pp. 611–670.

Salas-Pacheco, J.M., Setlow, B., Setlow, P., and Pedraza-Reyes, M. (2005). Role of the Nfo (YqfS) and ExoA apurinic/apyrimidinic endonucleases in protecting *Bacillus subtilis* spores from DNA damage. J. Bacteriol. *187*, 7374–7381.

Salas-Pacheco, J.M., Urtiz-Estrada, N., Martínez-Cadena, G., Yasbin, R.E., and Pedraza-Reyes, M. (2003). YqfS from *Bacillus subtilis* is a spore protein and a new functional member of the type IV apurinic/apyrimidinic-endonuclease family. J. Bacteriol. *185*, 5380–5390.

Sanchez-Salas, J.L., Santiago-Lara, M.L., Setlow, B., Sussman, M.D., and Setlow, P. (1992). Properties of mutants of *Bacillus megaterium* and *Bacillus subtilis* which lack the protease that degrades small, acid-soluble proteins during spore germination. J. Bacteriol. *174*, 807–814.

Sapru, V., and Labuza, T.P. (1993). Glassy state in bacterial spores predicted by polymer glass-transition theory. J. Food Sci. *58*, 445–448.

Setlow, P. (1988). Small, acid-soluble spore proteins of *Bacillus* species: structure, synthesis, genetics, function, and degradation. Annu. Rev. Microbiol. *42*, 319–338.

Setlow, P. (1992). I will survive: protecting and repairing spore DNA. J. Bacteriol. *174*, 2737–2741.

Setlow, P. (1995). Mechanisms for the prevention of damage to the DNA in spores of *Bacillus* species. Annu. Rev. Microbiol. *49*, 29–54.

Setlow, P. (2000). Resistance of bacterial spores. In Bacterial Stress Responses, Storz, G., and Hengge-Aronis, R., eds. (American Society for Microbiology, Washington D.C.), pp. 217–230.

Setlow, P. (2001). Resistance of spores of *Bacillus* species to ultraviolet light. Environ. Mol. Mutagen. *38*, 97–104.

Setlow, P. (2003). Spore germination. Curr. Opin. Microbiol. *6*, 550–556.

Setlow, P. (2006). Spores of *Bacillus subtilis*: their resistance to and killing by radiation, heat and chemicals. J. Appl. Microbiol. *101*, 514–525.

Setlow, P. (2007). I will survive: DNA protection in bacterial spores. Trends Microbiol. *15*, 172–180.

Setlow, P., and Kornberg, A. (1970). Biochemical studies of bacterial sporulation and germination. XXII. Energy metabolism in early stages of germination of *Bacillus megaterium* spores. J. Biol. Chem. *245*, 3637–3645.

Setlow, B., and Setlow, P. (1995). Small, acid-soluble proteins bound to DNA protect *Bacillus subtilis* spores from killing by dry heat. Appl. Environ. Microbiol. *61*, 2787–2790.

Setlow, B., and Setlow, P. (1998). Heat killing of *Bacillus subtilis* spores in water is not due to oxidative damage. Appl. Environ. Microbiol. *64*, 4109–4112.

Setlow, B., McGinnis, K.A., Ragkousi, K., and Setlow, P. (2000). Effects of major spore-specific DNA binding proteins on *Bacillus subtilis* sporulation and spore properties. J. Bacteriol. *182*, 6906–6912.

Setlow, B., Melly, E., and Setlow, P. (2001). Properties of spores of *Bacillus subtilis* blocked at an intermediate stage in spore germination. J. Bacteriol. *183*, 4894–4899.

Setlow, B., Loshon, C.A., Genest, P.C., Cowan, A.E., Setlow, C., and Setlow, P. (2002). Mechanisms of killing of spores of *Bacillus subtilis* by acid, alkali and ethanol. J. Appl. Microbiol. *92*, 362–375.

Setlow, B., Tautvydas., K.J., and Setlow, P. (1998). Small, acid-soluble spore proteins of the alpha/beta type do not protect the DNA in *Bacillus subtilis* spores against base alkylation. Appl. Environ. Microbiol. *64*, 1958–1962.

Shapiro, M., Setlow, B., and Setlow, P. (2004). Studies on the killing of spores of *Bacillus subtilis* by a modified Fenton reagent containing $CuCl_2$ and ascorbic acid. Appl. Environ. Microbiol. *70*, 2535–2539.

Slieman, T.A., and Nicholson, W.L. (2000). Artificial and solar UV radiation induces strand breaks and cyclobutane pyrimidine dimers in *Bacillus subtilis* spore DNA. Appl. Environ. Microbiol. *66*, 199–205.

Sousa, J.C., Silva, M.T., and Balassa, G. (1978a). Spore control (*sco*) mutations in *Bacillus subtilis*. Mol. Gen. Genet. *163*, 285–291.

Sousa, J.C., Silva, M.T., and Balassa, G. (1978b). Ultrastructure and development of an exosporium-like outer spore envelope in *Bacillus subtilis*. Ann. Microbiol. (Paris) *129*, 339–362.

Sunde, E.P., Setlow, P., Hederstedt, L., and Halle, B. (2009). The physical state of water in bacterial spores. Proc. Natl. Acad. Sci. U.S.A. *106*, 19334–19339.

Tennen, R., Setlow, B., Davis, K.L., Loshon, C.A., and Setlow, P. (2000). Mechanisms of killing of spores of *Bacillus subtilis* by iodine, glutaraldehyde and nitrous acid. J. Appl. Microbiol. *89*, 330–338.

Tipper, D.J., and Linnet, P.E. (1976). Distribution of peptidoglycan synthetase activities between sporangia and forespores in sporulating cells of *Bacillus sphaericus*. J. Bacteriol. *126*, 213–221.

Todd, S.J., Moir, A.J., Johnson, M.J., and Moir, A. (2003). Genes of *Bacillus cereus* and *Bacillus anthracis* encoding proteins of the exosporium. J. Bacteriol. *185*, 3373–3378.

Tovar-Rojo, F., Chander, M., Setlow, B., and Setlow, P. (2002). The products of the *spoVA* operon are involved in dipicolinic acid uptake into developing spores of *Bacillus subtilis*. J. Bacteriol. *184*, 584–587.

Tyrrell, R.M. (1978). Solar dosimetry with repair deficient bacterial spores: action spectra, photoproduct measurements and a comparison with other biological systems. Photochem. Photobiol. *27*, 571–579.

Urtiz-Estrada, N., Salas-Pacheco, J.M., Yasbin, R.E., and Pedraza-Reyes, M. (2003). Forespore-specific expression of *Bacillus subtilis yqfS*, which encodes type IV apurinic/apyrimidinic endonuclease, a component of the base excision repair pathway. J. Bacteriol. *185*, 340–348.

Varghese, A.J. (1970). 5-Thyminyl-5,6-dihydrothymine from DNA irradiated with ultraviolet light. Biochem. Biophys. Res. Commun. *38*, 484–490.

Vepachedu, V.R., and Setlow, P. (2007). Role of SpoVA proteins in the release of dipicolinic acid during germination of *Bacillus subtilis* spores triggered by dodecylamine or lysozyme. J. Bacteriol. *189*, 1565–1572.

Waller, L.N., Fox, N., Fox, K.F., Fox, A., and Price, R.L. (2004). Ruthenium red staining for ultrastructural visualization of a glycoprotein layer surrounding the spore of *Bacillus anthracis* and *Bacillus subtilis*. J. Microbiol. Methods *58*, 23–30.

Wang, S.T., Setlow, B., Conlon, E.M., Lyon, J.L., Imamura, D., Sato, T., Setlow, P., Losick, R., and Eichenberger, P. (2006). The forespore line of gene expression in *Bacillus subtilis*. J. Mol. Biol. *358*, 16–37.

Warth, A.D. (1978). Relationship between the heat resistance of spores and the optimum temperature and maximum growth temperatures of *Bacillus* species. J. Bacteriol. *131*, 699–705.

Warth, A.D. (1980). Heat stability of *Bacillus cereus* enzymes within spores and in extracts. J. Bacteriol. *143*, 27–34.

Warth, A.D., Ohye, D.F., and Murrell, W.G. (1963). The composition and structure of bacterial spores. J. Cell Biol. *16*, 579–592.

Warth, A.D., and Strominger, J.L. (1969). Structure of the peptidoglycan of bacterial spores: occurrence of the lactam of muramic acid. Proc. Natl. Acad. Sci. U.S.A. *64*, 528–535.

Warth, A.D., and Strominger, J.L. (1972). Structure of the peptidoglycan from spores of *Bacillus subtilis*. Biochemistry *11*, 389–1396.

Weller, G.R., Kysela, B., Roy, R., Tonkin, L.M., Scanlan, E., Della, M., Devine, S.K., Day, J.P., Wilkinson, A., d'Adda di Fagagna, F., *et al.* (2002). Identification of a DNA nonhomologous end–joining complex in bacteria. Science *297*, 1686–1689.

Wilson, T.E., Topper, L.M., and Palmbos, P.L. (2003). Non-homologous end-joining: bacteria join the chromosome breakdance. Trends Biochem. Sci. *28*, 62–66.

Yasbin, R.E., and Pedraza-Reyes, M. (2004). Stationary phase-induced mutagenesis: is directed mutagenesis alive and well within neo Darwinian theory? In Microbial Evolution: Gene Establishment, Survival, and Exchange, Miller, R.V., and Day, M.J., eds. (ASM Press, Washington D.C.), pp. 181–191.

Yasbin, R.E., Cheo, D., and Bol, D. (1993). DNA repair systems. In *Bacillus subtilis* and Other Gram-Positive Bacteria: Biochemistry, Physiology, and Molecular Genetics, Sonenshein, A.L., Hoch, J.A., and Losick, R., eds. (ASM Press, Washington D.C.), pp. 529–537DC.

Young, S.B., and Setlow, P. (2003). Mechanisms of killing of *Bacillus subtilis* spores by hypochlorite and chlorine dioxide. J. Appl. Microbiol. *95*, 54–67.

Young, S.B., and Setlow, P. (2004a). Mechanisms of killing of *Bacillus subtilis* spores by Decon and OxoneTM, two general decontaminants for biological agents. J. Appl. Microbiol. *96*, 289–301.

Young, S.B., and Setlow, P. (2004b). Mechanisms of *Bacillus subtilis* spore resistance to and killing by aqueous ozone. J. Appl. Microbiol. *96*, 1133–1142.

Initiation of Germination in *Bacillus* and *Clostridium* Spores

7

Graham Christie

Abstract

Dormant bacterial spores of the genera *Bacillus* and *Clostridium* employ receptors belonging to the GerA family as environmental sensors. The recognition and binding of appropriate small molecule ligands, termed germinants, irreversibly commits the spore to initiate the cascade of biophysical and hydrolytic reactions that constitute the germination process. Germination is a crucial event in the lifecycle of spore forming organisms, since in virtually all cases the detrimental properties associated with the bacterium are manifested during vegetative growth. However, despite intensive research spanning several decades, few details concerning the precise function of the nutrient germinant receptors (other than their ability to trigger spore germination in response to specific germinants) or the molecular mechanisms that underpin the earliest stages of the process, are known. The current chapter describes genetic, structural and functional aspects of spore germinant receptors, highlighting diverse functional roles and mechanistic strategies associated with GerA-type receptors. Indeed, present outputs from mutational and bioinformatic analyses seem to support the notion that, whereas the general mechanism of triggering germination may be largely conserved between *Bacillus* and *Clostridium* species, a number of different mechanisms involving GerA-type receptor proteins, and perhaps unrelated proteins, also exist, and are employed by at least by some pathogenic *Clostridia*.

Introduction

Bacterial cells of the genera *Bacillus* and *Clostridium* have the remarkable ability upon sensing nutrient limitation to differentiate and form metabolically dormant spores that, owing to their unique composition and structure, can persist as survival structures in the environment for perhaps millennia (Setlow, 2007). Equally remarkable, considering the absence of any perceptible metabolic processes, is the ability of the spore to continually monitor its environment, and, upon sensing that conditions are once again conducive to growth, to rapidly exit dormancy and form a new vegetative cell. The latter is achieved via the process of spore germination, which, from an applied perspective, is a crucial event in the lifecycle of spore forming organisms, since in virtually all cases the deleterious properties of the bacterium, whether food spoilage, toxigenesis, or disease, are manifested typically during vegetative growth.

A vast scientific literature concerning the spore germination process has accrued since the late 1940s – driven in part by being a topic of considerable biological interest, but also by the promise of benefits that improved control over the spore germination process might offer to various commercial, environmental and security/military-related sectors. The purpose of the current chapter is not to conduct a comprehensive review of this literature – the reader is referred instead to several review articles (Johnstone,

1994; Setlow, 2003; Moir, 2006; Paredes-Sabja *et al.*, 2011) – nor indeed to concentrate on physiological events that occur late in germination, such as cortical depolymerization (which is covered elsewhere in this book), but to focus instead on current knowledge of the earliest events of the germination process. Whereas in the recent past such an undertaking would have focused on information derived principally from studies conducted with *B. subtilis*, the availability of numerous spore-forming bacterial genomes and improved molecular genetic tools has permitted, over the past decade or so, a large increase in germination related research to be conducted with other species of *Bacillus*. Similarly, issues with genetic intractability and fastidious growth requirements, which have hindered germination-associated research in *Clostridium* species, has to a degree been circumvented by the introduction of improved methodologies for targeted disruption of genes of interest (Heap *et al.*, 2010), which allied to the availability of genomic sequences, has resulted in rapid advances in identifying proteins and mechanisms involved in germination of spores of this genus.

This expansion of molecular genetic and bioinformatic-led gene-mining approaches to encompass a wide range of *Bacillus* and *Clostridium* species has proved particularly beneficial to the development of molecular models of spore germination. In particular, the assumption of a common germination mechanism – which seems logical since spores of both genera share several defining characteristics – has been challenged by recent insights derived from different species. Features common to all spores include a multilayered structure, spore-specific chemical moieties (e.g. dipicolinic acid, muramic-delta lactam containing peptidoglycan), stimulation of germination by exposure to a similar range of germinant molecules, and similarities in the sequence of physiological and morphological events that constitute the germination process. However, differences do exist – certainly at least between some species of *Clostridium* and *Bacillus* – in the structure, organization, function and perhaps even identity of receptor proteins employed to initiate spore germination in these genera. This chapter is ordered therefore to include a relatively brief synopsis of the overall germination process, and to provide a more detailed overview of the current state of knowledge concerning the structure and function of the spore nutrient germinant receptors. Mechanistic aspects of germinant receptor function are considered by selecting key examples of the different classes of receptor from diverse species, highlighting apparent differences in strategies employed by spores to initiate germination and variance in the perceived mode of action of germinant receptors. Experimental approaches that have been used to investigate germinant receptor function are also described, albeit briefly. Additionally, a section describing non-nutrient routes to germination is also included towards the end of the chapter.

Physiology of spore germination

Spores of *Bacillus* and *Clostridium* species rely upon a diverse range of small molecules, termed germinants, which include amino acids, ribosides, sugars, ions and peptidoglycan fragments, to indicate that environmental conditions are conducive to the resumption of vegetative growth. Despite perhaps having lain dormant for many years, exposure of the spore to appropriate germinant molecules, for as little as a few seconds in some cases, irreversibly commits the spore to undergo the process of germination resulting ultimately in the emergence of a new vegetative cell. The germination process takes place in the absence of germinant (or general spore) metabolism and essentially comprises a series of exergonic biophysical and degradative reactions that lead to the rapid loss of properties of resistance to environmental stress factors (Foster and Johnstone, 1990; Johnstone, 1994).

Two pathways to germination that are probably most relevant under physiological conditions have been identified in spores of *Bacillus* species. The best-studied of these – which has been shown experimentally to apply also in the *Clostridia* – is mediated by a family of membrane bound receptors, termed germinant receptors, that serve to initiate germination upon sensing appropriate germinant ligands (Ross and Abel-Santos, 2010a). In contrast, the second germination pathway that is probably of physiological relevance was discovered relatively recently, and is mediated by a eukaryotic-like serine/threonine kinase (PrkC), which, upon binding of defined peptidoglycan fragments, is responsible for initiating subsequent germination events. 'Non-nutrient' routes to germination, including the PrkC pathway, are discussed further towards the end of this chapter.

The nutrient induced germination process can be divided into two distinct stages: stage 1 encompasses the earliest germination events, which comprise receptor-mediated changes to the permeability of the spore inner membrane that result in the movement of select ions and other small molecules through the membrane from the spore core, with water moving in the opposite direction. Stage 2 encompasses later germination events, in particular the degradation of spore-specific structures (cortex and coat), which permits further hydration and concomitant swelling of the spore core.

The germinant receptors themselves are buried deep on the inside of the spore, localized to the inner-membrane that separates the protoplast and protective integument layers of the spore (Hudson *et al.*, 2001; Paidhungat and Setlow, 2001; Alberto *et al.*, 2005). Therefore, in order to stimulate germination, germinant molecules have to traverse the protective outer layers of the spore and the cortex to bind at cognate receptor sites at the inner membrane. This interaction, in a process that is poorly understood at the molecular level, somehow activates the receptor and irreversibly commits the spore to proceed through the germination process. The process of commitment, which is the earliest measurable event during spore germination (Stewart *et al.*, 1981; Yi and Setlow, 2010) (and see below) is followed by the release of monovalent cations (H^+, Na^+ and K^+) from the spore core. This has the effect of increasing the internal pH of the core to levels that are commensurate with the re-initiation of metabolism when hydration levels are sufficient to support this (Swerdlow *et al.*, 1981). Cation release is followed by the release of the spore core's vast pool of pyridine-2, 6-dicarboxylic acid (dipicolinic acid, or DPA), predominantly chelated with Ca^{2+} and other divalent metal ions (Johnstone *et al.*, 1982). The dormant spore's reserves of free amino acids, including arginine and glutamic acid, and additional unidentified small molecules, are released in parallel with Ca^{2+}-DPA (Setlow *et al.*, 2008). The spore core is partially rehydrated with water at or around the same time as DPA release, and although this is insufficient for any enzyme activity, the heat resistance properties of these partially germinated spores is reduced (Setlow *et al.*, 2001; Cowan *et al.*, 2003). This latter point indicates that the local mobility of at least some vital protein(s) are such that they can be denatured irreversibly [see Sunde *et al.* (2009) for information concerning loss of heat resistance], although in general protein mobility and diffusion in the core remains extremely restricted at this point.

The major physiological event that defines stage 2 of the germination process is the depolymerization of the spore's thick layer of cortical peptidoglycan by germination-specific peptidoglycan hydrolases (Setlow, 2003). Two semi-redundant cortex-lytic enzymes (CLEs), SleB and CwlJ, are responsible for initiating this process in *Bacillus* species, which is facilitated by another protein, SleL, whose activity appears to vary between different species of *Bacillus* (Ishikawa *et al.*, 1998; Lambert and Popham, 2008; Heffron *et al.*, 2009; Setlow *et al.*, 2009; Christie *et al.*, 2010b). Structural genes encoding homologues of SleB and CwlJ are also observed in the genomes of several *Clostridium* species (Paredes-Sabja *et al.*, 2011), although another enzyme, SleC, appears to have an essential role in cortex hydrolysis in many – but not all – species of this genus (Miyata *et al.*, 1995; Paredes-Sabja *et al.*, 2009c; Burns *et al.*, 2010). Whereas CLEs are present as inactive but mature enzymes in *Bacillus* spores, SleC is present in *Clostridium* spores as an inactive zymogen that is proteolytically cleaved to a functional form during germination (Miyata *et al.*, 1995). Intriguingly, *C. perfringens* spores that have a null mutation in the subtilisin-like protease (CspB) that is responsible for activating pro-SleC, do not release Ca^{2+}-DPA upon exposure to germinants, whereas SleC mutant spores (and SleB CwlJ double mutants in *Bacillus*) do, revealing perhaps a fundamental difference between some *Clostridium* and *Bacillus* species in the sequence of physiological events that constitute the germination process

(Paredes-Sabja *et al.*, 2009a). Regardless of the identity of the enzymes involved, degradation of the cortex by CLEs removes a physical constraint on the spore core, permitting further uptake of water, and expansion of the core (and therefore the inner membrane and germ cell wall). Protein mobility can resume at this point, permitting enzyme action, marking the end of germination and the transition to vegetative metabolism and cellular outgrowth.

A genetic perspective on germinant receptors

Prior to its identification, the stereospecific requirement of spores of some species to germinate in response to amino acids indicated that the germinant receptor, whose presence had long been postulated and indeed incorporated into models of spore germination, must be a protein that could be activated allosterically (Wolgamott and Durham, 1971). Despite intensive biochemical research from the 1950s onwards, the identification of the proteins that constitute spore nutrient germinant receptors was not revealed until genetic studies designed to map loci associated with *B. subtilis* germination mutants identified putative receptor genes at the turn of the 1980s (Moir *et al.*, 1979; Moir and Smith, 1990). A series of random mutants defective in the L-alanine stimulated germinative response were isolated and the defect mapped to the *gerA* locus, which upon cloning and sequencing was revealed to constitute three genes, *gerAA*, *gerAB* and *gerAC*, organized as a single transcription unit (Feavers *et al.*, 1985; Zuberi *et al.*, 1985, 1987). Expression of the *gerA* operon is under the control of the forespore specific sigma factor sigma G in the developing forespore (Feavers *et al.*, 1990). Expression from the *gerA* locus is at very low levels, presumably as a result of relatively weak promoter, ribosomal binding site, and translation initiation sequences. Expression levels are further down-regulated by the SpoVT protein, and are also dependent on PrpE phosphatase activity (Bagyan *et al.*, 1996; Hinc *et al.*, 2006). The low abundance of germinant receptors in spores, probably in the order of tens of molecules per spore (Paidhungat and Setlow, 2001), may be an evolutionary strategy to ensure that germination is only triggered when environmental conditions are conducive to sustaining vegetative growth.

Similar work conducted with *B. subtilis* mutants with defects in the asparagine (or alanine)–glucose–fructose–potassium ion (AGFK)-mediated germination pathway, mapped mutations to two loci, *gerB* and *gerK*, which were predicted to encode proteins homologous to the GerA proteins (Corfe *et al.*, 1994; Irie *et al.*, 1996). Of the proteins that constitute the receptor, only some receptor B proteins show any detectable homology to protein sequences in non-spore-forming bacterial orders, forming a branch of the single component amino acid transporter family (Jack *et al.*, 2000; and see below), indicating that the spore germinant receptors represent a unique class of sensory transducers.

Several other genes encoding orthologues of GerA-type proteins, classified now as the GerA receptor family, were identified subsequently using molecular genetic techniques, and examined experimentally for function in other species of *Bacillus* (Clements and Moir, 1998; Barlass *et al.*, 2002; Ireland and Hanna, 2002; Christie and Lowe, 2007). More recently, bioinformatic analyses of the considerable number of spore-forming bacterial genomic sequences now available have revealed that most species encode a number of such operons, typically from one to seven, which appear to have evolved to respond to different germinants by a process of gene duplication and divergence (Moir *et al.*, 2002; Ross and Abel-Santos, 2010a; Paredes-Sabja *et al.*, 2011). Genes encoding germinant receptor proteins are predominantly arranged in tricistronic operons, although the order of individual genes within the operon is variable. However, the relatively conserved genetic organization and cotranscription of genes within *gerA*-type operons indicates that each germinant receptor is a complex of all three proteins. Indeed, where examined mutation of any cistron within the operon results in inactivation of the respective receptor function (Zuberi *et al.*, 1985). However, mutational analysis of the divergon-like *C. perfringens gerK* operon appears to challenge the requirement for all three subunits for functionality since receptors lacking the B subunit appear to retain a large part of their function (see below) (Paredes-Sabja *et al.*, 2009b).

Molecular genetic experiments conducted with *B. subtilis* and *B. megaterium* have conferred indirect evidence for physical interaction between the respective subunits (Igarashi and Setlow, 2005; Christie and Lowe, 2007). These studies have been complemented by Yeast-2-hybrid and far-Western analyses, which indicate that the A protein of the *B. subtilis* GerA receptor interacts with the respective protein B and C receptor subunits (Vepachedu and Setlow, 2007a). Perhaps the strongest evidence yet that the germinant receptor consists of a complex of interacting protein subunits can be inferred from a very recent mutagenesis study conducted on the *B. subtilis* GerA receptor (Cooper and Moir, 2011; Mongkolthanaruk *et al.*, 2011). In this work, spores bearing various single amino acid substitutions in either the A or B receptor subunits were revealed, by probing with antisera raised against the C subunit, to have severely reduced or a complete absence of the receptor C subunit. This observation, when considered with the defective germinative responses to L-alanine of these spores, indicates that perturbation of the structure of either the A or B proteins by mutation of key residues can lead to the destabilization of perhaps the entire receptor complex.

The rapid expansion of spore-forming bacterial genomic information has revealed that numerous orphan monocistronic and bicistronic Ger-receptor genes also abound in *Bacillus* and *Clostridium* genomes (Paredes-Sabja *et al.*, 2011). A number of these genes have been shown experimentally to encode functional proteins (see below), presumably by associating with the products of other *gerA*-type genes to form functional receptors. Additionally, some operons are characterized by an unusual genetic organization, containing duplicate or multiple copies of genes encoding individual receptor subunits. Proteolytic *C. botulinum* strains, for example, contain two conventional tricistronic GerA-type receptor operons, but also a pentacistronic operon comprising three very closely related B genes that are separated by subunit A and C genes respectively. Bioinformatic analyses have also revealed recently the presence of a fourth gene (*gerAD*), predicted to encode a small (50–80 residues) membrane-associated protein, within some *Bacillus* and *Clostridium gerA*-family operons (Paredes-Sabja *et al.*, 2011). The precise role of these duplicate gene products, and of the novel D protein, has yet to be determined experimentally, but it seems probable that they contribute to the functional diversity of the associated germinant receptors.

An attempt to introduce a more logical and systematic approach to the annotation of germinant receptors based on evolutionary relationships has been conducted recently in order to remove the considerable inconsistencies associated with current nomenclature procedures (Ross and Abel-Santos, 2010b). Phylogenetic-based analyses employing representative species for which complete genomes are available have revealed that germinant receptor proteins of *Bacillus* species cluster into four distinct clades. This is evident regardless of the receptor subunit protein used as the query sequence, further substantiating the hypothesis concerning co-evolution of proteins encoded within specific receptor operons. Additionally, proteins from each cluster tend to be more closely related to orthologous proteins from other *Bacillus* species than they are to receptor proteins that cluster in other groups from the same organism. A limited degree of functional similarity in terms of germinants recognized is also evident for the respective clades, facilitating the assignment of putative function to newly identified, or as yet uncharacterized receptors. Thus, a number of experimentally characterized receptors that cluster in the GerA/L group are cognate for L-alanine, those in the GerH(I)/Y clade recognize aromatic and related compounds, and those that cluster in the GerK clade appear to be associated with the recognition of glucose. Interestingly, the plasmid-encoded GerX receptor from *B. anthracis*, for which no cognate germinant has been identified, does not cluster with any other *Bacillus* receptor, forming its own clade, and perhaps indicating that it is the product of horizontal gene transfer (Ross and Abel-Santos, 2010b). *Clostridium* receptor proteins also appear to cluster in four well-defined clades; however, the lack of functionally characterized receptors in this genus precludes the assignment, however approximate, of potential germinant recognition profiles to the respective groups of receptors.

Structural and functional insights to germinant receptor proteins

As detailed above, orthologous proteins belonging to the GerA family form the receptors through which spores of *Bacillus* species sense their environment. All three proteins have been localized to the spore inner membrane, which is consistent with hydropathy profile analyses that predict that the A and B proteins are integral membrane proteins. However, whereas the receptor B protein is predicted to be entirely membrane associated, comprising typically of ten transmembrane helices, receptor A-proteins are predicted to consist of a large hydrophilic N-terminal domain, a central domain comprising five or six transmembrane helices, and a relatively small hydrophilic C-terminal domain (Fig. 7.1). The C protein, on the other hand, is a lipoprotein that is probably located on the outward facing aspect of the spore inner-membrane, anchored by a lipid moiety that is attached by a diacylglycerol-transferase (encoded by *gerF*) to a conserved N-terminal cysteine residue (Moir *et al.*, 2002).

Despite considerable research effort, however, the precise function of the germinant receptors – other than their ability to trigger spore germination in response to specific germinants – and the role, stoichiometry and oligomerization status of the individual protein subunits that constitute the receptor, remain unknown. The germinant receptors may have a role, for example, in transporting small molecules to or from the spore core very early in the germination process. However, an early study investigating the possible incorporation of radiolabelled germinant molecules into metabolic by-products of germination did not find any evidence for bulk transport of [3H]-alanine during the period of initiation of spore germination of *B. megaterium* KM (Scott and Ellar, 1978). Similarly, we know nothing of how the presumed binding of a germinant stimulates receptor function, or indeed, how many receptors have to be activated to trigger the germination cascade. Molecular details regarding integration of germinant binding events, and subsequent transduction of the signal(s) from germinant-activated receptor proteins, which are present only in extremely low abundance, to presumed Ca^{2+}-DPA channels (and possibly other ion channel proteins) that are thought to be significantly more plentiful (Vepachedu and Setlow, 2005), are also entirely unknown.

One area where progress has been made recently concerning the role of germinant receptors, taking advantage of strains engineered to have variable numbers of receptors and progress with analytical techniques, concerns the so called commitment reaction. Early biochemical-led

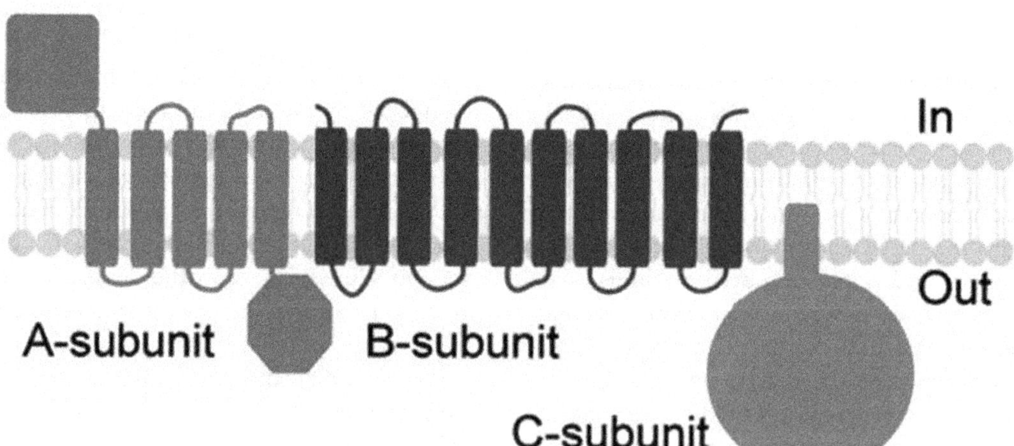

Figure 7.1 Schematic representation of the three proteins that constitute a typical GerA-type receptor. The number of predicted transmembrane helices in the A-subunit is not clear – probably five or six – hence, the hydrophilic domains may reside on the same side of the inner membrane.

investigations revealed that exposure of spores to nutrient germinants causes a reaction that irreversibly commits the spore to germinate, even if the germinant is subsequently removed or inhibited from binding to its cognate receptor (Stewart *et al.*, 1981). Although the precise nature of the commitment reaction is not understood, recent work has demonstrated that the time taken for a defined proportion of the spore population to undergo the commitment reaction correlates inversely with the average number of germinant receptors per spore (Yi and Setlow, 2010). Furthermore, average lag periods between addition of germinants, or between commitment, and subsequent release of DPA, which are highly heterogeneous in wild-type spore populations, is reduced in mutant strains designed to over-express defined germinant receptors. Thus, the abundance of 'activated' receptors per spore appears to be a major determinant in the rate of commitment, and indeed, in germination heterogeneity in general (Yi and Setlow, 2010; Zhang *et al.*, 2010). Additionally, activated germinant receptor levels also appear to play a role in post-commitment events, influencing the timing of subsequent DPA release, and perhaps even having some direct role in DPA release itself.

Despite these insights to the underlying causes of heterogeneity in rates of spore germination, progress in determining the precise function of spore germinant receptors, from either a biochemical or mechanistic/structural perspective, remains slow. Reasons for this are at least twofold. First, the receptors are present in the spore at extremely low abundance, localized to the inner membrane environment that is extremely well protected by the outer integument layers of the spore. As such, there are no published accounts of successful purification of functional germinant receptor complexes from this challenging and extremely complex environment. Second, like numerous other integral membrane proteins, the A and B subunit proteins have proved to be extremely recalcitrant to overexpression in several heterologous expression hosts, presenting a severe bottleneck to the initiation of biochemical and structural studies on these proteins.

However, several relatively hydrophilic C-protein subunits can be overexpressed in *E. coli*, and in a major advance for the field, the crystal structure of the C-protein from the *B. subtilis* GerB receptor (GerBC) has been solved recently, providing the first structural view of a spore germinant receptor subunit (Li *et al.*, 2010). The structure itself reveals GerBC to comprise three distinct domains, each of which adopts a protein fold that has not been observed previously. Bioinformatic analyses indicate also that the overall GerBC fold is probably adopted by all receptor C-subunits across the *Bacillaceae* and *Clostridiales*. However, while the structure represents the foundation of a molecular framework for understanding the architecture, conservation and mechanism(s) of spore germinant receptors, it has not, as yet, yielded information regarding the specific function of the C-protein, nor insight to its role within the receptor complex. It seems probable, therefore, that the answers to these questions will have to await high resolution structures of appropriate receptor A and B proteins, or indeed of an intact receptor complex.

Germinant recognition and location of the binding site

As alluded to in the previous section, the precise function of the individual protein subunits that constitute the receptor complex has not been elucidated, however, clearly one or more subunits must present cognate site(s) for the binding of specific germinant molecules. The location of these binding sites, definitive identification of amino acid residues that form receptor-ligand contacts, the effect that germinant binding has on receptor conformation, and the nature of subsequent signal transduction events, remain outstanding questions in the field.

The feasibility of using a bioinformatic approach to identify ligand-binding motifs seems logical since spores of several species of *Bacillus* and *Clostridium* have receptors that mutually recognize the same germinant molecules, with L-alanine probably being the most common. However, while the *B. subtilis* and *B. cereus* alanine-responsive receptors are relatively closely related and cluster in the same phylogenetic clade, primary amino acid sequences between these, and receptors from other species identified as being cognate for alanine, are too dissimilar to identify consensus motifs involved in germinant binding (Ross and Abel-Santos, 2010a,b). Indeed, the observation that L-alanine can stimulate relatively distantly related germinant receptors indicates perhaps that cognate binding sites for alanine have arisen several times in spore-forming lineages due to the abundance of this amino acid in diverse environmental niches (Ross and Abel-Santos, 2010b).

The possibility that the B protein of the receptor might be the site for germinant binding was suggested a number of years ago after point mutations in two *B. subtilis* strains that required increased concentrations of L-alanine to stimulate germination were mapped to the gene encoding the B subunit (GerAB) of the alanine receptor (Sammons *et al.*, 1981). This notion was supported further by phylogenetic analyses that revealed that the germinant receptor B proteins are related (albeit distantly) to microbial single component membrane transporter proteins, forming a branch of the amino acid, polyamine, and organocation (APC) superfamily of secondary transporters [spore germination protein (SGP) family; transporter classification no. 2.A.3.9; Moir *et al.*, 1994; Jack *et al.*, 2000].

Experimental evidence that the receptor B protein may be responsible for germinant recognition was conferred recently when *B. megaterium* QM B1551 spores bearing germinant receptors engineered to carry either of two closely related B-protein subunits (GerUB and GerVB) were observed to have different germination recognition profiles (Christie and Lowe, 2007). Additionally, *B. megaterium* spores bearing a B-protein cross-homologue chimeric receptor, comprising transmembrane (TM) helices one to eight of GerUB and TM helices nine and ten from GerVB, were shown to have gained a germinative response to proline, conferring further evidence that the B-subunit is the site of ligand binding (Christie *et al.*, 2008). Subsequent site-directed mutagenesis experiments, complemented with kinetic analyses of mutant spore germinative responses, have facilitated the identification of key regions, and indeed individual residues, within B-protein homologues of the *B. megaterium* GerU receptor that may participate directly in the binding of germinants to the receptor (Christie and Lowe, 2008; Christie *et al.*, 2008, 2010a). These include a number of conserved charged, aromatic and proline residues that cluster towards the middle of predicted TM domains 6 and 10, indicating that the germinant binding site may be located deep within the inner membrane (Fig. 7.2A). Alternatively, these residues may only be indirectly involved in receptor function, and perhaps have a structural role or influence germinant

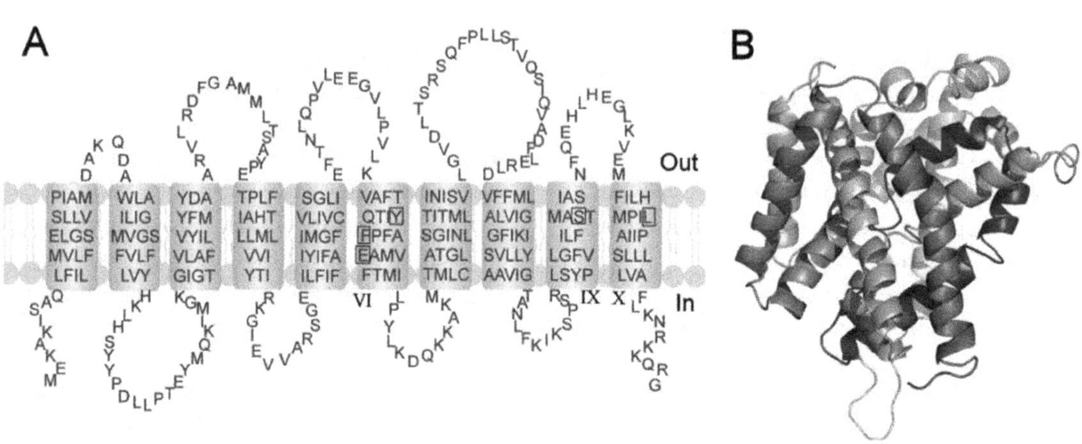

Figure 7.2 (A) Secondary structure prediction of *B. megaterium* GerVB, showing residues of functional or structural importance in putative TM6, TM9 and TM10 as determined by site-directed mutagenesis and kinetic analyses. (B) Structural model of GerVB obtained by comparative modelling, and based on the protein fold that is common to APC transporter proteins for which crystal structures have been determined (courtesy of Dr R.N. Miguel, University of Cambridge).

accessibility to the binding site. Improved discrimination between residues of functional and structural importance has been achieved recently by using antisera raised against the C protein of the *B. subtilis* GerA receptor to relay information, albeit indirectly, on the effect of substitutions to the A or B subunits on the apparent integrity of the entire receptor complex (Cooper and Moir, 2011; Mongkolthanaruk *et al.*, 2011).

The absence of high-resolution structural information and supporting biochemical analyses render the definitive identification of receptor domains and residues involved directly in ligand binding difficult to ascertain with any degree of certainty. However, residues in several prokaryotic APC transporters that are located at analogous positions to key residues identified in GerVB, for which high resolution structures have recently been solved, have been identified as being involved directly in ligand binding or as participants in substrate-binding cavities (Fang *et al.*, 2009; Gao *et al.*, 2009; Shaffer *et al.*, 2009). Further, despite undetectable homology at the amino acid level, the overall fold of the APC transporter proteins is very similar to the core structure observed in four other families of unrelated transport proteins (Lolkema and Slotboom, 2008; Fang *et al.*, 2009). Although only tentative in the absence of any firm structural data, collectively these observations indicate that germinant-receptor B-protein subunits might adopt structural, and perhaps mechanistic, similarities to some APC transporter proteins (Fig. 7.2B). This idea may be reinforced further by the observation that some bacteria utilize various transporter proteins, including at least one member of the APC superfamily of secondary transporters, in a sensory capacity to transduce environmental signals to the inside of the cell in a receptor-like fashion (Tetsch and Jung, 2009).

While molecular genetic and kinetic analyses have revealed that the receptor B protein exerts a significant influence over the selectivity and apparent affinity of the spore germinant receptor, indicating that it is the site for ligand binding, observations derived from germinative studies conducted with *B. subtilis* and *C. perfringens* spores, coupled with bioinformatic analyses, indicate that this may not necessarily be the case. Wild-type *B. subtilis* spores exposed to AGFK, for example, have an absolute requirement for the presence of both the GerB and GerK receptors in order to somehow cooperatively stimulate the germination response (Moir and Smith, 1990). However, a number of strains bearing point mutations in the GerB locus – two with single residue changes in the A protein (G297S and P326S) and another with a single substitution (F269I) in the B protein – have dispensed with the requirement for the GerK receptor, and can germinate via the variant GerB receptors in response to L-asparagine alone (Paidhungat and Setlow, 1999). Thus, while amino acid changes in this case do not alter the range of germinants recognized by the GerB receptor (Atluri *et al.*, 2006), since it recognizes the amino acid component in the AGFK mixture, they evidently have the capacity to introduce structural perturbations that lead to gain-of-function modifications to the receptor complex.

Similarly, *C. perfringens* spores bearing a null mutation in the structural gene for the B protein (GerKB) in the divergon-like GerK operon retain essentially wild-type germinative responses to nutrient (asparagine) and non-nutrient (KCl, Ca^{2+}-DPA) germinants identified as stimulating the GerK receptor (Paredes-Sabja *et al.*, 2008, 2009b). Since GerKB is the only B-protein homologue identifiably encoded in the *C. perfringens* genome, these analyses indicate that the A and C proteins of the receptor are responsible for germinant recognition and subsequent triggering of germination. Alternatively, another protein – perhaps a receptor B-protein whose sequence has diverged sufficiently from archetypal receptor B-proteins to preclude detection by genomic searches – might interact with the A and C proteins to form a functional germinant receptor. Intriguingly, proteins encoded by the *gerK* operon plus the associated GerKB protein all appear to have a crucial role in post-germinative outgrowth (Paredes-Sabja *et al.*, 2009b), contrasting with the situation in the bacilli, where the germinant receptors appear only to have a role in germination. Whether this role in outgrowth applies to other species of *Clostridia*, perhaps illustrating a fundamental distinction in receptor functionality between the major endospore forming genera, remains to be established.

Elsewhere, bioinformatic analyses have

revealed that the genomes of several *Clostridium* species encode only a single protein-A receptor subunit (Paredes-Sabja *et al.*, 2011). However, the germinative properties of spores of these species, which include the opportunistic pathogen *C. butyricum*, have not been characterized, and it may be that they germinate via alternative pathways such as the PrkC pathway (discussed below), or additional proteins interact with the A protein to form functional germinant receptor units. Indeed, *C. difficile*, a major causative agent of nosocomial-acquired pseudomembranous colitis, has no identifiable Ger-A type receptors encoded within its genome at all, yet spores of this species germinate specifically and display saturation-type kinetics when titrated with variable levels of glycine and the bile salt taurocholate (Sorg and Sonenshein, 2008; Ramirez *et al.*, 2010). These data indicate that alternative receptor-type proteins fulfil the requirement for environmental sensing in spores of this, and perhaps other species.

Thus, the location of the germinant binding site in nutrient germinant receptors awaits definitive identification. Indeed, it may be that both the A and B proteins individually form binding sites, particularly in cases where a range of structurally diverse germinants are recognized. However, combined site-directed mutagenesis and kinetic analyses conducted with *B. megaterium* spores, where the GerU receptor recognizes glucose, proline and leucine, indicate that the binding sites are close in proximity, and in the case of proline and glucose, probably overlap, indicating that a single subunit forms a binding pocket in this receptor (Christie *et al.*, 2008, 2010a). It is also formally possible that there is some variance in the receptor protein responsible for germinant binding, with the A protein serving this purpose in some receptors (perhaps only in some *Clostridium* species) and the B protein being tasked with this function in others.

Functional diversity and mechanistic insights to defined germination pathways

The multiplicity of receptors encoded in the genomes of most spore-forming bacteria, and differences in apparent function, appear not only to extend the range of germinants recognized by the spore, but to offer increased flexibility in germinant pathways, and routes to optimization of the germinative response in diverse environmental niches. Flexibility in germinative responses is exemplified in *B. anthracis*, where the five different germinant receptors – several of which can recognize multiple germinants – can variously interact to confer at least four general germination pathways, each of which requires the coordinated binding to two different germinant molecules for activation (Fisher and Hanna, 2005). Furthermore, the activation of defined pathways in pathogenesis appears to be determined by the route of infection, reflecting variance in intracellular conditions in different regions of the host (Carr *et al.*, 2010).

A recent Boolean-type analysis of the function of germinant receptors from various species identified six general strategies employed by *Bacillus* and *Clostridium* spores when integrating germinant signals (Ross and Abel-Santos, 2010a). The simplest of these involves the interaction of a germinant with a cognate receptor that can function independently of other receptors to initiate germination. Examples include the *B. subtilis* GerA receptor and the GerU receptor in *B. megaterium* QM B1551. However, although superficially similar, the structural diversity of ligands recognized by the GerU receptor is much greater than that of GerA, which responds primarily to L-alanine and a range of analogues, the effectiveness of which correlates essentially with the length of the alkyl side chain (Yasuda and Tochikubo, 1985). This apparent difference in structural flexibility of ligand binding sites between the receptors is reflected also in their propensity to inhibition by binding of germinant analogues. Thus, while stereospecific inversion of the side chain of L-ala to D-ala confers an extremely potent antagonist that binds to the GerA receptor with greater affinity than the agonist (Yasuda and Tochikubo, 1985), neither stereoisomers of amino acid germinants, nor several other germinant analogues, exert any measurable inhibitory effect on GerU receptor function (G. Christie, unpublished results).

Elsewhere, the use of germinant analogues as molecular probes to investigate receptor structure–activity relationships has provided fresh insights to the germinative strategy of *B. cereus* 569 spores (Dodatko *et al.*, 2010), which require both the GerI and GerQ receptors to stimulate germination in response to inosine as a sole germinant (Clements and Moir, 1998; Barlass *et al.*, 2002). Germinative analyses conducted with defined null mutant spores and several nucleoside analogues indicate that the GerI receptor has a well-ordered binding site that can bind productively with only a very narrow range of agonists, but can be inhibited in logical patterns by a number of nucleoside analogues. The GerQ receptor, on the other hand, can interact productively with several purine nucleosides, but is inhibited only by relatively hydrophobic analogues. These observations indicate differential mechanisms of nucleoside recognition by the respective receptors, with GerI displaying specific protein–substrate interactions, whereas GerQ has a more flexible binding site that interacts with the germinant via non-specific interactions (Dodatko *et al.*, 2010). This germinative pathway is notable also in that inosine binding results in the spore releasing L-alanine from the spore core, which, presumably by stimulating cognate receptors, facilitates the efficiency of the germinative response (Dodatko *et al.*, 2009). However, while these studies relay mechanistic and structural information concerning probable differences in the binding sites of receptors that recognize the same germinant, they do not explain the spore's prerequisite for both receptors to stimulate a germinative response.

The requirement for two different receptors to stimulate a germinative response is more commonly associated with the recognition of multiple germinant signals. As detailed previously, the GerB and GerK receptors are required to initiate germination in response to AGFK in *B. subtilis*. Despite the AGFK acronym, a detailed investigation of this germinative pathway – originally identified via the observation that *B. subtilis* spores are stimulated to germinate in caramel (Wax *et al.*, 1967) – revealed that the GerB receptor is in fact cognate for a diverse range of amino acids (and fructose), and therefore shares similarities with previously described receptors with apparently flexible binding sites (Atluri *et al.*, 2006). As might be expected for a receptor of this class, GerB function is not inhibited by D-alanine, indicating that differences exist in the structural organization of the binding sites of the *B. subtilis* alanine receptors.

It is in these areas that the application of kinetic-type analyses have proved to be very useful; particularly as increased knowledge of the proteins involved in mediating germination, and the availability of molecular tools with which they can specifically be manipulated, has conferred the potential to gain fresh mechanistic insights to receptor function. At the simplest level, the calculation of Michaelis–Menten parameters associated with spores in which the GerU receptor had been modified by site-directed mutagenesis permitted the putative identification of several residues that appear to contribute to the formation of a binding pocket that accommodates, in close proximity, structurally dissimilar ligands that trigger germination in *B. megaterium* QM B1551 (Christie *et al.*, 2010a). Elsewhere, detailed kinetic analyses of the *B. anthracis* germinative response to a mixture of alanine and inosine revealed a sequential cooperative mechanism, in which binding of the first germinant to its receptor increases the affinity for binding of the second germinant to its cognate receptor (Akoachere *et al.*, 2007). Furthermore, analyses of double reciprocal plots indicate strongly that the observed synergy in the germinative response is achieved by both germinants being present simultaneously on their respective receptors, although the precise molecular and biochemical details that underpin this cooperative response are not clear.

A similar sequential binding mechanism is evident from kinetic analysis of the spore germinative response of *C. difficile* spores, even although the protein(s) that mediate this response have yet to be identified (Ramirez *et al.*, 2010). These analyses indicate that only the receptor site cognate for taurocholate is available for binding in the dormant spore. Binding and subsequent saturation of taurocholate sites then reveal the glycine binding sites, which are rapidly occupied, resulting in the formation of putative hetero-receptor complexes that trigger initiation of spore germination.

Non-nutrient stimulation of germination

As mentioned previously, a second pathway for the initiation of germination that may be of physiological relevance, and which by-passes the germinant receptor mediated pathway, has been identified in *Bacillus* spores fairly recently (Shah *et al.*, 2008). This pathway involves a eukaryotic-like serine/threonine kinase (PrkC), which like the nutrient germinant receptors is localized to the spore inner membrane. This protein, homologues of which appear to be present in the genomes of all *Bacillus* and *Clostridium* species, has a peptidoglycan-binding domain that recognizes peptidoglycan fragments that contain a *meso*-diaminopimelic acid residue at the third position of the stem peptide. Binding of appropriate muropeptides – which are recycled during vegetative growth and may therefore signify the presence of growing bacteria in the environment – results in the PrkC-kinase domain phosphorylating the ribosomal GTPase, EF-G. However, the precise molecular details of how phosphorylation and presumed activation of EF-G in the dehydrated environment of the spore core results in triggering of subsequent germination events have yet to be established.

In addition to the two recognized physiological routes to germination, breakage of spore dormancy can also be stimulated by various physicochemical agents that in some cases can bypass individual components of the nutrient- or muropeptide-induced germination pathways. In addition to information that may be of use in an applied sense, research in this area has led to important insights concerning fundamental aspects of spore germination. The realization that exogenous Ca-DPA, for example, which had long since been recognized as a potent non-nutrient germinant, was responsible for the activation of the cortex lytic enzyme CwlJ, led to the first identification of a signalling pathway between stage 1 and stage 2 germination events (Paidhungat *et al.*, 2001).

Driven by potential application in the food industry, where reduced thermal processing offers potential improvements from economic, energy and food quality/organoleptic perspectives, the use of high pressures to stimulate spore germination is perhaps the best studied of the non-nutrient germinant pathways. Exposure of spores to moderate hydrostatic pressures, in the region of 100–150 MPa, has been shown to stimulate germination of *B. subtilis* and *B. cereus* spores (and by inference, all spore formers) by somehow activating the germinant receptors (Wuytack *et al.*, 2000; Paidhungat *et al.*, 2002). Further, individual germinant receptors appear to show differing levels of sensitivity to pressure treatment, with the hierarchy in responsiveness mirroring that observed when *B. subtilis* spores are germinated in a mixture of nutrients, i.e. GerA is more sensitive to pressure than GerB, which is more sensitive than GerK (Black *et al.*, 2005). The observation that spores bearing the variant GerB* receptor, which are > 5 times more responsive to pressure than spores bearing the native GerB receptor, despite the receptors being present at similar levels, indicates that differences in responsiveness to pressure are, at least to a degree, intrinsic to the structure and/or function of specific receptor complexes, and not just to the relative abundance of the respective receptors within the spore. Similarly, *B. subtilis* spores prepared at higher temperatures are demonstrably more responsive to moderate pressures compared with those prepared at lower temperatures, despite probably having lower numbers of germinant receptors, as adjudged from receptor–*lacZ* fusion assays.

It seems probable therefore that factors other than receptor abundance, probably concerning the fluidity or other aspects of the composition of the spore inner membrane, influence the spore germinative response to moderate pressures (Black *et al.*, 2005). Indeed, spores of the important food pathogen *B. cereus* bearing a null mutation in the GerN antiporter protein, which is required for efficient germination in response to inosine as a sole germinant, show a defective germination response to moderate high pressures when compared with wild-type spores. This effect is enhanced when the homologous GerT protein is also null, indicating that spore-associated monovalent cation transporters are also involved significantly in the moderate high pressure germinative response (Wei *et al.*, 2009).

Exposure of spores of various species to high hydrostatic pressures, i.e. upwards of 500 MPa, also stimulates germination, as one might expect from observations with moderate pressures. However, research conducted with various *B. subtilis* mutant strains has demonstrated that high pressure germination is not mediated by the spore germinant receptors, as evidenced by the efficient high pressure germination of receptorless spores (Paidhungat *et al.*, 2002). Instead, exposure to high pressures stimulates the release of DPA from the spore core, which results subsequently in the activation of cortex lytic enzymes and other stage 2 germination events. A number of lines of evidence that can be inferred from work in this area indicate that the release of DPA, and other small molecules from the spore core, is via the opening of specific channels in the spore inner membrane as opposed to physical disruption to the membrane. The release from the spore core of DPA, arginine and glutamic acid, for example, but not the slightly larger adenosine monophosphate (AMP), indicates that a proteinaceous channel is involved in this process, as do rate measurements that correlate with a temperature profile curve that has a defined optimum value (Vepachedu *et al.*, 2007). Further, rates of DPA release from spores engineered to have either increased levels of SpoVA proteins – which are required for the uptake (but not synthesis) of DPA into the developing forespore during sporulation (Tovar-Rojo *et al.*, 2002) – or temperature-sensitive mutations in a number of these proteins, are commensurate with SpoVA protein abundance and/or functionality of temperature sensitive variants (Vepachedu *et al.*, 2007).

While the precise function of the SpoVA proteins is unknown, their involvement in the transport of DPA during both sporulation and germination has been suspected for a number of years (Errington, 1993; Setlow, 2003). Further evidence that they are involved in the movement of DPA out of the spore during germination, perhaps as components of a DPA channel, has been provided by research aimed at investigating the germinative properties of cationic surfactants, as exemplified by dodecylamine. Although first identified in the 1960s (Rode and Foster, 1960), more recent work conducted with a range of defined *B. subtilis* germination mutants has provided insights to the mechanism by which dodecylamine (and by inference other cationic surfactants) stimulates germination. Indeed, dodecylamine-stimulated germination shares many parallels with the route to high pressure-induced germination described previously, whereby exposure to relatively low (50 μM) concentrations of surfactant by-passes the germinant receptors to stimulate the release of DPA from spores, which in turn activates the cortex lytic enzymes (Setlow *et al.*, 2003). Insight to the mode of action of dodecylamine in stimulating germination was achieved by comparing the germinative responses of wild-type spores and mutants that are incapable of degrading the cortex. Thus, whereas the action of dodecylamine ultimately ruptures the inner membrane in wild-type spores, resulting in the non-specific release of core-associated small molecules, spores engineered to lack cortex lytic enzymes release only DPA and free amino acids. The retention of AMP in these latter spores, in which the inner membrane is presumably resistant to rupture by surfactant due to the constraining effect of the cortex, indicates that dodecylamine causes the opening of specific DPA channels as opposed to more general pore formation in the membrane that is independent of proteins (Setlow *et al.*, 2003). This view is reinforced by the response to dodecylamine of spores with enhanced levels (or temperature sensitive) SpoVA proteins, where, as observed with high pressure experiments, the rate of the germinative response is commensurate with the abundance of native or functionality of temperature-sensitive SpoVA proteins (Vepachedu and Setlow, 2007b).

However, while strong evidence has been attained for the opening of specific DPA pores by either dodecylamine or high hydrostatic pressures, and it is evident that spore germinant receptors are activated by moderate pressures, it has not yet been established whether these agents act upon these proteins directly (i.e. as agonists), or whether they exert a more general influence on the properties of the inner membrane [or unidentified component(s) thereof], resulting indirectly in the activation of receptors or opening of DPA channels. Intriguingly, however, germination of spores by high pressure is strongly inhibited by prior exposure of spores to subgerminative concentrations of dodecylamine, which is consistent with high pressure acting upon protein(s) rendered non-responsive by dodecylamine that, from a DPA-release perspective, has bound non-productively (Black *et al.*, 2007). Additionally, the inhibition of dodecylamine-induced DPA release by Hg^{2+} ions is indicative that this process is induced by some action on proteins, although conversely, Hg^{2+} does not inhibit high pressure stimulated germination.

Research conducted with *B. subtilis* spores bearing enhanced levels of SpoVA proteins or temperature sensitive variants has also revealed their involvement in the release of DPA, and presumably other small molecules from the spore core, during lysozyme-mediated spore germination (Vepachedu and Setlow, 2007b). Lysozyme is capable of initiating germination in spores in which the integrity of the spore coat has been compromised, either by mutation or by chemical removal, to permit access to the cortical peptidoglycan. As with dodecylamine and high pressure germination, the germinant receptors are not required in this lytic route to germination. It is not clear, however, why the removal of the constraining cortex and germ cell wall should lead to the opening of DPA channels, unless perhaps the channel has mechanosensitive properties. However, the viability of lysozyme germinated spores when rescued on hypertonic medium does indicate that the release of DPA and other small molecules under these circumstances is in a controlled manner, presumably through the opening of designated channels, as opposed to leakage through a physically disrupted membrane. Activation of the endogenous cortex lytic enzyme CwlJ by relatively high concentrations of Ca^{2+}-DPA also leads to cortex depolymerization, with subsequent stimulation of DPA release and completion of spore germination, although in this case the germ cell wall remains in place (Paidhungat *et al.*, 2001). The removal of the cortex by lytic enzymes, however, does not necessarily lead to core rehydration and swelling, since dodecylamine germinated spores release DPA and degrade the cortex but hydrate only to a level equivalent to Stage one germinated spores (Setlow *et al.*, 2003). The reason for this inhibition of core swelling by dodecylamine has not been established, but clearly factors other than, or in addition to, the release of cortical pressure by lytic enzymes are involved in the uptake of water into germinating spores. Ultimately, further work is required in this area, with the aim of clarifying the identity of the proteins that constitute the DPA channel, to ascertain how this channel is regulated during nutrient and non-nutrient stimulated germination pathways, and to elucidate the mechanism by which water enters the spore core during germination.

A model for germinant receptor-initiated spore germination

Despite intensive research and considerable progress since the identification of the proteins that constitute the spore germinant receptor, several key unanswered questions concerning the mechanism of spore germination at the molecular level preclude the establishment of a definitive mechanistic model of the process. In all likelihood this will only be achieved with the availability of high resolution structures of representative receptor subunits, coupled with insights to the molecular basis of multigerminant synergy and hetero-receptor cooperativity. Determination of the role and molecular functions of accessory proteins such as GerD in *Bacillus*, the GerN-type antiporters, and Csp proteases in *Clostridium* species will also be required. As mentioned previously, definitive identification of the proteins that form putative DPA channels, and indeed their mode of gating and activation, is critical to the understanding of the molecular mechanisms involved in spore germination. Clearly much remains to be achieved, especially when apparent differences in overall receptor function and molecular mechanisms involved in spore germination of important pathogenic species, including *C. perfringens* and *C. difficile*, are also considered.

However, since the release of ions and amino acids from the spore core constitutes the earliest biochemical event detectable during spore germination, it seems plausible that the receptor may function at the molecular level in a transporter-type capacity. Although speculative, a model based on this hypothesis for initiation of spore germination in *Bacilli* that would apply to a receptor that can function independently, might consist of germinant molecule(s) binding to a receptor B-protein, inducing a conformational change that leads to the opening of a ligand-gated channel. Such a channel may comprise transmembrane regions of the A and B-subunit proteins, or a single protein, since both A and B proteins have sufficient predicted membrane-spanning domains to form a channel (Pebay-Peyroula, 2008). In this scenario, productive binding of germinant molecules resulting in channel opening, which may only be at low frequency, would define the commitment reaction. The selective movement of ions and/or amino acids through the putative channel, probably by facilitated diffusion, ultimately has the effect of irreversibly breaching the integrity of the spore inner membrane, committing the spore to proceed through the germination process. This could be achieved by the movement of some critical level of select small molecules from the spore core through the receptor channel, which either directly or indirectly stimulates opening of Ca^{2+}-DPA channels, and subsequent influx of water and commencement of stage 2 germination reactions. In this model, the time taken for the movement of a critical concentration of small molecules across the inner membrane would account for the lag period between commitment and rapid release of Ca^{2+}-DPA.

This model for initiation of spore germination does not encompass a defined role for accessory proteins such as GerD or GerN-type antiporter proteins, since at least in some receptor-mediated pathways proteins of the latter class do not appear to be required (Thackray *et al.*, 2001). Similarly, although GerD clearly facilitates the germinative response where examined in *Bacillus* spores (Pelczar *et al.*, 2007), it appears only to play an auxiliary role, particularly in non-optimal conditions, and indeed a homologue of this protein is not encoded within the genomes of the *Clostridia*. This latter point illustrates further apparently fundamental differences in molecular mechanisms involved in initiating germination in *Bacillus* and *Clostridium* species, and it is anticipated that the clarification of these, and other major unknown issues concerning the molecular details that constitute spore germination, will be the focus of future research in this field.

References

Akoachere, M., Squires, R.C., Nour, A.M., Angelov, L., Brojatsch, J., and Abel-Santos, E. (2007). Identification of an *in vivo* inhibitor of *Bacillus anthracis* spore germination. J. Biol. Chem. *282*, 12112–12118.

Alberto, F., Botella, L., Carlin, F., Nguyen-The, C., and Broussolle, V. (2005). The *Clostridium botulinum* GerAB germination protein is located in the inner membrane of spores. FEMS Microbiol. Lett. *253*, 231–235.

Atluri, S., Ragkousi, K., Cortezzo, D.E., and Setlow, P. (2006). Cooperativity between different nutrient receptors in germination of spores of *Bacillus subtilis* and reduction of this cooperativity by alterations in the GerB receptor. J. Bacteriol. *188*, 28–36.

Bagyan, I., Hobot, J., and Cutting, S. (1996). A compartmentalized regulator of developmental gene expression in *Bacillus subtilis*. J. Bacteriol. *178*, 4500–4507.

Barlass, P.J., Houston, C.W., Clements, M.O., and Moir, A. (2002). Germination of *Bacillus cereus* spores in response to L-alanine and to inosine: the roles of gerL and gerQ operons. Microbiology *148*, 2089–2095.

Black, E.P., Koziol-Dube, K., Guan, D., Wei, J., Setlow, B., Cortezzo, D.E., Hoover, D.G., and Setlow, P. (2005). Factors influencing germination of *Bacillus subtilis* spores via activation of nutrient receptors by high pressure. Appl. Environ. Microbiol. *71*, 5879–5887.

Black, E.P., Wei, J., Atluri, S., Cortezzo, D.E., Koziol-Dube, K., Hoover, D.G., and Setlow, P. (2007). Analysis of factors influencing the rate of germination of spores of *Bacillus subtilis* by very high pressure. J. Appl. Microbiol. *102*, 65–76.

Burns, D.A., Heap, J.T., and Minton, N.P. (2010). SleC is essential for germination of *Clostridium difficile* spores in nutrient-rich medium supplemented with the bile salt taurocholate. J. Bacteriol. *192*, 657–664.

Carr, K.A., Lybarger, S.R., Anderson, E.C., Janes, B.K., and Hanna, P.C. (2010). The role of *Bacillus anthracis* germinant receptors in germination and virulence. Mol. Microbiol. *75*, 365–375.

Christie, G., and Lowe, C.R. (2007). Role of chromosomal and plasmid-borne receptor homologues in the response of *Bacillus megaterium* QM B1551 spores to germinants. J. Bacteriol. *189*, 4375–4383.

Christie, G., and Lowe, C.R. (2008). Amino acid substitutions in transmembrane domains 9 and 10 of GerVB that affect the germination properties of *Bacillus megaterium* spores. J. Bacteriol. *190*, 8009–8017.

Christie, G., Lazarevska, M., and Lowe, C.R. (2008). Functional consequences of amino acid substitutions to GerVB, a component of the *Bacillus megaterium* spore germinant receptor. J. Bacteriol. *190*, 2014–2022.

Christie, G., Gotzke, H., and Lowe, C.R. (2010a). Identification of a receptor subunit and putative ligand-binding residues involved in the *Bacillus megaterium* QM B1551 spore germination response to glucose. J. Bacteriol. *192*, 4317–4326.

Christie, G., Ustok, F.I., Lu, Q., Packman, L.C., and Lowe, C.R. (2010b). Mutational analysis of *Bacillus megaterium* QM B1551 cortex-lytic enzymes. J. Bacteriol. *192*, 5378–5389.

Clements, M.O., and Moir, A. (1998). Role of the gerI operon of *Bacillus cereus* 569 in the response of spores to germinants. J. Bacteriol. *180*, 6729–6735.

Cooper, G.R., and Moir, A. (2011). Amino Acid Residues in the GerAB Protein Important in the Function and Assembly of the Alanine Spore Germination Receptor of *Bacillus subtilis* 168. J. Bacteriol. *193*, 2261–2267.

Corfe, B.M., Sammons, R.L., Smith, D.A., and Mauel, C. (1994). The *gerB* region of the *Bacillus subtilis* 168 chromosome encodes a homologue of the *gerA* spore germination operon. Microbiology *140 (Pt 3)*, 471–478.

Cowan, A.E., Koppel, D.E., Setlow, B., and Setlow, P. (2003). A soluble protein is immobile in dormant spores of *Bacillus subtilis* but is mobile in germinated spores: implications for spore dormancy. Proc. Natl. Acad. Sci. U.S.A. *100*, 4209–4214.

Dodatko, T., Akoachere, M., Muehlbauer, S.M., Helfrich, F., Howerton, A., Ross, C., Wysocki, V., Brojatsch, J., and Abel-Santos, E. (2009). *Bacillus cereus* spores release alanine that synergizes with inosine to promote germination. PLoS One *4*, e6398.

Dodatko, T., Akoachere, M., Jimenez, N., Alvarez, Z., and Abel-Santos, E. (2010). Dissecting interactions between nucleosides and germination receptors in *Bacillus cereus* 569 spores. Microbiology *156*, 1244–1255.

Errington, J. (1993). *Bacillus subtilis* sporulation: regulation of gene expression and control of morphogenesis. Microbiol. Rev. *57*, 1–33.

Fang, Y., Jayaram, H., Shane, T., Kolmakova-Partensky, L., Wu, F., Williams, C., Xiong, Y., and Miller, C. (2009). Structure of a prokaryotic virtual proton pump at 3.2 A resolution. Nature *460*, 1040–1043.

Feavers, I.M., Miles, J.S., and Moir, A. (1985). The nucleotide sequence of a spore germination gene (gerA) of *Bacillus subtilis* 168. Gene *38*, 95–102.

Feavers, I.M., Foulkes, J., Setlow, B., Sun, D., Nicholson, W., Setlow, P., and Moir, A. (1990). The regulation of transcription of the gerA spore germination operon of *Bacillus subtilis*. Mol. Microbiol. *4*, 275–282.

Fisher, N., and Hanna, P. (2005). Characterization of *Bacillus anthracis* germinant receptors *in vitro*. J. Bacteriol. *187*, 8055–8062.

Foster, S.J., and Johnstone, K. (1990). Pulling the trigger: the mechanism of bacterial spore germination. Mol. Microbiol. *4*, 137–141.

Gao, X., Lu, F., Zhou, L., Dang, S., Sun, L., Li, X., Wang, J., and Shi, Y. (2009). Structure and mechanism of an amino acid antiporter. Science *324*, 1565–1568.

Heap, J.T., Cartman, S.T., Kuehne, S.A., Cooksley, C., and Minton, N.P. (2010). ClosTron-targeted mutagenesis. Methods Mol. Biol. *646*, 165–182.

Heffron, J.D., Orsburn, B., and Popham, D.L. (2009). Roles of germination-specific lytic enzymes CwlJ and SleB in *Bacillus anthracis*. J. Bacteriol. *191*, 2237–2247.

Hinc, K., Nagorska, K., Iwanicki, A., Wegrzyn, G., Seror, S.J., and Obuchowski, M. (2006). Expression of genes coding for GerA and GerK spore germination receptors is dependent on the protein phosphatase PrpE.J. Bacteriol. *188*, 4373–4383.

Hudson, K.D., Corfe, B.M., Kemp, E.H., Feavers, I.M., Coote, P.J., and Moir, A. (2001). Localization of GerAA and GerAC germination proteins in the *Bacillus subtilis* spore. J. Bacteriol. *183*, 4317–4322.

Igarashi, T., and Setlow, P. (2005). Interaction between individual protein components of the GerA and GerB nutrient receptors that trigger germination of *Bacillus subtilis* spores. J. Bacteriol. *187*, 2513–2518.

Ireland, J.A., and Hanna, P.C. (2002). Amino acid- and purine ribonucleoside-induced germination of *Bacillus anthracis* DeltaSterne endospores: *gerS* mediates responses to aromatic ring structures. J. Bacteriol. *184*, 1296–1303.

Irie, R., Fujita, Y., and Kobayashi, M. (1996). Nucleotide sequence and gene organisation of the *gerK* spore germination locus of *Bacillus subtilis* 168. J. Gen. Appl. Microbiol. *42*, 141–153.

Ishikawa, S., Yamane, K., and Sekiguchi, J. (1998). Regulation and characterization of a newly deduced cell wall hydrolase gene (*cwlJ*) which affects germination of *Bacillus subtilis* spores. J. Bacteriol. *180*, 1375–1380.

Jack, D.L., Paulsen, I.T., and Saier, M.H. (2000). The amino acid/polyamine/organocation (APC) superfamily of transporters specific for amino acids, polyamines and organocations. Microbiology *146 (Pt 8)*, 1797–1814.

Johnstone, K. (1994). The trigger mechanism of spore germination: current concepts. Soc. Appl. Bacteriol. Symp. Ser. *23*, 17S-24S.

Johnstone, K., Stewart, G.S., Scott, I.R., and Ellar, D.J. (1982). Zinc release and the sequence of biochemical events during triggering of *Bacillus megaterium* KM spore germination. Biochem. J. *208*, 407–411.

Lambert, E.A., and Popham, D.L. (2008). The *Bacillus anthracis* SleL (YaaH) protein is an *N*-acetylglucosaminidase involved in spore cortex depolymerization. J. Bacteriol. *190*, 7601–7607.

Li, Y., Setlow, B., Setlow, P., and Hao, B. (2010). Crystal structure of the GerBC component of a *Bacillus subtilis* spore germinant receptor. J. Mol. Biol. *402*, 8–16.

Lolkema, J.S., and Slotboom, D.J. (2008). The major amino acid transporter superfamily has a similar core structure as Na+-galactose and Na+-leucine transporters. Mol. Membr. Biol. *25*, 567–570.

Miyata, S., Moriyama, R., Miyahara, N., and Makino, S. (1995). A gene (*sleC*) encoding a spore-cortex-lytic enzyme from *Clostridium perfringens* S40 spores; cloning, sequence analysis and molecular characterization. Microbiology *141 (Pt 10)*, 2643–2650.

Moir, A. (2006). How do spores germinate? J. Appl. Microbiol. *101*, 526–530.

Moir, A., and Smith, D.A. (1990). The genetics of bacterial spore germination. Annu. Rev. Microbiol. *44*, 531–553.

Moir, A., Lafferty, E., and Smith, D.A. (1979). Genetics analysis of spore germination mutants of *Bacillus subtilis* 168: the correlation of phenotype with map location. J. Gen. Microbiol. *111*, 165–180.

Moir, A., Kemp, E.H., Robinson, C., and Corfe, B.M. (1994). The genetic analysis of bacterial spore germination. J. Appl. Bacteriol. *77*, 9S-16S.

Moir, A., Corfe, B.M., and Behravan, J. (2002). Spore germination. Cell Mol. Life Sci. *59*, 403–409.

Mongkolthanaruk, W., Cooper, G.R., Mawer, J.S., Allan, R.N., and Moir, A. (2011). Effect of amino acid substitutions in the GerAA protein on the function of the alanine-responsive germinant receptor of *Bacillus subtilis* spores. J. Bacteriol. *193*, 2268–2275.

Paidhungat, M., and Setlow, P. (1999). Isolation and characterization of mutations in *Bacillus subtilis* that allow spore germination in the novel germinant D-alanine. J. Bacteriol. *181*, 3341–3350.

Paidhungat, M., and Setlow, P. (2001). Localization of a germinant receptor protein (GerBA) to the inner membrane of *Bacillus subtilis* spores. J. Bacteriol. *183*, 3982–3990.

Paidhungat, M., Ragkousi, K., and Setlow, P. (2001). Genetic requirements for induction of germination of spores of *Bacillus subtilis* by Ca^{2+}-dipicolinate. J. Bacteriol. *183*, 4886–4893.

Paidhungat, M., Setlow, B., Daniels, W.B., Hoover, D., Papafragkou, E., and Setlow, P. (2002). Mechanisms of induction of germination of *Bacillus subtilis* spores by high pressure. Appl. Environ. Microbiol. *68*, 3172–3175.

Paredes-Sabja, D., Torres, J.A., Setlow, P., and Sarker, M.R. (2008). *Clostridium perfringens* spore germination: characterization of germinants and their receptors. J. Bacteriol. *190*, 1190–1201.

Paredes-Sabja, D., Setlow, P., and Sarker, M.R. (2009a). The protease CspB is essential for initiation of cortex hydrolysis and dipicolinic acid (DPA) release during germination of spores of *Clostridium perfringens* type A food poisoning isolates. Microbiology *155*, 3464–3472.

Paredes-Sabja, D., Setlow, P., and Sarker, M.R. (2009b). Role of GerKB in germination and outgrowth of *Clostridium perfringens* spores. Appl Environ Microbiol. *75*, 3813–3817.

Paredes-Sabja, D., Setlow, P., and Sarker, M.R. (2009c). SleC is essential for cortex peptidoglycan hydrolysis during germination of spores of the pathogenic bacterium *Clostridium perfringens*. J. Bacteriol. *191*, 2711–2720.

Paredes-Sabja, D., Setlow, P., and Sarker, M.R. (2011). Germination of spores of *Bacillales* and *Clostridiales* species: mechanisms and proteins involved. Trends Microbiol. doi:10.1016/j.tim.2010.10.004.

Pebay-Peyroula, E. (2008). High Resolution Structures of Membrane Proteins: From X-ray Crystallography to an Integrated Approach of Membranes. In Biophysical Analysis of Membrane Proteins, Pebay-Peyroula, E., ed. (Wiley-VCH, Berlin), pp. 1–27.

Pelczar, P.L., Igarashi, T., Setlow, B., and Setlow, P. (2007). Role of GerD in germination of *Bacillus subtilis* spores. J. Bacteriol. *189*, 1090–1098.

Ramirez, N., Liggins, M., and Abel-Santos, E. (2010). Kinetic evidence for the presence of putative germination receptors in *Clostridium difficile* spores. J. Bacteriol. *192*, 4215–4222.

Rode, L.J., and Foster, J.W. (1960). The action of surfactants on bacterial spores. Arch. Mikrobiol. *36*, 67–94.

Ross, C., and Abel-Santos, E. (2010a). The Ger receptor family from sporulating bacteria. Curr. Issues Mol. Biol. *12*, 147–158.

Ross, C.A., and Abel-Santos, E. (2010b). Guidelines for nomenclature assignment of Ger receptors. Res. Microbiol. *161*, 830–837.

Sammons, R.L., Moir, A., and Smith, D.A. (1981). Isolation and properties of spore germination mutants of *Bacillus subtilis* 168 defective in the initiation of germination. J. Gen. Microbiol. *124*, 229–241.

Scott, I.R., and Ellar, D.J. (1978). Metabolism and the triggering of germination of *Bacillus megaterium*. Use of L-[3H]alanine and tritiated water to detect metabolism. Biochem. J. *174*, 635–640.

Setlow, B., Melly, E., and Setlow, P. (2001). Properties of spores of *Bacillus subtilis* blocked at an intermediate stage in spore germination. J. Bacteriol. *183*, 4894–4899.

Setlow, B., Cowan, A.E., and Setlow, P. (2003). Germination of spores of *Bacillus subtilis* with dodecylamine. J. Appl. Microbiol. *95*, 637–648.

Setlow, B., Wahome, P.G., and Setlow, P. (2008). Release of small molecules during germination of spores of *Bacillus* Species. J. Bacteriol. *190*, 4759–4763.

Setlow, B., Peng, L., Loshon, C.A., Li, Y.Q., Christie, G., and Setlow, P. (2009). Characterization of the germination of *Bacillus megaterium* spores lacking enzymes that degrade the spore cortex. J. Appl. Microbiol. *107*, 318–328.

Setlow, P. (2003). Spore germination. Curr. Opin. Microbiol. *6*, 550–556.

Setlow, P. (2007). I will survive: DNA protection in bacterial spores. Trends Microbiol. *15*, 172–180.

Shaffer, P.L., Goehring, A., Shankaranarayanan, A., and Gouaux, E. (2009). Structure and mechanism of a Na+-independent amino acid transporter. Science *325*, 1010–1014.

Shah, I.M., Laaberki, M.H., Popham, D.L., and Dworkin, J. (2008). A eukaryotic-like Ser/Thr kinase signals bacteria to exit dormancy in response to peptidoglycan fragments. Cell *135*, 486–496.

Sorg, J.A., and Sonenshein, A.L. (2008). Bile salts and glycine as cogerminants for *Clostridium difficile* spores. J. Bacteriol. *190*, 2505–2512.

Stewart, G.S., Johnstone, K., Hagelberg, E., and Ellar, D.J. (1981). Commitment of bacterial spores to germinate. A measure of the trigger reaction. Biochem. J. *198*, 101–106.

Sunde, E.P., Setlow, P., Hederstedt, L., and Halle, B. (2009). The physical state of water in bacterial spores. Proc. Natl. Acad. Sci. U.S.A. *106*, 19334–19339.

Swerdlow, B.M., Setlow, B., and Setlow, P. (1981). Levels of H^+ and other monovalent cations in dormant and germinating spores of *Bacillus megaterium*. J. Bacteriol. *148*, 20–29.

Tetsch, L., and Jung, K. (2009). The regulatory interplay between membrane-integrated sensors and transport proteins in bacteria. Mol. Microbiol. 73, 982–991.

Thackray, P.D., Behravan, J., Southworth, T.W., and Moir, A. (2001). GerN, an antiporter homologue important in germination of *Bacillus cereus* endospores. J. Bacteriol. *183*, 476–482.

Tovar-Rojo, F., Chander, M., Setlow, B., and Setlow, P. (2002). The products of the *spoVA* operon are involved in dipicolinic acid uptake into developing spores of *Bacillus subtilis*. J. Bacteriol. *184*, 584–587.

Vepachedu, V.R., and Setlow, P. (2005). Localization of SpoVAD to the inner membrane of spores of *Bacillus subtilis*. J. Bacteriol. *187*, 5677–5682.

Vepachedu, V.R., and Setlow, P. (2007a). Analysis of interactions between nutrient germinant receptors and SpoVA proteins of *Bacillus subtilis* spores. FEMS Microbiol. Lett. *274*, 42–47.

Vepachedu, V.R., and Setlow, P. (2007b). Role of SpoVA proteins in release of dipicolinic acid during germination of *Bacillus subtilis* spores triggered by dodecylamine or lysozyme. J. Bacteriol. *189*, 1565–1572.

Vepachedu, V.R., Hirneisen, K., Hoover, D.G., and Setlow, P. (2007). Studies of the release of small molecules during pressure germination of spores of *Bacillus subtilis*. Lett. Appl. Microbiol. *45*, 342–348.

Wax, R., Freese, E., and Cashel, M. (1967). Separation of two functional roles of L-alanine in the initiation of *Bacillus subtilis* spore germination. J. Bacteriol. *94*, 522–529.

Wei, J., Setlow, P., and Hoover, D.G. (2009). Effects of moderately high pressure plus heat on the germination and inactivation of *Bacillus cereus* spores lacking proteins involved in germination. Lett. Appl. Microbiol. *49*, 646–651.

Wolgamott, G.D., and Durham, N.N. (1971). Initiation of spore germination in *Bacillus cereus*: a proposed allosteric receptor. Can. J. Microbiol. *17*, 1043–1048.

Wuytack, E.Y., Soons, J., Poschet, F., and Michiels, C.W. (2000). Comparative study of pressure- and nutrient-induced germination of *Bacillus subtilis* spores. Appl. Environ. Microbiol. *66*, 257–261.

Yasuda, Y., and Tochikubo, K. (1985). Germination-initiation and inhibitory activities of L- and D-alanine analogues for *Bacillus subtilis* spores. Modification of methyl group of L- and D-alanine. Microbiol. Immunol. *29*, 229–241.

Yi, X., and Setlow, P. (2010). Studies of the commitment step in the germination of spores of *Bacillus* species. J. Bacteriol. *192*, 3424–3433.

Zhang, P., Garner, W., Yi, X., Yu, J., Li, Y.Q., and Setlow, P. (2010). Factors affecting variability in time between addition of nutrient germinants and rapid dipicolinic acid release during germination of spores of *Bacillus* species. J. Bacteriol. *192*, 3608–3619.

Zuberi, A.R., Feavers, I.M., and Moir, A. (1985). Identification of three complementation units in the gerA spore germination locus of *Bacillus subtilis*. J. Bacteriol. *162*, 756–762.

Zuberi, A.R., Moir, A., and Feavers, I.M. (1987). The nucleotide sequence and gene organization of the gerA spore germination operon of *Bacillus subtilis* 168. Gene *51*, 1–11.

Germination of *Bacillus anthracis* Spores

8

Jonathan D. Giebel, Katherine A. Carr and Philip C. Hanna

Abstract

Bacillus anthracis exists in two morphologically distinct states, the spore and the vegetative cell. *B. anthracis* spores, not the vegetative bacilli, are the etiologic agent of the disease anthrax. Anthrax infections result from these inert spores entering a suitable host, recognizing that the environment is suitable for growth, and then transitioning to the rapidly growing, toxin- and capsule-producing vegetative cell through a programmed germination process whose proteins are produced during spore formation and remain inactive until triggered by specific cues. The process is, generally, conserved among spore-forming bacteria. In this chapter we outline some of the unique attributes of *B. anthracis* spore germination, such as the spatial and temporal regulation of spore germination inside a host, the repertoire of small molecule germinants recognized by the spore, their concomitant germinant receptors and the group of lytic enzymes necessary to complete the cortex degradation that allows for vegetative cell outgrowth and discuss how the proper functioning of each of these steps contributes directly to the pathogenesis of this organism.

Introduction

The bacterial spore, and its concomitant sporulation and germination capabilities, are critical to the life cycle of the *Bacillus* and *Clostridium* genera. For all firmicute species capable of forming spores the developmental processes of spore formation and germination appear conserved at both genetic and mechanistic levels. The resistance properties and extreme longevities of dormant bacterial spores allow bacteria to survive, in a dormant form, long periods of nutrient deprivation and challenging physical and chemical stresses (Nicholson *et al.*, 2000b). Within the large group of spore-forming species there are several that are severe human and animal pathogens and, as such, are the focus of biomedical research efforts. In this chapter we discuss the contribution of the *B. anthracis* spore to the disease anthrax and outline important differences between *Bacillus anthracis* and other related spore-forming bacteria in regards to this spore's contributions to disease.

Bacillus anthracis has been the subject of renewed research interest in the last decade. *B. anthracis* is the causative agent of anthrax and is identified as a Select Agent by the Centers for Disease Control and Prevention (http://www.selectagents.gov) and the United Stated Department of Agriculture (http://www.aphis.usda.gov/) for its potential use as an agent of bioterrorism. Like other spore-forming species, *B. anthracis* spores are very stable in the environment (Gest and Mandelstam, 1987; Cano and Borucki, 1995). Decontamination of effected areas is extremely difficult and costly (Turnbull, 1996; WHO, 2008). In addition to their persistence in nature, *B. anthracis* spores are highly infectious and are the etiologic agents for all variations of the disease, with each variation dictated by their point of entry into the body (Druett *et al.*, 1953a; Navacharoen *et al.*, 1985a; Sirisanthana *et al.*, 1988; Dixon *et al.*, 1999; Friedlander, 1999a; Sirisanthana and Brown, 2002b). In the early twentieth century, *B. anthracis* was used as a weapon of war, perhaps its first well-documented use as a bioweapon. During the First World War, Germany intentionally infected the horses of neutral countries that were aiding the allied powers (Wheelis, 1999). In the mid-twentieth century, especially in the years following the Second World War, *B. anthracis* research interest increased with several countries including the United States, the UK and the USSR conducting research into the defensive measures against, and offensive capabilities of, this bacterium (Hammond and Carter, 2002). The 1979 accidental release of *B. anthracis* spores from a Soviet weapons plant in Sverdlovsk and the Amerithrax attacks through the US mail in 2001 each promoted intense periods of research interest into this pathogen (Meselson *et al.*, 1994; Jernigan *et al.*, 2001, 2002). Here we will discuss the unique traits of the *B. anthracis* spore that potentially contribute to its pathogenicity including, recognition of spores by host immune cells, role of redundant germinant receptors in host infection, and the array of germination-specific lytic enzymes used to breakdown spore cortex during the breaking of dormancy and return to its vegetative, virulent, infectious cycle.

Anthrax

As initially described by Robert Koch, *B. anthracis* is the aetiological agent of the disease anthrax (Koch, 1877). *B. anthracis* exists in two morphologically distinct forms, the spore and the vegetative bacillus. Koch went on to show that in fact it is the spore form of the bacterium that initiates disease once it has entered a suitable host (Koch, 1881). Although grazing ruminants such as cattle and sheep are the most common hosts of *B. anthracis* infection it is believed that most mammals are susceptible to the bacterium (Dixon *et al.*, 1999). The most common form of the disease in natural veterinary outbreaks is known as oropharyngeal anthrax. This occurs when spores in the soil enter the body of domesticated or wild animals while grazing through abrasions in the mouth or pharynx (Navacharoen *et al.*, 1985b; Quinn and Turnbull, 1998; Glomski *et al.*, 2007). A second form of the disease, gastrointestinal anthrax, occurs when spores are ingested, usually from the consumption of contaminated meat (Sirisanthana *et al.*, 1984; Dixon *et al.*, 1999; Sirisanthana and Brown, 2002a; Watts *et al.*, 2008, 2009). The most common natural form of the disease in human cases is cutaneous anthrax. This form of the disease occurs when spores cross the epidermis through small cuts or abrasions (Dixon *et al.*, 1999; Friedlander, 1999b). Cutaneous anthrax results in a self-contained black eschar where spores have entered the skin. This is considered the least severe form of anthrax and is often resolved without medical intervention (Smego *et al.*, 1998). In some cases bacteria from a cutaneous anthrax infection can enter the blood stream causing toxaemia and death to the host. The final and most severe form of the disease is inhalational anthrax. This occurs when spores are taken up into the lungs and is usually fatal (Druett *et al.*, 1953b; Dixon *et al.*, 1999). Because this form of anthrax is the most severe and is therefore the most probable targeted pathway for bioterrorism applications of *B. anthracis*, it has received the most attention from researchers in the community and will be the focus of the remainder of this chapter.

The spore is the infectious form of the bacterium (Koch, 1881; Bergman *et al.*, 2006; Fisher *et al.*, 2006). It has been a long held belief in the field that spores entering the lungs associate with phagocytes and are trafficked to nearby lymph nodes where the bacteria escape causing a widespread infection. This model was first suggested in a study published by Ross in 1957 that used guinea pigs as a model organism to study inhalation anthrax (Fig. 8.1) (Ross, 1957). Since then, several groups have attempted to confirm this theory and to determine when in the internalization process spore germination occurs. In her initial study, Ross relied on two new methods for accurately determining the fate of *B. anthracis* spores entering the lung. Ross developed a differential staining technique that distinguished dormant spores from those that had begun germination. Vegetative cells were identified by both their morphology as well as their staining properties. Ross also employed a new method for rapid fixation of pulmonary tissue in formaldehyde. Following infection, animals were anesthetized then euthanized by removing the blood and perfusing the tissue with 4% formaldehyde. This effectively halted spore germination immediately providing a snapshot of the

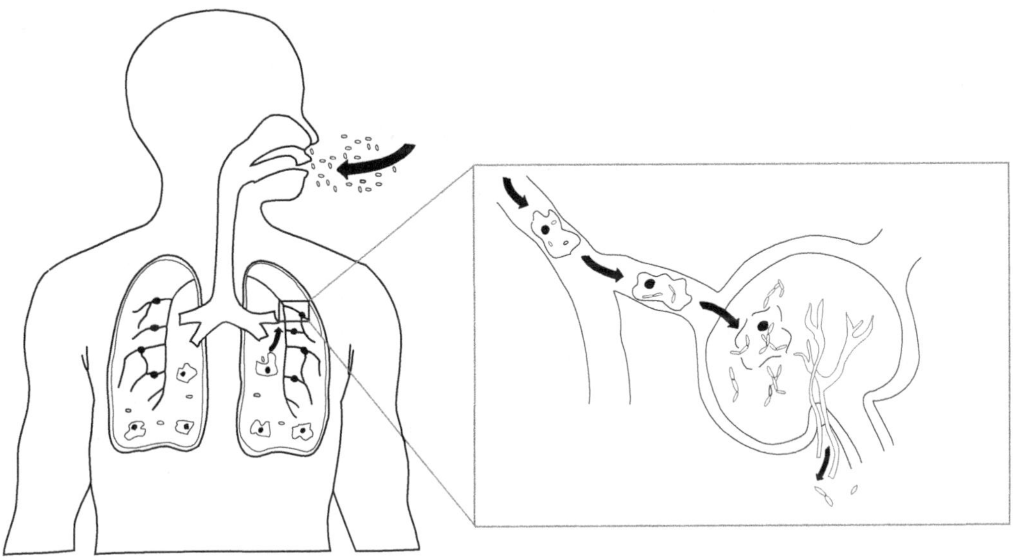

Figure 8.1 Model of inhalation anthrax progression based on the 1957 study by Ross. (1) Aerosolized spores are inhaled. (2) Alveolar macrophages phagocytose spores in the lungs. (3) Macrophages containing spores are trafficked through the lymphatic system to a mediastinal lymph node. (4) Spores germinate within the macrophage while en route and vegetative cells begin to divide. (5) Once inside the lymph node, vegetative bacilli are released and bacterial growth continues. (6) Vegetative cells escape the lymph node into the circulatory system causing bacteraemia, toxaemia and eventually death. Artwork by Nicole Biltz.

infection. Using these techniques, Ross found that *B. anthracis* spores in the lungs are rapidly taken up (less than 35 minutes) by alveolar phagocytes that she termed macrophages. Although no vegetative growth or spore germination was observed in the extracellular lung tissue (there is typically no acute pneumonia during inhalation anthrax until very late in the disease, if ever), macrophages were seen to contain dormant spores, germinating spores and bacilli at various time points between 0 and 18 hours post infection. After 18 hours, vegetative cells were predominantly present and by 24 hours vegetative cell division was seen in the lymph nodes, free of the macrophages. *B. subtilis* spores did not show signs of germination following infection and no vegetative cells were found. Ross concluded from this study that anthrax was not a traditional pneumonia, but that spores were engulfed by lung phagocytes, and only once bacteria were trafficked to lymph nodes, and released, did they cause a general bacteraemia. The model established by Ross in 1957 has been further studied and refined by many groups using the newest technologies available.

In 2007, Cleret *et al.* used fluorescently activated cell sorting and confocal microscopy to further define the interaction between *B. anthracis* spores and lung phagocytes following infection (Cleret *et al.*, 2007). In agreement with Ross, Cleret *et al.* found that spores were engulfed by alveolar macrophages within 10 minutes of infection. However, these cells were not the phagocytes responsible for transport to the lymph nodes. Within 30 minutes of infection, dendritic cells in the interstitial space of the lungs were able to sample spores without crossing the lung epithelial barrier. These cells then trafficked the spores to the draining lymph nodes. This resolution of lung phagocytic cell type was not available to Ross in the 1950s.

Work done by Guidi-Rontani and colleagues also found that *B. anthracis* spores rapidly associate with mononuclear phagocytes in the lungs (Guidi-Rontani *et al.*, 1999b). Guidi-Rontani *et al.* developed polyclonal antibodies against spores and vegetative bacilli to distinguished dormant and germinated spores using immunofluorescence confocal microscopy. Using these tools, they found that germinated spores were associated with phagolysosomes within phagocytes collected from bronchoalveolar lavage (BAL). They also showed that spores only germinated when associated with phagocytes, not in the presence of BAL fluid alone. This further supports the idea that spores do not germinate in the extracellular space of the lungs, but only inside alveolar phagocytes.

Despite the abundance of evidence supporting the theory of macrophage/dendritic cell-associated germination and trafficking to lymph nodes, there remain many studies that find conflicting results. Glomski and colleagues have seen that germination can occur at the initial site of inoculation (Glomski *et al.*, 2007, 2008). This was done using a capsulated non-toxigenic strain in which a bioluminescence cassette was expressed under the *pag* promoter. Therefore, only cells that had resumed metabolic activity, or had active expression from the *pag* promoter would produce light. This difference cannot be solely attributed to the new technique employed. Sanz *et al.* used a similar model in which the bioluminescence cassette was expressed under the *sspB* promoter or the *pag* promoter to produce light in germinating spores or vegetative cells, respectively (Sanz *et al.*, 2008). In this study spores were seen to germinate in the lung tissue, but only when associated with macrophages. Additionally, Sanz *et al.* found no germination within the nearby lymph nodes. This study suggests that spore germination occurs primarily within the phagocyte while in transit to the lymph node. Still other groups have found that spores remain dormant in the lungs for several days or even weeks (Henderson *et al.*, 1956; Friedlander *et al.*, 1993; Pickering *et al.*, 2004; Cote *et al.*, 2006; Heine *et al.*, 2007; Loving *et al.*, 2007).

In addition to *in vivo* studies tracking the movements of *B. anthracis* spores within a host, several groups have used *in vitro* cell co-culture techniques with primary phagocytes as well as transformed cell lines to study the interaction between spores and immune cells (Dixon *et al.*, 2000; Ireland and Hanna, 2002a; Weiner and Hanna, 2003; Kang *et al.*, 2005; Ribot *et al.*, 2006; Basu *et al.*, 2007; Hu *et al.*, 2007; Gut *et al.*, 2011). RAW 264.7 cells, a murine macrophage-like cell line, are capable of rapidly phagocytosing spores of several bacillus species including *B. subtilis*, *B. megaterium*, and *B. anthracis*. Association of *B. anthracis* spores with phagocytic cells leads to germination, while the spores of the closely related but non-pathogenic species are not triggered to germinate (Dixon *et al.*, 2000; Ireland and Hanna, 2002a; Weiner and Hanna, 2003). In the germination model originally described by Ross, spores germinate within phagocytes not in the extracellular spaces of the lungs. However it has been shown *in vitro* that macrophages can induce germination without direct contact with the spores (Weiner and Hanna, 2003; Gut *et al.*, 2011). The ability of cell-free conditioned media to induce germination is heat stable, but is lost following dialysis, suggesting that this phenotype maybe caused by the release of small molecule germinants supplied by the macrophages themselves.

Germinant receptors

Upon entry into the host, germination of dormant spores is dependent upon an appropriate germinant/germinant receptor interaction (Corfe *et al.*, 1994b; Moir *et al.*, 1979; Sammons *et al.*, 1981; Zuberi *et al.*, 1987; Ross and Abel-Santos, 2010). These interactions allow for both sensitive and on-going environmental 'sensing' by the dormant spore *and* trigger a very rapid return to vegetative growth even after very extended periods of dormancy when environmental conditions are favourable. Germinant receptors localize to the spore inner membrane, where they sense the presence of discrete germinant signals in their environment. The detection of germinants triggers an irreversible cascade of physical and biochemical changes, leading to the outgrowth of a metabolically active bacillus, with the ability to cause disease in the host (Stewart *et al.*, 1981). Between species, these receptors recognize a variety of small molecules that are often unique in their chemistry but all rapidly, and with high efficiency, initiate breaking of spore dormancy and onset of vegetative growth. In order to appreciate the level of control required for spore germination it is useful to compare related species in terms of their germinants, germinant-specific receptors and germination machinery genes.

Bacillus subtilis

Bacillus subtilis encodes for three germinant receptors, GerA, GerB, and GerK. The germinant receptors of *Bacillus* species are encoded by tricistronic operons under the control of a σ^G promoter (Feavers *et al.*, 1990; Corfe *et al.*, 1994a; Liu *et al.*, 2004). These receptors respond to different defined germinants, and disruption of these receptors results in the inability to respond to those germinants (Sammons *et al.*, 1981; Corfe *et al.*, 1994b; McCann *et al.*, 1996). GerA can respond to L-alanine alone (Paidhungat and Setlow, 2000). While no single germinant has the ability to stimulate the other two germinant receptors, GerB recognizes fructose and amino acids, while GerK has shown responsiveness to glucose (Sammons *et al.*, 1981; Corfe *et al.*, 1994b; McCann *et al.*, 1996). GerB and GerK also can act cooperatively to stimulate germination in the presence of glucose, fructose, and potassium ions (GFK) combined with either L-asparagine or L-alanine (Atluri *et al.*, 2006).

Bacillus cereus

The *Bacillus cereus* strain ATCC 14579 encodes seven putative germinant receptor operons, including *germ*, *gerI*, *gerK*, *gerL*, *gerQ*, *gerR*, and *gerS* (Ivanova *et al.*, 2003). GerL has been implicated in L-alanine-based germination. GerG has been shown to respond to L-glutamine, while GerR stimulates germination in response to L-alanine, L-cysteine, and L-threonine. GerI, GerQ, and GerR play a role in inosine-mediated germination. GerI also responds to L-glutamine and phenylalanine. Within the *B. cereus* classification, different strains appear to possess different germinant specificities. In strain ATCC 14579 GerI, GerQ, and GerR behave redundantly, as all three can independently recognize inosine. In contrast, strain 569 requires the presence of both GerI and GerQ for inosine recognition. To date, no germinants have been identified for GerK, GerL, or GerS, although evidence suggests that the *gerK*, *gerL*, and *gerS* operons are transcribed during sporulation of strain ATCC 14579 (Hornstra *et al.*, 2006).

Clostridia

The germinant receptors of *Clostridia* species exhibit the same genetic organization as those of the *Bacilli*, with the exception of *Clostridium difficile*. Although *C. difficile* has been shown to germinate in the presence of glycine and the bile salt taurocholate, no germinant receptor-like genes have been identified during genome sequencing efforts. This is intriguing, and suggests the possibility of a novel set of proteins or pathways required for germinant recognition (Sorg and Sonenshein, 2008; Ramirez *et al.*, 2010). Proteolytic *Clostridium botulinum* encodes for three highly conserved germinant receptors. Additionally, some proteolytic *C. botulinum* strains also possess additional 'B' protein subunits, or additional *ger* receptor operons, but the significance of these is yet to be fully elucidated. One possibility is that the extra germinant receptor components may expand the germinant or host range of the organism, resulting in the variability seen in germinant specificities between strains. This theory has been given validity in *Bacillus megaterium*. This *Bacillus* species contains a tricistronic receptor operon, *gerU*, on a plasmid, in addition to the single *gerVB* and *gerWB* genes located on the chromosome. The GerVB and GerWB proteins provide additional germinant specificities to the GerUA and GerUC proteins present, probably by forming additional germinant receptor complexes (Christie and Lowe, 2007; Christie *et al.*, 2010). Some of the germinants for these proteolytic strains include L-alanine, L-alanine/L-lactate, L-alanine/bicarbonate, inosine, AGFK (L-asparagine, D-glucose, D-fructose, KCl) and a variety of hydrophobic L-amino acids (Alberto *et al.*, 2003). In contrast, non-proteolytic *C. botulinum* strains encode for only one germinant receptor, which may explain its narrower germinant specificities (Peck and Robert, 2009). Non-proteolytic *C. botulinum* responds to the defined germinant pairs L-alanine/L-lactate, L-cysteine/L-lactate and L-serine/L-lactate (Plowman and Peck, 2002). All *Clostridium perfringens* tested respond to the germinants KCl/L-asparagine. *C. perfringens* strains encoding for enterotoxin in their chromosome (C-cpe), often associated with food borne diseases, can also germinate in the presence of KCl or L-asparagine alone (Paredes-Sabja *et al.*, 2008). Additionally, some of these strains can germinate in the presence of KCl and sodium phosphate (NaPi) (Paredes-Sabja *et al.*, 2008, 2009). In contrast, strains expressing enterotoxin on a plasmid (P-cpe), responsible for non-food borne gastrointestinal diseases, can respond to L-alanine or L-valine. The differences in germination specificities between these two categories of enterotoxin expression are surprising, as both encode for the same single germinant receptor, GerK (Paredes-Sabja *et al.*, 2008). Whether additional germinant receptors remain to be identified or additional spore components contribute to this phenomenon remains to be determined.

B. anthracis

The germinant receptors of *B. anthracis* share the same organization as other receptors of the *Bacillus* genera, but recognize a somewhat different array of germinants (Fig. 8.2). *B. anthracis* encodes for four known functional germinant receptor operons on the chromosome, *gerH*, *gerK*, *gerL*, and *gerS* (Ireland and Hanna, 2002b; Read *et al.*, 2003; Weiner *et al.*, 2003; Fisher and Hanna, 2005). An additional putative receptor operon, *gerX*, is located on the pXO1 virulence plasmid, although the function of the GerX receptor remains unclear (Read *et al.*, 2003). Mutational analyses indicate that at least one germinant receptor must be present in order to initiate disease in a murine model of anthrax infection, although there are some differences in receptor specificity depending upon route of infection (Carr *et al.*, 2010). Two additional receptor-like operons, *gerA* and *gerY*, were identified on the chromosome, but probably do not play a role in germination, as the *gerAA*, *gerYB*, and *gerYC* open reading frames contain frameshift mutations (Read *et al.*, 2003). Additionally, the *gerY* operon is expressed at a reduced level, and the *gerA* genes are not expressed at all.

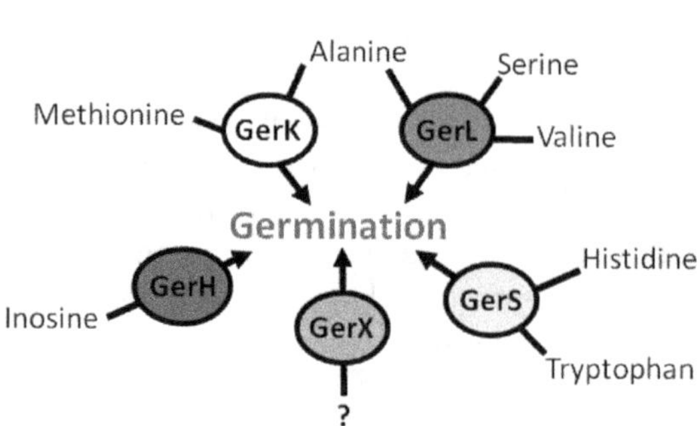

Figure 8.2 The germinant specificities of *B. anthracis* germinant receptors.

Finally, disruption of these loci did not result in any measurable germination defect (Fisher and Hanna, 2005).

The five receptors of *B. anthracis* have distinct germinant requirements, although redundancy and promiscuity in these requirements exist. Germinants consist of either a purine nucleoside (typically adenosine or inosine) or L-alanine paired with an L-amino acid co-germinant. These L-amino acids include L-alanine, L-serine, L-valine, L-histidine, L-methionine, L-proline, and L-tryptophan (Ireland and Hanna, 2002b; Weiner *et al.*, 2003; Fisher and Hanna, 2005). L-alanine at high concentrations can also germinate spores in the absence of an additional germinant. Inosine, however requires the co-stimulation provided by the L-amino acid co-germinant, although the mechanism behind this process is not yet clear (Fisher and Hanna, 2005; Ireland and Hanna, 2002b; Weiner *et al.*, 2003; Carr *et al.*, 2010). Interestingly, spores of *B. cereus* are able to respond to inosine alone. It appears that L-alanine is still required for this germination event to occur in *B. cereus* 569, but that the spore releases its own L-alanine stores to provide the co-germinant upon recognition of inosine (Dodatko *et al.*, 2009).

The GerH receptor of *B. anthracis* appears to be an orthologue of the GerI receptor of *B. cereus* 569, in that it recognizes purine nucleosides and L-amino acids (Clements and Moir, 1998; Barlass *et al.*, 2002; Carr *et al.*, 2010). Mutant strains of *B. anthracis* expressing single germinant receptor operons demonstrated that, in the presence of inosine, GerH responded at wild-type levels to all amino acids tested, including L-alanine, L-serine, L-valine, L-histidine, L-methionine, and L-tryptophan. Although it was previously thought that germination required at least two germinant receptors in *B. anthracis*, evidence now suggests that individual germinant receptors can respond to germinants independent of other receptors (Carr *et al.*, 2010). In an inhalational model of infection, the presence of any one of the chromosomally encoded receptors, GerH, GerK, GerL, or GerS, was enough to elicit a fully virulent infection. This suggests the presence of a wide range of germinants at the site of germination. In contrast, full virulence following subcutaneous inoculation requires the presence of GerH (Carr *et al.*, 2010).

These different receptor requirements suggest a possible explanation for the presence of so many different, yet partially overlapping, germinant specificities. These receptors may show cooperativity in the complex germinant environment of a mammalian host, resulting in accelerated germination. In fact, recent work *in vitro* has demonstrated that different germinant receptor pathways can either synergize or block each other depending on the germinants present (Luu *et al.*, 2011). Additionally, a large amount of germinant specificities provides a wider range of hosts and infectious routes available to *B. anthracis* in order to optimize its ability to germinate and cause disease.

Promiscuity in germinant specificity also exists, to some extent, for GerK, GerL, and GerS (Ireland and Hanna, 2002b; Fisher and Hanna, 2005; Carr *et al.*, 2010). GerK and GerL both respond to L-alanine in the absence of other germinants, when present at high enough concentration, and these two receptors appear to act cooperatively to stimulate this response when both are present in the spore (Fisher and Hanna, 2005; Carr *et al.*, 2010). Additionally, GerK shows a weak response to L-histidine, L-methionine, L-serine, and L-valine when present alone. In the absence of GerK, L-methionine germination is decreased. GerL also shows varying degrees of weak specificity for the amino acids listed above. Its absence results in decreased germination in L-serine or L-valine. GerS is unable to stimulate full germination in any inosine/ L-amino acid combination tested to date, although it appears to play a role in L-histidine and L-tryptophan mediated germination.

As mentioned above, *B. anthracis* also encodes for a putative germinant receptor on the pXO1 virulence plasmid. Studies have suggested that this receptor may play a role in L-alanine mediated germination, and possibly virulence (Guidi-Rontani *et al.*, 1999a). However, when GerX is present alone in the spore, germination is extremely limited, resulting in severe attenuation in a mouse model of infection (Carr *et al.*, 2010). Several possible functions for GerX have been proposed, but additional work is needed to fully appreciate the role this receptor plays in germination. GerX specific germinants may be found in niches yet to be tested, such as in other mammalian hosts, organ sites, or the environment. GerX may also play a role in augmenting the germination responses of other receptors. By playing an additive role when present with other receptors, GerX may allow for a more rapid or efficient germination response. This response may be specific to a limited number of germinants, or it may be a global mediator of germination upon the sensing of nutrients, depending on whether GerX actually recognizes germinants or facilitates some other germination event.

Germination-specific lytic enzymes

A critical step in the process of spore germination is the breakdown of the cortex. The spore cortex is a thick layer of modified peptidoglycan located just outside the germ cell wall of the spore (Setlow, 2003). The cortex contributes greatly to the environmental resistance of the spore but in order for vegetative cell growth to begin it must be selectively broken down leaving the cell wall intact (Setlow, 2006). This is accomplished through the action of a group of enzymes known as germination-specific cortex lytic enzymes (GSLEs). Genes encoding GSLEs are expressed during sporulation (Popham, 2002b). As the spore is being formed, GSLEs are packaged into the outer spore layers but remain inactive. Spores in the environment may be subjected to an extended period of dormancy, possibly as long as several decades (Gest and Mandelstam, 1987; Cano and Borucki, 1995; Nicholson *et al.*, 2000a). GSLEs must remain silent during this dormant period to maintain the environmental resistance of the spore. Therefore, tightly regulating the activity of these enzymes is critical for the viability of the spore but the processes by which this is accomplished in the various spore-forming bacterial species remains largely a mystery. For example, the GSLE SleB in *Clostridium perfringens* as well as *Bacillus megaterium* is packaged as an inactive proenzyme that is proteolytically cleaved during germination in order to be activated (Foster and Johnstone, 1988; Miyata *et al.*, 1995; Moriyama *et al.*, 1999). However, homologues of this enzyme in *Bacillus subtilis* and *Bacillus cereus* are packaged in their active forms and do not require proteolytic processing for activation. The need for this type of processing in *B. anthracis* spores remains unknown.

One common feature of all GSLEs identified to date is their ability to preferentially degrade cortex peptidoglycan over cell wall peptidoglycan. This is accomplished through modifications in the sugar residues of the cortex. Bacterial cell wall peptidoglycan comprises repeating subunits of the saccharides *N*-acetylglucosamine and *N*-acetylmuramic acid (Atrih *et al.*, 1996; Popham, 1996, 2002a). These two sugars are linked in an alternating pattern to form chains. Polysaccharide chains are then cross-linked by peptide side chains found on the *N*-acetylmuramic acid residues giving the peptidoglycan its structure. In cortex peptidoglycan, 50% of the *N*-acetylmuramic acid residues are replaced by muramic δ-lactam residues (alternating every other residue) (Warth and Strominger, 1969). This modification is somehow recognized by active GSLEs allowing for preferential breakdown of the cortex.

In *B. anthracis* the process of cortex degradation has been shown to involve four enzymes, SleB, CwlJ1, CwlJ2, and SleL (Lambert and Popham, 2008; Giebel *et al.*, 2009; Heffron *et al.*, 2010). The enzymes SleB, CwlJ1 and CwlJ2 are predicted members of the group known as spore cortex lytic enzymes (SCLEs). These enzymes degrade intact peptidoglycan. The enzyme SleL is a predicted member of the cortex fragment lytic enzyme (CFLE) family (Makino and Moriyama, 2002). This type of enzyme further hydrolyses partially degraded peptidoglycan. The predicted roles of these enzymes have been confirmed by analysing muropeptide fragments released from germinating spores.

SleB and CwlJ are the primary GSLEs involved in cortex degradation of *Bacillus* species. In *B. subtilis* and *B. anthracis* the presence of at least one of these genes is required for efficient spore germination indicating that these enzymes have partially redundant roles (Chirakkal *et al.*, 2002; Giebel *et al.*, 2009; Heffron *et al.*, 2010). In both *B. subtilis* and *B. anthracis* the *sleB* gene is found in a bicistronic operon with the gene *ypeB*. The specific function of *ypeB* is not known; however, mutant spores lacking this gene do not contain SleB (Boland *et al.*, 2000). This suggests that YpeB is necessary for proper expression, localization or stability of the SleB protein. *B. anthracis*, unlike *B. subtilis* and the even more closely related *B. cereus*, contains two homologues of *cwlJ* termed *cwlJ-1* and *cwlJ-2* (Heffron *et al.*, 2009). The *B. anthracis* gene *cwlJ-1*, like the single copy of *cwlJ* found in the other species, is located in a bicistronic operon with the gene *gerQ*. Much as is the case for *ypeB* and *sleB*, the presence of *gerQ* is required for CwlJ function (Ragkousi *et al.*, 2003; Ragkousi and Setlow, 2004). The unique *cwlJ2*, which is 63% identical to *cwlJ1* at the amino acid level, is not found in an operon with a *gerQ* homologue.

B. anthracis spores lacking SleB or CwlJ1 are not significantly impaired in germination rate or colony forming efficiency. However, when both of these enzymes are removed 99.9% of spores are no longer able to complete germination and initiate outgrowth (Giebel *et al.*, 2009; Heffron *et al.*, 2010). When incubated in the presence of rich nutrients normally suitable to induce germination, these mutant spores begin to lose their characteristic phase bright appearance and become phase grey suggesting that germination is triggered through the intact germinant receptors, but the process is arrested at the cortex degradation step preventing outgrowth. There are conflicting reports regarding the role of CwlJ2 in germination. One of two independent studies performed at about the same time found that CwlJ2 had no effect on spore germination (Heffron *et al.*, 2010). Deletion of *cwlJ2* alone, or in combination with other GSLE mutations had no additional effect on germination rate or colony forming efficiency. The second study also found that a single deletion of *cwlJ2* had no defect compared with wild-type spores. However, when *cwlJ2* was removed from a strain already lacking *cwlJ1* and *sleB*, colony forming efficiency was decreased from 1500 spores/cfu for Δ*sleB*Δ*cwlJ1* spores to 8.2×10^7 spores/cfu for Δ*sleB*Δ*cwlJ1*Δ*cwlJ2* spores (Giebel *et al.*, 2009). The results from this study suggest that CwlJ2 alone is not sufficient to complete cortex degradation in the majority of spores, but in the absence of the primary GSLEs, SleB and CwlJ1, CwlJ2 is able to function in a subset of the population. The difference between the two data sets remains unclear. Both groups used the same genetic system to construct their mutants and as reported followed the same assay protocols.

SleB or CwlJ1 is also required for virulence (Giebel *et al.*, 2009). Spores lacking both proteins are highly attenuated (80-fold increase in LD_{50} compared with wild type) in a mouse model of inhalation anthrax. Spores lacking all three proteins, SleB, CwlJ1 and CwlJ2, are further attenuated in mice (550-fold increase in LD_{50} compared with wild type). An additive defect in virulence corresponding to the lack of CwlJ2 supports the finding that this protein has limited functionality in the absence of SleB or CwlJ1. Although *ΔsleBΔcwlJ1ΔcwlJ2* spores are highly attenuated they were found to have a higher rate of mortality than would be predicted based on *in vitro* germination rates in BHI media. Although *ΔsleBΔcwlJ1ΔcwlJ2* spores do not germinate in response to nutrients, they can be germinated by incubation in purified lysozyme. Presumably lysozyme is able to access the cortex through holes in the coats of a small number of spores. When mice deficient in lysozyme M production are challenged with *ΔsleBΔcwlJ1ΔcwlJ2* spores they are protected compared with wild-type mice (Giebel *et al.*, unpublished results). This finding confirms a role for host lysozyme in the germination of *ΔsleBΔcwlJ1ΔcwlJ2* spores and may suggest that a very small percentage of spores in a natural infection are germinated in this way, although it is undoubtedly not a contributing factor to disease in a natural infection.

There are conflicting reports suggesting that SleL acts as either a glucosaminidase (Chen *et al.*, 2000; Lambert and Popham, 2008) or an epimerase (Chirakkal *et al.*, 2002). *B. anthracis* mutants lacking SleL do not show defects in the speed with which they escape the cortex and initiate outgrowth suggesting that this enzyme is not required for cortex hydrolysis. However, mutants lacking SleL do retain ~65% more *N*-acetylmuramic acid than wild-type spores. This finding is consistent with the role of SleL as a CFLE that acts on partially degraded peptidoglycan (Lambert and Popham, 2008). Presumably, cortex degradation, at a level sufficient for outgrowth, is carried out by other GSLEs such as SleB and CwlJ. SleL then acts on the large peptidoglycan fragments produced by these enzymes, further refining them to smaller muropeptide subunits. It is possible that this secondary processing of cortex degradation products plays a role in the interaction between *B. anthracis* spores and the host immune system during infection. A second possibility is that additional processing is involved in signalling neighbouring spores in the population to begin germination. It has been shown that cell-free *B. subtilis* peptidoglycan fragments can act as a germinant for *B. anthracis* spores (Shah *et al.*, 2008). However, the relative efficacy of peptidoglycan fragments collected from *sleB*, *cwlJ1* or *sleL* mutants to trigger this germination pathway has not been tested.

Conclusion

The successful and timely germination of spores is the essential first step in the deadly disease anthrax. Proper *B. anthracis* spore germination is a strictly regulated process that is programmed during sporulation and only triggered upon entry into a suitable host, possibly decades later. This process requires coordination of both germinant receptors and germination-specific lytic enzymes. Dormant spores passively sense their environment through a group of five germinant receptors. When at least one of these receptors is triggered dormant enzymes are activated and the cortex is degraded. Only after these steps have been successfully completed can the spore resume metabolic activity and transform into its actively dividing vegetative cell morphology. The redundancy of these critical steps in spore germination and subsequent disease may represent the evolutionary adaptation that allows *B. anthracis* to infect such a broad host range or may be involved in producing the proper germination response based on the microenvironments specific to the unique points of entry into the host. Ongoing research into the uptake and trafficking of spores in the body, the function of characterized and uncharacterized germinant receptors, and the mode of silencing and activation of germination-specific lytic enzymes will greatly increase our understanding of spore germination and how it impacts disease.

References

Alberto, F., Broussolle, V., Mason, D.R., Carlin, F., and Peck, M.W. (2003). Variability in spore germination response by strains of proteolytic *Clostridium botulinum* types A, B and F. Lett. Appl. Microbiol. *36*, 41–45.

Atluri, S., Ragkousi, K., Cortezzo, D.E., and Setlow, P. (2006). Cooperativity between different nutrient receptors in germination of spores of *Bacillus subtilis* and reduction of this cooperativity by alterations in the GerB receptor. J. Bacteriol. *188*, 28–36.

Atrih, A., Zöllner, P., Allmaier, G., and Foster, S.J. (1996). Structural analysis of *Bacillus subtilis* 168 endospore peptidoglycan and its role during differentiation. J. Bacteriol. *178*, 6173–6183.

Barlass, P.J., Houston, C.W., Clements, M.O., and Moir, A. (2002). Germination of *Bacillus cereus* spores in response to L-alanine and to inosine: the roles of *gerL* and *gerQ* operons. Microbiology *148*, 2089–2095.

Basu, S., Kang, T.J., Chen, W.H., Fenton, M.J., Baillie, L., Hibbs, S., and Cross, A.S. (2007). Role of *Bacillus anthracis* spore structures in macrophage cytokine responses. Infect. Immun. *75*, 2351–2358.

Bergman, N.H., Anderson, E.C., Swenson, E.E., Niemeyer, M.M., Miyoshi, A.D., and Hanna, P.C. (2006). Transcriptional profiling of the *Bacillus anthracis* life cycle *in vitro* and an implied model for regulation of spore formation. J. Bacteriol. *188*, 6092–6100.

Boland, F.M., Atrih, A., Chirakkal, H., Foster, S.J., and Moir, A. (2000). Complete spore-cortex hydrolysis during germination of *Bacillus subtilis* 168 requires SleB and YpeB. Microbiology *146*, 57–64.

Cano, R.J., and Borucki, M.K. (1995). Revival and identification of bacterial spores in 25- to 40-million-year-old Dominican amber. Science *268*, 1060–1064.

Carr, K.A., Lybarger, S.R., Anderson, E.C., Janes, B.K., and Hanna, P.C. (2010). The role of *Bacillus anthracis* germinant receptors in germination and virulence. Mol. Microbiol. *75*, 365–375.

Chen, Y., Fukuoka, S., and Makino, S. (2000). A novel spore peptidoglycan hydrolase of *Bacillus cereus*: biochemical characterization and nucleotide sequence of the corresponding gene, sleL. J. Bacteriol. *182*, 1499–1506.

Chirakkal, H., O'Rourke, M., Atrih, A., Foster, S.J., and Moir, A. (2002). Analysis of spore cortex lytic enzymes and related proteins in *Bacillus subtilis* endospore germination. Microbiology *148*, 2383–2392.

Christie, G., Gotzke, H., and Lowe, C.R. (2010). Identification of a receptor subunit and putative ligand-binding residues involved in the *Bacillus megaterium* QM B1551 spore germination response to glucose. J. Bacteriol. *192*, 4317–4326.

Christie, G., and Lowe, C.R. (2007). Role of chromosomal and plasmid-borne receptor homologues in the response of *Bacillus megaterium* QM B1551 spores to germinants. J. Bacteriol. *189*, 4375–4383.

Clements, M.O., and Moir, A. (1998). Role of the *gerI* operon of *Bacillus cereus* 569 in the response of spores to germinants. J. Bacteriol. *180*, 6729–6735.

Cleret, A., Quesnel-Hellmann, A., Vallon-Eberhard, A., Verrier, B., Jung, S., Vidal, D., Mathieu, J., and Tournier, J.N. (2007). Lung dendritic cells rapidly mediate anthrax spore entry through the pulmonary route. J. Immunol. *178*, 7994–8001.

Corfe, B.M., Moir, A., Popham, D., and Setlow, P. (1994a). Analysis of the expression and regulation of the *gerB* spore germination operon of *Bacillus subtilis* 168. Microbiology *140*, 3079–3083.

Corfe, B.M., Sammons, R.L., Smith, D.A., and Mauel, C. (1994b). The *gerB* region of the *Bacillus subtilis* 168 chromosome encodes a homologue of the *gerA* spore germination operon. Microbiology *140*, 471–478.

Cote, C.K., Van Rooijen, N., and Welkos, S.L. (2006). Roles of macrophages and neutrophils in the early host response to *Bacillus anthracis* spores in a mouse model of infection. Infect. Immun. *74*, 469–480.

Dixon, T.C., Fadl, A.A., Koehler, T.M., Swanson, J.A., and Hanna, P.C. (2000). Early *Bacillus anthracis*–macrophage interactions: intracellular survival survival and escape. Cell Microbiol. *2*, 453–463.

Dixon, T.C., Meselson, M., Guillemin, J., and Hanna, P.C. (1999). Anthrax. N. Engl. J. Med. *341*, 815–826.

Dodatko, T., Akoachere, M., Muehlbauer, S.M., Helfrich, F., Howerton, A., Ross, C., Wysocki, V., Brojatsch, J.R., and Abel-Santos, E. (2009). *Bacillus cereus* spores release alanine that synergizes with inosine to promote germination. PLoS ONE *4*, e6398.

Druett, H.A., Henderson, D.W., Packman, L., and Peacock, S. (1953a). Studies on respiratory infection. I. The influence of particle size on respiratory infection with anthrax spores. J. Hyg. *51*, 359–371.

Druett, H.A., Henderson, D.W., Packman, L., and Peacock, S. (1953b). Studies on respiratory infection. I. The influence of particle size on respiratory infection with anthrax spores. J. Hyg. *51*, 359–371.

Feavers, I.M., Foulkes, J., Setlow, B., Sun, D., Nicholson, W., Setlow, P., and Moir, A. (1990). The regulation of transcription of the *gerA* spore germination operon of *Bacillus subtilis*. Mol. Microbiol. *4*, 275–282.

Fisher, N., and Hanna, P. (2005). Characterization of *Bacillus anthracis* germinant receptors *in vitro*. J. Bacteriol. *187*, 8055–8062.

Fisher, N., Shetron-Rama, L., Herring-Palmer, A., Heffernan, B., Bergman, N., and Hanna, P. (2006). The dltABCD operon of *Bacillus anthracis* sterne is required for virulence and resistance to peptide, enzymatic, and cellular mediators of innate immunity. J. Bacteriol. *188*, 1301–1309.

Foster, S.J., and Johnstone, K. (1988). Germination-specific cortex-lytic enzyme is activated during triggering of *Bacillus megaterium* KM spore germination. Mol. Microbiol. *2*, 727–733.

Friedlander, A. (1999a). Clinical aspects, diagnosis and treatment of anthrax. J. Appl. Microbiol. *87*, 303.

Friedlander, A.M. (1999b). Anthrax. In Vaccines, Plotkin, S., and Orenstein, W., eds. (WB Saunders Company, Philadelphia), pp. 629–637.

Friedlander, A.M., Bhatnagar, R., Leppla, S.H., Johnson, L., and Singh, Y. (1993). Characterization of macrophage sensitivity and resistance to anthrax lethal toxin. Infect. Immun. *61*, 245–252.

Gest, H., and Mandelstam, J. (1987). Longevity of microorganisms in natural environments. Microbiol. Sci. *4*, 69–71.

Giebel, J.D., Carr, K.A., Anderson, E.C., and Hanna, P.C. (2009). The germination-specific lytic enzymes SleB, CwlJ1, and CwlJ2 each contribute to *Bacillus anthracis* spore germination and virulence. J. Bacteriol. *191*, 5569–5576.

Glomski, I.J., Corre, J.P., Mock, M., and Goossens, P.L. (2007). Noncapsulated toxinogenic *Bacillus anthracis* presents a specific growth and dissemination pattern in naive and protective antigen-immune mice. Infect. Immun. *75*, 4754–4761.

Glomski, I.J., Dumetz, F., Jouvion, G., Huerre, M.R., Mock, M., and Goossens, P.L. (2008). Inhaled non-capsulated *Bacillus anthracis* in A/J mice: nasopharynx and alveolar space as dual portals of entry, delayed dissemination, and specific organ targeting. Microbes Infect. *10*, 1398–1404.

Guidi-Rontani, C., Pereira, Y., Ruffie, S., Sirard, J.-C., Weber-Levy, M., and Mock, M. (1999a). Identification and characterization of a germination operon on the virulence plasmid pXO1 of *Bacillus anthracis*. Mol. Microbiol. *33*, 407–414.

Guidi-Rontani, C., Weber-Levy, M., Labruyere, E., and Mock, M. (1999b). Germination of *Bacillus anthracis* spores within alveolar macrophages. Mol. Microbiol. *31*, 9–17.

Gut, I.M., Tamilselvam, B., Prouty, A.M., Stojkovic, B., Czeschin, S.N., van der Donk, W.A., and Blanke, S.R. (2011). *Bacillus anthracis* spore interactions with mammalian cells: relationship between germination state and the outcome of *in vitro* infections. BMC Microbiol. *11*, 46.

Hammond, P.M., and Carter, G.B. (2002). From biological warfare to healthcare: Porton Down 1940–2000 (Palgrave, Basingstoke).

Heffron, J.D., Lambert, E.A., Sherry, N., and Popham, D.L. (2010). Contributions of four cortex lytic enzymes to germination of *Bacillus anthracis* spores. J. Bacteriol. *192*, 763–770.

Heffron, J.D., Orsburn, B., and Popham, D.L. (2009). Roles of germination-specific lytic enzymes CwlJ and SleB in *Bacillus anthracis*. J. Bacteriol. *191*, 2237–2247.

Heine, H.S., Bassett, J., Miller, L., Hartings, J.M., Ivins, B.E., Pitt, M.L., Fritz, D., Norris, S.L., and Byrne, W.R. (2007). Determination of antibiotic efficacy against *Bacillus anthracis* in a mouse aerosol challenge model. Antimicrob. Agents Chemother. *51*, 1373–1379.

Henderson, D.W., Peacock, S., and Belton, F.C. (1956). Observations on the prophylaxis of experimental pulmonary anthrax in the monkey. J. Hyg. *54*, 28–36.

Hornstra, L.M., de Vries, Y.P., de Vos, W.M., and Abee, T. (2006). Influence of sporulation medium composition on transcription of *ger* operons and the germination response of spores of *Bacillus cereus* ATCC 14579. Appl. Environ. Microbiol. *72*, 3746–3749.

Hu, H., Emerson, J., and Aronson, A.I. (2007). Factors involved in the germination and inactivation of *Bacillus anthracis* spores in murine primary macrophages. FEMS Microbiol. Lett. *272*, 245–250.

Ireland, J., and Hanna, P. (2002a). Macrophage-enhanced germination of *Bacillus anthracis* endospores requires gerS. Infect. Immun. *70*, 5870–5872.

Ireland, J.A., and Hanna, P.C. (2002b). Amino acid- and purine ribonucleoside-induced germination of *Bacillus anthracis* DeltaSterne endospores: *gerS* mediates responses to aromatic ring structures. J. Bacteriol. *184*, 1296–1303.

Ivanova, N., Sorokin, A., Anderson, I., Galleron, N., Candelon, B., Kapatral, V., Bhattacharyya, A., Reznik, G., Mikhailova, N., Lapidus, A., *et al.* (2003). Genome sequence of *Bacillus cereus* and comparative analysis with *Bacillus anthracis*. Nature *423*, 87–91.

Jernigan, D.B., Raghunathan, P.L., Bell, B.P., Brechner, R., Bresnitz, E.A., Butler, J.C., Cetron, M., Cohen, M., Doyle, T., Fischer, M., *et al.* (2002). Investigation of bioterrorism-related anthrax, United States, 2001: epidemiologic findings. Emerging Infect. Dis. *8*, 1019–1028.

Jernigan, J.A., Stephens, D.S., Ashford, D.A., Omenaca, C., Topiel, M.S., Galbraith, M., Tapper, M., Fisk, T.L., Zaki, S., Popovic, T., *et al.* (2001). Bioterrorism related inhalational anthrax: the first 10 cases reported in the United States. Emerging Infect. Dis. *7*, 933–944.

Kang, T.J., Fenton, M.J., Weiner, M.A., Hibbs, S., Basu, S., Baillie, L., and Cross, A.S. (2005). Murine macrophages kill the vegetative form of *Bacillus anthracis*. Infect. Immun. *73*, 7495–7501.

Koch, R. (1877). The aetiology of anthrax based on the ontogeny of the anthrax bacillus. Beitrage zur Biologie der Phlanzen *2*, 277–282.

Koch, R. (1881). On the etiology of anthrax. In Essays of Robert Koch, Codell Cartes, K., ed. (Greenwood Press, New York), pp. 57–81.

Lambert, E.A., and Popham, D.L. (2008). The *Bacillus anthracis* SleL (YaaH) protein is an *N*-acetylglucosaminidase involved in spore cortex depolymerization. J. Bacteriol. *190*, 7601–7607.

Liu, H., Bergman, N.H., Thomason, B., Shallom, S., Hazen, A., Crossno, J., Rasko, D.A., Ravel, J., Read, T.D., Peterson, S.N., *et al.* (2004). Formation and composition of the *Bacillus anthracis* endospore. J. Bacteriol. *186*, 164–178.

Loving, C.L., Kennett, M., Lee, G.M., Grippe, V.K., and Merkel, T.J. (2007). Murine aerosol challenge model of anthrax. In Infect Immun, pp. 2689–2698.

Luu, H., Akoachere, M., Patra, M., and Abel-Santos, E. (2011). Mechanistic insights into *Bacillus anthracis* spore germination. J. Bacteriol. *193*, 4192–4198.

Makino, S., and Moriyama, R. (2002). Hydrolysis of cortex peptidoglycan during bacterial spore germination. Med. Sci. Monit. *8*, RA119–127.

McCann, K.P., Robinson, C., Sammons, R.L., Smith, D.A., and Corfe, B.M. (1996). Alanine germination receptors of *Bacillus subtilis*. Lett. Appl. Microbiol. *23*, 290–294.

Meselson, M., Guillemin, J., Hugh-Jones, M., Langmuir, A., Popova, I., Shelokov, A., and Yampolskaya, O. (1994). The Sverdlovsk anthrax outbreak of 1979. Science *266*, 1202–1208.

Miyata, S., Moriyama, R., Miyahara, N., and Makino, S. (1995). A gene (sleC) encoding a spore-cortex-lytic enzyme from *Clostridium perfringens* S40 spores; cloning, sequence analysis and molecular characterization. Microbiology *141*, 2643–2650.

Moir, A., Lafferty, E., and Smith, D.A. (1979). Genetics analysis of spore germination mutants of *Bacillus subtilis* 168: the correlation of phenotype with map location. J. Gen. Microbiol. *111*, 165–180.

Moriyama, R., Fukuoka, H., Miyata, S., Kudoh, S., Hattori, A., Kozuka, S., Yasuda, Y., Tochikubo, K., and Makino, S. (1999). Expression of a germination-specific amidase, SleB, of Bacilli in the forespore compartment of sporulating cells and its localization on the exterior side of the cortex in dormant spores. J. Bacteriol. *181*, 2373–2378.

Navacharoen, N., Sirisanthana, T., Navacharoen, W., and Ruckphaopunt, K. (1985). Oropharyngeal anthrax. J. Laryngol. Otol. *99*, 1293–1295.

Nicholson, W.L., Munakata, N., Horneck, G., Melosh, H.J., and Setlow, P. (2000). Resistance of *Bacillus* endospores to extreme terrestrial and extraterrestrial environments. Microbiol. Mol. Biol. Rev. *64*, 548–572.

Paidhungat, M., and Setlow, P. (2000). Role of *ger* proteins in nutrient and non-nutrient triggering of spore germination in *Bacillus subtilis*. J. Bacteriol. *182*, 2513-2519.

Paredes-Sabja, D., Torres, J.A., Setlow, P., and Sarker, M.R. (2008). *Clostridium perfringens* spore germination: characterization of germinants and their receptors. J. Bacteriol. *190*, 1190–1201.

Paredes-Sabja, D., Udompijitkul, P., and Sarker, M.R. (2009). Inorganic phosphate and sodium ions are cogerminants for spores of *Clostridium perfringens* type A food poisoning-related isolates. Appl. Environ. Microbiol. 75, 6299–6305.

Peck, M.W., and Robert, K.P. (2009). Biology and genomic analysis of *Clostridium botulinum*. In Advances in Microbial Physiology, Poole, R.K., ed. (Academic Press), pp. 183–265, 320.

Pickering, A.K., Osorio, M., Lee, G.M., Grippe, V.K., Bray, M., and Merkel, T.J. (2004). Cytokine response to infection with *Bacillus anthracis* spores. Infect. Immun. 72, 6382–6389.

Plowman, J., and Peck, M.W. (2002). Use of a novel method to characterize the response of spores of non-proteolytic *Clostridium botulinum* types B, E and F to a wide range of germinants and conditions. J. Appl. Microbiol. *92*, 681–694.

Popham, D.L. (2002). Specialized peptidoglycan of the bacterial endospore: the inner wall of the lockbox. Cell Mol. Life Sci. *59*, 426–433.

Popham, D.L., Helin, J., Costello, C.E., and Setlow, P. (1996). Analysis of the peptidoglycan structure of *Bacillus subtilis* endospores. J. Bacteriol. *178*, 6451–6458.

Quinn, C., and Turnbull, P. (1998). Anthrax. In Topley and Wilson's Microbiology and Microbial Infections, Collier, L., Balows, A., Sussman, M., and Hausler, W., eds. (Arnold, London), pp. 799–818.

Ragkousi, K., Eichenberger, P., van Ooij, C., and Setlow, P. (2003). Identification of a new gene essential for germination of *Bacillus subtilis* spores with Ca2+-dipicolinate. J. Bacteriol. *185*, 2315–2329.

Ragkousi, K., and Setlow, P. (2004). Transglutaminase–mediated cross-linking of GerQ in the coats of *Bacillus subtilis* spores. J. Bacteriol. *186*, 5567–5575.

Ramirez, N., Liggins, M., and Abel-Santos, E. (2010). Kinetic evidence for the presence of putative germination receptors in *Clostridium difficile* spores. J. Bacteriol. *192*, 4215–4222.

Read, T.D., Peterson, S.N., Tourasse, N., Baillie, L.W., Paulsen, I.T., Nelson, K.E., Tettelin, H., Fouts, D.E., Eisen, J.A., Gill, S.R., *et al.* (2003). The genome sequence of *Bacillus anthracis* Ames and comparison to closely related bacteria. Nature 423, 81–86.

Ribot, W.J., Panchal, R.G., Brittingham, K.C., Ruthel, G., Kenny, T.A., Lane, D., Curry, B., Hoover, T.A., Friedlander, A.M., and Bavari, S. (2006). Anthrax lethal toxin impairs innate immune functions of alveolar macrophages and facilitates *Bacillus anthracis* survival. Infect. Immun. *74*, 5029–5034.

Ross, C., and Abel-Santos, E. (2010). The Ger receptor family from sporulating bacteria. Curr. Issues Mol. Biol. *12*, 147–158.

Ross, J.M. (1957). The pathogenesis of anthrax following the administration of spores by the respiratory route. J. Pathol. Bacteriol. 73, 485–494.

Sammons, R.L., Moir, A., and Smith, D.A. (1981). Isolation and properties of spore germination mutants of *Bacillus subtilis* 168 deficient in the initiation of germination. J. Gen. Microbiol. *124*, 229–241.

Sanz, P., Teel, L.D., Alem, F., Carvalho, H.M., Darnell, S.C., and O'Brien, A.D. (2008). Detection of *Bacillus anthracis* spore germination *in vivo* by bioluminescence imaging. Infect. Immun. *76*, 1036–1047.

Setlow, P. (2003). Spore germination. Curr. Opin. Microbiol. *6*, 550–556.

Setlow, P. (2006). Spores of *Bacillus subtilis*: their resistance to and killing by radiation, heat and chemicals. J. Appl. Microbiol. *101*, 514–525.

Shah, I.M., Laaberki, M.-H., Popham, D.L., and Dworkin, J. (2008). A eukaryotic-like Ser/Thr kinase signals bacteria to exit dormancy in response to peptidoglycan fragments. Cell *135*, 486–496.

Sirisanthana, T., and Brown, A.E. (2002). Anthrax of the gastrointestinal tract. Emerg. Infect. Dis. *8*, 649–651.

Sirisanthana, T., Navachareon, N., Tharavichitkul, P., Sirisanthana, V., and Brown, A.E. (1984). Outbreak of oral-oropharyngeal anthrax: an unusual manifestation of human infection with *Bacillus anthracis*. Am. J. Trop. Med. Hyg. 33, 144–150.

Sirisanthana, T., Nelson, K.E., Ezzell, J.W., and Abshire, T.G. (1988). Serological studies of patients with cutaneous and oral-oropharyngeal anthrax from northern Thailand. Am. J. Trop. Med. Hyg. *39*, 575–581.

Smego, R.A., Jr., Gebrian, B., and Desmangels, G. (1998). Cutaneous manifestations of anthrax in rural Haiti. Clin. Infect. Dis. *26*, 97–102.

Sorg, J.A., and Sonenshein, A.L. (2008). Bile salts and glycine as cogerminants for *Clostridium difficile* spores. J. Bacteriol. *190*, 2505–2512.

Stewart, G.S., Johnstone, K., Hagelberg, E., and Ellar, D.J. (1981). Commitment of bacterail spores to germinate: a measure of the trigger reaction. Biochem. J. *198*, 101–106.

Turnbull, P. (1996). Stubborn contamination with anthrax spores. Environmental Health Series *106*, 171.

Warth, A.D., and Strominger, J.L. (1969). Structure of the peptidoglycan of bacterial spores: occurrence of the lactam of muramic acid. Proc. Natl. Acad. Sci. U.S.A. *64*, 528–535.

Watts, C.J., Hahn, B.L., and Sohnle, P.G. (2008). Progressive and destructive hair follicle infections in a murine cutaneous anthrax model. Microb. Pathog. *44*, 363–369.

Watts, C.J., Hahn, B.L., and Sohnle, P.G. (2009). Resistance of athymic nude mice to experimental cutaneous *Bacillus anthracis* infection. J. Infect. Dis. *199*, 673–679.

Weiner, M.A., and Hanna, P.C. (2003). Macrophage-mediated germination of *Bacillus anthracis* endospores requires the gerH operon. Infect. Immun. *71*, 3954–3959.

Weiner, M.A., Read, T.D., and Hanna, P.C. (2003). Identification and characterization of the *gerH* operon of *Bacillus anthracis* endospores: a differential role for purine nucleosides in germination. J. Bacteriol. *185*, 1462–1464.

Wheelis, M. (1999). Biological sabotage in World War I. In Biological and Toxin Weapons: Research, Developments and Use from the Middle Ages to 1945, Geissler, E., and van Courtland Moon, J.E., eds. (Oxford University Press, Oxford), pp. 35–62.

WHO. (2008). Anthrax in humans and animals, 4th edn. (World Health Organization, Geneva).

Zuberi, A., Moir, A., and Feavers, I.M. (1987). The nucleotide sequence and gene organization of the *gerA* spore germination operon of *Bacillus subtilis* 168. Gene *51*, 1–11.

Degradation of Spore Peptidoglycan During Germination

9

David L. Popham, Jared D. Heffron and Emily A. Lambert

Abstract

During spore germination, several spore components are broken down and are discarded or recycled. The first major degradative step in germination is depolymerization of the spore cortex peptidoglycan. This is essential to allow full hydration of the spore core and the resumption of cellular metabolism. The spore cortex is a thick layer of peptidoglycan with structural modifications that differentiate it from vegetative cell wall material. Germination-specific cortex lytic enzymes exhibit specificity for the muramic-δ-lactam modification of the peptidoglycan strands. As no protein synthesis can take place during germination prior to cortex breakdown, the cortex lytic enzymes must be produced during spore formation and packaged within the dormant spore in an inactive and highly stable form. A mechanism(s) must then exist for the activation of lytic enzymes during germination. This chapter will cover the current knowledge concerning the identities of cortex lytic enzymes in bacilli and clostridia, their expression and incorporation into dormant spores, the mechanisms that hold them inactive during spore dormancy and their activation during germination, and their specific lytic activities, substrates, and products.

Introduction

Bacterial endospores are highly resistant to most treatments that rapidly kill vegetative cells. Spores owe their extreme resistance properties to a variety of cell modifications, including alteration of the nucleoprotein complex, relative dehydration and absolute dormancy of the spore cytoplasm or 'core' with accumulation of a large depot of calcium-dipicolinate (Ca^{2+}-DPA), and encasement of the spore within multiple highly stable coat protein layers (Setlow, 2006). During the return to vegetative growth via the processes of germination and outgrowth, degradative processes are required to reverse some or all of these cellular modifications. Germination-triggered reactions result in several spore components being broken down and discarded or recycled. These reactions fall into three categories based upon the stage of sporulation at which they are required. The first major degradative step in germination is depolymerization of the spore cortex peptidoglycan. This is essential to allow full hydration of the spore core and the resumption of cellular metabolism (Popham *et al.*, 1996b; Setlow *et al.*, 2001). Upon spore core rehydration, the second major degradative process commences, which is hydrolysis of a significant subset of the spore core proteins (Setlow *et al.*, 2001). As the germinated spore resumes metabolism and growth, it must break out of the spore coat layers, which may involve specific degradative reactions within this structure.

The spore cortex is a thick layer of peptidoglycan with structural modifications that differentiate it from vegetative cell wall material (Warth and Strominger, 1969, 1972; Atrih *et al.*, 1996; Popham *et al.*, 1996a). Synthesis of a sufficient amount of cortex peptidoglycan is essential for the formation of stable, fully resistant dormant spores (Imae and Strominger, 1976). Similarly, treatments that damage the cortex of fully dormant spores can result in the loss of dehydration and resistance properties (Koshikawa *et al.*, 1984). Upon initiation of germination and release of the core Ca^{2+}-DPA, partial core rehydration can take place in the absence of any cortex breakdown (Popham *et al.*, 1996b; Setlow *et al.*, 2001). However, this rehydration is not sufficient to allow resumption of cellular metabolism and spore outgrowth (Setlow *et al.*, 2001). To begin these processes, the cortex must be sufficiently depolymerized to allow core swelling and additional water uptake. As no protein synthesis can take place during germination prior to cortex breakdown, the cortex lytic enzymes must be produced during spore formation and packaged within the dormant spore in an inactive and highly stable form. A mechanism(s) must then exist for the activation of lytic enzymes during germination.

As metabolism resumes within the germinating spore core, a significant percentage of core proteins are hydrolysed. The greatest proportion of this turnover, and the most well studied component, is hydrolysis of the small acid-soluble spore proteins (SASP) by the germination protease Gpr (Setlow, 2007). Within the dormant spore, the SASP play an important role in protection of the DNA, changing its chemistry and greatly increasing resistance to UV irradiation and to other processes that damage DNA. During germination, the SASP must be removed from the DNA in order to allow resumption of transcription. In addition, hydrolysis of the large mass of SASP contributes to the pool of metabolic precursors used during spore outgrowth. Studies of the hydrolysis of SASP have been recently reviewed (Setlow, 2007) and little has been published concerning this process in the ensuing years.

Outgrowing spores must escape from the highly stable spore coat protein structures. This escape may require the action of protease(s) or other enzymes on coat layers during germination or outgrowth, but to date no activity of this type has been demonstrated. In one case, a localized, programmed discontinuity in a spore exosporium layer has been shown to define the site of rupture during outgrowth (Steichen *et al.*, 2007).

This chapter will cover the current knowledge concerning the identities of cortex lytic enzymes in Bacilli and Clostridia, their expression and incorporation into dormant spores, the mechanisms that hold them inactive during spore dormancy and allow activation during germination, and their specific lytic activities, substrates, and products.

Spore cortex synthesis, structure, and relationships to other spore components.

Spore core dehydration is the major factor determining spore heat resistance (Nakashio and Gerhardt, 1985; Beaman and Gerhardt, 1986; Popham *et al.*, 1995) and is dependent on the presence of a sufficient spore peptidoglycan cell wall (Imae and Strominger, 1976). The spore peptidoglycan can be differentiated into two layers based on structural and functional characteristics. The innermost layer, the germ cell wall, more closely resembles vegetative wall structure (Tipper and Linnet, 1976; Atrih *et al.*, 1998; Meador-Parton and Popham, 2000). This layer is stable during spore germination and serves as the initial cell wall of the outgrowing spore (Atrih *et al.*, 1998; Meador-Parton and Popham, 2000; Dowd *et al.*, 2008). The outer layer, the cortex, constitutes > 80% of the spore peptidoglycan, has major structural modifications, and is rapidly degraded during spore germination. Cortex structure is highly conserved across genera (Warth and Strominger, 1972; Atrih and Foster, 2001a; Orsburn *et al.*, 2008). The primary structural characteristic of cortex peptidoglycan is the removal of up to 50% of the peptide side chains, with conversion of the corresponding muramic acid residues to muramic-δ-lactam (Fig. 9.1) (Warth and Strominger, 1969). This modification takes place during spore formation. One study suggested that further muramic-δ-lactam formation takes place during spore germination (Johnstone and Ellar, 1982), but the validity of these data will be discussed below.

During spore formation, the peptidoglycan is synthesized between the two membranes surrounding the developing forespore (Fig. 9.2). In the dormant spore, the inner membrane is clearly an intact permeability barrier that resides just inside the peptidoglycan. The outer membrane is at least partially permeable in the dormant spore,

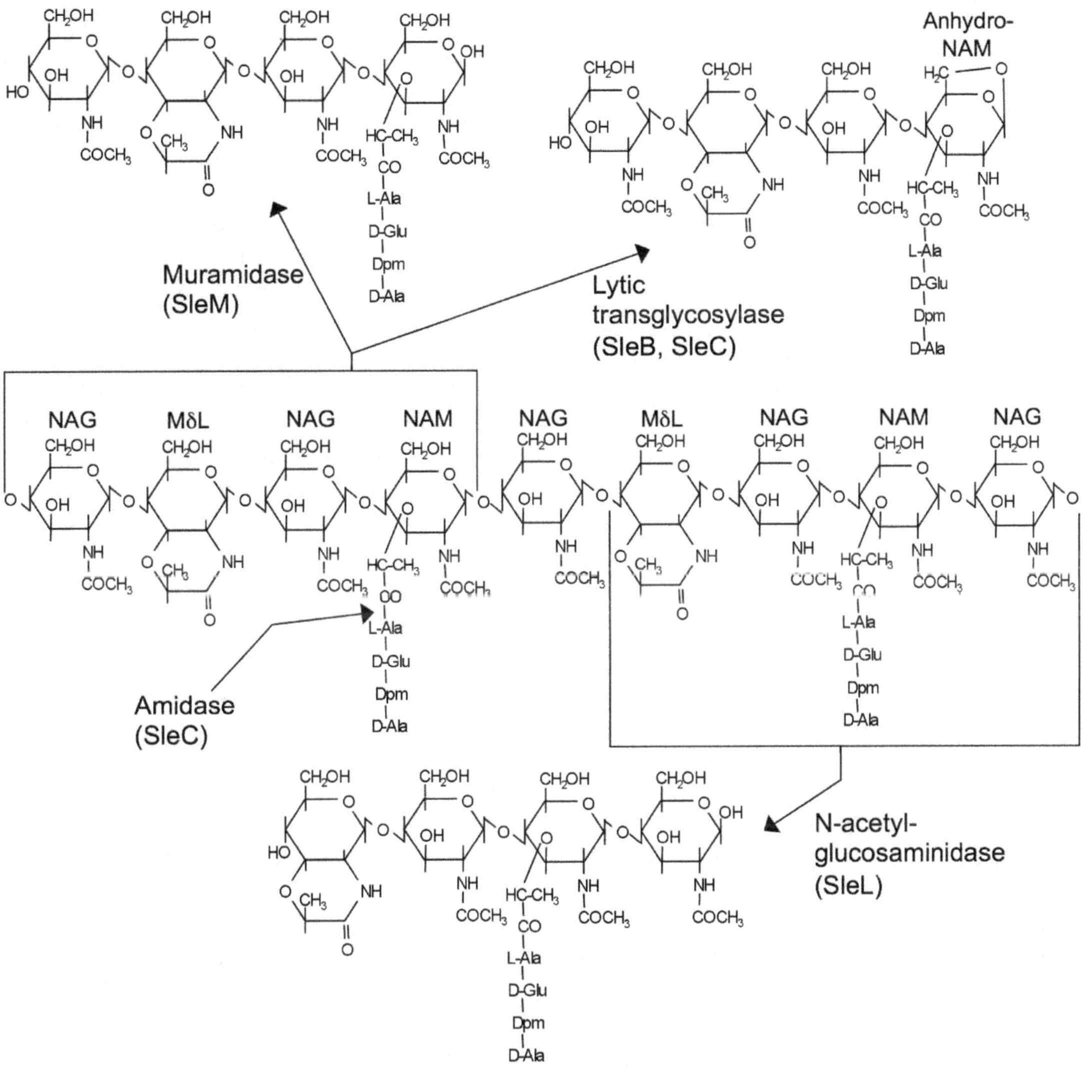

Figure 9.1 Spore peptidoglycan structure and products of lytic enzymes. The central structure represents a section of one peptidoglycan strand composed of *N*-acetyl glucosamine (NAG), muramic-δ-lactam (MδL), and *N*-acetyl-muramic acid (NAM) with peptide side chains. Some of these peptide side chains are used to form cross-links with other strands, and some are cleaved to single L-ala side chains. Potential cleavage sites and products of known GSLEs acting on the glycan strands are indicated, as well as the cleavage site of an amidase activity. Muramidases and lytic transglycosylases cleave in the same location but their products differ in the structure of the terminal NAM residue.

allowing the passage of nutrient germinants and other small molecules. The spore coat protein layers are outside of the cortex and outer membrane, and serve as a further barrier, blocking entry of proteins such as lysozyme and contributing to resistance to some small molecules such as chloroform (Driks, 2002). However, the coat layer permeability must be somewhat selective as other small molecules, for instance nutrient germinants, can pass through. The structure of the spore coats is variable between species and genera, differing in thickness, visual appearance, protein complement, and potentially in permeability. Some species carry an outer layer called the exosporium, which also varies significantly in appearance between species.

While spores can be stably dormant for years, they can be triggered into rapid germination upon

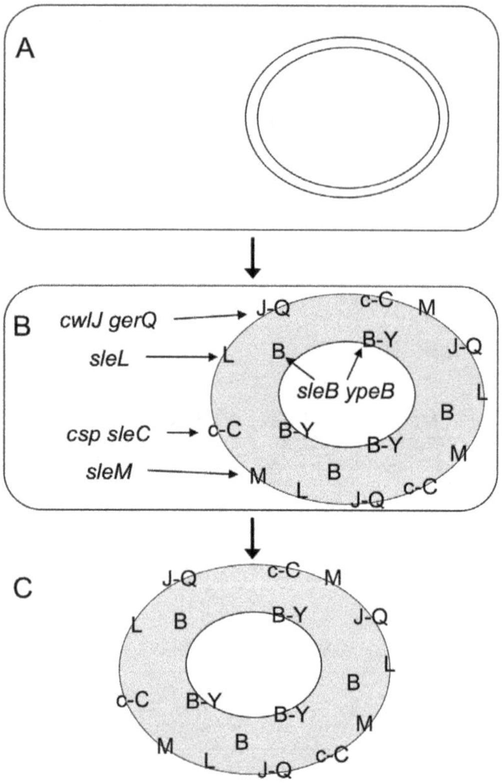

Figure 9.2 Incorporation of GSLEs into spores. Following engulfment of the developing forespore by the mother cell (A), synthesis of the cortex peptidoglycan commences in the intermembrane space, and the genes encoding GSLEs and associated proteins are expressed in the mother cell and forespore compartments (B). As described in the text, Csp-SleC (c-C) and SleM (M) are primarily found in the *Clostridiales*, while SleB-YpeB (B-Y), CwlJ-GerQ (J-Q), and SleL (L) are commonly in the *Bacillales*. SleB and YpeB traverse the inner forespore membrane and remain associated with that membrane or enter the intermembrane space. All of the relevant proteins expressed in the mother cell appear to associate with the spore coats (not shown for clarity), with the outer forespore membrane, or with the outer edge of the peptidoglycan. The Csp proteases have been shown to associate with SleC, and spore incorporation of SleB and CwlJ depend on YpeB and GerQ, respectively, and thus these protein pairs are depicted as complexes. Lysis of the mother cell releases the dormant spore (C) with the GSLE proteins in stable inactive forms that can be activated upon germination.

introduction into favourable growth conditions. Spores can be induced to germinate by a variety of treatments, but the most 'natural' germination pathways utilize specific nutrient germinant receptors. These receptors reside in the inner spore membrane and respond to single or combinations of small molecule nutrient or chemical germinants (reviewed in Moir, 2006; Setlow, 2003). The first step in germination following triggering of these receptors is a rapid release of spore core solutes, including essentially the entire spore Ca^{2+}-DPA depot. A simultaneous uptake of water results in partial rehydration of the spore cytoplasm, but apparently not enough to allow full resumption of metabolism. The next major, essential step in germination is the depolymerization of the cortex peptidoglycan, which allows full core rehydration and the resumption of cell growth.

General properties of cortex lytic enzymes

Degradation of spore peptidoglycan was recognized before extensive genetic systems were developed for sporulating bacteria, and the earliest studies of cortex lytic enzymes were therefore based upon their biochemical identification. These studies generally involved concentration and/or purification of proteins from the exudates of germinating spores or from broken spores, with subsequent demonstration of activity in stimulating germination of dormant spores. In some cases the studies continued to the identification of chemical changes in the cortex PG and prediction of a particular enzymatic activity.

Cortex breakdown is carried out by a class of enzymes referred to as germination-specific lytic enzymes (GSLEs). Two characteristics have been used to define these enzymes. First, GSLEs must be synthesized during sporulation and incorporated into the developing spore in an inactive form, because they act during germination before any protein synthesis takes place. Second, GSLEs must be specific for cortex peptidoglycan, presumably in order to leave the germ cell wall intact. This specificity has frequently been demonstrated as a lack of activity on the peptidoglycan of *cwlD* mutant spores, which lack the cortex structural element and GSLE specificity determinant muramic-δ-lactam (Chen *et al.*, 1997, 2000; Kumazawa *et al.*, 2007). Studies have suggested that GSLEs can be further subdivided based on the order of their action during germination (Miyata *et al.*, 1995a; Chen *et al.*, 1997). Spore cortex lytic enzymes (SCLEs) can act on intact spore peptidoglycan, and must act first during germination, whereas cortical fragment lytic enzymes (CFLEs) can act only on fragments of spore peptidoglycan and thus presumably act in a secondary step on the products of SCLE activity.

Identification of GSLEs has been carried out in several species using a range of methodologies, from direct biochemical analysis to classical and reverse genetic analyses. Coupling these studies to analyses of cortex peptidoglycan structure has provided information on the enzymatic specificities of the enzymes. Along the way a number of contradictory findings have arisen, and some potentially unfounded assertions have been perpetuated in the literature. Localization of some GSLEs in the spore has been studied using fluorescence and immuno-electron microscopy. Locations identified have ranged from immediately adjacent to the inner membrane to within the cortex layer to associations with the spore coats. Localization has also been tied to or inferred from sensitivity of GSLEs to chemical coat permeabilization treatments that remove some components of the spore coats and probably some proteins near the outer surface of the cortex. The following sections examine the history of studies of cortex lytic activities, from biochemical analyses to genomic studies. The origins of certain conclusions carried in the literature, and the strength of findings supporting or refuting them, are traced. Potential reasons for some incorrect conclusions, and data required to clarify the results, are presented. In the end, the state of the field will hopefully be clarified for future studies. We will first examine the history and state of knowledge concerning cortex lytic enzymes in *Bacillus* species and then progress to studies in *Clostridium* species.

Detection and characterization of lytic proteins in *Bacillus* spores

Early work on the presence of spore lytic enzymes utilized *Bacillus* spores (Strange and Dark, 1957a,b; Dring and Gould, 1971; Hsieh and Vary, 1975). While these studies left little doubt that spores contained lytic enzymes that could be involved in cortex degradation during germination, the strains and methods used, and the data produced do not lend themselves to direct correlation with activities identified in later studies. Later studies using *Bacillus megaterium* by Ellar, Foster, and Johnstone had major effects on the development of the literature in this field to this day.

Johnstone and Ellar (1982) carried out chemical analyses of cortex structure and detected additional sugar reducing groups as germination progressed. These reducing groups were attributed to muramic-δ-lactam. With the idea that additional muramic-δ-lactam was produced during germination, and with published ideas on the mechanism of muramic-δ-lactam formation (Warth and Strominger, 1969), they postulated that an amidase must be functioning to remove peptide side chains and thus peptidoglycan cross-links during germination (Fig. 9.1). This was the origin of all future references to amidase activity during *Bacillus* species spore germination. Outside of this, there appears to be no direct evidence for peptidoglycan amidase activity during germination of *Bacillus* spores. Unfortunately, the assertion of amidase activity appears to have been based on an improperly controlled assay for muramic-δ-lactam (see below).

Foster and Johnstone purified a germination-specific lytic enzyme from germinating *B. megaterium* spores (Foster and Johnstone, 1987). This enzyme could drive germination of coat-permeabilized spores and caused the previously observed increase in muramic-δ-lactam reduction products. It was thus proposed to be an amidase. Further characterization of this enzyme (Foster and Johnstone, 1988) suggested that it was processed from a phosphorylated 63 kDa precursor to a 29 kDa enzyme during germination. The identity of this lytic enzyme is not known, and later studies of *B. megaterium* cortex lysis have revealed neither amidase activity nor processing of a lytic enzyme (Setlow *et al.*, 2009; Christie *et al.*, 2010).

Major progress in identifying specific proteins and genes involved in cortex lysis was made in studies of *B. subtilis* and *B. cereus* in the 1990s. Makino *et al.* purified an enzyme from germinating *B. cereus* spores that was able to stimulate germination of coat-permeabilized spores (Makino *et al.*, 1994). They determined the amino terminal sequence of this protein and found that it was what we now call SleB (Moriyama *et al.*, 1996a,b). Treatment of spore PG with SleB resulted in the generation of free amino groups but no increase in sugar reducing groups (Makino *et al.*, 1994). This suggested that the activity was cleaving the peptide cross-links rather than the glycan strands. Based upon this data and on the previous proposal of an amidase involved in muramic-δ-lactam production during germination (Johnstone and Ellar, 1982), it was proposed that SleB was an amidase or a peptidase.

Later detailed genetic and biochemical studies of SleB have demonstrated conclusively that it functions as a lytic transglycosylase that cleaves glycan strands (Fig. 9.1) rather than as an amidase (See below). How can these results be reconciled with the earlier increased measures of free amino groups and of muramic-δ-lactam reducing groups in some cases, and of no increase in sugar reducing groups in another case, following SleB function? First, when a lytic transglycosylase cleaves the glycan strand, its product is a non-reducing 1,6-anhydro-muramic acid rather than reducing end muramic acid (Fig. 9.1). Second, increased detection of muramic-δ-lactam and reducing ends can result simply from a loosening of the cortex structure by any lytic activity. Muramic-δ-lactam was quantified based upon its reduction by ^{3}H-NaBH$_4$ (Johnstone and Ellar, 1982; Warth and Strominger, 1969). It was later found that in isolated cortex and decoated spores, of the type used in these assays, a majority of the muramic-δ-lactam is not reduced by NaBH$_4$ treatment (Atrih *et al.*, 1999; Popham and Setlow, unpublished). It appears that a significant increase in muramic-δ-lactam reduction occurs during germination because the PG structure is opened up and the muramic-δ-lactam becomes more accessible. This increase in reduction can also be achieved by digestion of the cortex using lysozyme, which should have no ability to produce additional muramic-δ-lactam. Similarly, loosening of the cortex structure using lysozyme or other glycan-cleaving enzymes can allow greater detection of free amino groups (Popham, unpublished), indicating that merely detection of more amino groups is not clear evidence for amidase or peptidase activity. In agreement with this, complete analyses of muropeptides generated from dormant and germinated spore peptidoglycan has not demonstrated any increase in muramic-δ-lactam or in free amino groups during germination (Atrih *et al.*, 1998, 1999; Dowd *et al.*, 2008; Heffron *et al.*, 2009, 2010). In short, in no case has a *Bacillus* GSLE been directly demonstrated to carry out amidase or peptidase activity or to cause the production of additional muramic-δ-lactam during germination.

Chen *et al.* (2000) purified a different cortex lytic enzyme from germinating *B. cereus* spores and identified it as SleL. They demonstrated that this enzyme functioned as an *N*-acetylglucosaminidase (Fig. 9.1) and that it could act on cortex fragments but not on intact cortex. They thus called it a CFLE and correctly proposed that SleL acted in a secondary step of cortex lysis.

Cortex structural modification during *Bacillus* spore germination

The structures of intact cortex peptidoglycan in several species were first determined in the 1970s (Warth and Strominger, 1969, 1971, 1972), and the application of modern methods to these analyses in the 1990s allowed more rapid and detailed studies of multiple samples (Atrih *et al.*, 1996; Popham *et al.*, 1996a,b).

Atrih *et al.* analysed cortex structural changes during germination of *B. subtilis* spores and produced evidence of two types of lytic activities (Atrih *et al.*, 1998). The appearance of anhydromuropeptides indicated the action of a lytic transglycosylase, and muropeptides with *N*-acetylglucosamine reducing ends indicated the action of an *N*-acetyl-glucosaminidase (Fig. 9.1). In addition, they observed the production of a large quantity of muropeptides with some alteration of their stereochemistry that was not associated with any particular lytic activity. They proposed the action of an epimerase probably functioning to alter muramic-δ-lactam. Analyses of cortex structural changes in germinating *B. megaterium* spores similarly revealed the production of *N*-acetyl-glucosaminidase, lytic transglycosylase, and epimerase products (Atrih *et al.*, 1999). Importantly, these analyses in *B. subtilis* and *B. megaterium* did not reveal any evidence for significant amidase, peptidase, or muramidase activities.

Cortex structure analyses in *B. anthracis* similarly revealed the actions of *N*-acetyl-glucosaminidase (Dowd *et al.*, 2008) and lytic transglycosylase during germination (Heffron *et al.*, 2009). However, in this species epimerase products were not present, and instead *N*-acetyl-glucosaminidase products were highly abundant. This difference from other species probably reflects differences in their SleL (YaaH) proteins discussed below.

In the late 1990s and early 2000s, the genetics of *Bacillus* cortex lytic enzymes caught up with the chemical analyses of cortex structure. Here we will briefly describe associations made between specific genes and the various identified lytic activities. Detailed analyses of the genes and proteins are covered later in the chapter.

SleB was the first cortex lytic enzyme for which a gene was identified and disrupted (Moriyama *et al.*, 1996a,b). Boland *et al.* demonstrated that a *B. subtilis sleB* mutant produced no lytic transglycosylase products (Boland *et al.*, 2000), implicating SleB in producing or regulating this activity. Similar losses of lytic transglycosylase activity were later observed for *sleB* mutants of *B. anthracis* (Heffron *et al.*, 2009) and *B. megaterium* (Christie *et al.*, 2010).

A *B. subtilis sleL (yaaH)* mutant had a complete lack of epimerase products, but still produced a small number of *N*-acetyl-glucosaminidase products (Chirakkal *et al.*, 2002). The same situation was observed for a *B. megaterium sleL* mutant (Christie *et al.*, 2010). To the contrary, a *B. anthracis sleL* mutant lacked all *N*-acetyl-glucosaminidase products, and this species never produced any epimerase products to begin with (Lambert and Popham, 2008). This result is consistent with the demonstration of *N*-acetyl-glucosaminidase activity for *B. cereus* SleL (Chen *et al.*, 2000). The highly similar *sleL* products in different species can act to produce significantly different muropeptide products.

The largest mystery in the association of a particular enzymatic activity with a protein acting in spore cortex lysis concerns CwlJ. The *cwlJ* product is clearly involved in spore germination in multiple species, and the phenotypic properties of *cwlJ* mutants indicate a direct role in cortex lysis (Ishikawa *et al.*, 1998; Heffron *et al.*, 2009, 2010; Setlow *et al.*, 2009). However, *cwlJ* mutants exhibit no differences in germination muropeptide profiles relative to wild-type strains of *B. subtilis* (Chirakkal *et al.*, 2002), *B. anthracis* (Heffron *et al.*, 2009, 2010), or *B. megaterium* (Christie *et al.*, 2010). It has been suggested that CwlJ might have *N*-acetyl-glucosaminidase (Christie *et al.*, 2010) or muramidase (Chirakkal *et al.*, 2002; Heffron *et al.*, 2010) activity, but no data are available to support either assertion. In fact, some unpublished data has been cited to suggest that CwlJ is not a muramidase (Chirakkal *et al.*, 2002).

Phenotypic analyses of mutants lacking GSLEs

The phenotypic effects of mutations affecting cortex lytic enzymes have been remarkably consistent across *Bacillus* species. In general, mutations eliminating any single enzyme have had minimal effects on the progression of nutrient-triggered germination. This reflects some redundancy in the roles of lytic enzymes in particular steps of cortex degradation as well as the fact that while greater levels of activity may accelerate the rate of germination, only minimal lytic activity is required to allow the completion of germination.

Loss of SleB results in only a slight delay in the loss of spore optical density during germination and little or no delay in progression of the spores into outgrowth (Boland *et al.*, 2000; Giebel *et al.*, 2009; Heffron *et al.*, 2009; Christie *et al.*, 2010). In addition, *sleB* mutant spores release Ca^{2+}-DPA as rapidly as wild-type spores. The slight delay in loss of optical density appears to correspond directly with the rate of cortex degradation, as the rate of release of cortex fragments is reduced in a *B. anthracis sleB* mutant (Heffron *et al.*, 2009). These mutant spores are, however, fully competent to complete outgrowth and produce colonies (Boland *et al.*, 2000; Giebel *et al.*, 2009; Heffron *et al.*, 2009; Christie *et al.*, 2010).

Similar results are found with *cwlJ* mutants (Ishikawa *et al.*, 1998; Giebel *et al.*, 2009; Setlow *et al.*, 2009 Heffron *et al.*, 2010), including *B. anthracis* strains that lack both copies of the *cwlJ* gene (Giebel *et al.*, 2009; Heffron *et al.*, 2010). In response to nutrient germinants, *cwlJ* mutants are slightly delayed in loss of optical density, release of cortex fragments, and progression into outgrowth, but the overall viability of the spores is not affected. In some cases, the rate of release of Ca^{2+}-DPA has been shown to be slower in *cwlJ* mutants than in the wild type (Ishikawa *et al.*, 1998; Setlow *et al.*, 2009). The major unique property of *cwlJ* mutants is their failure to respond to Ca^{2+}-DPA as a germinant (Paidhungat *et al.*, 2001; Setlow *et al.*, 2009; Heffron *et al.*, 2010). Ca^{2+}-DPA -triggered germination appears to result directly from activation of CwlJ activity in cortex degradation, leading eventually to spore core rehydration and resumption of metabolism (Paidhungat *et al.*, 2001). In *B. anthracis*, the products of both *cwlJ* genes appear to contribute to this process, with CwlJ1 playing the dominant role and CwlJ2 a more minor role (Heffron *et al.*, 2010).

A dramatic effect on spore germination rate and spore viability is only observed in strains that lack both CwlJ and SleB (Ishikawa *et al.*, 1998; Paidhungat *et al.*, 2001; Giebel *et al.*, 2009; Heffron *et al.*, 2009, 2010). Such spores initiate germination with the loss of approximately 25% of their optical density and the release of their Ca^{2+}-DPA pool, but they are unable to achieve the > 50% optical density loss seen in spores containing at least one of these two lytic enzymes. The spores achieve partial rehydration, but apparently not enough to allow resumption of metabolism (Setlow *et al.*, 2001, 2009). The colony-forming ability of *cwlJ sleB* double mutants are reduced at least 10^4-fold, and in some cases 10^8-fold (Ishikawa *et al.*, 1998; Giebel *et al.*, 2009; Heffron *et al.*, 2009, 2010; Setlow *et al.*, 2009). Differences in the degree of this viability phenotype may be due to the presence of small amounts of lytic enzymes carried over or introduced during spore purification and assay (Heffron *et al.*, 2010). The fact that the addition of exogenous lysozyme can rescue the colony-forming ability of these spores verifies that their defect is in cortex degradation. The overall conclusion is that CwlJ and SleB play redundant roles as SCLEs in an essential early step in cortex depolymerization.

Kodama reported that a *B. subtilis* strain lacking *sleL* was defective in L-alanine triggered germination (Kodama *et al.*, 1999), though this phenotype was not reproduced in other studies of *B. subtilis* or *B. anthracis* (Chirakkal *et al.*, 2002; Lambert and Popham, 2008). A *B. anthracis* strain lacking SleL exhibited a minimal delay in the loss of spore optical density and progression of the spores into outgrowth, but it released Ca^{2+}-DPA at a normal rate (Lambert and Popham, 2008). As noted above and discussed further below, the apparent difference in enzymatic activities of the SleL proteins of these species affects the rate of cortex fragment release. Germinating *B. subtilis* spores release cortex fragments very slowly (Atrih *et al.*, 1998), and this is unaffected by an *sleL* mutation (Popham, unpublished). *B. anthracis* releases cortex fragments to the medium very rapidly (Dowd *et al.*, 2008), and the rate of this release is dramatically reduced by the loss of SleL (Lambert and Popham, 2008). The introduction of *sleL* mutations into *B. anthracis* strains lacking *cwlJ* and/or *sleB* had minimal effects on phenotypic properties, aside from slightly increased delays in optical density loss and resumption of growth (Heffron *et al.*, 2010). This added delay is tied to an even slower rate of release of cortex fragments (Heffron *et al.*, 2010). The inability of SleL to initiate cortex hydrolysis and allow completion of germination, and its major role in hydrolysis of SleB and CwlJ reaction products into smaller fragments, classifies SleL as a CFLE.

SleB and YpeB

In the species in which an *sleB* gene has been shown to be involved in spore germination, it is in an operon with a conserved downstream gene that was named *ypeB* in *B. subtilis* (Moriyama *et al.*, 1996a; Boland *et al.*, 2000; Giebel *et al.*, 2009; Heffron *et al.*, 2009; Christie *et al.*, 2010). In *B. anthracis* and closely related species, there appears to be a third gene in the *sleB* operon that is predicted to encode a lipoprotein that is similar to other proteins involved in spore formation or germination (Bagyan *et al.*, 1998; Kuwana *et al.*, 2002). A *B. anthracis* strain lacking this third gene does not exhibit a dramatic alteration in its germination phenotypes (Popham, unpublished).

The *B. subtilis sleB* operon is expressed in the forespore compartment of sporulating cells (Fig. 9.2) under the control of sigma factor G (Moriyama *et al.*, 1999), and examination of the timing of expression and predicted promoter sequences suggests that this is true in other *Bacillus* species

(Liu *et al.*, 2004; Heffron *et al.*, 2009). SleB possesses a classic signal sequence (Fig. 9.3) that is cleaved during spore formation (Moriyama *et al.*, 1996a,b; Moriyama *et al.*, 1999; Hu *et al.*, 2007), indicating that SleB is translocated into the intermembrane space of the developing spore (Fig. 9.2). The ultimate location of SleB within the dormant spore is subject to some disagreement. Immuno-electron microscopy indicated that SleB was localized within the cortex peptidoglycan matrix (Masayama *et al.*, 2006a), while in a spore fractionation study some SleB appeared to be membrane associated with the remainder potentially in the cortex space (Chirakkal *et al.*, 2002). It may be important to remember that whatever the location of SleB within the dormant spore, this localization could be altered dramatically upon initiation of germination, and physical or chemical treatments involved in EM preparation or fractionation could trigger early germination events. The fact that SleB is not affected by chemical spore coat permeabilization procedures that inactivate lytic enzymes associated with the coat layers (Paidhungat *et al.*, 2001, Atrih and Foster, 2001b; Heffron *et al.*, 2009), suggests that at least some SleB, which could theoretically represent a functional fraction of the total SleB pool, is located at a more interior protected location. Efforts at protein localization within viable dormant spores may resolve conflicting results concerning SleB localization.

B. subtilis SleB is not released from germinating spores (Moriyama *et al.*, 1999; Chirakkal *et al.*, 2002), whereas at least some of the SleB of *B. cereus* spores is found in the germination exudate (Moriyama *et al.*, 1996b). SleB is stable during germination (Chirakkal *et al.*, 2002). SleB that was purified from *B. cereus* germination exudate was capable of causing loss of optical density of coat-stripped spores but not of purified spore peptidoglycan (Makino *et al.*, 1994), which led to a proposal that the relative turgor-induced stress on the cortex is important for SleB substrate recognition. However, *B. anthracis* SleB expressed and purified from *E. coli* was active on both intact and broken PG but not on vegetative cell or *cwlD* spore PG (Heffron *et al.*, 2011). This *in vitro* detection of SleB activity on cortex fragments may have been possible due to a higher enzyme specific activity and more detailed analysis of products than those used in earlier studies that proposed a role for turgor pressure.

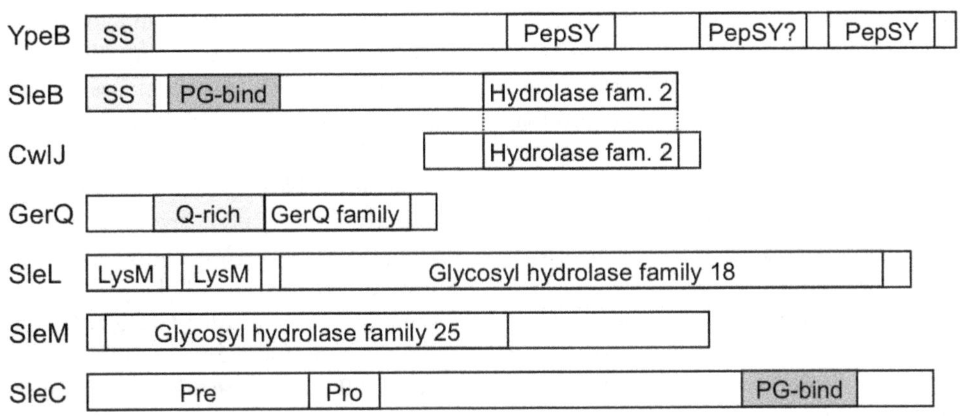

Figure 9.3 Domain structures of GSLEs and associated proteins. Conserved and predicted protein domains are indicated by boxes and are drawn to scale. Signal sequences (SS) for protein export are present in YpeB and SleB. The SleB signal peptide is cleaved while that of YpeB does not possess a predicted cleavage site. A peptidoglycan-binding domain (PG-bind)(pfam01471) is shared between SleB and SleC, and LysM domains (pfam01476) of SleL are involved in peptidoglycan binding. Hydrolase family 2 (pfam07486) and glycosyl hydrolase family 18 (pfam00704) and 25 (pfam01183) domains are the locations of glycan-cleaving active sites. Glutamine-rich (Q-rich) and highly conserved domains of GerQ are indicated. The pre and pro regions of SleC that are removed during spore incorporation and germination activation, respectively, are indicated.

Mature SleB possesses an N-terminal peptidoglycan-binding domain (pfam01471) and a C-terminal enzymatic domain (pfam07486) (Fig. 9.3). The peptidoglycan-binding motif is shared with *C. perfringens* SleC (Moriyama *et al.*, 1996b) and a number of other peptidoglycan lytic proteins. Localization of a fusion of this domain to GFP suggested that it exhibited specificity for peptidoglycan containing muramic-δ-lactam (Masayama *et al.*, 2006a). To the contrary, *in vitro* assays of SleB proteins indicated that specificity for muramic-δ-lactam was present in the enzymatic domain, and that the enzymatic domain could function as a lytic transglycosylase (Fig. 9.1) in the absence of the PG-binding domain (Heffron *et al.*, 2011). An *in vivo* analysis of altered SleB proteins in *B. megaterium* suggested that either domain could be deleted without abrogating SleB function (Christie *et al.*, 2010), however, this study may have been compromised by the presence of multiple functional SleB homologues in that organism.

The *ypeB* product possesses a potential uncleaved signal peptide (Fig. 9.3) that could serve to drive YpeB translocation across and anchoring to the inner spore membrane (Fig. 9.2) (Boland *et al.*, 2000; Chirakkal *et al.*, 2002). Indeed, spore fractionation studies revealed YpeB localization to both the membrane and the cortex space, similar to that of SleB (Chirakkal *et al.*, 2002). This suggested colocalization of SleB and YpeB is consistent with the observation that YpeB is required for stable incorporation of SleB in the dormant spore (Chirakkal *et al.*, 2002). In the absence of YpeB, SleB disappears from the developing spore, suggesting that it is subject to proteolysis. YpeB contains either two or three copies of the PepSY motif (pfam03413) (Fig. 9.3), which is found in numerous, frequently excreted, proteins and in some cases functions in inhibiting peptidase activity (Yeats *et al.*, 2004; Das *et al.*, 2010). The inhibitory action of PepSY domains has been suggested to result from a chaperone-type interaction. Interesting possibilities for the role of YpeB include direct inhibition of SleB prior to germination or inhibition of a protease that can degrade SleB during spore formation.

CwlJ and GerQ

In *B. subtilis, cwlJ* is expressed from a monocistronic locus in the mother cell (Fig. 9.2) compartment under the control of sigma factor E (Ishikawa *et al.*, 1998). Various studies suggest that *cwlJ* expression in *B. anthracis* is also sporulation-specific (Liu *et al.*, 2004; Bergman *et al.*, 2006; Heffron *et al.*, 2009). Genome sequences reveal that in several other species, *cwlJ* is expressed in a bicistronic operon with the downstream *gerQ* gene. In *B. anthracis*, the *cwlJ1* locus contains *gerQ* while *cwlJ2* is monocistronic, and *cwlJ1* is expressed at significantly higher levels than *cwlJ2* (Heffron *et al.*, 2009). In *B. subtilis, gerQ* is not adjacent to *cwlJ*, but is also expressed under the control of sigma factor E (Ragkousi *et al.*, 2003).

Both CwlJ1 and GerQ have been demonstrated to be associated with spore coat layers (Fig. 9.2) (Paidhungat *et al.*, 2001; Bagyan and Setlow, 2002; Chirakkal *et al.*, 2002; Lai *et al.*, 2003; Ragkousi *et al.*, 2003; Heffron *et al.*, 2009). This is consistent with the fact that all of the spore coat proteins are expressed and assembled within the mother cell cytoplasm surrounding the developing spore (Driks, 2002). Similar to the role played by YpeB in SleB incorporation into the spore, GerQ is required for the presence of CwlJ in the spore whereas GerQ is stable in the absence of CwlJ (Ragkousi *et al.*, 2003). In *B. subtilis*, GerQ is cross-linked to the spore coat structure via isopeptide bonds, and this cross-linking is dependent on the Tgl transglutaminase (Ragkousi and Setlow, 2004). Cross-linking can involve multiple lysine residues in the N-terminal region of GerQ, though the partner protein(s) involved in the cross-linking is unknown (Monroe and Setlow, 2006). Furthermore, this cross-linking is not required for incorporation of either GerQ or CwlJ into the spore (Ragkousi and Setlow, 2004; Monroe and Setlow, 2006). The *B. subtilis* GerQ lysine residues involved in cross-linking are not conserved in the homologous proteins of some other species such as *B. anthracis*, though glutamine residues in the same region could play this role (Fig. 9.3). There is no current evidence for or against cross-linking of GerQ in other species.

Extraction of spores with SDS and DTT can strip CwlJ (Bagyan and Setlow, 2002) but not

GerQ (Ragkousi and Setlow, 2004) from *B. subtilis* spores, consistent with cross-linking of GerQ but not CwlJ. Chemical extractions can also eliminate CwlJ function from *B. anthracis* spores (Heffron *et al.*, 2009), though it is not clear if this results from removal or simply denaturation of the CwlJ. The fate of CwlJ during germination is unclear. Conflicting reports indicate that *B. subtilis* CwlJ is stable (Chirakkal *et al.*, 2002) or disappears (Bagyan and Setlow, 2002) during germination. *B. subtilis* GerQ would appear to remain stably associated with the spore coats during germination (Ragkousi *et al.*, 2003).

There is currently no evidence that any spore lytic activity previously recovered from spores was due to CwlJ, and attempts to express and purify CwlJ in an active form have met with failure (Bagyan and Setlow, 2002; Chirakkal *et al.*, 2002). Such problems have been encountered with multiple spore coat proteins and could be a function of their inherent requirement for long-term aggregation and stability. The CwlJ protein sequence bears high similarity to that of the SleB lytic transglycosylase domain (Fig. 9.3) (Ishikawa *et al.*, 1998). This suggests that CwlJ would also act to cleave the peptidoglycan strands. As both of these proteins lie within a large family that includes both lytic transglycosylases and muramidases (pfam07486), and that these two activities share a similar reaction mechanism (Blackburn and Clarke, 2001; Scheurwater *et al.*, 2008), muramidase activity of CwlJ is a strong potential. The fact that *B. anthracis* spores possessing CwlJ1 as their sole GSLE can germinate efficiently but release very little cortex material to the medium (Heffron *et al.*, 2010) suggests that CwlJ need only carry out minimal cortex hydrolysis. The few products of a putative CwlJ muramidase activity would be impossible to detect using the standard method of muropeptide fractionation that involves the use of an exogenous muramidase (Atrih *et al.*, 1996; Popham *et al.*, 1996a), and such minimal products might be below the limit of detection of the alternative methods employed for this purpose (Chirakkal *et al.*, 2002). Clear understanding of CwlJ lytic activity may await its successful *in vitro* demonstration, and additional genetic studies may reveal the details of CwlJ incorporation into the spore coat and the mechanism of its activation during germination.

SleL (YaaH)

In *B. subtilis*, *sleL (yaaH)* is expressed from a monocistronic locus in the mother cell compartment (Fig. 9.2) under the control of sigma factor E (Kodama *et al.*, 1999) and genome sequences suggest that this locus is monocistronic in other species. Various types of expression studies suggest that *sleL* expression is also sporulation-specific in *B. anthracis* (Bergman *et al.*, 2006; Lambert and Popham, 2008; Liu *et al.*, 2004). Proteomic (Lai *et al.*, 2003) and fluorescence microscopy (Imamura *et al.*, 2010; McKenney *et al.*, 2010) studies indicate that SleL is associated with the spore coats (Fig. 9.2), and this is consistent with the demonstration that SleL can be removed and/or inactivated by chemical coat permeabilization procedures (Chen *et al.*, 2000; Heffron *et al.*, 2009; Lai *et al.*, 2003). Studies of the dependence of SleL localization on other gene products indicate that it localizes to the inner coat, and this localization is dependent on SafA and SpoVID as is true for other proteins in this coat layer (McKenney *et al.*, 2010), however no direct interactions between SleL and other proteins have been demonstrated. Positioning in this layer would presumably prevent direct contact between SleL and the intact cortex peptidoglycan. Action of SleL on the cortex would therefore require either a change in SleL position during germination, or that it act on fragments of peptidoglycan that have been rendered small enough to penetrate into the coat structure. Consistent with a potential relocalization, SleL has been identified within the exudate of germinating *B. cereus* spores (Chen *et al.*, 2000), though what fraction of the total SleL this represents is unknown. In *B. anthracis*, less than 10% of the SleL is found in the exudate of germinating spores (Lambert and Popham, unpublished).

SleL proteins comprise two N-terminal LysM domains (pfam01476) (Buist *et al.*, 2008) and a C-terminal glycosyl hydrolase family 18 domain (pfam00704) (Fig. 9.3) (Terwisscha van Scheltinga *et al.*, 1995; van Aalten *et al.*, 2001). The

LysM motif has been demonstrated in several proteins to be involved in association of proteins with peptidoglycan structures (Buist *et al.*, 2008), and *in vitro* studies of *B. anthracis* SleL with and without these motifs supports their role in binding of SleL to cortex peptidoglycan (Lambert and Popham, unpublished). LysM domains are found in several spore proteins where they are implicated in association with the developing spore (Kodama *et al.*, 2000; Costa *et al.*, 2006; Buist *et al.*, 2008 Wang *et al.*, 2009), though in cases where protein localization has been examined, these proteins are found to be coat-associated (Kodama *et al.*, 2000; Lai *et al.*, 2003; Costa *et al.*, 2006; Wang *et al.*, 2009; Imamura *et al.*, 2010; McKenney *et al.*, 2010). In addition, these proteins do not possess clear signal sequences for membrane translocation, and during the time of their assembly to the developing spore (Kodama *et al.*, 2000), the outer forespore membrane separates the cortex from the coats. This raises the interesting question of whether peptidoglycan binding via LysM domains positions the proteins for coat incorporation or if this domain can play a dual role in both coat assembly and later peptidoglycan association.

The enzymatic activity of the SleL glycosyl hydrolase domain (Fig. 9.3) appears to vary between species. The proteins from *B. anthracis* (Lambert and Popham, 2008) and *B. cereus* (Chen *et al.*, 2000) function as *N*-acetyl-glucosaminidases (Fig. 9.1), while those of *B. megaterium* (Christie *et al.*, 2010) and *B. subtilis* (Chirakkal *et al.*, 2002) have epimerase activity. The epimerase activity of the latter proteins has not been demonstrated *in vitro*, but proteins from the two former species have been either purified from the exudate of germinating spores (Chen *et al.*, 2000) or expressed in *E. coli* and purified in active form (Lambert and Popham, unpublished). *B. anthracis* SleL was produced in multiple forms with and without one or both LysM domains. The isolated glycosyl hydrolase domain can function *in vitro* as an *N*-acetyl-glucosaminidase independently of the LysM motifs, though both the efficiency of peptidoglycan hydrolysis and the affinity for peptidoglycan binding are reduced (Lambert and Popham, unpublished).

This interesting dichotomy of enzyme activities between such similar proteins in the various *Bacillus* species may be explained by a slight alteration in the reaction chemistry of the enzymes. This protein falls within a large family of glycosyl hydrolases, and within this family, some of the closest neighbours are chitinases (pfam00704). The similarity with chitinases is consistent with the observed *N*-acetyl-glucosaminidase activity, as chitin is a polymer of *N*-acetyl-glucosamine. Studies of the related *Serratia marcescens* chitinase A have suggested a reaction mechanism that involves an initial epimerization of a glucosamine residue followed by cleavage of the polysaccharide chain (Papanikolau *et al.*, 2001). If such a reaction mechanism is utilized by SleL, then perhaps the *B. subtilis* and *B. megaterium* SleL proteins possess a structural modification that allows the initial epimerization step but not the strand hydrolysis.

As the presence or absence of SleL activity has little or no effect on the rate or efficiency of spore germination, one could ask which of these two activities might have been present in a progenitor *Bacillus* species and which activity appeared later in the evolution of the genus. This raises questions about potential natural selection for one or the other activity. The major effect of SleL *N*-acetyl-glucosaminidase activity is the rapid production and release into the medium of large amounts of muropeptides (Lambert and Popham, 2008). A species that possesses only epimerase activity releases cortex fragments much more slowly (Atrih *et al.*, 1998). What effect might this differential release have on the success of the germinating spores? In the case of *B. anthracis*, which releases *N*-acetyl-glucosaminidase products, germination is generally taking place within a host organism. Muropeptides can have significant immunostimulatory effects (Guan and Mariuzza, 2007; Kanneganti *et al.*, 2007), which would appear to be counterproductive with regard to the establishment of an infection. Immune response to peptidoglycan during a mouse infection with *B. anthracis* spores has been documented, though the data suggest that the majority of this response is to vegetative cell peptidoglycan rather than that of spores (Loving *et al.*, 2009). Could reduced release of cortex muropeptides provide an advantage to *B. subtilis* and *B. megaterium*? As these compounds could be significant nutrient sources, their recycling or at least preventing nearby competitor species from using them could provide the germinating spore with an advantage. There is no available evidence that outgrowing spores recycle any cortex PG fragments. However, while it has been generally accepted in the field that Bacilli did not recycle peptidoglycan fragments, a recycling system has recently been demonstrated to function in *B. subtilis* stationary phase cells (Litzinger *et al.*, 2010). It therefore remains to be seen if such a recycling process could function during spore outgrowth.

Cortex depolymerization in *Clostridium perfringens*

Release of spore lytic activity from germinating *C. perfringens* spores was demonstrated in 1979 (Ando, 1979) and two different enzyme fractions were subsequently characterized (Gombas and Labbe, 1981, 1985). A 31 kDa lytic protein that was active on intact cortex peptidoglycan but not cortex fragments was purified from the exudate of germinating *C. perfringens* spores and was suggested to act as an amidase, cleaving peptides from the peptidoglycan backbone (Miyata *et al.*, 1995b). The size of this enzyme is similar to a 29.5 kDa protein identified by Gombas and Labbe (1985), but their studies suggested that this protein did not function as an amidase. The gene encoding the 31 kDa amidase was identified and named *sleC* (Miyata *et al.*, 1995a). A second *C. perfringens* cortex lytic enzyme was found to be active only on cortex fragments rather than intact spore cortex, was shown to function as a muramidase, and was named SleM (Chen *et al.*, 1997). SleC and SleM therefore fall into the categories of SCLE and CFLE, respectively.

SleC

The initially identified 31-kDa SleC enzyme is produced as a 50-kDa pre-pro-protein, is both N-terminally and C-terminally processed, is incorporated into the spore as a 35-kDa pro-protein, and is finally processed to the active 31-kDa form during germination (Fig. 9.3) (Miyata *et al.*, 1995a; Urakami *et al.*, 1999). Cloning of *sleC* revealed that it is present in a locus along with the *cspABC* genes, which encode serine proteases involved in SleC processing (Miyata *et al.*, 1995a; Shimamoto *et al.*, 2001). Significant variability in the structure of the *csp-sleC* genetic locus is found within the *C. perfringens* species. Some strains possess three tandem serine protease-encoding genes (*cpsABC*)(Shimamoto *et al.*, 2001), while others have a single *cspB* gene upstream of *sleC* (Myers *et al.*, 2006; Paredes-Sabja *et al.*, 2010). The *sleC* gene is transcribed separately from the *csp* genes, and all are expressed in the mother cell compartment of the sporangium (Fig. 9.2) (Masayama *et al.*, 2006b; Paredes-Sabja *et al.*, 2009a). The CspA, CspB, and CspC proteases copurify as a large complex called Germination-Specific Protease (GSP) and are able to carry out the final SleC processing step *in vitro* (Shimamoto *et al.*, 2001). It appears that SleC is localized to an outer layer of the dormant spore, presumably just outside the cortex or with the inner coats, where it is associated with its cleaved propeptide (Okamura *et al.*, 2000) and the Csp protease(s) (Fig. 9.2) (Shimamoto *et al.*, 2001; Masayama *et al.*, 2006b).

A *C. perfringens* mutant strain lacking SleC is severely impaired in completion of germination and spore outgrowth in response to KCl or nutrient germinants (Paredes-Sabja *et al.*, 2009b). For an unknown reason, these spores are also slow in releasing Ca^{2+}-DPA during germination. *C. perfringens* spores are also subject to germination by application of exogenous Ca^{2+}-DPA (Paredes-Sabja *et al.*, 2008). However, unlike the case in spores of Bacilli, Ca^{2+}-DPA does not appear to trigger germination via activation of a GSLE (Paredes-Sabja *et al.*, 2009b). A similar essential role of an *sleC* homologue in germination of *C. difficile* spores has recently been demonstrated (Burns *et al.*, 2010).

A *C. perfringens* mutant strain lacking its sole *csp* gene exhibited a germination defect similar to that of an *sleC* mutant, and was unable to process SleC during germination (Paredes-Sabja *et al.*, 2009a). This *cspB* mutant appeared to have an additional defect, in that it failed to release any Ca^{2+}-DPA when triggered to germinate. No mechanism for an effect of CspB on Ca^{2+}-DPA release is obvious.

Initiation of germination results in the processing and activation of SleC, which then acts on the cortex. Further *in vitro* study of SleC action indicated that it possesses both lytic transglycosylase and amidase activities (Fig. 9.1) (Kumazawa *et al.*, 2007). Both of these activities would appear to be carried within the 31 kDa mature form of the protein. This domain of SleC possesses some similarity to the region of SleB predicted to be a peptidoglycan-binding domain (pfam01471) (Fig. 9.3), but this would not appear to include the SleB enzymatic domain (Heffron *et al.*, 2011). A different region of SleC exhibits similarity to a small region of the SpoIID protein (pfam08486), which functions as a lytic transglycosylase, but this similarity does not include an essential active site glutamate residue of SpoIID (Morlot *et al.*, 2010). Clearly, many details remain to be determined about these enzymatic functions of SleC.

SleM

SleM is expressed from a monocistronic locus in the mother cell compartment and localizes to the outer surface of the cortex or to the inner coat layers (Fig. 9.2) (Chen *et al.*, 1997; Masayama *et al.*, 2006b). Consistent with an SleM role as a CFLE, a strain lacking *sleM* was not impaired in germination and colony formation (Paredes-Sabja *et al.*, 2009b). However, the absence of *sleM* exacerbated some aspects of the germination defect of an *sleC* mutant (Paredes-Sabja *et al.*, 2009b), suggesting that SleM may assist in a cortex hydrolysis pathway that is independent of the primary pathway involving SleC. Different from the situation with the *Bacillus* CFLE SleL, loss of SleM did not reduce the overall release of cortex fragments from germinating spores (Paredes-Sabja *et al.*, 2009b). However, release was assayed only after 2 hours of germination, so an effect of SleM might be apparent at earlier time points.

No other protein has yet been implicated in the recruitment of SleM to the spore, and no processing of the protein is required to allow lytic activity (Chen *et al.*, 1997). It would therefore appear that SleM is present in the spore in an active form and simply requires the action of an SCLE such as SleC to produce a proper substrate. Active SleM can be recovered from both broken dormant spores and from germinating spore exudate (Chen *et al.*, 1997), indicating that it can be released from a coat association during germination or that it is associated with the cortex and is released as it hydrolyses that structure. The muramidase activity of SleM is consistent with its sequence similarity to a family of muramidases and lysozymes (pfam01183) (Fig. 9.3) (Chen *et al.*, 1997).

Conservation of GSLEs across genera and species

As a general rule, bacterial species possess genes encoding a large number of peptidoglycan lytic activities. For example, the *B. subtilis* genome encodes at least 35 lytic proteins (Smith *et al.*, 2000). While some of these proteins can be identified based on sequence similarities to well-studied homologues, novel lytic proteins are identified in the literature on a consistent basis (examples are in Litzinger *et al.*, 2010; Morlot *et al.*, 2010; Sudiarta *et al.*, 2010). Classification of lytic proteins is further complicated by domain mixing among this large group of proteins. Several peptidoglycan-binding domains that do not participate directly in the lytic reaction have been identified, and these can be found associated with enzymes with significantly different enzymatic activities. This had led to a great deal of confusion in annotation of genes identified in genome sequencing projects. An additional factor to be considered is that similar lytic enzymes can play extremely different physiological roles depending on the timing of their expression within the cell cycle or process of differentiation, or due to specific localization in the cell.

Three clear examples of problems with identification of GSLEs based on sequence information alone are illustrated here, though others are present in the literature. As described above, SleB has been clearly characterized as a GSLE in *B. subtilis* and other Bacilli. The CwlJ protein of *B. subtilis* has 30% amino acid sequence identity with the enzymatic domain of SleB (Ishikawa *et al.*, 1998), illustrating that a relationship of this distance could potentially signal conservation of GSLE function. The *B. subtilis* genome also encodes the YkvT protein, which has 31% amino acid sequence identity with the enzymatic domain of SleB, but studies show that *ykvT* is not expressed during spore formation and plays no role in germination (Chirakkal *et al.*, 2002). Similarly, in *B.*

subtilis, YdhD and YvbX are 31% and 23% identical to SleL, respectively, both are expressed during spore formation (Chirakkal *et al.*, 2002), and YdhD is present in the dormant spore (Kodama *et al.*, 2000), yet neither plays a significant role in cortex modification (Chirakkal *et al.*, 2002). A protein with 45% sequence identity with *B. subtilis* SleB is encoded in the *C. difficile* genome (Burns *et al.*, 2010). Disruption of the encoding gene resulted in no alteration of germination, consistent with SleC acting as a sole SCLE in this species (Burns *et al.*, 2010). Presumably each of these GSLE homologues plays an alternative role in peptidoglycan hydrolysis within the life histories of these species. Identification of GSLEs based simply on genome sequence information must be done very carefully, and genetic and/or biochemical verification of predictions will generally be required.

Below is an attempt to organize the information about GSLEs that can be extracted from the sequenced genomes of spore-forming bacteria. Similar information was presented in a recent review (Paredes-Sabja *et al.*, 2010), and in most cases the following analysis is consistent with that publication. Parameters used to identify probable GSLEs and associated proteins are overall amino acid sequence identity and similarity, sequence alignment over the entire length of a known GSLE, and shared genomic context.

Genes encoding homologues of SleB and YpeB are found almost universally in the genomes of *Bacillus* species, including *Geobacillus*, *Oceanobacillus*, and *Paenibacillus*. These two genes are almost always adjacent to one another, with an exception to this rule in *Paenibacillus polymyxa*. SleB homologues are also encoded in a number of *Clostridiales*, and in many of these species YpeB homologues are also present. This observation of the abundance of these YpeB homologues differs with that of Paredes-Sabja *et al.*, who identified very few (Paredes-Sabja *et al.*, 2010). While the sequence similarities between the *Bacillus* and *Clostridium* YpeB-type proteins are not particularly strong, they do span the length of the protein, leading to their identification as potential homologues. In most *Clostridiales* species, these *sleB* and *ypeB* homologues are not adjacent to one another on the chromosome, though they are adjacent in *C. thermocellum* and in several more distant *Clostridiales* such as *Alkaliphilus oremlandii*, *Desulfotomaculum reducens*, and *Thermoanaerobacter tencongensis*. In no case have any of these *Clostridiales* genes been shown to function in spore germination. In the one case where a putative *C. difficile sleB/cwlJ* homologue was disrupted, no effect on germination was observed (Burns *et al.*, 2010). The length of this protein is reminiscent of a CwlJ homologue, while the sequence is more similar to the SleB enzymatic domain. It should be noted that this *C. difficile* strain does not possess YpeB or GerQ homologues, suggesting that the disrupted *sleB/cwlJ* homologous gene has an alternative role.

Genes encoding CwlJ and GerQ homologues are also almost universal among the *Bacillales* genomes examined. However, while these genes are adjacent on the chromosomes of many species, they are separated in ~30% of species, as they are in *B. subtilis* (Ragkousi *et al.*, 2003). The GerQ sequences among these species are well conserved in the C-terminal 50% (~90 residues) of the proteins, while the N-terminal halves of the proteins are highly divergent (Fig. 9.3). This divergence includes the previously cited lack of the residues involved in *B. subtilis* GerQ cross-linking (Monroe and Setlow, 2006) in the *B. anthracis* protein. CwlJ and GerQ homologues are encoded in only two of the sequenced *Clostridiales* genome, those of *Heliobacterium modesticaldum* and *Symbiobacterium thermophilum*. In both cases the genes are adjacent on the chromosome, supporting the possibility that they function together in a germination role in these species.

SleL homologues are encoded in virtually all *Bacillales* genomes, though the genome context of these genes varies greatly. Reasonably similar genes can also be found in some non-spore-forming species, and the glycosyl hydrolase family 18 and LysM domains of SleL are shared with several other protein families, making attribution of a role in cortex hydrolysis to these SleL homologues difficult. SleL homologues are also encoded in about 40% of the examined *Clostridiales* genomes. The genome contexts of these genes are also highly variable, and there is no current data to support a role for any in cortex hydrolysis. As the function of a CFLE such as SleL or SleM has never been found to be essential for germination, it is certainly possible that such proteins could be absent from sporulating species. Consistent with this, SleM homologues are encoded in only a small subset of *Clostridium* species and are absent from the *Bacillales* and more diverse *Clostridiales* that were examined.

SleC homologues are encoded in about 50% of the sequenced *Clostridiales* genomes, and some species have more than one copy. Aside from two clear exceptions, *Faecalibacterium prausnitzii* and *Subdoligranulum variabile,* all genomes with an *sleC* gene also contained one or more *csp* gene homologues. As subtilisin-type serine proteases with some similarity to Csp protein are fairly common in Gram-positive species, it is possible that SleC in these two species is processed by an alternative protease. The *csp* gene(s) were adjacent to *sleC* in the genomes of some species, as in *C. perfringens,* but were separated from *sleC* in other cases, such as in *C. difficile.* Again, function of SleC has only been demonstrated genetically in *C. difficile* (Burns *et al.,* 2010) and *C. perfringens* (Paredes-Sabja *et al.,* 2009b).

A comprehensive model of spore cortex hydrolysis

Based on the sum of the data described above, a model emerges for assembly, regulation, and function of the spore cortex lytic machinery (Fig. 9.4). In the absence of cortex degradation, germinating spores are not capable of achieving sufficient rehydration to resume metabolism, and thus transcription and translation cannot be initiated. The cortex lytic enzymes must therefore be synthesized during sporulation and incorporated into the dormant spore in an inactive, stable form (Fig. 9.2). Some of these enzymes such as CwlJ, SleL, SleC, and SleM are produced in the mother cell compartment and become associated with the spore coat layers surrounding the outer surface of the cortex. A common characteristic of these enzymes is their inactivation and/or removal by chemical treatments designed to disrupt the coat layers. The inherent stability of spore coat protein layers may play a role in the long term stability of the cortex lytic enzymes located in this region and could play a role in holding the enzymes inactive

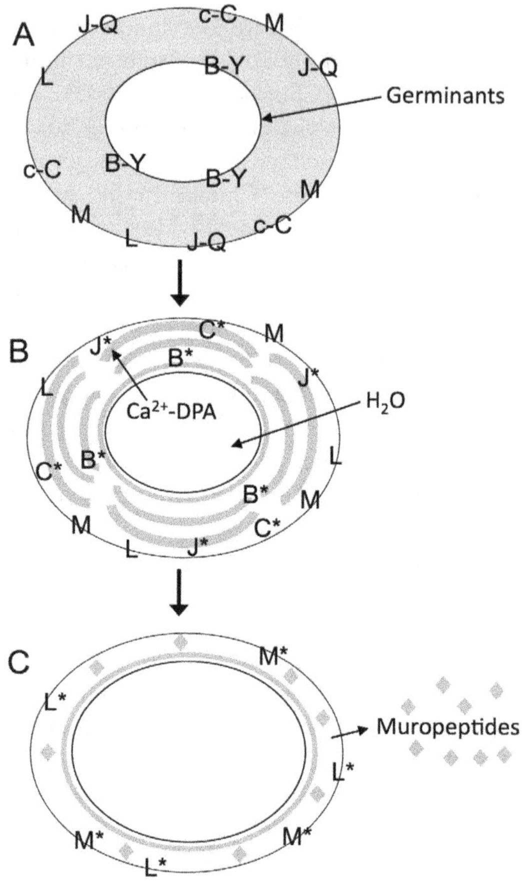

Figure 9.4 Model for GSLE activation and spore cortex hydrolysis. Proteins are indicated by letters as described in Fig. 9.1, and activation of enzymes is indicated with asterisks. While all of the GSLEs described in the text are shown in a single spore, it should be noted that most spores will only have a subset of these proteins, and that certain GSLEs are commonly restricted to the *Clostridiales* and the *Bacillales*. Germination is triggered by the binding of nutrient germinants to Ger receptors in the inner spore membrane (A). This results in flux of ions, including the large depot of Ca^{2+}DPA, out of the spore core with a corresponding influx of water (B). The presence of Ca^{2+}DPA in the cortex/coat region results in activation of CwlJ. SleB is activated via an unknown mechanism. SleC is activated via cleavage of its pro sequence by the Csp protease(s), but the mechanism controlling this cleavage is unknown. Initial cortex depolymerization by any single one of these enzymes (B) is sufficient to allow completion of spore germination and outgrowth. The thin, inner layer of spore peptidoglycan, the germ cell wall, is not degraded (B) because it contains little or no muramic-δ-lactam. SleL and SleM act as CFLEs (C) to cleave the products of the initial depolymerization into smaller muropeptides that are more readily released from the spore.

until germination is initiated. The fact that the coats are formed on the opposite side of the outer forespore membrane from the cortex would suggest that this physical separation prevents lytic activity during spore formation. This membrane does not remain a significant permeability barrier in the dormant spore but the physical separation between the lytic enzymes and the cortex established during spore assembly may preclude lysis even after the membrane is breached.

Cortex lytic enzymes that are expressed within the forespore compartment, such as SleB, must cross the inner forespore membrane in order to reach the cortex. This membrane remains an intact barrier, with regard to protein diffusion, throughout spore formation, dormancy, and germination. SleB possesses a classic signal peptide that drives translocation across the membrane and that is cleaved during spore formation (Fig. 9.2). It is then held in an inactive state either associated with the inner forespore membrane or in the intermembrane space with the cortex. SleB localized at the membrane could be held inactive via association with another protein (Fig. 9.4). An obvious candidate for this is YpeB, which is required to prevent SleB proteolysis during spore formation. Such a model provides an obvious method for preventing cortex hydrolysis by SleB during spore formation. Any SleB that is exported into the intermembrane space and that escapes its inhibitory mechanism could cleave peptidoglycan and lead to a failure in achieving spore dormancy. If SleB is highly sensitive to proteolysis once it escapes from the inner membrane, this would provide a failsafe mechanism for SleB inhibition. It has been suggested that the cortex peptidoglycan within the dormant spore is not in a suitable substrate conformation for SleB cleavage. The initial ion and water exchange and partial spore core rehydration during the initial stage of germination could alter the stress on the peptidoglycan, bringing it into an SleB substrate conformation. In support of this idea, SleB extracted from germinating spores was not found to be active on intact cortex PG of decoated spores, whereas it was active on fragmented cortex. While this is a viable theory, there are at least two points that argue against it. First, SleB is produced and its signal peptide is cleaved while cortex synthesis is in progress and prior to the attainment of spore core dehydration. A model would therefore have to include some alternative mechanism of SleB inhibition or a different state of osmotic stress on the cortex during spore formation than during germination. One could perhaps cite the fact that the developing spore is contained within the mother cell cytoplasm to postulate such an altered osmotic condition. Second, contrary to results achieved with SleB extracted from germinating spores, SleB expressed in *E. coli* was found to be active on intact cortex peptidoglycan as well as cortex fragments. The significantly greater amount of SleB available for assay may have allowed the observation of this activity. Alternatively one could argue that the SleB produced in a heterologous host might have altered properties. Clearly there is much left to be learned about the state of cortex lytic enzymes that are held in an inactive and highly stable form within the spore.

Tied to our gaps in knowledge of the states of lytic enzymes in the dormant spore are questions concerning the mechanism by which they are activated during germination. For the cortex fragment lytic enzymes SleL and SleM, the questions of inhibition and activation are slightly simpler. As these enzymes can only act on peptidoglycan that has been previously cleaved by another enzyme, their substrate only becomes available during germination (Fig. 9.4). Whether the localization or function of these enzymes are subject to other levels of control is unknown. There is also a clearly defined trigger regulating the initiation of CwlJ activity. This enzyme becomes active following exposure to high concentrations of Ca^{2+}-DPA (Fig. 9.4); either that released from the germinating spore or that provided exogenously in the lab. Numerous questions about the mechanism of this CwlJ activation remain. It is not known if Ca^{2+}-DPA interacts with CwlJ or with some other molecule associated with CwlJ. It is not known if Ca^{2+}–DPA interaction results in conversion of CwlJ to an active form or if CwlJ is already active and Ca^{2+}-DPA alters its availability to the substrate. It is not known if CwlJ localization changes upon Ca^{2+}-DPA exposure. Some of these questions will only be answered when the hurdle of direct detection of CwlJ enzymatic activity is cleared.

Unlike CwlJ, which can be activated by exogenous Ca^{2+}-DPA, activation of SleB or SleC depends on triggering of germination via the Ger receptors (Fig. 9.4). It is not clear which outcome of Ger receptor activation is key to lytic enzyme activation. It could be a direct interaction between the Ger complexes and the lytic enzyme, or with another protein or structure that is inhibiting the lytic enzyme. It could be via flux of ions out of the spore core in some way that has not been reproduced by exogenous application of ions. It could be a result of recovery of inner membrane fluidity upon core rehydration (Cowan *et al.*, 2004). It could be via alteration of the osmotic stress exerted by the core upon the cortex. It is clear that proteolytic processing of SleC by a GSP results in SleC activation, whereas SleB does not require processing during germination. However, this simply moves the regulatory key in the Clostridia one link further up a chain. How activation of the Ger receptor response regulates GSP activity is an open question.

The lytic transglycosylase, amidase, or other activity of a GSLE on the cortex peptidoglycan decreases its ability to withstand the turgor of the partially rehydrated spore core, and the core can swell and reach a greater state of hydration. In each case examined, the action of any single SCLE has been sufficient to allow the completion of germination and outgrowth, though release of cortex fragments to the medium may be minimal (Fig. 9.4). The action of a CFLE can speed the completion of germination and can increase the rate and amount of release of cortex fragments (Fig. 9.4). The *N*-acetyl-glucosaminidase activity of *B. anthracis* SleL produces the majority of the products of cortex degradation and greatly increases muropeptide excretion.

Part of the definition of a GSLE is that it has some specificity for hydrolysis of the cortex during germination, rather than being a more generic peptidoglycan lytic enzyme that has the potential to assist in germination. To drive the most efficient germination process, GSLEs should be highly active during a short time frame early in germination. This high activity raises the challenge of preventing degradation of all of the spore peptidoglycan, as the outgrowing spore may require an initial cell wall for stability or as a template for efficient synthesis of new wall material. All GSLEs studied to date have demonstrated specificity for peptidoglycan containing muramic-δ-lactam. This structural modification is found in abundance and is evenly distributed in the cortex peptidoglycan, but is lacking in the thinner germ cell wall. GSLE action therefore removes > 80% of the spore peptidoglycan but leaves behind an intact cell wall for the outgrowing spore.

New challenges

A remarkable characteristic of endospores is the maintenance of enzymatic activities during a long period of dormancy and through periods of elevated temperature. For proteins within the spore core, this stability is achieved by removal of water and the presence of a high concentration of Ca^{2+}-DPA. The GSLEs on the other hand, are held in the relatively hydrated environment of the spore cortex or coat layers. This hydration is indicated by the ability of nutrient germinants to diffuse rapidly to the inner spore membrane. The GSLEs that have been examined *in vitro* exhibit no particular heat stability and must therefore be held in a stable state via interaction with other spore components. For some GSLEs, it is probable that this same mechanism is responsible for preventing premature lytic activity in the absence of initiation of germination. Detailed characterization of the locations and binding partners of the GSLEs will shed light on the stabilization and inhibition of these enzymatic activities.

Once germination is triggered, the GSLEs must be released from inhibition. The mechanisms of activation appear to be varied, and could include physical release from an inhibitory molecule, physical release from a site separated from the cortex, proteolytic activation of the GSLE, or alteration of the cortex to produce an appropriate substrate. Again, detailed knowledge of GSLE localization and binding partners will inform the likelihood of each mechanism. An understanding of the required circumstances for activation of each GSLE will then allow dissection of the signal transduction pathway leading to activation. Several specific outstanding questions are presented below.

For SleB, is the functional protein associated

with the spore membrane or with the cortex peptidoglycan? If it is at the membrane, what other proteins is it associated with, and upon triggering of germination, what is the signalling pathway leading from the Ger receptors to SleB? For CwlJ, does it interact directly with Ca^{2+}-DPA, and if so, must this association be maintained to allow enzymatic activity or just for triggering of CwlJ function? Further work to demonstrate CwlJ function *in vitro* is required to answer these questions. If another spore structure is the true receptor for Ca^{2+}-DPA, then does this interaction release CwlJ from an inhibitory interaction? For SleC, what is the signal transduction pathway leading from Ger receptor at the inner membrane to Csp protease action at the spore coat layer? Is it a biophysical signal such as ion flux or altered turgor, or could there be a protein link across the cortex space?

Understanding the activities of GSLEs and mechanisms that control them will make these enzymes better targets for sporicidal agents. Cortex degradation is an essential step in the return to vegetative growth, so irreversible inhibitors of GSLEs or their activators could be lethal to spores. For species with multiple redundant GSLEs, it may be challenging to achieve the high level of spore inactivation desired for blocking infection by a human pathogen, but for decreasing the abundance of a food spoilage organism, GSLE inhibition might be a viable strategy. An alternative approach is the development of gratuitous GSLE activators. Highly efficient triggering of GSLE function by compounds that interact with the GSLEs or with a component of their regulatory functions will result in cortex degradation, spore rehydration, and the accompanying loss of most spore resistance properties. Subsequent application of antibacterial agents that are less toxic, corrosive, and/or expensive than current sporicidal agents will provide for a simpler and safer decontamination.

References

van Aalten, D.M., Komander, D., Synstad, B., Gaseidnes, S., Peter, M.G., and Eijsink, V.G. (2001). Structural insights into the catalytic mechanism of a family 18 exo-chitinase. Proc. Natl. Acad. Sci. U.S.A. *98*, 8979–8984.

Ando, Y. (1979). Spore lytic enzyme released from *Clostridium perfringens* spores during germination. J. Bacteriol. *140*, 59–64.

Atrih, A., and Foster, S.J. (2001a). Analysis of the role of bacterial endospore cortex structure in resistance properties and demonstration of its conservation amongst species. J. Appl. Microbiol. *91*, 364–372.

Atrih, A., and Foster, S.J. (2001b). *In vivo* roles of the germination-specific lytic enzymes of *Bacillus subtilis* 168. Microbiology *147*, 2925–2932.

Atrih, A., Zollner, P., Allmaier, G., and Foster, S.J. (1996). Structural analysis of *Bacillus subtilis* 168 endospore peptidoglycan and its role during differentiation. J. Bacteriol. *178*, 6173–6183.

Atrih, A., Zollner, P., Allmaier, G., Williamson, M.P., and Foster, S.J. (1998). Peptidoglycan structural dynamics during germination of *Bacillus subtilis* 168 endospores. J. Bacteriol. *180*, 4603–4612.

Atrih, A., Bacher, G., Korner, R., Allmaier, G., and Foster, S.J. (1999). Structural analysis of *Bacillus megaterium* KM spore peptidoglycan and its dynamics during germination. Microbiology *145*, 1033–1041.

Bagyan, I., and Setlow, P. (2002). Localization of the cortex lytic enzyme CwlJ in spores of *Bacillus subtilis*. J. Bacteriol. *184*, 1219–1224.

Bagyan, I., Noback, M., Bron, S., Paidhungat, M., and Setlow, P. (1998). Characterization of *yhcN*, a new forespore-specific gene of *Bacillus subtilis*. Gene *212*, 179–188.

Beaman, T.C., and Gerhardt, P. (1986). Heat resistance of bacterial spores correlated with protoplast dehydration, mineralization, and thermal adaptation. Appl. Environ. Microbiol. 52, 1242–1246.

Bergman, N.H., Anderson, E.C., Swenson, E.E., Niemeyer, M.M., Miyoshi, A.D., and Hanna, P.C. (2006). Transcriptional profiling of the *Bacillus anthracis* life cycle *in vitro* and an implied model for regulation of spore formation. J. Bacteriol. *188*, 6092–6100.

Blackburn, N.T., and Clarke, A.J. (2001). Identification of four families of peptidoglycan lytic transglycosylases. J. Mol. Evol. 52, 78–84.

Boland, F.M., Atrih, A., Chirakkal, H., Foster, S.J., and Moir, A. (2000). Complete spore-cortex hydrolysis during germination of *Bacillus subtilis* 168 requires SleB and YpeB. Microbiology *146*, 57–64.

Buist, G., Steen, A., Kok, J., and Kuipers, O.P. (2008). LysM, a widely distributed protein motif for binding to (peptido)glycans. Mol. Microbiol. *68*, 838–847.

Burns, D.A., Heap, J.T., and Minton, N.P. (2010). SleC is essential for germination of *Clostridium difficile* spores in nutrient-rich medium supplemented with the bile salt taurocholate. J. Bacteriol. *192*, 657–664.

Chen, Y., Miyata, S., Makino, S., and Moriyama, R. (1997). Molecular characterization of a germination-specific muramidase from *Clostridium perfringens* S40 spores and nucleotide sequence of the corresponding gene. J. Bacteriol. *179*, 3181–3187.

Chen, Y., Fukuoka, S., and Makino, S. (2000). A novel spore peptidoglycan hydrolase of *Bacillus cereus*: biochemical characterization and nucleotide sequence of the corresponding gene, *sleL*. J. Bacteriol. *182*, 1499–1506.

Chirakkal, H., O'Rourke, M., Atrih, A., Foster, S.J., and Moir, A. (2002). Analysis of spore cortex lytic enzymes and related proteins in *Bacillus subtilis* endospore germination. Microbiology *148*, 2383–2392.

Christie, G., Ustok, F.I., Lu, Q., Packman, L.C., and Lowe, C.R. (2010). Mutational analysis of *Bacillus megaterium* QM B1551 cortex-lytic enzymes. J. Bacteriol. *192*, 5378–5389.

Costa, T., Isidro, A.L., Moran, C.P., Jr., and Henriques, A.O. (2006). Interaction between coat morphogenetic proteins SafA and SpoVID. J. Bacteriol. *188*, 7731–7741.

Cowan, A.E., Olivastro, E.M., Koppel, D.E., Loshon, C.A., Setlow, B., and Setlow, P. (2004). Lipids in the inner membrane of dormant spores of *Bacillus* species are largely immobile. Proc. Natl. Acad. Sci. U.S.A. *101*, 7733–7738.

Das, D., Finn, R.D., Carlton, D., Miller, M.D., Abdubek, P., Astakhova, T., Axelrod, H.L., Bakolitsa, C., Chen, C., Chiu, H.J., *et al.* (2010). The structure of BVU2987 from *Bacteroides vulgatus* reveals a superfamily of bacterial periplasmic proteins with possible inhibitory function. Acta Crystallogr. Sect. F Struct. Biol. Cryst. Commun. *66*, 1265–1273.

Dowd, M.M., Orsburn, B., and Popham, D.L. (2008). Cortex peptidoglycan lytic activity in germinating *Bacillus anthracis* spores. J. Bacteriol. *190*, 4541–4548.

Driks, A. (2002). Maximum shields: the assembly and function of the bacterial spore coat. Trends Microbiol. *10*, 251–254.

Dring, G.J., and Gould, G.W. (1971). Sequence of events during rapid germination of spores of *Bacillus cereus*. J. Gen. Microbiol. *65*, 101–104.

Foster, S.J., and Johnstone, K. (1987). Purification and properties of a germination-specific cortex-lytic enzyme from spores of *Bacillus megaterium* KM. Biochem. J. *242*, 573–579.

Foster, S.J., and Johnstone, K. (1988). Germination-specific cortex-lytic enzyme is activated during triggering of *Bacillus megaterium* KM spore germination. Mol. Microbiol. *2*, 727–733.

Giebel, J.D., Carr, K.A., Anderson, E.C., and Hanna, P.C. (2009). Germination-specific lytic enzymes SleB, CwlJ1, and CwlJ2 each contribute to *Bacillus anthracis* spore germination and virulence. J. Bacteriol. *191*, 5569–5576.

Gombas, D.E., and Labbe, R.G. (1981). Extraction of spore-lytic enzyme from *Clostridium perfringens* spores. J. Gen. Microbiol. *126*, 37–44.

Gombas, D.E., and Labbe, R.G. (1985). Purification and properties of spore-lytic enzymes from *Clostridium perfringens* type A spores. J. Gen. Microbiol. *131*, 1487–1496.

Guan, R., and Mariuzza, R.A. (2007). Peptidoglycan recognition proteins of the innate immune system. Trends Microbiol. *15*, 127–134.

Heffron, J.D., Lambert, E.A., Sherry, N., and Popham, D.L. (2010). Contributions of four cortex lytic enzymes to germination of *Bacillus anthracis* spores. J. Bacteriol. *192*, 763–770.

Heffron, J.D., Orsburn, B., and Popham, D.L. (2009). Roles of germination-specific lytic enzymes CwlJ and SleB in *Bacillus anthracis*. J. Bacteriol. *191*, 2237–2247.

Heffron, J.D., Sherry, N., and Popham, D.L. (2011). *In vitro* studies of peptidoglycan binding and hydrolysis by the *Bacillus anthracis* germination-specific lytic enzyme SleB. J. Bacteriol. *193*, 125–131.

Hsieh, L.K., and Vary, J.C. (1975). Germination and peptidoglycan solubilization in *Bacillus megaterium* spores. J. Bacteriol. *123*, 463–470.

Hu, K., Yang, H., Liu, G., and Tan, H. (2007). Cloning and identification of a gene encoding spore cortex-lytic enzyme in *Bacillus thuringiensis*. Curr. Microbiol. *54*, 292–295.

Imae, Y., and Strominger, J.L. (1976). Relationship between cortex content and properties of *Bacillus sphaericus* spores. J. Bacteriol. *126*, 907–913.

Imamura, D., Kuwana, R., Takamatsu, H., and Watabe, K. (2010). Localization of proteins to different layers and regions of *Bacillus subtilis* spore coats. J. Bacteriol. *192*, 518–524.

Ishikawa, S., Yamane, K., and Sekiguchi, J. (1998). Regulation and characterization of a newly deduced cell wall hydrolase gene (*cwlJ*) which affects germination of *Bacillus subtilis* spores. J. Bacteriol. *180*, 1375–1380.

Johnstone, K., and Ellar, D.J. (1982). The role of cortex hydrolysis in the triggering of germination of *Bacillus megaterium* KM spores. Biochim. Biophys. Acta. *714*, 185–191.

Kanneganti, T.D., Lamkanfi, M., and Nunez, G. (2007). Intracellular NOD-like receptors in host defense and disease. Immunity *27*, 549–559.

Kodama, T., Takamatsu, H., Asai, K., Kobayashi, K., Ogasawara, N., and Watabe, K. (1999). The *Bacillus subtilis yaaH* gene is transcribed by SigE RNA polymerase during sporulation, and its product is involved in germination of spores. J. Bacteriol. *181*, 4584–4591.

Kodama, T., Takamatsu, H., Asai, K., Ogasawara, N., Sadaie, Y., and Watabe, K. (2000). Synthesis and characterization of the spore proteins of *Bacillus subtilis* YdhD, YkuD, and YkvP, which carry a motif conserved among cell wall binding proteins. J. Biochem. *128*, 655–663.

Koshikawa, T., Beaman, T.C., Pankratz, H.S., Nakashio, S., Corner, T.R., and Gerhardt, P. (1984). Resistance, germination, and permeability correlates of *Bacillus megaterium* spores successively divested of integument layers. J. Bacteriol. *159*, 624–632.

Kumazawa, T., Masayama, A., Fukuoka, S., Makino, S., Yoshimura, T., and Moriyama, R. (2007). Mode of action of a germination-specific cortex-lytic enzyme, SleC, of *Clostridium perfringens* S40. Biosci. Biotechnol. Biochem. *71*, 884–892.

Kuwana, R., Kasahara, Y., Fujibayashi, M., Takamatsu, H., Ogasawara, N., and Watabe, K. (2002). Proteomics characterization of novel spore proteins of *Bacillus subtilis*. Microbiology *148*, 3971–3982.

Lai, E.M., Phadke, N.D., Kachman, M.T., Giorno, R., Vazquez, S., Vazquez, J.A., Maddock, J.R., and Driks, A. (2003). Proteomic analysis of the spore coats of *Bacillus subtilis* and *Bacillus anthracis*. J. Bacteriol. *185*, 1443–1454.

Lambert, E.A., and Popham, D.L. (2008). The *Bacillus anthracis* SleL (YaaH) protein is an *N*-acetylglucosaminidase involved in spore cortex depolymerization. J. Bacteriol. *190*, 7601–7607.

Litzinger, S., Duckworth, A., Nitzsche, K., Risinger, C., Wittmann, V., and Mayer, C. (2010). Muropeptide rescue in *Bacillus subtilis* involves sequential hydrolysis by beta-*N*-acetylglucosaminidase and *N*-acetylmuramyl-L-alanine amidase. J. Bacteriol. *192*, 3132–3143.

Liu, H., Bergman, N.H., Thomason, B., Shallom, S., Hazen, A., Crossno, J., Rasko, D.A., Ravel, J., Read, T.D., Peterson, S.N., *et al.* (2004). Formation and composition of the *Bacillus anthracis* endospore. J. Bacteriol. *186*, 164–178.

Loving, C.L., Osorio, M., Kim, Y.G., Nunez, G., Hughes, M.A., and Merkel, T.J. (2009). Nod1/Nod2-mediated recognition plays a critical role in induction of adaptive immunity to anthrax after aerosol exposure. Infect. Immun. 77, 4529–4537.

Makino, S., Ito, N., Inoue, T., Miyata, S., and Moriyama, R. (1994). A spore-lytic enzyme released from *Bacillus cereus* spores during germination. Microbiology *140*, 1403–1410.

Masayama, A., Fukuoka, H., Kato, S., Yoshimura, T., Moriyama, M., and Moriyama, R. (2006a). Subcellular localization of a germination-specific cortex-lytic enzyme, SleB, of Bacilli during sporulation. Genes Genet. Syst. *81*, 163–169.

Masayama, A., Hamasaki, K., Urakami, K., Shimamoto, S., Kato, S., Makino, S., Yoshimura, T., Moriyama, M., and Moriyama, R. (2006b). Expression of germination-related enzymes, CspA, CspB, CspC, SleC, and SleM, of *Clostridium perfringens* S40 in the mother cell compartment of sporulating cells. Genes Genet. Syst. *81*, 227–234.

McKenney, P.T., Driks, A., Eskandarian, H.A., Grabowski, P., Guberman, J., Wang, K.H., Gitai, Z., and Eichenberger, P. (2010). A distance-weighted interaction map reveals a previously uncharacterized layer of the *Bacillus subtilis* spore coat. Curr. Biol. *20*, 934–938.

Meador-Parton, J., and Popham, D.L. (2000). Structural analysis of *Bacillus subtilis* spore peptidoglycan during sporulation. J. Bacteriol. *182*, 4491–4499.

Miyata, S., Moriyama, R., Miyahara, N., and Makino, S. (1995a). A gene (*sleC*) encoding a spore-cortex-lytic enzyme from *Clostridium perfringens* S40 spores; cloning, sequence analysis and molecular characterization. Microbiology *141*, 2643–2650.

Miyata, S., Moriyama, R., Sugimoto, K., and Makino, S. (1995b). Purification and partial characterization of a spore cortex-lytic enzyme of *Clostridium perfringens* S40 spores. Biosci Biotechnol Biochem *59*, 514–515.

Moir, A. (2006). How do spores germinate? J. Appl. Microbiol. *101*, 526–530.

Monroe, A., and Setlow, P. (2006). Localization of the transglutaminase cross-linking sites in the *Bacillus subtilis* spore coat protein GerQ. J. Bacteriol. *188*, 7609–7616.

Moriyama, R., Hattori, A., Miyata, S., Kudoh, S., and Makino, S. (1996a). A gene (*sleB*) encoding a spore cortex-lytic enzyme from *Bacillus subtilis* and response of the enzyme to L-alanine-mediated germination. J. Bacteriol. *178*, 6059–6063.

Moriyama, R., Kudoh, S., Miyata, S., Nonobe, S., Hattori, A., and Makino, S. (1996b). A germination-specific spore cortex-lytic enzyme from *Bacillus cereus* spores: cloning and sequencing of the gene and molecular characterization of the enzyme. J. Bacteriol. *178*, 5330–5332.

Moriyama, R., Fukuoka, H., Miyata, S., Kudoh, S., Hattori, A., Kozuka, S., Yasuda, Y., Tochikubo, K., and Makino, S. (1999). Expression of a germination-specific amidase, SleB, of Bacilli in the forespore compartment of sporulating cells and its localization on the exterior side of the cortex in dormant spores. J. Bacteriol. *181*, 2373–2378.

Morlot, C., Uehara, T., Marquis, K.A., Bernhardt, T.G., and Rudner, D.Z. (2010). A highly coordinated cell wall degradation machine governs spore morphogenesis in *Bacillus subtilis*. Genes Dev. *24*, 411–422.

Myers, G.S., Rasko, D.A., Cheung, J.K., Ravel, J., Seshadri, R., DeBoy, R.T., Ren, Q., Varga, J., Awad, M.M., Brinkac, L.M., *et al.* (2006). Skewed genomic variability in strains of the toxigenic bacterial pathogen, *Clostridium perfringens*. Genome Res. *16*, 1031–1040.

Nakashio, S., and Gerhardt, P. (1985). Protoplast dehydration correlated with heat resistance of bacterial spores. J. Bacteriol. *162*, 571–578.

Okamura, S., Urakami, K., Kimata, M., Aoshima, T., Shimamoto, S., Moriyama, R., and Makino, S. (2000). The N-terminal prepeptide is required for the production of spore cortex-lytic enzyme from its inactive precursor during germination of *Clostridium perfringens* S40 spores. Mol. Microbiol. *37*, 821–827.

Orsburn, B., Melville, S.B., and Popham, D.L. (2008). Factors contributing to heat resistance of *Clostridium perfringens* endospores. Appl. Environ. Microbiol. *74*, 3328–3335.

Paidhungat, M., Ragkousi, K., and Setlow, P. (2001). Genetic requirements for induction of germination of spores of *Bacillus subtilis* by Ca(2+)-dipicolinate. J. Bacteriol. *183*, 4886–4893.

Papanikolau, Y., Prag, G., Tavlas, G., Vorgias, C.E., Oppenheim, A.B., and Petratos, K. (2001). High resolution structural analyses of mutant chitinase A complexes with substrates provide new insight into the mechanism of catalysis. Biochemistry *40*, 11338–11343.

Paredes-Sabja, D., Torres, J.A., Setlow, P., and Sarker, M.R. (2008). *Clostridium perfringens* spore germination: characterization of germinants and their receptors. J. Bacteriol. *190*, 1190–1201.

Paredes-Sabja, D., Setlow, P., and Sarker, M.R. (2009a). The protease CspB is essential for initiation of cortex hydrolysis and dipicolinic acid (DPA) release during germination of spores of *Clostridium perfringens* type A food poisoning isolates. Microbiology *155*, 3464–3472.

Paredes-Sabja, D., Setlow, P., and Sarker, M.R. (2009b). SleC is essential for cortex peptidoglycan hydrolysis during germination of spores of the pathogenic bacterium *Clostridium perfringens*. J. Bacteriol. *191*, 2711–2720.

Paredes-Sabja, D., Setlow, P., and Sarker, M.R. (2010). Germination of spores of Bacillales and Clostridiales species: mechanisms and proteins involved. Trends Microbiol. *19*, 85–94.

Popham, D.L., Illades-Aguiar, B., and Setlow, P. (1995). The *Bacillus subtilis dacB* gene, encoding penicillin-binding protein 5*, is part of a three-gene operon required for proper spore cortex synthesis and spore core dehydration. J. Bacteriol. *177*, 4721–4729.

Popham, D.L., Helin, J., Costello, C.E., and Setlow, P. (1996a). Analysis of the peptidoglycan structure of *Bacillus subtilis* endospores. J. Bacteriol. *178*, 6451–6458.

Popham, D.L., Helin, J., Costello, C.E., and Setlow, P. (1996b). Muramic lactam in peptidoglycan of *Bacillus subtilis* spores is required for spore outgrowth but not for spore dehydration or heat resistance. Proc. Natl. Acad. Sci. U.S.A. *93*, 15405–15410.

Ragkousi, K., and Setlow, P. (2004). Transglutaminase-mediated cross-linking of GerQ in the coats of *Bacillus subtilis* spores. J. Bacteriol. *186*, 5567–5575.

Ragkousi, K., Eichenberger, P., van Ooij, C., and Setlow, P. (2003). Identification of a new gene essential for germination of *Bacillus subtilis* spores with Ca2+-dipicolinate. J. Bacteriol. *185*, 2315–2329.

Scheurwater, E., Reid, C.W., and Clarke, A.J. (2008). Lytic transglycosylases: bacterial space-making autolysins. Int. J. Biochem. Cell. Biol. *40*, 586–591.

Setlow, B., Melly, E., and Setlow, P. (2001). Properties of spores of *Bacillus subtilis* blocked at an intermediate stage in spore germination. J. Bacteriol. *183*, 4894–4899.

Setlow, B., Peng, L., Loshon, C.A., Li, Y.Q., Christie, G., and Setlow, P. (2009). Characterization of the germination of *Bacillus megaterium* spores lacking enzymes that degrade the spore cortex. J. Appl. Microbiol. *107*, 318–328.

Setlow, P. (2003). Spore germination. Curr. Opin. Microbiol. *6*, 550–556.

Setlow, P. (2006). Spores of *Bacillus subtilis*: their resistance to and killing by radiation, heat and chemicals. J. Appl. Microbiol. *101*, 514–525.

Setlow, P. (2007). I will survive: DNA protection in bacterial spores. Trends Microbiol. *15*, 172–180.

Shimamoto, S., Moriyama, R., Sugimoto, K., Miyata, S., and Makino, S. (2001). Partial characterization of an enzyme fraction with protease activity which converts the spore peptidoglycan hydrolase (SleC) precursor to an active enzyme during germination of *Clostridium perfringens* S40 spores and analysis of a gene cluster involved in the activity. J. Bacteriol. *183*, 3742–3751.

Smith, T.J., Blackman, S.A., and Foster, S.J. (2000). Autolysins of *Bacillus subtilis*: multiple enzymes with multiple functions. Microbiology *146*, 249–262.

Steichen, C.T., Kearney, J.F., and Turnbough, C.L., Jr. (2007). Non-uniform assembly of the *Bacillus anthracis* exosporium and a bottle cap model for spore germination and outgrowth. Mol. Microbiol. *64*, 359–367.

Strange, R.E., and Dark, F.A. (1957a). A cell-wall lytic enzyme associated with spores of *Bacillus* species. J. Gen. Microbiol. *16*, 236–249.

Strange, R.E., and Dark, F.A. (1957b). Cell-wall lytic enzymes at sporulation and spore germination in *Bacillus* species. J. Gen. Microbiol. *17*, 525–537.

Sudiarta, I.P., Fukushima, T., and Sekiguchi, J. (2010). *Bacillus subtilis* CwlQ (previous YjbJ) is a bifunctional enzyme exhibiting muramidase and soluble-lytic transglycosylase activities. Biochem. Biophys. Res. Commun. *398*, 606–612.

Terwisscha van Scheltinga, A.C., Armand, S., Kalk, K.H., Isogai, A., Henrissat, B., and Dijkstra, B.W. (1995). Stereochemistry of chitin hydrolysis by a plant chitinase/lysozyme and X-ray structure of a complex with allosamidin: evidence for substrate assisted catalysis. Biochemistry *34*, 15619–15623.

Tipper, D.J., and Linnet, P.E. (1976). Distribution of peptidoglycan synthetase activities between sporangia and forespores in sporulating cells of *Bacillus sphaericus*. J. Bacteriol. *126*, 213–221.

Urakami, K., Miyata, S., Moriyama, R., Sugimoto, K., and Makino, S. (1999). Germination-specific cortex-lytic enzymes from *Clostridium perfringens* S40 spores: time of synthesis, precursor structure and regulation of enzymatic activity. FEMS Microbiol. Lett. *173*, 467–473.

Wang, K.H., Isidro, A.L., Domingues, L., Eskandarian, H.A., McKenney, P.T., Drew, K., Grabowski, P., Chua, M.H., Barry, S.N., Guan, M., *et al.* (2009). The coat morphogenetic protein SpoVID is necessary for spore encasement in *Bacillus subtilis*. Mol. Microbiol. *74*, 634–649.

Warth, A.D., and Strominger, J.L. (1969). Structure of the peptidoglycan of bacterial spores: occurrence of the lactam of muramic acid. Proc. Natl. Acad. Sci. U.S.A. *64*, 528–535.

Warth, A.D., and Strominger, J.L. (1971). Structure of the peptidoglycan from vegetative cell walls of *Bacillus subtilis*. Biochemistry *10*, 4349–4358.

Warth, A.D., and Strominger, J.L. (1972). Structure of the peptidoglycan from spores of *Bacillus subtilis*. Biochemistry *11*, 1389–1396.

Yeats, C., Rawlings, N.D., and Bateman, A. (2004). The PepSY domain: a regulator of peptidase activity in the microbial environment? Trends Biochem. Sci. *29*, 169–172.

Water and Cations Flux During Sporulation and Germination

10

Daniela Bassi, Fabrizio Cappa and Pier Sandro Cocconcelli

Abstract

Bacterial spore water uptake and solute permeability are hypothesized to play a central role during the sporulation and germination processes but the leading mechanisms are at now poorly understood. In this chapter, a state of the art and the last experiments to better study this topic are presented. In the first part, major questions are addressed to the role of water content in spore dormancy and resistance, to the mechanisms which lead to spore dehydration and to the permeability of the spore layers to water and solutes. In the second part, the attention is focused on the ion composition of spores and their importance for the spore life cycle, with particular attention to the leading mechanisms acting during spore formation in the mother cell and during the germination process and outgrowth.

Water in the bacterial spore

In both prokaryotic and eukaryotic biological systems, a decrease in water content is fundamental in the shift from active cell life to dormancy and it contributes to resistance towards adverse conditions. Dehydration is therefore a key factor in the extraordinary resistance of bacterial spores. Dormancy induced by dehydration is reversible, with rehydration restoring normal cell life (Powell and Hunter, 1955).

Initial attempts to study bacterial spores raised numerous doubts about spore permeability. Preliminarily, spores were considered non-stainable, highly refractive and often 'non-wetable' (Fischer, 1897). Then, towards the end of the 1880s, on the basis of water vapour sorption by bacterial spores, researchers attributed their heat resistance to the fact that spores had a lower moisture content than vegetative cells (Drymont, 1886). Subsequently, different methods were used to measure water content in spores and in vegetative cells. Based on hygroscopic properties, Virtanen and Pulkky (1933), stated that there were no differences between spores and vegetative cells. Some years later, Friedman and Henry (1938), in their pioneering work, used cryoscopic methods to study bound water in the vegetative and spore forms of *Bacillus subtilis*, *Bacillus megatherium* and *Bacillus mycoides*. They affirmed that spores had a stronger water-binding capacity than vegetative cells, so that their higher resistance to heat was due to the amount of bound water. It was only in 1950–1960, that numerous scientists highlighted the difference between the hydration state of spores and cells (Stewart and Halvorson, 1953; Waldham and Halvorson, 1954) demonstrating that the vegetative cells were more hygroscopic than spores. Murrell and Scott (1958) also supported this theory; using D2O, they measured a 96% free isotope exchange with spore water, demonstrating that almost all the water in spores was exchangeable with external water in the environment. They also estimated that 30–40% of the spore volume could be dry space. It was studying the heat resistance of spores of *Bacillus* and *Clostridium* that some hypotheses were made:

1 Spores, unlike vegetative cells, contain thermoresistant enzymes (Green and Sadoff, 1965).

2 Spores have a high refractive index, suggesting that they contain a large amount of refractile material or that the water content of spores is low (Ross and Billing, 1957; Leman, 1973).
3 Owing to spore apparent gravity, it was postulated that these dormant cells present a condensed structure with an atypically low water content, compared with the vegetative cell (Lewis *et al.*, 1965).

Most of the pioneering studies in the 1960s and 1970s addressed three main questions: first, the role of water content in spore dormancy and resistance; secondly, the mechanisms which lead to spore dehydration; and, thirdly, the permeability of the spore envelope to water and solutes.

Precise determination of the total water content of the fully hydrated spores has been, since many years, a challenge for scientists. Black and Gerhardt (1962) demonstrated in *B. cereus* that the water content of dormant spores was lower than in germinated spores, being, respectively, 66.6% and 75.6% of the total spore weight. Other studies in *Bacillus* indicated that the total water content depended on the species and varied from 0.45 to 0.65 g of water per grams of wet spore (Marshall and Murrell, 1970; Watt, 1981). Initially, such a water content did not seem to correlate with the high refractive index and apparent gravity measured for spores. However, if water distribution was unequal within the spores, with more in an exterior compartment and less in an interior compartment, then the values could be justified. Using laser diffractometry, Ulanowski *et al.* (1987) and De Pieri *et al.* (1992) measured water content and size of two layers of *Bacillus sphaericus* spores, which they recognized as the cortex and the core, respectively in a total spore crops. They found a water content of ≈ 0.6 g/ml in the cortex and only ≈ 0.3 g/ml in the core.

These observations led to the assumption that spores are partially dry and this feature is prerequisite to dormancy and resistance (Gould, 1977). Most of the investigations into water properties and water content in spores were made on members of the *Bacillus* genus (Lewis *et al.*, 1960; Black and Gerhardt, 1961; Neihof *et al.*, 1967; Setlow and Setlow, 1980; Rubel, 1997; Leuschner and Lillford, 2000; Westphal *et al.*, 2003; Plomp *et al.*, 2005; Sunde *et al.*, 2009). Ishihara and colleagues (1994) studied the free and bound water in vegetative cells and spores of *B. subtilis* using time-domain reflectometry, and demonstrated that spores contained only two thirds of the water contained in vegetative cells.

The mechanism which leads to a low water content was first investigated by Lewis *et al.* (1960), who proposed the mechanical pressure created by spore cortex contraction during sporulation as being responsible for forcing water out of the core. Alderton and Schnell (1963) meanwhile hypothesized that it was the mechanical force exerted by cortex expansion that lead to core dehydration. Though presenting contrasting mechanisms, both theories relied on the presence of specific envelopes that are highly impermeable to water (Lewis *et al.*, 1960), an idea that was refuted when different authors demonstrated that dormant spores are permeable to water and small solutes, and that this is influenced by their molecular weight, charge and lipophilicity (Black *et al.*, 1960; Lewis *et al.*, 1960; Black and Gerhardt, 1961; Gerhardt and Black, 1961b; Black and Gerhardt, 1962). Later, Marshall and Murrell (1970) in experiments of H_2O/D_2O exchange demonstrated that almost 97% of spore water is exchangeable with the environment. Therefore, the impermeability theory was completely excluded, but the nature of water distribution in the spore structure became complicated and remained a wide and interesting area of study. In the following paragraphs these themes are discussed in detail.

Core water content

The unusual capacity of spores to remain dormant for years and to resist physical and chemical insults, was mostly attributed to the dehydration of the spore protoplast, which appeared to remain encased within a robust integument, even when spores were exposed to excess water. This could help to understand why spores can stay dormant for long times. The driving evidence for this idea was the historical light microscopic image of the bacterial spore with a refractive core and a non/refractive surroundings that were assumed to correspond with the protoplast and the exterior integuments, respectively (Beaman *et al.*, 1984).

Initial attempts to measure the distribution of water content were difficult because the protoplast could not be isolated without loss of its intrinsic *in situ* properties. Moreover, having discovered that the spore possessed, not only an inner membrane surrounding the cytoplasm, but also an outer membrane protecting the cortex (Crafts-Lighty and Ellar, 1980), the question became more complex.

The high resistance of spores, when compared with vegetative cells, was traditionally ascribed to a sort of molecular stabilization in the core caused by the presence of specific compounds (such as dipicolinic acid) and a low water. The core was shown to have a low water content even in full hydrated spores (Beaman and Gerhardt, 1986). It has been estimated that the water content of the spore protoplast varies among *Bacillus* species from 26% to 55% on a total wet weight basis (Beaman *et al.*, 1984; Lindsay *et al.*, 1985; Nakashio and Gerhardt, 1985; Beaman and Gerhardt, 1986; Gerhardt and Marquis, 1989; Leuschner and Lillford, 2000). The quantity of free water in the spore core is particularly low and, as a consequence, the movement of macromolecules is strongly inhibited (Cowan *et al.*, 2004). A low water content is essential for a spore's metabolic dormancy (Gerhardt and Marquis, 1989), but little is known about the cell's behaviour during sporulation, when there is a rapid reduction in core water. During the germination process, water uptake in *B. subtilis* spores happens in the first minutes, so that macromolecular dynamics and enzymatic activity are restored. Since the core dormancy is maintained by cortical pressure, cortex degradation causes a relaxation of the inner membrane with a consequent rehydration of the spore core, triggering germination (Cowan *et al.*, 2003). Sonenshein and colleagues (2001) proposed a model according to the fact that any event allowing the redistribution of water and ions across the inner membrane would induce germination; particularly, an intake of water into the spore core and an efflux of Ca^{2+}-DPA are the main events, triggered by germinants binding and consequent activation of germinant receptors and which results in a modulation of specific channels. After water and ions redistribution, the core expands applying a stress on the cortex which, at the same time, activates the cortex lytic enzymes (CLEs) and the cortex breakdown. The complete release of cortical pressure produces further exchanges of water and ions until a total core rehydration. More evidences are needed to open also new scenarios beside this model.

Other than water, the physical space in the core is quite completely occupied by the genome, about half of the spore's proteins, ribosomes, tRNAs and a large depot of calcium and other divalent ions bound by pyridine-2,6-dicarboxylic acid (DPA) (Gerhardt and Marquis, 1989; Setlow, 2006). Dipicolinic acid has been considered a major constituent of the spore core, representing 5–20% of the dry weight of spores of both *Bacillus* and *Clostridium* species, where it is most probably chelated 1:1 with divalent cations, largely Ca^{2+} (Murrell, 1969; Leanz and Gilvarg, 1973; Kozuka and Tochikubo, 1985; Gerhardt and Marquis, 1989; Setlow, 2006). Presently, the precise state of DPA in the core is still unknown, however it seems clear that the accumulation of a high amount of DPA in the spore core is responsible for some of the reduction in core water content during sporulation, and that DPA also plays a significant role in the UV photochemistry of spore DNA (Setlow, 2006).

Gould in 1986 proposed that the core moisture content, low enough to be comparable to a vitrified solid state (glassy), is responsible for the high resistance of spores. This hypothesis was supported by differential scanning calorimetric (DSC) and NMR studies (Ablett *et al.*, 1999) which provided evidence that the core region of spore is in a low moisture glassy state. This theory was consistent with the mechanism developed by anhydrobiotes, such as plants, seeds, pollen and fungal spores. In these organisms dormancy is achieved by increasing the cellular viscosity and transforming the cytoplasm in a glassy state, conferring long term survival in a dry condition. The formation of intracellular glasses is the result of the accumulation of disaccharides and other compatible osmolytes (Crowe *et al.*, 1998; Buitink and Leprince, 2008). If so, spore dormancy could be explained as a consequent suppression of chemical reactions and extreme slowness in denaturation and metabolic processes. According to this, in 1979, experiments determining the dielectric permittivity and the electron paramagnetic resonance spectra demonstrated that many ions in the core are immobilized (Carstensen *et al.*, 1979) and further experiments in *B. subtilis*, expressing green fluorescent protein, verified that in the spore core the proteins are four orders of magnitude less mobile than in the cytoplasm of the vegetative cell (Cowan *et al.*, 2003; Newsome, 2003; Gould, 2006). Other attempts to study the core water state using NMR spectroscopy have not been exhaustive (Bradbury *et al.*, 1961; Maeda *et al.*, 1968).

More recently, Ghosal *et al.* (2010), in agreement with Black and Gerhardt (1962) and Leuschner and Lillford (2000), clearly showed that the core is permeable and behaves as an insoluble heat-stable gel which exchanges water with the external environment, being, particularly in this last point, in contradiction with the previous theory of the glassy state.

Recent studies using NMR have enabled the dynamics of water mobility in the dense spore core of *B. subtilis* to be determined. Water mobility in the different spore compartments has been quantified using the dynamic perturbation factor (DPF), which is the rotational correlation time averaged over all water molecules in the spore region considered, divided by the same quantity in bulk water (Sunde *et al.*, 2009). The DPF value permitted the evaluation of the water molecules external to proteins and other solutes but not of water inside polar cavities of proteins or at the interfaces of associated macromolecules (Mattea *et al.*, 2008). In addition, a symmetric flip motion of water molecules bounded to Ca^{2+} divalent ions in the core has to be considered in order to obtain a proper water mobility evaluation (Denisov and Halle, 1995). According to the 'core glassy state' hypothesis (Gould, 1986; Sapru and Labuza, 1993; Ablett *et al.*, 1999), the core water should be a 'rigid lattice' limiting molecular diffusion. Sunde *et al.* (2009) demonstrated instead an high water mobility through the spore. In fact, it was found that the core water was only 30-fold less mobile than bulk water, an observation that lead to the conclusion that the core is not in a glassy state.

From these recent advances, it has been concluded that water in the core preferentially hydrates non-protein molecules. As a consequence, core proteins are surrounded, at the most, by one layer of water, and that this dehydrated state is probably sufficient to alter their active folding and stop their functionality (Setlow, 1994; Gregory, 1995). In this way, spores could be resistant to environmental adverse conditions, simply stabilizing proteins, avoiding their thermal denaturation and maintaining their unfolded state (Rupley and Careri, 1991; Gregory, 1995) However, minor changes in the water content could affect this state (Eijlander *et al.*, 2010; Mafart *et al.*, 2010). When the level of hydration is high enough to allow intermolecular disulfide exchanges, proteins unfolding translates to irreversible aggregation that affects the spore viability (Klibanov and Schefiliti, 2004). This last position agrees also with Raman spectroscopy experiments made on *B. subtilis* spores (Zhang *et al.*, 2009), in which a heat activation at 70°C has been followed by partial but reversible protein denaturation. Since wet heat kills spores by protein denaturation and enzyme inactivation, this suggests that, as long as protein denaturation is reversible, it should not kill the spore (Sunde *et al.*, 2009).

Some issues relevant to core water content still remain unclear; we know that the low water content in the core is responsible of spore's heat resistance but we do not know the exact quantity of free water in the core and we ignore to what extent core proteins are hydrated, and via which mechanism wet heat is able to kill spores. Since the core could not be defined as a solid glassy environment, major understanding of this internal spore 'viscosity' is absolutely requested.

Water mobility across the inner membrane

Numerous hypotheses suggest that the spore inner membrane (IM) is the major permeability barrier selecting the flux of chemicals and small molecules into and out of the core (Gerhardt *et al.*, 1971; Setlow and Setlow, 1980; Swerdlow and Setlow, 1984; Setlow *et al.*, 2000; Paidhungat and Setlow, 2002; Westphal *et al.*, 2003). The rare permeability of the spore IM to hydrophilic and hydrophobic molecules is, in this way, one of the factor that influences spore chemical resistance (Gerhardt *et al.*, 1971). The IM appears to be in a gel state (Cowan *et al.*, 2004) and could have a protective role towards the core DNA, due to its low permeability to ionic species and even to small non-ionic solutes (Setlow, 2006).

A small molecule such as methylamine has been used to study the IM permeability, observing that its passage through the membrane was very slow; also, water crossed the IM very slowly (Setlow and Setlow, 1980; Swerdlow *et al.*, 1981; Westphal *et al.*, 2003; Cortezzo *et al.*, 2004; Cortezzo and Setlow, 2005).

The unusual low permeability of IM was initially attributed to some abnormalities in its lipid composition, but deeper studies demonstrated that the IM lipid structure was very similar to that of the plasma membrane of growing cells (Cortezzo *et al.*, 2004; Cortezzo and Setlow, 2005). The IM in the dormant spore is very tightly packed and it is composed by lipid molecules which are largely immobile and create a crystalline structure (Cowan *et al.*, 2004). Although the majority of lipids in the IM are immobile, there is an important ratio that remains mobile. Cowan *et al.* (2004) proposed that the low diffusion value (*D*) of the spore inner membrane is strictly related to the organization of mobile and immobile lipids. The first hypothesis proposes that the two leaflets of the IM present a different mobility. The variation of movement could be due to the low water content in the core compared with the cortex area (Gerhardt and Marquis, 1989). This leads to a lower mobility of lipids from the inner part of the membrane towards the core. When germination occurs, the core water grows, restoring the normal lipids mobility (Gerhardt and Marquis, 1989; Popham *et al.*, 1996; Setlow *et al.*, 2001). A second hypothesis is that the IM is organized in different domains that can, respectively contain, respectively mobile and immobile phospholipids. This view is supported by studies on lipid and protein composition of *B. subtilis* and *B. megaterium* IM. Ellar *et al.* (1978) reported that the IM has a strongly higher protein content than the growing cell and this might contribute to establish large areas of ordered domains in condition of a special dehydrated core state. These domains have been distinguished in growing cells of *B. subtilis* (Vanounou *et al.*, 2003), while in *B. megaterium* the IM has been seen to be composed by two fractions of similar lipids but with different protein composition (Swerdlow and Setlow, 1984).

Additionally, the spore IM has the particular capacity to expand at different steps of the spore-former's cell cycle; in fact, the volume it encompasses, decreases as much as 2-fold late in sporulation and increases up to two-fold in the first minutes of spore germination, when the spore peptidoglycan cortex is degraded and the outgrowing cell expands (Gould, 1969; Warth, 1978; Popham *et al.*, 1996). This increment in spore volume happens without either new membrane synthesis or ATP production, and restores the permeability to the germinated spore's plasma membrane (Setlow and Kornberg, 1970; Gerhardt *et al.*, 1971; Swerdlow *et al.*, 1981; Setlow *et al.*, 2001).

The major germinant receptors are thought to be located in the spore IM. The inner membrane is in fact the place where the germination cascade of events begins. Germinants cross the outer spore layers to interact with the receptors proteins located in the IM; this binding causes membrane-associated changes and bulk ion movements, such as release of dipicolinic acid, through specific channels. Immediately after, water goes into the core followed by the activation of cortex and coat lytic enzymes by an unknown signal molecule (Moir, 2002).

Considering that the spore IM is (i) the major barrier to the passage of small molecules such as DPA that are essential for spore resistance, (ii) where a number of proteins involved in the triggering of spore germination are found and (iii) where membrane-active agents such as cationic surfactants act to trigger spore germination (Setlow, 2003), it seems probable that a greater knowledge of the structure of the inner membrane will give us new insights, not only on spore resistance, but also on spore dormancy and spore germination.

The role of cortex

The cortex, a layer composed by a cross-linked peptidoglycan matrix, is the most hydrated region of the spore. Gould and Dring (1974, 1975a) underlined that the polymer in this spore layer is water-filled and highly expanded if compared with the dense water-depleted core. Such a structure is in line both with the high refractive index and peculiar gravity of the bacterial spores, and the total high water content of these dormant cells, compensating the water-deprived state of the core. The cortex structure has been found to be generally preserved among species but its role and its implication in heat resistance is not completely clear (Atrih and Foster, 2001; Popham, 2002).

The spore core dehydration state is in part dependent on the inner forespore membrane but is also greatly dependent on the spore cortex (Popham, 2002). What processes enables a perfect and equal distribution to be maintained inside the spore's integuments? The first studies postulated that the leading force in maintaining this distribution was osmosis and the cortex was designated as an osmoregulatory organelle (Gould and Dring, 1975a). Since the cortex is mainly composed of electronegative peptidoglycan, it is also supposed to contain also positive counter-ions that confer an osmotic potential of around 2 MPa or more. In this way, the expanded cortex should be in osmotic equilibrium with the dehydrated enclosed core, with the inner membrane regulating the passage of water from the core to the cortex or vice versa until the total osmotic pressure is balanced. The equilibrium between the two compartments is also dependent on the osmotic activity of the core molecules. Since molecules of high molecular mass, or salts such as calcium dipicolinate or calcium salts of weak acids (glutamic, sulpholactic or phosphoglyceric) usually reside in the cortex, the osmotic balance is in the direction of a high water content cortex surrounded by a lower water content protoplast. If spores are suspended in water, they absorb it into the cortex; successively, the hydrostatic pressure inside the spore will balance the osmotic pressure to set the system in perfect equilibrium. Different experiments have been conducted to validate this hypothesis. Spores with modified coats have been produced using chemical treatments so that they were permeable to multivalent cations; all these spores were heat sensitive if high concentrations of calcium or other multivalent cation salts were present in the medium (Gould and Dring, 1975a). The effect caused by the divalent cations that succeed in passing through the spore coat was a strong interaction with the electronegative peptidoglycan of the cortex giving rise to its collapse, the replacement of mobile counterions, and a consequent rehydration of the core that then becomes heat sensitive. This theory was validated by confirming the opposite reaction of full heat resistance when spores are suspended in solutions of sugars concentrated enough to re-establish protoplast dehydration via osmosis (Gould,1977). Though the heat sensitivity of rehydrated spores was not demonstrated, it should be assumed that rehydration is at the basis of the germination process, and it is known that spores that have entered germination become more susceptible to heat. Water flux into the core can occur within minutes, in which time the spore does not synthesize new plasma membrane, and so the inner membrane has to expand its surface area almost 4-fold (Dawes and Halvorson, 1972).

Using the information gained until now, the cortex can be defined as a motionless structure able to assure core dehydration (Popham *et al.*, 1996; Popham, 2002). The different level of peptidoglycan cross-linkage in the varied cortex regions is responsible of the different degree of pressure that the cortex exercises on the core (Meador-Parton and Popham, 2000; Popham, 2002). Owing to the cortex peptidoglycan composition, water cannot be trapped inside it for more than 10 ns (Vaca Chavez *et al.*, 2006), transferring rapidly to the near coat, where it is better bound by its protein layers.

Water dynamics through the spore coat

Proceeding towards the exterior, the spore's first line of defence is the coat. This structure, which is composed of multiple protein layers, has a central role in the mechanism of protection, contributing to exclude degradative enzymes and to defend the spore against reactive molecular species (Young and Setlow, 2003). In the coat of *B. subtilis* at least 50 different individual proteins have been found in two major layers(Driks, 1999, 2002; Lai *et al.*, 2003). Only a small subset of coat proteins play essential roles in the determination of coat morphology (Driks and Setlow, 2000; Driks, 2002; Henriques *et al.*, 2004). Several studies have assigned coat proteins an important role in defending spores against a number of chemicals, including sodium hypochlorite, chlorine dioxide, ozone, organic solvents, and lytic enzymes that can damage the spore cortex. It was observed that chemically decoated wild-type spores were much more sensitive to these agents than intact wild-type spores (Setlow *et al.*, 2000; Young and Setlow, 2003, 2004; Setlow, 2006). However, the coat is not responsible for protection against heat, radiations and some other chemicals (Driks, 1999; Nicholson *et al.*, 2000; Setlow *et al.*, 2000; Klobutcher *et al.*, 2006). No specific coat protein has been discovered to be responsible for spore resistance to these chemicals and the role of the coat seemed to be that of a 'reactive armour' (Setlow, 2006) acting to filter these chemicals when they enter into the spore.

Studies conducted on functions of *Bacillus* spore coat proteins revealed an unexpected grade of disorder in this structure (Kim *et al.*, 2006). This opens the discussion on how a discrete biological disorder in the network of coat proteins can not only be tolerate, but may also help to maintain an important function such as the ability of the spore to flex and move. This means that every coat protein could be able to bind to each other, presumably through relatively low-affinity, low-specificity interactions and this may directly contribute to function, by permitting structural flexibility.

If compared with cortex and core compartments, first studies revealed that little water was present in the spore coat, as demonstrated in *B. stearothermophilus* by calculating the dry mass, dry volume, wet mass, and wet volume of the spore and its components (Algie and Watt, 1984). This view was confirmed by similar results achieved in *B. subtilis* and *B. cereus* T (Algie, 1984) as shown in Table 10.1.

More recently, the water content of coat has been investigated by means of an NMR approach called magnetic relaxation dispersion (MRD) profile (Sunde *et al.*, 2009), a method which determines the water ^{2}H and ^{17}O spin relaxation rates in D_2O-exchanged *B. subtilis* spores. This study aimed to calculate the water distribution in the different spore compartments, and demonstrated that the internal water fraction of the coat is limited, accounting for 0.14% of spore water trapped inside proteins. This value may indicate a lower internal water content of coat proteins or a larger fraction of internal water molecules with residence times $> 10\,\mu s$ in the extensively cross-linked coat proteins.

With regard to permeability, the spore coat is a semipermeable layer allowing the passage of molecules with a molecular mass < 5 kDa, including water, and the passage of molecules to deeper layers, particularly through the inner membrane that contains specific receptors (Driks, 1999; Setlow, 2003; Henriques and Moran, 2007). Westphal *et al.* (2003) measured how humidity influences spore response of *B. thuringiensis* in terms of water flux and swelling of dried spores exposed to water vapour. Using an automated scanning microscope and observing the kinetics of size changes at different values of relative humidity, they demonstrated a rapid diffusion of water into the spore coat and cortex layers with a time scale close to 50 seconds. In the same experiments, the time for diffusion of water into the

Table 10.1 Masses, water contents, densities and water activities of spores and their components for *B. subtilis* and *B. cereus T* (Algie, 1984)

Species	Quantity	Spore	Coat	Cortex	Core	Ca2+-DPA
B. subtilis	Dry mass (fg)	292	146	98	49	33
	Water content (g H2O/g dry wt)	0.34	0.24	0.48	0.23	0.05
	Dry density (g/ml)	1.38	1.30	1.40	1.66	1.8
	aw	–	–	0.85	0.85	–
B. cereus T	Dry mass (fg)	693	200	253	240	71
	Water content (g H2O/g dry wt)	0.50	0.35	0.57	0.55	0.08
	Dry density (g/ml)	1.40	1.29	1.40	1.51	1.8
	aw	–	–	–	0.89	–

spore core was found to be much longer, being approximately 8 minutes. Taking into account the time needed for water diffusion and the estimated thickness of coat, cortex and core, these authors proposed a simplified model where water diffusion time through the spore layers is proportional to square root of the summed radial distance of the different spore's layers.

Clear explanation of the precise function of the spore coat has been prevented by the difficulty to obtain spores without most coat layer. Ghosh *et al.* (2008) studied a *B. subtilis* mutant strain lacking in both CotE and GerE proteins. CotE is a protein essential for assembly of many coat proteins as well as the outer coat layer; GerE is a DNA binding protein acting in the mother cell compartment late in sporulation and which positively regulates the expression of genes coding for a number of proteins in the spore coat. These spores, which present a modified coat, had a normal core water content supporting the hypothesis of the limited action of coat in reducing core water during spore development, which is the primary cause of spore resistance to wet heat. It was unexpectedly found that the water permeability of the IM is greatly enhanced in these coat-deficient spores.

The coat only seems to act as a detoxifying filter for chemicals with its large amount of proteins, but it has no involvement in either heat resistance or water flux through the spore.

Exosporium: a semipermeable barrier

Being the outermost layer of the spore, the exosporium is the primary site of contact with the environment, including host defences in the case of pathogenic spore-formers (Guidi-Rontani *et al.*, 2001). This envelope acts as a source of surface antigens (Gerhardt and Ribi, 1964; Steichen *et al.*, 2003)and as a semipermeable barrier that excludes large, potentially harmful molecules such as antibodies and hydrolytic enzymes (Gerhardt and Black, 1961a; Gerhardt, 1967). It has a wide surface and, in some cases, a hirsute appearance, and it occupies a great part of the spore's volume (Fig. 10.1).

Some strains of *Bacillus* lack this outermost layer (Setlow, 2006). The exosporium was firstly studied in *Bacillus cereus* T, a strain frequently used in basic studies of spore resistance (Gerhardt and Ribi, 1964; Gould, 1977; Beaman *et al.*, 1982). This strain has a superficial, loose-fitting, bilayered exosporium which expands when spores are suspended in solution, and tens to collapse when spores are packed. This can be explained by the presence of liquid, macromolecules, and in some cases protein crystals in the space between the exosporium and the spore coat, suggesting a free flow of water through the exosporium.

Figure 10.1 Scanning electron microscopy analysis of *C. sporogenes* spores reveals the presence of exosporium with and without hirsute appearance.

An additional factor which may influence water permeability is surface hydrophobicity of spores, which depends mainly on its exosporium and its components. It is well described that hydrophobic features of spores are strictly related to the presence or not of the exosporium (Takubo *et al.*, 1988; Koshikawa *et al.*, 1989). However, both dimensions and hydrophobicity of the exosporium were shown to be strain dependent (Tauveron *et al.*, 2006). For example, spores of *Bacillus subtilis*, *B. licheniformis*, and *B. macerans* do not have distinct exosporium layers and are less hydrophobic than exosporium-containing *B. cereus*, *B. brevis*, and *B. thuringiensis* spores (Koshikawa *et al.*, 1989). Specific information on structural component(s) responsible for the hydrophobic behaviour of spores is limited. In *B. anthracis* the collagen-like protein (BclA) present in the exosporium, markedly influenced the spore hydrophobicity, probably affecting the access of molecules to the internal spore compartments (Brahmbhatt *et al.*, 2007). The same authors also observed that deletion mutants lacking the *bclA* gene were less water repellent than the wild-type. Contrary to this, recent studies in *B. thuringiensis* subsp. *israeliensis* spores, analysing the uptake of water by exposing dry spores to D_2O vapour, demonstrate that the exosporium is highly permeable to water (Ghosal *et al.*, 2010).

Data available to date indicate that the exosporium, due to its composition, elasticity and the fact that it loosely surrounds the spore periphery, often dissociating from the spore itself (Zaman *et al.*, 2005), has a limited role in water and ions fluxes.

Ions in spores

Spore electrical conductivity

Spore resistance to environmental stresses has been the focus of extensive research since the 1950s, with particular attention having been paid to the involvement of dipicolinic acid and calcium content and distribution, as well as to the polymeric components of the *Firmicutes* spore structure and their relation to physical and chemical stress resistance. Overall, research has shown that spore resistance to environmental stress is certainly related to metal ions, mostly Ca^{2+}, and dipicolinic acid (Murrell, 1969; Tipper and Gauthier, 1972). Particularly, the ion composition of spores and their distribution in different layers is considered of fundamental relevance to spore formation, resistance, dormancy and germination.

Chemical analysis of spores demonstrated that the spore integuments have different compositions, and minerals are distributed more in the protoplast than in the other sectors (Scherrer and Gerhardt, 1972; Scherrer and Shull, 1987). The minerals in spores do not seem to constitute a crystalline phase and it has been proposed that they form complexes with spore polymers and low-molecular-mass anions (Marquis and Shin, 1994).

In the 1980s, scientists considered that the state of spore dormancy was strictly associated to the low electrical conductivity of the core, where ions are found tightly bound and immobile. However, this state of electrical stasis seems to engage not only the core mineral composition, but also ions in the other spore compartments (Stewart, M., *et al.*, 1980). When germination and outgrowth take place, a new ion flux in the cytoplasm starts, restoring the normal cell conductivity of the vegetative state. Being the spore a multilayered structure, enzymatically rich and chemically complex, it is quite difficult to attribute a proper electrical function to each component. In spores of *B. cereus* and in *B. megaterium*, Carstensen (1979) detected the presence of two dielectrically effective membranes, the plasma membrane and the outer membrane, as well as the presence of tightly bound ions in the spore core. The inner membrane, that in dormant spore forms a close dielectrically homogeneous complex with the core, was considered a good electrical insulator that affects dispersions when germination starts and ions become mobile. The cortex is a sort of 'reservoir' of self mobile ions independent from the environmental ionic charge.

Moreover, In decoated spores, an increase in conductivity levels has been observed in response to environmental ion flux from outside the spore. This indicates that the coat has the attribute of being an insulating layer able to maintain its own fixed charge concentration (with an order of 10 to 20 milliequivalents per litre and a corresponding number of mobile counterions).

Spore mineral content

Bacterial spores contain a discrete amount of metals and mineral elements that are known to have a fundamental role in the spore development, but also in its singular resistance to heat. The major mineral constituents are calcium, potassium, magnesium and manganese, while the minor components refer to sodium, iron, aluminium, silicon, and phosphorus (Curran *et al.*, 1943; Powell and Strange, 1956; Vinter, 1957; Thomas, 1964). Sometimes also copper is reported. A scheme of the major inorganic elements in *Bacillus*, *Clostridium* and *Sporosarcina* species is represented in Table 10.2 (Murrell, 1969).

Attempts were made during years to study elements localization in the spore interior.

Depending on the different phases of spore cycle, calcium is generally the most abundant and 'participant' mineral. During spore formation, we assist to a passive diffusion of Ca^{2+} into the prespore, particularly inside the cortex and the core.

Table 10.2 Levels of monovalent and divalent ions in spores of various species. Values are reported in per cent of dry weight. *B. species* corresponds to the average of 13 species and 14 unidentified strains (from Murrel, 1969; Warth ,1978; Stewart, D.S., *et al.*, 1980; Loshon and Setlow, 1993)

	Major inorganic elements in spores % dry weight								
Species	Ca	K	Mg	Mn	Na	Fe	Al	Si	P
Bacillus spp.	2.27	0.65	0.45	0.23	0.08	0.07	0.026	0.53	1.10
C. bifermentans	6.4	1.6	0.25	0.01	–	–	–	–	–
Bacillus cereus	1.68	0.09	0.29	0.18	–	–	–	–	–
Bacillus megaterium	1.52	0.12	0.46	0.30	0.01	–	–	–	–
Bacillus subtilis	3.64	0.55	0.21	0.15	0.1	–	–	–	–
Clostridium perfringens	3.04	0.09	0.26	0.01	0.02	–	–	–	–
Sporosarcina halophila	1.28	0.02	0.32	0.01	0.03	–	–	–	–
Sporosarcina ureae	1.36	0.04	0.26	1.86	0.19	–	–	–	–

In mature spores calcium can be found in a molar ratio of 1:1 with dipicolinic acid (Murrell, 1967) and a calcium complex with DPA has been found to constitute 10% of the total spore mass (Scully *et al.*, 2002; De Vries, 2004). The calcium content is variable and the amount in the spore depends on the availability of calcium in the environment outside the cell. Results obtained in *B. cereus* demonstrated that almost all the calcium, magnesium, phosphorous and manganese were confined to the core, though some residual was also found at the cortex/coat interface (Stewart, G.S., *et al.*, 1980). This distribution was consistent with a study performed by X-ray microanalysis in *B. coagulans* that detected calcium, manganese, and phosphorus in the core and in the coat (Stewart *et al.*, 1981).

In mature spores, phosphorous was found to be present in a significant amount as 3-phosphoglyceric acid, a useful acid in ATP production during the first step of germination (Loshon and Setlow, 1993). The mature spore has a storage reservoir of 3-phosphoglyceric acid (3-PGA) that is metabolized when the germination process starts. The accumulation of 3-PGA is achieved during spore formation following an important drop of pH in the intracellular compartment with a contemporary dissociation of the Mn^{2+} cofactor from 3-phosphoglycerate mutase enzyme (Kuhn *et al.*, 1995; Jakubovics and Jenkinson, 2001).

Cells of *B. subtilis* contain high level of potassium to satisfy the mineral requirement for the growing (Lubin, 1964) and germination (Eisenstadt and Silver, 1972b) phases. Since it is required for protein synthesis, potassium is also implied in sporulation in some *Bacillus* species, and its levels in sporulating cells is quite similar to that present in growing cells. Nevertheless, high amounts of potassium are not sufficient to stimulate sporulation without the support of manganese (Eisenstadt, 1972c).

Magnesium seems to be essential for sporulation but not for heat resistance, as it interferes with Ca^{2+}–DPA complex formation in the core (Murrell, 1969; Roberts and Hitchins, 1969). Manganese plays a relevant role in the spore cycle because its homeostasis is fundamental in initiating spore-forming activity in vegetative cells. Mn^{2+} ion is, in fact, necessary for septum formation in the process regulated by pre-spore transcription factor σ^F and it then accumulates in the new spore. In addition, Mn^{2+} is able to stimulate up to 25-fold the activity of phosphoglycerate phosphomutase enzyme in the forespore or in dormant spore extracts (Singh and Setlow, 1979). In the core, the high content of mineral ions–DPA complex together with the effect of protection of the thick coat, the low level of water content and the presence of α/β-type small, acid soluble proteins (SASP) are very relevant in protecting DNA from damage desiccation, wet heat, dry heat, H_2O_2 and gamma radiation (Setlow, 2006; Moeller *et al.*, 2008; Magge *et al.*, 2009). Recently, Granger *et al.* (2011) studied the role of Mn^{2+} and other divalent cations to understand the mechanisms that increase spore resistance to dry heat, gamma irradiation and UV. These agents can damage the spore by generation of ROS, including H_2O_2. High content of Mn^{2+} alone was not sufficient to protect the spore from the damaging effects of ionizing radiation and H_2O_2, while Mn^{2+} in complex with DPA and with or without ortophosphate was demonstrated to be effective, acting in the protection of restriction enzymes. Furthermore, the relevance of Mn^{2+} and superoxide dismutase (SOD) during sporulation was shown in *B. subtilis*. It is well known that the role of the enzyme SOD in bacteria is to defend the cell from reactive oxygen species (ROS). By using SOD negative mutants and a medium depleted of Mn^{2+}, the sensitivity of sporulating cells and premature spores to O_2^- was observed. The complex Mn-superoxidismutase has been shown to increase the resistance of vegetative cells, sporulating cells and spores against oxidative stress (Inaoka *et al.*, 1999).

Iron was revealed in the coat but not in the core of *B. coagulans* spores, while in *C. botulinum* it is considered necessary for heat resistance (Sugiyama, 1951).

Silicon, one of the most common minerals on earth, appears also in spores (Rouf *et al.*, 1964). In 1980, using scanning transmission electron microscopy (STEM), the unexpected finding of silicon in spores of *B. cereus* and *B. megaterium*, later assumed as some sort of 'lab contamination', was demonstrated in two different scientific papers (Johnstone *et al.*, 1980; Stewart, G.S., *et al.*, 1980). The silicon content in bacterial spores became the focus of scientists' attention again in 2001 during the *Bacillus anthracis* U.S. bioterrorist attack (Matsumoto, 2003); when analysing the killer spores, researchers found them rich *in silicon* and it was argued whether its presence was related to a natural behaviour (Bhattacharjee and Enserink, 2008). Last year, a *Bacillus cereus* strain capable of growing on a silicate-rich medium was selected and studied (Hirota *et al.*, 2010). It was observed that the silicon uptake happened during the last stage of sporulation when the spores is already heat resistant. The passage of the mineral from the mother cell to the new spore leads to its final storage in the spore coat layer. The exact role of silicon has to be re-evaluated but what it has revealed is that high-Si spores are highly resistant to acidic conditions.

Copper was detected in *B. subtilis* and *B. megaterium* spores in greater amounts than in vegetative cells (Windle and Sacks, 1963) and it has been suggested its possible function in inducing sporulation and increasing heat resistance (Charney *et al.*, 1951; Roberts and Hitchins, 1969; Weinberg, 1964).

It is clear that the metal content of spores depends on the cultural conditions and on the metal availability in the environment during growth and sporulation (Slepecky and Foster, 1959), though specifically the higher the concentration of metals in the sporulation medium, the greater the incorporation of divalent cations (Ca^{2+}, Ni^{2+}, Zn^{2+}, Cu^{2+}, Co^{2+}, Mn^{2+}) and minerals into the spores. At high concentrations, Zn^{2+}, Mn^{2+} and Ni^{2+} compete for Ca^{2+} sites in the spores, thus reducing the final Ca^{2+} content (Murrell, 1967; Bender and Marquis, 1985; Gerhardt and Marquis, 1989). Another factor that reduces Ca^{2+} incorporation is growing spores at high temperatures. In both scenarios, the Ca^{2+}-less spores have been shown to be more susceptible to heat stress. What is important to note, however, is that when Ca^{2+} competition occurred, the competing divalent ions were not incorporated with the same efficiency onto the Ca^{2+} binding sites. Thus, these replacing ions do not confer the same efficiency of heat resistance as calcium. Differently, when a medium composed of multiple metal ions, Ca^{2+} included, was used during the sporulation process, a consistent increase in heat resistance was observed in different *Bacillus* spores (*B. subtilis, B. coagulans, B. licheniformis, B. cereus*; Cazemier *et al.*, 2001).

Monovalent and divalent cations can be found in spores of *Clostridium* and *Bacillus* species in similar amounts, suggesting that the levels of these small molecules could be comparable in all spore formers belonging to *Firmicutes*. This observation reflects important and conserved characteristics of the sporulation process and dormant spores (Loshon and Setlow, 1993). The importance and significance of ions for the spore life cycle has been underlined in many studies over the years (Fleming and Ordal, 1964; Swerdlow *et al.*, 1981) but many mechanisms are still unclear. Regarding wet-heat resistance, Ca^{2+} spores are the more tolerant, while H^{+} spores are the least resistant; generally, spores which carry divalent cations are more heat stable than the counterpart binding monovalent cations (Bender and Marquis, 1985; Gerhardt and Marquis, 1989).

Calcium is the most abundant cation in *Bacillus* spores as calcium dipicolinate in the core (Scherrer and Gerhardt, 1972; Stewart, G.S., *et al.*, 1980). However, when Ca^{2+} is not abundant in the spore, other elements present in the sporulating medium can replace it (Slepecky and Foster, 1959; Foerster and Foster, 1966). Calcium can also be removed from spores by acid treatments and thereafter substituted by other elements (Marquis and Bender, 1985). Chemical elements can permeate without any special treatment after the spore has been produced in the mother cell (Ghosal *et al.*, 2010). The permeability to water vapour and solutes depends on the molecular weight, the charge of ions and lipophilicity (Gerhardt and Black, 1961a). The efficiency of certain bivalent cations to permeate and form a complex with dipicolinic acid was first studied in *B. megaterium* by using the technique of converting spore in hydrogen form (H-spores) by titration to pH 4 with HNO_3. Calcium, strontium and barium were the interchangeable bivalent cations involved in the formation of endotrophic spores. On the contrary, magnesium and sodium were not so efficient in the spore formation process (Black *et al.*, 1960; Lewis *et al.*, 1960; Black and Gerhardt, 1961, 1962; Gerhardt and Black, 1961a).

Another important aspect of cations in spores is their exchange through the different spore layers, membranes and structures (core, cortex and coat). There are cation fluxes driven by vapour water permeation and there is cation diffusion through channels. The uptake of solvated ions into the spore is guided by two balanced processes; the first is a chemical gradient, called diffusion, or an hydration gradient, called advection; the second is the adsorption at particular sites via an exchange mechanism.

When the spore loses water, ions, that are not bound to an exchange site, diffuse out of the spore and the charge balance is maintained by a movement of cations together with anions. When the diffusion occurs more rapidly than the exchange at active sites, the gradient does not form and the elemental abundance is only in function of binding-sites saturation. Diffusion motion is quite rapid, and cations permeate the wall spore, including the core, on a time scale of seconds to minutes (Ghosal *et al.*, 2010). The flux produced in spores leads us to imagine a peculiar distribution of cation exchange sites in the integuments rather than in the core. Bacterial cell walls are in fact known to adsorb cations (Foerster and Foster, 1966; Beveridge and Murray, 1976, 1980; Doyle *et al.*, 1980). In *Bacillus* spores, all of the layers contain potential exchange sites for cation uptake, while phosphate and amino groups associated with the spore DNA, SASPs, and dipicolinic acid are example of cation binding sites in the core. Ghosal and colleagues (2010) applied successfully high-resolution secondary ion mass spectrometry (NMR) to analyse the spatial distribution of permeating water vapour and ions in *B. thuringiensis* subsp. *israeliensis*. The experiments were performed by exposure of spores to cations in dehydrated and aqueous conditions. Li^{+} and Cs^{+} were selected because these cations are usually not present in the spores. They observed that the Li^{+} ions permeated the entire layers of the spore, showing that the core of dormant spores was subjected to a flux of water and small ions. The amount of Li^{+} after three days of exposure increased by 10^4 compared with spores that were not exposed. The NMR analysis showed high Li^{+} concentrations in the spore structures (exosporium, coat, and cortex) and a reasonable amount also in the core. The permeation is more rapid than the actual incorporation and the check point for ion distribution is the accessibility of binding sites. In addition, elements diffused in the spore on the basis of the chemistry and composition of the outer environment in which they find themselves.

Ion fluxes during sporulation, dormancy and germination

Roles of ions during sporulation

The very first hypotheses about the specific functions of metal ions in sporulation described their probable role as catalytic activators of the enzyme systems necessary for the sporulation process (Kolodziej and Slepecky, 1962). Kolodziej and Slepecky, in 1964, analysed the sporulation cycle in *Bacillus megaterium* finding traces of metal cations connected to the phenomenon. Using a purified medium supplemented with copper for sporulation they found traces of the following metals in the spores: iron, 0.5 μg/ml ferrous ion; zinc, 1.1 μg/ml zinc ion; manganese, 0.037 μg/ml manganese ion; and calcium, 0.9 μg/ml calcium ion.

Several experiments have shown that there is a significant variability in ions incorporated into spores depending on the mineral composition of the medium (El-Bisi and Ordal, 1956; Amaha and Ordal, 1957; Slepecky and Foster, 1959; Levinson and Hyatt, 1964; Murrell, 1967; Aoki and Slepecky, 1973; Krueger and Kolodziej, 1978; Mallidis and Scholefield, 1987; Marquis, 1989). Sporulating cultures have a strong tendency to accumulate metals, including Ca, Zn, Mn, Ni and Cu, and the ion content affects the heat resistance of the spores (Levinson and Hyatt, 1964; Cazemier *et al.*, 2001).

In spore-forming bacteria calcium was shown to be a prerequisite for the formation of refractile and heat-resistant spores, and it was necessary at the time of spore formation (Kolodziej and Slepecky, 1964; Lee *et al.*, 1978). During the onset of sporulation, in fact, a specific Ca^{2+} transport system begins to develop. This is in close agreement with the demonstration that the calcium content of cells increased, during the formation of the spores, within the spore itself (Vinter, 1956, 1957). As the spores matured, they progressively accumulated calcium until they contained several times more calcium than their corresponding vegetative cells. The proper mechanism of transport has not yet been elucidated. Usually, in the cell, microorganisms move calcium through secondary transport mechanisms such as Ca^{2+}/H^+ exchangers which use the energy accumulated in the gradient of other ions (Smith, 1995; Norris *et al.*, 1996). A similar apparatus could be postulated for sporulating cells where a transporter accumulates Ca^{2+} in the cytosol of the mother cell, followed by passive diffusion of Ca^{2+} into the prespore, particularly inside the cortex and the core; here the level of free calcium is low due to the complex formation with DPA (Stewart, G.S., *et al.*, 1980; Seto-Young and Ellar, 1981; Gerhardt and Marquis, 1989; Hintze and Nicholson, 2010). Eisenstadt and Silver in 1971 proposed an active transport system of Ca^{2+} from the environment to the mother cell, with internal accumulation following saturation kinetics, and a membrane-situated calcium transport system was suggested. In *Bacillus subtilis* and *Bacillus megaterium* the carrier-mediated transporter of Ca^{2+} has been suggested as being a Ca^{2+}/H^+ exchanger driven by the electrical gradient over the plasma membrane (Seto-Young and Ellar, 1981) and in accordance with these results, an antiporter Ca^{2+}/H^+ has been characterized in *B. subtilis* (Matsushita *et al.*, 1986). Some works have hypothesized the presence of a putative P-type Ca^{2+} transport ATPases in *B. subtilis* codified by the *yloB* gene (Raeymaekers *et al.*, 2002). This protein has only been seen to be active during the induction of sporulation and it is possible that it catalyses also the transport of Mn^{2+} and other cations. Calcium inside the prespore plays a relevant role in affecting gene expression during the sporulation process. Oomes *et al.* (2009), using a transcriptome approach, individuated 305 genes influenced by calcium level in sporulating *B. subtilis* cells. Particularly, other than genes involved in sporulation, *spoVFA* and *spoVFB*, responsible of DPA synthesis, were induced, together with spore coat polysaccharide synthesis genes (*sps* operon), germination genes and pathway of genes responsible of flagella and biofilm matrix production.

Manganese is the most firmly established metal requirement for sporulation (Charney *et al.*, 1951; Grelet, 1952; Curran and Evans, 1954; Amaha *et al.*, 1956; Powell, 1957). Recent genetic works in *B. subtilis* have demonstrated that Mn^{2+} homeostasis is fundamental not only for an efficient sporulation (Que and Helmann, 2000), but also for several steps of its developmental cycle (Jakubovics and Jenkinson, 2001). During vegetative exponential growth initial Mn, together with Mg uptake, was rapid and ended by mid-log phase. Secondary uptake followed immediately and continued through stage V of sporulation (Krueger and Kolodziej, 1978). The manganese accumulated during growth and early sporulation is exchangeable and therefore relatively 'free';

intracellular manganese is converted later during sporulation into a bound form that cannot be released (Eisenstadt *et al.*, 1973). In sporulation Mn^{2+} seems to be a cofactor of SpoIIE serine phosphatase (Schroeter *et al.*, 1999), implied in the formation of the spore polar septum through interactions with FtsZ (King *et al.*, 1999) and, at the same time, involved in the activation of σ^F in the forespore by dephosphorylation of SpoIIAA (Kroos *et al.*, 1999).

Magnesium active transport is of particular interest since this cation is involved in the functioning of the nucleic acid and protein synthesis apparatus, as well as being necessary for the functioning of numerous individual enzymes. Magnesium is therefore considered essential for sporulation (Murrell, 1969), though not for heat resistance (Roberts and Hitchins, 1969). Magnesium, in fact, has been shown to decrease the heat resistance of *Bacillus* species by interfering with the formation of the Ca–DPA complex (Murrell, 1969). Mg^{2+}, phosphate, and an energy source in minimal concentrations were required in order that early stationary-phase cells become committed (stage II) (Greene and Slepecky, 1972). In *Clostridium perfringens* Mg seemed to be important for growth of vegetative cells and maintenance of normal cellular morphology (Lee *et al.*, 1978). With reference to magnesium uptake and transport system, Scribner *et al.* (1973) stated that active transport of magnesium in *B. subtilis* is indicated by the energy dependence of the transport system and by the substrate saturation kinetics of magnesium uptake. They noted that, at the very beginning of sporulation, there was a sudden decline in the Mg accumulation, and they observed that 4.5 or 8 hours after the end of log-phase growth, the rate of magnesium uptake had decreased almost 80% and seemed to be independent of substrate concentration. In *B. subtilis*, two magnesium carriers have been postulated, and so that the Mg decline during sporulation could be associated with the fact that one of two magnesium transport systems can no longer function late in sporulation. Such specificity of regulation would not be surprising, since it has been already shown that potassium transport in sporulating *B. subtilis* continues at essentially the same rate as with log-phase cells (Eisenstadt *et al.*, 1973) and that manganese transport increases in rate during sporulation in response to the external growth conditions. This kinetics could be explained by the fact that Mg^{2+} ion is implied in the four components phospho-relay system. As already known, the first two steps of this system involve induced auto-phosphorylation of either of two kinases, KinA or KinB, and transfer of the phosphoryl group to the response regulator, Spo0F. A phosphotransferase, Spo0B, then catalyses the transfer of the phosphoryl group from the Spo0F protein to the transcriptional activator/repressor, Spo0A. Spo0A-P plays a dual role; it represses expression of proteins that regulate the vegetative phase and activates gene expression important for sporulation (Hoch, 1993). During sporulation KinA-mediated Spo0F-P formation also displayed both activation and inhibition by Mg^{2+}. The concentration dependence was that of Mg^{2+} ions stoichiometry with ATP produced full activity, while higher levels caused inhibition (Grimshaw *et al.*, 1998).

Few are the hypotheses regarding the role of potassium in sporulation. In normal growth, *B. subtilis* accumulates potassium against a concentration gradient. Eisenstadt (1972) was the first to study K^+ involvement in *B. subtilis*; he found that sporulating cells had a potassium content similar to that of vegetative cells, and since this mineral is fundamental for protein synthesis, its activity is enhanced not only in normal growth but also in sporulating cells (Eisenstadt, 1972). Moreover, it was observed that addition of manganese to stationary phase cells before sporulation produced a rapid intake of potassium ions. This relationship was quite unusual, but it indicated that during the sporulation phase there is an overall activation of different ion transport systems. This concept could also explain the role of manganese in stimulating magnesium accumulation in stationary cells (Eisenstadt *et al.*, 1973). A final note on potassium is that unlike calcium, it is not firmly bound inside the sporulating cells (Eisenstadt and Silver, 1972).

Slepecky and Foster determining the approximate time required by the cell to use ions, postulated that copper could be one possible requirement for sporulation in *B. megaterium* and then in *B. mycoides* and *B. cereus* subsp. *albolactis*; this role could be highlighted by the previous demonstration of its presence in vegetative cells and spores using spectrochemical analysis (Curran *et al.*, 1943), colorimetric means (Slepecky and Foster, 1959), diphenylcarbazone staining (Troger, 1959), and electro paramagnetic resonance studies (Windle and Sacks, 1963). Moreover, Powell and Strange (1956) demonstrated, with their data, a strong chelation between DPA and copper. Copper behaves, in fact, as a cofactor or a metal ion activator in many animals, plants, and bacterial systems (McElroy and Glass, 1950; Underwood, 1959; Stiles, 1961). They observed that a $CuSO_4{\cdot}5H_2O$ concentration within the range 0.025–0.10 µg/ml stimulated maximum sporulation. Lower concentrations limited sporulation, while greater concentrations resulted in incomplete sporulation after 62 hours. Contrary to what was observed until that time, Kihm *et al.* in 1990 conducted the first studies on *Clostridium botulinum* sporulation influenced by transition metals, particularly Cu and Zn. During metabolism, clostridia produce organic acids that assist in solubilizing metals and making them available for growth and incorporation into spores. It was found that Cu incorporation into spores during sporulation resulted in increased thermal sensitivity. Cu inside *Clostridium* spores can react with essential macromolecules catalysing hydrolytic reactions (Williams, 1981) and spontaneously producing free radicals, greatly toxic to DNA and cellular components (Loeb *et al.*, 1988). As a matter of fact, the possible use of copper to limit *Clostridium* sporulation, germination and growth of vegetative cell is now being studied (Mato Rodriguez and Alatossava, 2010); the model supposes the interaction between copper ions and spore-specific molecules involved in sporulation and germination processes, such as spore membranes and coat proteins rich in cysteine residues that are involved in disulphide bridge formation (Chada *et al.*, 2003; Henriques and Moran, 2000). Cu^{2+}, therefore, targets consequently sulphydryl groups of spore proteins with the formation of non-functional metal–protein complexes, resulting in a reduction of the sporulation process.

Sporulation requirements for other metals include molybdenum, that has to some extent been considered stimulatory, though not essential for sporulation (Kolodziej and Slepecky, 1964), iron and zinc, that have been established as being needed for *B. megaterium* sporulation, though their role is not certain.

Ions transport and germination

Breaking of dormancy and returning back to life is a peculiar and extraordinary behaviour of spore-forming bacteria that resume in the germination process. This complex mechanism, triggered by germinants, can be described as a biophysical and degradative process (Moir and Smith, 1990) where the IM begins to change upon the activation of a germinant receptor, making its state more fluid and permeable (Stewart, G.S., *et al.*, 1980; Skomurski *et al.*, 1983). At the same time, a flux of monovalent cations, such as K^+, H^+ and Na^+, move throughout the membrane and Ca^{2+}–DPA complexes are eliminated from the spore core through pores and channels. This leads to a restoration of ion composition, ionic strength and osmotic strength in the cells at the beginning of the outgrowing process.

The involvement of ions in the germination process has been under consideration ever since the first studies by Levinson and Sevag (1953) and Hyatt and Levinson (1961). Analyses were performed in the bacterial model *B. megaterium* where a varied group of organic and inorganic substances were signalled as part of the germination step (Rode and Foster, 1962). It was calcium and manganese cations that were first recognized as playing a role in germination (Murrell, 1961).

At the beginning of germination, the first step implies a rapid change of membrane permeability features with a consequent water uptake in the protoplast, a passage of ions and water across the spore layers and an induction of cortex lytic enzymes (Keynan, 1978). The suppression of ion channel activity is, in fact, critical for spore germination (Mitchell *et al.*, 1986). The release of monovalent ions in the first minutes of germination is, probably concomitant with the release of anions (Swerdlow *et al.*, 1981).

The first step of germination in a natural environment is triggered by a response to nutrients named germinants (Setlow *et al.*, 2003). These substances have to permeate the outer coat and cortex layers of the spore and encounter their specific cognate receptor located in the spore inner membrane (Paidhungat *et al.*, 2001). Germinants may be single nutrients or a combination of nutrients and they are functionally different depending on the species of the spore-forming bacteria. The addition of monovalent cations in combination with germinants seems to be more effective at triggering germination than when just germinants are present (Foerster and Foster, 1966). The second step of germination is the release of monovalent ions from the spore core, more frequently Na^+ and K^+ followed by the release of the large depot of calcium-DPA. It was observed that the release of calcium and DPA in the germinating spore is possible because of the presence in the spore of one or more channels for the ions to cross the spore layers; furthermore, the channels are opened to allow germinants to reach and bind with appropriate receptors (Setlow *et al.*, 2003).

Genes for Ger receptor homologues have been described in most bacilli and clostridia. Each Ger receptor can be recognized by a unique germinant such as amino acids, nucleosides, sugars, and cations (Ross and Abel-Santos, 2010).

In *B. megaterium* ATCC12872, Tani and co-workers (1996) found a sodium proton antiporter homologue, called GrmA, which exerted an important function in spore germination even when the spore was in contact with different kinds of germinants (glucose, proline, leucine or KNO_3). *B. megaterium* germination also needed receptors of the gerA type discovered for the first time in *B. subtilis* (Moir *et al.*, 2002) (Genbank, accession no. U61380). The role of Ca-DPA in germination is different among the studied species. In *B. subtilis* the release of Ca-DPA is the signal molecule for the activation of cortex-lytic enzyme CwlJ (Paidhungat *et al.*, 2001). In *C. perfringens* Ca-DPA has a different role; it acts as a signal molecule in GerKA and GerKC to trigger germination, but it has no effect on hydrolytic enzymes (Paredes-Sabja *et al.*, 2008a,b). The result of interaction of germinants with cognate receptors triggers the release of monovalent (H^+, Na^+, K^+) and divalent ions (Ca^{2+}, Mg^{2+}, Mn^{2+}). The ions' uptake and the hydrolysis of the cortex enable the spore to hydrate and expand, reassuming the water content found in growing cells, with the resumption of enzymatic activity and energy metabolism (Setlow *et al.*, 2003; Paredes-Sabja *et al.*, 2008a).

Recent works to analyse the gene expression in the different steps in the life cycle of the spore-forming *C. sporogenes* have demonstrated the presence of mRNA transcripts for ions transporters in dormant spores (Bassi *et al.*, unpublished). Genes codifying for sulphur, magnesium and iron regulation and transport proteins were particularly abundant. Sulphurtransferase family protein, sulphate permease *SulP* family, magnesium/cobalt transporter *CorA*, magnesium-translocating P-type ATPase, and ferrous iron transport protein B are some examples. This raises the hypothesis if these mRNA are residues of sporulation stages and they remain entrapped in the dormant spore, or if they have a functional role during germination. A wide expression of these genes belonging to the ions transport functional category has also been detected in spores undergoing germination. Ions carriers, including K^+, Mn^{2+}, Fe^{2+}, Zn^{2+} and Cu^{2+} transporters, are overexpressed in *C. sporogenes* spores induced with germinants. Particularly, ions transporters appear to be critical during the early stages of endospore germination and outgrowth, the moment in which significant outward fluxes of cations Na^+, K^+ and H^+ occur (Swerdlow *et al.*, 1981). Transcriptomic data indicate that within the germinating spore the rapid start of transport functions is fundamental in providing molecules the means to front a new life cycle (Bassi *et al.*, unpublished).

It was suggested that in the early stages of germination the release of K^+ and Na^+ occurred due to an increased permeability of spore layers. At the same time, the low levels of K^+ and Na^+ in germination medium enhanced the release and balance of K^+ and Na^+ as a consequence of gradient balancing and pH stabilization between the inside and the outside of the spore. The release of Na^+ and K^+ has been shown to be one of the first steps of germination, coming before the DPA release, and causing a rise of the spore core's pH; at least 80% of the spore's Na^+ and K^+ is released in the early germination steps and this efflux is followed later by the reuptake of K^+ by an energy dependent system (Swerdlow *et al.*, 1981).

Different $Na^+/H^+–K^+$ antiporters have been identified in spore-forming bacteria: GerN in *B.*

cereus and *B. megaterium*, GerA in *B. subtilis*, and GerO in *Clostridium perfringens* (Southworth *et al.*, 2001; Thackray *et al.*, 2001; Paredes-Sabja *et al.*, 2008a).

GerN, homologue of *B. megaterium* GrmA (Fujisawa *et al.*, 2007) has been studied also in *B. cereus* (Senior and Moir, 2008). This antiporter is important in the germination process mediated by inosine and, by using a *gerN* null mutant, the normal order of germination events was perturbed or drastically reduced producing an amount of asynchronous spores when inosine was used as the only germinant (Thackray *et al.*, 2001). In *E. coli*, GerN has been classified as a Na^+/H^+ and $Na^+/H^+–K^+$ antiport (Southworth *et al.*, 2001). Even if GerN is present in *B. cereus* ATCC 14579 (Ivanova *et al.*, 2003) and in other *Bacillus* species such as *B. anthracis* (Read *et al.*, 2003) and *B. thuringiensis* (Radnedge *et al.*, 2003), its germination function is not so sure. On the contrary, the role in germination seems to be fundamental for GerI, a germinant receptor whose activity is reforming local ion equilibrium. In *B. subtilis*, instead, any near homologue of GerN/GrmA has been found (Southworth *et al.*, 2001).

Differently to GerN and GerI, recent studies of GerO and GerQ, putative $Na^+/H^+–K^+$ antiporters in *C. perfringens*, have shown that they are involved in the transport of monovalent cations. The results indicated a possible role during sporulation and germination (Paredes-Sabja *et al.*, 2009a) and the involvement of GerO in germination in presence of Na^+ and inorganic phosphate as germinants (Paredes-Sabja *et al.*, 2009b). The role of Na^+ and sodium phosphate (100 mM, pH 6) has been demonstrated by using null mutants of germinant receptors. Germination tests were performed with *C. perfringens* spores mutants of *gerKA-KC*, *gerAA* and *gerO* and their function as a receptor of the sodium phosphate germinant was demonstrated. More precisely, the Na^+ acted as co-germinant with inorganic phosphate and the presence of the putative $Na^+/H^+–K^+$ antiporter was necessary for the relative transport (Paredes-Sabja *et al.*, 2009b). Previously, the same authors (Paredes-Sabja *et al.*, 2008b) studied the role of KCl as the most effective germinant in *C. perfringens*. The K^+ ions had a strong power, contributing to germination in strains of *C. perfringens* responsible of food poisoning. The response was different depending on the localization of the chromosomal enterotoxin *cpe* gene or plasmid-borne *cpe* gene. It has been hypothesized that the spore of food poisoning strains isolated from meat during ripening have a system to sense the ions that trigger the germination. This mechanism is not present in spores from non-food-poisoning strains. Differently from other germinant ions, the sodium phosphate requires all four Ger receptors (GerKA-KC, GerAA, GerKB) as well as the putative antiporters GerO and GerQ to have a normal germination. The receptor GerKA-KC is also involved in triggering germination in the presence of KCl and L-asparagine (Paredes-Sabja *et al.*, 2008a, 2009b).

The ion need for spore germination depends also on different germinants; in *B. cereus*, sodium ions are considered stimulating factors for inosine, while potassium ions act as strong inhibitors (Clements and Moir, 1998). The rapid release of calcium dipicolinate was studied in *Clostridium tyrobutyricum* by using X-ray microanalysis (Bassi *et al.*, 2009). The molecular mechanisms that regulate the germination of this clostridial species is yet unknown. Similarly to the data and observation in *C. perfringens*, combinations of ions and amino acids were more effective at triggering germination than a single germinant. Among all the conditions tested, the greatest drop in OD_{625} (57.4%) was obtained when spores were incubated in L-alanine/L-lactate $NaHCO_3$ buffer, pH 4.6. It was surprising that neither of these two compounds, when used alone in solution, caused a sharp decline in OD_{625}. Moreover, significant differences were obtained in germination measurements when phosphate, Na^+ or the amino acid alanine added to lactate were assayed in single to trigger germination. The greatest release of Ca-DPA was reached when germinants such as alanine or lactate were used in solution together with ions, $NaHCO_3$. In *C. sordelli* the $NaHCO_3$ has not a positive effect to stimulate germination when used alone as germinant and, similarly, germination was not triggered by using defined germination medium without $NaHCO_3$. The germination was induced normally when the experiment is carried out with either solution of a defined germination medium and $NaHCO_3$ or by replacing it with $KHCO_3$ (Ramirez and Abel-Santos, 2010).

References

Ablett, S., Darke, A., Lillford, P., and Martin, D. (1999). Glass formation and dormancy in bacterial spores. Int. J. Food Sci. Technol. *34*, 59–69.

Alderton, G., and Snell, N. (1963). Base exchange and heat resistance in bacterial spores. Biochem. Biophys. Res. Commun. *10*, 139–143.

Algie, J.E. (1984). Effect of internal water activity of bacterial spores on their heat resistance in water. Curr. Microbiol. *11*, 293–296.

Algie, J.E., and Watt, I.C. (1984). Calculation of mass and water content between the core, cortex and coat of *B. stearothermophilus* spores. Curr. Microbiol. *10*, 249–254.

Amaha, M., and Ordal, Z.J. (1957). Effect of divalent cations in the sporulation medium on the thermal death rate of *Bacillus coagulans* var. thermoacidurans. J. Bacteriol *74*, 596–604.

Amaha, M., Ordal, Z.J., and Touba, A. (1956). Sporulation requirements of *Bacillus coagulans* var. thermoacidurans in complex media. J. Bacteriol. *72*, 34–41.

Aoki, H., and Slepecky, R.A. (1973). Inducement of a heat-shock requirement for germination and production of increased heat resistance in *Bacillus fastidiosus* spores by manganous ions. J. Bacteriol. *114*, 137–143.

Atrih, A., and Foster, S.J. (2001). Analysis of the role of bacterial endospore cortex structure in resistance properties and demonstration of its conservation amongst species. J. Appl. Microbiol. *91*, 364–372.

Bassi, D., Cappa, F., and Cocconcelli, P.S. (2009). A combination of a SEM technique and X-ray microanalysis for studying the spore germination process of *Clostridium tyrobutyricum*. Res. Microbiol. *160*, 322–329.

Beaman, T.C., and Gerhardt, P. (1986). Heat resistance of bacterial spores correlated with protoplast dehydration, mineralization, and thermal adaptation. Appl. Environ. Microbiol. *52*, 1242–1246.

Beaman, T.C., Greenamyre, J.T., Corner, T.R., Pankratz, H.S., and Gerhardt, P. (1982). Bacterial spore heat resistance correlated with water content, wet density, and protoplast/sporoplast volume ratio. J. Bacteriol *150*, 870–877.

Beaman, T.C., Koshikawa, T., Pankratz, H.S., and Gerhardt, P. (1984). Dehydration partitioned within core protoplast accounts for heat resistance of bacterial spores. FEMS Microbiol. Lett. *24*, 47–51.

Bender, G.R., and Marquis, R.E. (1985). Spore heat resistance and specific mineralization. Appl. Environ. Microbiol. *50*, 1414–1421.

Beveridge, T.J., and Murray, R.G. (1976). Uptake and retention of metals by cell walls of *Bacillus subtilis*. J. Bacteriol. *127*, 1502–1518.

Beveridge, T.J., and Murray, R.G. (1980). Sites of metal deposition in the cell wall of *Bacillus subtilis*. J. Bacteriol. *141*, 876–887.

Bhattacharjee, Y., and Enserink, M. (2008). Anthrax investigation. FBI discusses microbial forensics – but key questions remain unanswered. Science *321*, 1026–1027.

Black, S.H., and Gerhardt, P. (1961). Permeability of bacterial spores. J. Bacteriol. *83*, 960–967.

Black, S.H., and Gerhardt, P. (1962). Permeability of bacterial spores. IV. Water content, uptake, and distribution. J. Bacteriol. *83*, 960–967.

Black, S.H., Hashimoto, T., and Gerhardt, P. (1960). Calcium reversal of the heat susceptibility and dipicolinate deficiency of spores formed 'endotrophically' in water. Can. J. Microbiol. *6*, 213–224.

Bradbury, J.H., Foster, J.R., Hammer, B., Lindsay, J., and Murrell, W.G. (1981). The source of the heat resistance of bacterial spores. Study of water in spores by NMR. Biochim. Biophys. Acta. *678*, 157–164.

Brahmbhatt, T.N., Janes, B.K., Stibitz, E.S., Darnell, S.C., Sanz, P., Rasmussen, S.B., and O'Brien, A.D. (2007). *Bacillus anthracis* exosporium protein BclA affects spore germination, interaction with extracellular matrix proteins, and hydrophobicity. Infect. Immun. *75*, 5233–5239.

Buitink, J., and Leprince, O. (2008). Intracellular glasses and seed survival in the dry state. C. R. Biol. *331*, 788–795.

Carstensen, E.L., Marquis, R.E., Child, S.Z., and Bender, G.R. (1979). Dielectric properties of native and decoated spores of *Bacillus megaterium*. J. Bacteriol. *140*, 917–928.

Cazemier, A.E., Wagenaars, S.F., and ter Steeg, P.F. (2001). Effect of sporulation and recovery medium on the heat resistance and amount of injury of spores from spoilage bacilli. J. Appl. Microbiol. *90*, 761–770.

Chada, V.G., Sanstad, E.A., Wang, R., and Driks, A. (2003). Morphogenesis of bacillus spore surfaces. J. Bacteriol. *185*, 6255–6261.

Charney, J., Fisher, W.P., and Hegarty, C.P. (1951). Managanese as an essential element for sporulation in the genus *Bacillus*. J. Bacteriol. *62*, 145–148.

Clements, M.O., and Moir, A. (1998). Role of the gerI operon of *Bacillus cereus* 569 in the response of spores to germinants. J. Bacteriol. *180*, 6729–6735.

Cortezzo, D.E., and Setlow, P. (2005). Analysis of factors that influence the sensitivity of spores of *Bacillus subtilis* to DNA damaging chemicals. J. Appl. Microbiol. *98*, 606–617.

Cortezzo, D.E., Koziol-Dube, K., Setlow, B., and Setlow, P. (2004). Treatment with oxidizing agents damages the inner membrane of spores of *Bacillus subtilis* and sensitizes spores to subsequent stress. J. Appl. Microbiol. *97*, 838–852.

Cowan, A.E., Koppel, D.E., Setlow, B., and Setlow, P. (2003). A soluble protein is immobile in dormant spores of *Bacillus subtilis* but is mobile in germinated spores: implications for spore dormancy. Proc. Natl. Acad. Sci. U.S.A. *100*, 4209–4214.

Cowan, A.E., Olivastro, E.M., Koppel, D.E., Loshon, C.A., Setlow, B., and Setlow, P. (2004). Lipids in the inner membrane of dormant spores of *Bacillus* species are largely immobile. Proc. Natl. Acad. Sci. U.S.A. *101*, 7733–7738.

Crafts-Lighty, A., and Ellar, D.J. (1980). The structure and function of the spore outer membrane in dormant and germinating spores of *Bacillus megaterium*. J. Appl. Bacteriol. *48*, 135–145.

Crowe, J.H., Carpenter, J.F., and Crowe, L.M. (1998). The role of vitrification in anhydrobiosis. Annu. Rev. Physiol. *60*, 73–103.

Curran, H.R., and Evans, F.R. (1954). The influence of iron or manganese upon the formation of spores by mesophilic aerobes in fluid organic media. J. Bacteriol. *67*, 489–497.

Curran, H.R., Brunstetter, B.C., and Myers, A.T. (1943). Spectrochemical analysis of vegetative cells and spores of bacteria. J. Bacteriol. *45*, 485–494.

Dawes, E.W., and H.0. Halvorson. (1972). Membrane synthesis during outgrowth of bacterial spores. In Spores V, Halvorson, H.O., Hanson, R., and Campbel, L.L., eds. (ASM Press, Washington, D.C.), pp. 449–455.

De Pieri, L.A., Ludlow, I.K., and Waites, W.M. (1993). The application of laser diffractometry to study the water content of spores of *Bacillus sphaericus* with different heat resistances. J. Appl. Bacteriol. *74*, 578–582.

De Vries, Y.P. (2004). The role of calcium in bacterial spore germination. Microbes Environ. *19*, 199–202.

Denisov, V.P., and Halle, B. (1995). Direct observation of calcium-coordinated water in calbindin D_{9k} by nuclear magnetic relaxation dispersion. Biochemistry *117*, 8456–8465.

Doyle, R.J., Matthews, T.H., and Streips, U.N. (1980). Chemical basis for selectivity of metal ions by the *Bacillus subtilis* cell wall. J. Bacteriol. *143*, 471–480.

Driks, A. (1999). *Bacillus subtilis* spore coat. Microbiol. Mol. Biol. Rev. *63*, 1–20.

Driks, A. (2002). Proteins of the spore core and coat. In *Bacillus subtilis* and its Closest Relatives: from Genes to Cells, Sonenshein, A.L., Hoch, J.A., and Losick, R., eds. (ASM Press, Washington, D.C.), pp. 527–535.

Driks, A., and Setlow, P. (2000). Morphogenesis and properties of the bacterial spore. In Prokaryotic Development, Brun, Y.V., and Shinkets, L.J., eds. (ASM Press, Washington, D.C.) pp. 191–218.

Drymont, A. (1886). Einige Beobachtungen tVber die Milzbrandbacillen. Arch. Exptl. Pathol. Pharmakol. *21*, 309–317.

Eijlander, R.T., Abee, T., and Kuipers, O.P. (2010). Bacterial spores in food: how phenotypic variability complicates prediction of spore properties and bacterial behavior. Curr. Opin. Biotechnol. *22*, 180–186.

Eisenstadt, E. (1972). Potassium content during growth and sporulation in *Bacillus subtilis*. J. Bacteriol. *112*, 264–267.

Eisenstadt, E., and Silver, S. (1972a). Calcium transport during sporulation in *Bacillus subtilis*. In Spores V, Halvorson, H.O., Hanson, R., and Campbell, L.L., eds. (ASM Press, Washington, D.C.), pp. 180–186.

Eisenstadt, E., and Silver, S. (1972b). Restoration of cation transport systems as an early physiologically-active event in germination. In Spores V, Halvorson, H.O., Hanson, R., and Campbell, L.L., eds. (ASM Press, Washington, D.C.), pp. 443–448.

Eisenstadt, E., Fisher, S., Der, C.L., and Silver, S. (1973). Manganese transport in *Bacillus subtilis* W23 during growth and sporulation. J. Bacteriol. *113*, 1363–1372.

El-Bisi, H.M., and Ordal, Z.J. (1956). The effect of certain sporulation conditions on the thermal death rate of *Bacillus coagulans* var. thermoacidurans. J. Bacteriol. *71*, 1–9.

Ellar, D.J. (1978). Spore specific structures and their function. Symp. Soc. Gen. Microbiol. *28*, 295–325.

Fischer, A. (1897). Vorlesungen uber Bakterien (Fischer, Jena).

Fleming, H.P., and Ordal, Z.J. (1964). Responses of *Bacillus subtilis* spores to ionic environments during sporulation and germination. J. Bacteriol. *88*, 1529–1537.

Foerster, H.F., and Foster, J.W. (1966). Response of *Bacillus* spores to combinations of germinative compounds. J. Bacteriol. *91*, 1168 1177.

Friedman, C.A., and Henry, B.S. (1938). Bound water content of vegetative and spore forms of bacteria. J. Bacteriol. *36*, 99–105.

Fujisawa, M., Ito, M., and Krulwich, T.A. (2007). Three two-component transporters with channel-like properties have monovalent cation/proton antiport activity. Proc. Natl. Acad. Sci. U.S.A. *104*, 13289–13294.

Gerhardt, P. (1967). Cytology of *Bacillus anthracis*. Fed. Proc. *26*, 1504–1517.

Gerhardt, P., and Black, S.H. (1961). Permeability of bacterial spores. II. Molecular variables affecting solute permeation. J. Bacteriol. *82*, 750–760.

Gerhardt, P., and Marquis, R.E. (1989). Spore thermoresistance mechanisms. In Regulation of Prokaryotic Development, Smith, I., Slepecky, R.A., and Setlow, P., eds. (ASM Press, Washington, DC), pp. 43–63.

Gerhardt, P., and Ribi, E. (1964). Ultrastructure of the exosporium enveloping spores of *Bacillus cereus*. J. Bacteriol. *88*, 1774–1789.

Gerhardt, P., Scherrer, R., and Black, S.H. (1971). Molecular sieving by dormant spore structures. In *Spores V*, Halvorson, H.O., Hanson, R.S., and Campbell, L.L., eds. (ASM Press, Washington, DC), pp. 68–74.

Ghosal, S., Leighton, T.J., Wheeler, K.E., Hutcheon, I.D., and Weber, P.K. (2010). Spatially resolved characterization of water and ion incorporation in *Bacillus* spores. Appl. Environ. Microbiol. *76*, 3275–3282.

Ghosh, S., Setlow, B., Wahome, P.G., Cowan, A.E., Plomp, M., Malkin, A.J., and Setlow, P. (2008). Characterization of spores of *Bacillus subtilis* that lack most coat layers. J. Bacteriol. *190*, 6741–6748.

Gould, G.W. (1969). Germination. In The Bacterial Spore, Gould, G.W., and Hurst, A., eds. (Academic Press, London) pp. 397–444.

Gould, G.W. (1977). Recent advances in the understanding of resistance and dormancy in bacterial spores. J. Appl. Bacteriol. *42*, 297–309.

Gould, G.W. (1986). Water and the survival of bacterial spores. In Membranes, Metabolism, and Dry Organisms, Leopold, A.C., ed. (Cornell University Press, Ithaca, N.Y.), pp. 143–156.

Gould, G.W. (2006). History of science – spores. J. Appl. Microbiol. *101*, 507–513.

Gould, G.W., and Dring, G.J. (1974). Mechanisms of spore heat resistance. Adv. Microbial Physiol. *11*, 137–164.

Gould, G.W., and Dring, G.J. (1975a). Heat resistance of bacterial endospores and concept of an expanded osmoregulatory cortex. Nature *258*, 402–405.

Gould, G.W., and Dring, G.J. (1975b). Role of expanded cortex in resistance of bacterial endospores. In Spores IV, Gerhardt, P., Sadoff, H.L., Costilow, R.N., eds. (ASM Press, Washington, DC), pp. 541–546.

Granger, A.C., Gaidamakova, E.K., Matrosova, V.Y., Daly, M.J., and Setlow, P. (2011). Effects of Mn and Fe levels on *Bacillus subtilis* spore resistance and effects of Mn2+, other divalent cations, orthophosphate, and dipicolinic acid on protein resistance to ionizing radiation. Appl. Environ. Microbiol. *77*, 32–40.

Green, J.H., and Sadoff, H.L. (1965). Comparision of soluble reduced nicotinamide adenine dinucleotide oxidases from cells and spores of *Clostridium botulinum*. J. Bacteriol. *89*, 1499–1505.

Greene, R.A., and Slepecky, R.A. (1972). Minimal requirements for commitment to sporulation in *Bacillus megaterium*. J. Bacteriol. *111*, 557–565.

Grelet, N. (1952). Le determinisme de la sporulation de *Bacillus megatherium*. II. L'effect de la penurie de constituants minoraux du milieu synthetique. Ann. Inst. Pasteur. *82*, 66–77.

Gregory, R.B. (1995). Protein hydration and glass transition behavior. In Protein-Solvent Interactions, Gregory, R.B., ed. (Dekker, New York), pp. 191–264.

Grimshaw, C.E., Huang, S., Hanstein, C.G., Strauch, M.A., Burbulys, D., Wang, L., Hoch, J.A., and Whiteley, J.M. (1998). Synergistic kinetic interactions between components of the phosphorelay controlling sporulation in *Bacillus subtilis*. Biochemistry *37*, 1365–1375.

Guidi-Rontani, C., Levy, M., Ohayon, H., and Mock, M. (2001). Fate of germinated *Bacillus anthracis* spores in primary murine macrophages. Mol. Microbiol. *42*, 931–938.

Henriques, A.O., and Moran, C.P., Jr. (2007). Structure, assembly, and function of the spore surface layers. Annu. Rev. Microbiol. *61*, 555–588.

Henriques, A.O., Costa, T., Martins, L.O., and Zilhão, R. (2004). Functional architecture and assembly of the spore coat. In Bacterial Spore Formers: Probiotics and Emerging Applications, Ricca, E., Henriques, A.O., and Cutting, S.M., eds. (Horizon Scientific Press, London, UK), pp. 34–52.

Henriques, A.O., and Moran, C.P., Jr. (2000). Structure and assembly of the bacterial endospore coat. Methods *20*, 95–110.

Hintze, P.E., and Nicholson, W.L. (2010). Single-spore elemental analyses indicate that dipicolinic acid-deficient *Bacillus subtilis* spores fail to accumulate calcium. Arch. Microbiol. *192*, 493–497.

Hirota, R., Hata, Y., Ikeda, T., Ishida, T., and Kuroda, A. (2010). The silicon layer supports acid resistance of *Bacillus cereus* spores. J. Bacteriol. *192*, 111–116.

Hoch, J.A. (1993). Regulation of the phosphorelay and the initiation of sporulation in *Bacillus subtilis*. Annu. Rev. Microbiol. *47*, 441–465.

Hyatt, M.T., and Levinson, H.S. (1961). Interaction of heat, glucose, L-alanine, and potassium nitrate in spore germination of *Bacillus megaterium*. J. Bacteriol. *81*, 204–211.

Inaoka, T., Matsumura, Y., and Tsuchido, T. (1999). SodA and manganese are essential for resistance to oxidative stress in growing and sporulating cells of *Bacillus subtilis*. J. Bacteriol. *181*, 1939–1943.

Ishihara, Y., Takano, J., Mashimo, S., and Yamamura, M. (1994). Determination of the water necessary for survival of *Bacillus subtilis* vegetative cells and spores. Thermochim. Acta *235*, 153–160.

Ivanova, N., Sorokin, A., Anderson, I., Galleron, N., Candelon, B., Kapatral, V., Bhattacharyya, A., Reznik, G., Mikhailova, N., Lapidus, A., *et al.* (2003). Genome sequence of *Bacillus cereus* and comparative analysis with *Bacillus anthracis*. Nature *423*, 87–91.

Jakubovics, N.S., and Jenkinson, H.F. (2001). Out of the iron age: new insights into the critical role of manganese homeostasis in bacteria. Microbiology *147*, 1709–1718.

Johnstone, K., Ellar, D.J., and Appleton, T.C. (1980). Location of metal ions in bacillus megaterium spores by high-resolution electron probe x-ray microanalysis. FEMS Microbiol. Lett. *7*, 97–101.

Keynan, A. (1978). Spore structure and its relations to resistance, dormancy and germination. In Spores VII, Chambliss, G., and Vary, J.C., eds. (ASM Press, Washington, DC), pp. 43–53.

Kihm, D.J., Hutton, M.T., Hanlin, J.H., and Johnson, E.A. (1990). Influence of transition metals added during sporulation on heat resistance of *Clostridium botulinum* 113B spores. Appl. Environ. Microbiol. *56*, 681–685.

Kim, H., Hahn, M., Grabowski, P., McPherson, D.C., Otte, M.M., Wang, R., Ferguson, C.C., Eichenberger, P., and Driks, A. (2006). The *Bacillus subtilis* spore coat protein interaction network. Mol. Microbiol. *59*, 487–502.

King, N., Dreesen, O., Stragier, P., Pogliano, K., and Losick, R. (1999). Septation, dephosphorylation, and the activation of sigmaF during sporulation in *Bacillus subtilis*. Genes Dev. *13*, 1156–1167.

Klibanov, A.M., and Schefiliti, J.A. (2004). On the relationship between conformation and stability in solid pharmaceutical protein formulations. Biotechnol. Lett. *26*, 1103–1106.

Klobutcher, L.A., Ragkousi, K., and Setlow, P. (2006). The *Bacillus subtilis* spore coat provides 'eat resistance' during phagocytic predation by the protozoan Tetrahymena thermophila. Proc. Natl. Acad. Sci. U.S.A. *103*, 165–170.

Kolodziej, B.J., and Slepecky, R.A. (1962). A copper requirement for the sporulation of *Bacillus megaterium*. Nature *194*, 504–505.

Kolodziej, B.J., and Slepecky, R.A. (1964). Trace metal requirements for sporulation of *Bacillus megaterium*. J. Bacteriol. *88*, 821–830.

Koshikawa, T., Yamazaki, M., Yoshimi, M., Ogawa, S., Yamada, A., Watabe, K., and Torii, M. (1989). Surface hydrophobicity of spores of *Bacillus* spp. J. Gen. Microbiol. *135*, 2717–2722.

Kozuka, S., and Tochikubo, K. (1985). Properties and origin of filamentous appendages on spores of *Bacillus cereus*. Microbiol. Immunol. *29*, 21–37.

Kroos, L., Zhang, B., Ichikawa, H., and Yu, Y.T. (1999). Control of sigma factor activity during *Bacillus subtilis* sporulation. Mol. Microbiol. *31*, 1285–1294.

Krueger, W.B., and Kolodziej, B.J. (1978). Divalent cation mobility throughout exponential growth and sporulation of *Bacillus megaterium*. Microbios *18*, 159–167.

Kuhn, N.J., Setlow, B., Setlow, P., Cammack, R., and Williams, R. (1995). Cooperative manganese (II) activation of 3-phosphoglycerate mutase of *Bacillus megaterium*: a biological pH-sensing mechanism in bacterial spore formation and germination. Arch. Biochem. Biophys. *320*, 35–42.

Lai, E.M., Phadke, N.D., Kachman, M.T., Giorno, R., Vazquez, S., Vazquez, J.A., Maddock, J.R., and Driks, A. (2003). Proteomic analysis of the spore coats of *Bacillus subtilis* and *Bacillus anthracis*. J. Bacteriol. *185*, 1443–1454.

Leanz, G., and Gilvarg, C. (1973). Dipicolinic acid location in intact spores of *Bacillus megaterium*. J. Bacteriol. *114*, 455–456.

Lee, K.Y., Juang, T.C., and Lee, K.C. (1978). Effect of metal ions on growth and sporulation of *Clostridium perfringens* in a synthetic medium. Zhonghua Min Guo Wei Sheng Wu Xue Za Zhi *11*, 50–61.

Leman, A. (1973). Interference microscopical determination of bacterial dry weight during germination and sporulation. Jena Rev. *5*, 263–270.

Leuschner, R.G., and Lillford, P.J. (2000). Effects of hydration on molecular mobility in phase-bright *Bacillus subtilis* spores. Microbiology *146* , 49–55.

Levinson, H.S., and Hyatt, M.T. (1964). Effect of sporulation medium on heat resistance, chemical composition, and germination of *Bacillus megaterium* spores. J. Bacteriol. *87*, 876–886.

Levinson, H.S., and Sevag, M.G. (1953). Stimulation of germination and respiration of the spores of *Bacillus megaterium* by manganese and monovalent anions. J. Gen. Physiol. *36*, 617–629.

Lewis, J., Snell, N.S., and Alderton, G. (1965). Dormancy and activation of bacterial spores. In Spores III, Campbell, L.L., and Halvorson, H.O., eds. (ASM Press, Washington, DC).

Lewis, J.C., Snell, N.S., and Burr, H.K. (1960). Water permeability of bacterial spores and the concept of a contractile cortex. Science *132*, 544–545.

Lindsay, J.A., Beaman, T.C., and Gerhardt, P. (1985). Protoplast water content of bacterial spores determined by buoyant density sedimentation. J. Bacteriol. *163*, 735–737.

Loeb, L.A., James, E.A., Waltersdorph, A.M., and Klebanoff, S.J. (1988). Mutagenesis by the autoxidation of iron with isolated DNA. Proc. Natl. Acad. Sci. U.S.A. *85*, 3918–3922.

Loshon, C.A., and Setlow, P. (1993). Levels of small molecules in dormant spores of Sporosarcina species and comparison with levels in spores of *Bacillus* and *Clostridium* species. Can. J. Microbiol. *39*, 259–262.

Lubin, M. (1964). Intracellular potassium and control of protein synthesis. Fed. Proc. *23*, 994–1001.

McElroy, W.D., and Glass, B. (1950). Copper Metabolism. A Symposium on Animal, Plant and Soil Relationships (Johns Hopkins Press, Baltimore).

Maeda, Y., Fujita, T., Sugiura, Y., and Koga, S. (1968). Physical properties of water in spores of *Bacillus megaterium*. J. Gen. Appl. Microbiol. *14*, 217–226.

Mafart, P., Leguerinel, I., Couvert, O., and Coroller, L. (2010). Quantification of spore resistance for assessment and optimization of heating processes: a never-ending story. Food Microbiol. *27*, 568–572.

Magge, A., Setlow, B., Cowan, A.E., and Setlow, P. (2009). Analysis of dye binding by and membrane potential in spores of *Bacillus* species. J. Appl. Microbiol. *106*, 814–824.

Mallidis, C.G., and Scholefield, J. (1987). Relation of the heat resistance of bacterial spores to chemical composition and structure. II. Relation to cortex and structure. J. Appl. Bacteriol. *63*, 207–215.

Marquis, R.E. (1989). Minerals and bacterial spores. In Metal Ions and Bacteria, Beveridge, T.J., and Doyle, R.J., eds. (Wiley, New York), pp. 247–274.

Marquis, R.E., and Bender, G.R. (1985). Mineralization and heat resistance of bacterial spores. J. Bacteriol. *161*, 789–791.

Marquis, R.E., and Shin, S.Y. (1994). Mineralization and responses of bacterial spores to heat and oxidative agents. FEMS Microbiol. Rev. *14*, 375–379.

Marshall, B.J., and Murrell, W.J. (1970). Symposium on bacterial spores: IX. Biophysical analysis of the spore. J. Appl. Bacteriol. 33, 103–129.

Mato Rodriguez, L., and Alatossava, T. (2010). Effects of copper on germination, growth and sporulation of *Clostridium tyrobutyricum*. Food Microbiol. *27*, 434–437.

Matsumoto, G. (2003). Bioterrorism. Anthrax powder: state of the art? Science *302*, 1492–1497.

Matsushita, T., Ueda, T., and Kusaka, I. (1986). Purification and characterization of Ca2+/H+ antiporter from *Bacillus subtilis*. Eur. J. Biochem. *156*, 95–100.

Mattea, C., Qvist, J., and Halle, B. (2008). Dynamics at the protein–water interface from 17O spin relaxation in deeply supercooled solutions. Biophys. J. 95, 2951–2963.

Meador-Parton, J., and Popham, D.L. (2000). Structural analysis of *Bacillus subtilis* spore peptidoglycan during sporulation. J. Bacteriol. *182*, 4491–4499.

Mitchell, C., Skomurski, J.F., and Vary, J.C. (1986). Effect of ion channel blockers on germination of

Bacillus megaterium spores. FEMS Microbiol. Lett. *34*, 211–214.

Moeller, R., Setlow, P., Horneck, G., Berger, T., Reitz, G., Rettberg, P., Doherty, A.J., Okayasu, R., and Nicholson, W.L. (2008). Roles of the major, small, acid-soluble spore proteins and spore-specific and universal DNA repair mechanisms in resistance of *Bacillus subtilis* spores to ionizing radiation from X rays and high-energy charged-particle bombardment. J. Bacteriol. *190*, 1134–1140.

Moir, A., and Smith, D.A. (1990). The genetics of bacterial spore germination. Annu. Rev. Microbiol. *44*, 531–553.

Moir, A., Corfe, B.M., and Behravan, J. (2002). Spore germination. Cell Mol. Life Sci. *59*, 403–409.

Murrell, W.G. (1961). Spore formation and germination as a microbial reaction to the environment. In Microbial Reaction to Environment: Symp. Soc. Gen. Microbiol. *11*, 100.

Murrell, W.G. (1967). The biochemistry of the bacterial endospore. In Advances in Microbial Physiology, Rose, A.H., and Wilkinson, J.F., eds. (Academic Press, New York), pp. 133–251.

Murrell, W.G. (1969). Chemical composition of spores and spore structures. In The Bacterial Spore, Gould, G.W., and Hurst, A., eds. (Academic Press, New York), pp. 215–273.

Murrell, W.G., and Scott, W.J. (1958). The permeability of bacterial spores to water. Paper presented at 7th Int. Congr. Microbiol. (Stockholm)

Nakashio, S., and Gerhardt, P. (1985). Protoplast dehydration correlated with heat resistance of bacterial spores. J. Bacteriol. *162*, 571–578.

Neihof, R., Thompson, J.K., and Deitz, V.R. (1967). Sorption of water vapour and nitrogen gas by bacterial spores. Nature *216*, 1304–1306.

Newsome, R. (2003). Dormant microbes: research needs. Foodtechnology *57*, 38–42.

Nicholson, W.L., Munakata, N., Horneck, G., Melosh, H.J., and Setlow, P. (2000). Resistance of *Bacillus* endospores to extreme terrestrial and extraterrestrial environments. Microbiol. Mol. Biol. Rev. *64*, 548–572.

Norris, V., Grant, S., Freestone, P., Canvin, J., Sheikh, F.N., Toth, I., Trinei, M., Modha, K., and Norman, R.I. (1996). Calcium signalling in bacteria. J. Bacteriol. *178*, 3677–3682.

Oomes, S.J.C.M., Jonker, M.J., Wittink, F.R.A., Hehenkamp, J.O., Breit, T.M., and Brul, S. (2009). The effect of calcium on the transcriptome of sporulating *B. subtilis* cells. Int. J. Food Microbiol. *133*, 234–242.

Paidhungat, M., and Setlow, P. (2002). Genes to Cells. In *Bacillus subtilis* and Its Relatives, Hoch, J.A., Losick, R., Sonenshein, A., eds. (ASM Press, Washington, DC), pp. 537–548.

Paidhungat, M., Ragkousi, K., and Setlow, P. (2001). Genetic requirements for induction of germination of spores of *Bacillus subtilis* by Ca(2+)-dipicolinate. J. Bacteriol. *183*, 4886–4893.

Paredes-Sabja, D., Bond, C., Carman, R.J., Setlow, P., and Sarker, M.R. (2008). Germination of spores of *Clostridium difficile* strains, including isolates from a hospital outbreak of *Clostridium difficile*-associated disease (CDAD). Microbiology *154*, 2241–2250.

Paredes-Sabja, D., Torres, J.A., Setlow, P., and Sarker, M.R. (2008a). *Clostridium perfringens* spore germination: characterization of germinants and their receptors. J. Bacteriol. *190*, 1190–1201.

Paredes-Sabja, D., Sarker, N., Setlow, B., Setlow, P., and Sarker, M.R. (2008b). Roles of DacB and spm proteins in clostridium perfringens spore resistance to moist heat, chemicals, and UV radiation. Appl. Environ. Microbiol. *74*, 3730–3738.

Paredes-Sabja, D., Setlow, P., and Sarker, M.R. (2009a). GerO, a putative Na^+/H^+–K^+ antiporter, is essential for normal germination of spores of the pathogenic bacterium *Clostridium perfringens*. J. Bacteriol. *191*, 3822–3831.

Paredes-Sabja, D., Udompijitkul, P., and Sarker, M.R. (2009b). Inorganic phosphate and sodium ions are cogerminants for spores of *Clostridium perfringens* type A food poisoning-related isolates. Appl. Environ. Microbiol. *75*, 6299–6305.

Plomp, M., Leighton, T.J., Wheeler, K.E., and Malkin, A.J. (2005). The high-resolution architecture and structural dynamics of *Bacillus* spores. Biophys. J. *88*, 603–608.

Popham, D.L. (2002). Specialized peptidoglycan of the bacterial endospore: the inner wall of the lockbox. Cell Mol. Life Sci. *59*, 426–433.

Popham, D.L., Helin, J., Costello, C.E., and Setlow, P. (1996). Muramic lactam in peptidoglycan of *Bacillus subtilis* spores is required for spore outgrowth but not for spore dehydration or heat resistance. Proc. Natl. Acad. Sci. U.S.A. *93*, 15405–15410.

Powell, J.F. (1957). Effect of manganese on sporulation of laboratory strains of *Bacillus cereus*, *Bacillus subtilis*, and *Bacillus megaterium*. In Spores V, Halvorson, H.O., ed. (American Institute of Biological Sciences, Washington, DC), pp. 163.

Powell, J.F., and Hunter, J.R. (1955). The sporulation of *Bacillus sphaericus* stimulated by association with other bacteria: an effect of carbon dioxide. J. Gen. Microbiol. *13*, 54–58.

Powell, J.F., and Strange, R.E. (1956). Biochemical changes occurring during sporulation in *Bacillus* species. Biochem. J. *63*, 661–668.

Que, Q., and Helmann, J.D. (2000). Manganese homeostasis in *Bacillus subtilis* is regulated by MntR, a bifunctional regulator related to the diphtheria toxin repressor family of proteins. Mol. Microbiol. *35*, 1454–1468.

Radnedge, L., Agron, P.G., Hill, K.K., Jackson, P.J., Ticknor, L.O., Keim, P., and Andersen, G.L. (2003). Genome differences that distinguish *Bacillus anthracis* from *Bacillus cereus* and *Bacillus thuringiensis*. Appl. Environ. Microbiol. *69*, 2755–2764.

Raeymaekers, L., Wuytack, E., Willems, I., Michiels, C.W., and Wuytack, F. (2002). Expression of a P-type Ca(2+)-transport ATPase in *Bacillus subtilis* during sporulation. Cell Calcium *32*, 93.

Ramirez, N., and Abel-Santos, E. (2010). Requirements for germination of *Clostridium sordellii* spores *in vitro*. J. Bacteriol. *192*, 418–425.

Read, T.D., Peterson, S.N., Tourasse, N., Baillie, L.W., Paulsen, I.T., Nelson, K.E., Tettelin, H., Fouts, D.E., Eisen, J.A., Gill, S.R., *et al.* (2003). The genome sequence of *Bacillus anthracis* Ames and comparison to closely related bacteria. Nature *423*, 81–86.

Roberts, T.A., and Hitchins, A.D. (1969). Resistance of spores. In The Bacterial Spore, Gould, G.W., and Hurst, A., eds. (Academic Press Inc., New York), pp. 611–670.

Rode, L.J., and Foster, J.W. (1962). Ions and the Germination of Spores of *Bacillus cereus* T. Nature *194*, 1300–1301.

Ross, C., and Abel-Santos, E. (2010). The Ger receptor family from sporulating bacteria. Curr. Issues Mol. Biol. *12*, 147–158.

Ross, K.F., and Billing, E. (1957). The water and solid content of living bacterial spores and vegetative cells as indicated by refractive index measurements. J. Gen. Microbiol. *16*, 418–425.

Rouf, M.A. (1964). Spectrochemical analysis of inorganic elements in bacteria. J. Bacteriol. *88*, 1545–1549.

Rubel, G.O. (1997). A non-intrusive method for the measurement of water vapour sorption by bacterial spores. J. Appl. Microbiol. *83*, 243–247.

Rupley, J.A., and Careri, G. (1991). Protein hydration and function. In Advances in Protein Chemistry, Anfinsen, C.B., Richards, F.M., Edsall, J.T., and Eisenberg, D.S., eds. (Academic Press, London), pp. 37–172.

Sapru, V., and Labuza, T.P. (1993). Glassy state in bacterial spores predicted by polymer glass-transition theory. J. Food Sci. *58*, 445–448.

Scherrer, R., and Gerhardt, P. (1972). Location of calcium within *Bacillus* spores by electron probe x-ray microanalysis. J. Bacteriol. *112*, 559–568.

Scherrer, R., and Shull, V.E. (1987). Microincineration and elemental X-ray microanalysis of single *Bacillus cereus* T spores. Can. J. Microbiol. 33, 304–313.

Schroeter, R., Schlisio, S., Lucet, I., Yudkin, M., and Borriss, R. (1999). The *Bacillus subtilis* regulator protein SpoIIE shares functional and structural similarities with eukaryotic protein phosphatases 2C. FEMS Microbiol. Lett. *174*, 117–123.

Scribner, H., Eisenstadt, E., and Silver, S. (1974). Magnesium transport in *Bacillus subtilis* W23 during growth and sporulation. J. Bacteriol. *117*, 1224–1230.

Scully, M.O., Kattawar, G.W., Lucht, R.P., Opatrny, T., Pilloff, H., Rebane, A., Sokolov, A.V., and Zubairy, M.S. (2002). FAST CARS: engineering a laser spectroscopic technique for rapid identification of bacterial spores. Proc. Natl. Acad. Sci. U.S.A. *99*, 10994–11001.

Senior, A., and Moir, A. (2008). The *Bacillus cereus* GerN and GerT protein homologs have distinct roles in spore germination and outgrowth, respectively. J. Bacteriol. *190*, 6148–6152.

Setlow, B., and Setlow, P. (1980). Measurements of the pH within dormant and germinated bacterial spores. Proc. Natl. Acad. Sci. U.S.A. *77*, 2474–2476.

Setlow, B., McGinnis, K.A., Ragkousi, K., and Setlow, P. (2000). Effects of major spore-specific DNA binding proteins on *Bacillus subtilis* sporulation and spore properties. J. Bacteriol. *182*, 6906–6912.

Setlow, B., Melly, E., and Setlow, P. (2001). Properties of spores of *Bacillus subtilis* blocked at an intermediate stage in spore germination. J. Bacteriol. *183*, 4894–4899.

Setlow, P. (1994). Mechanisms which contribute to the long-term survival of spores of *Bacillus* species. Soc. Appl. Bacteriol. Symp. Ser. *23*, 49S-60S.

Setlow, P. (2003). Spore germination. Curr. Opin. Microbiol. *6*, 550–556.

Setlow, P. (2006). Spores of *Bacillus subtilis*: their resistance to and killing by radiation, heat and chemicals. J. Appl. Microbiol. *101*, 514–525.

Setlow, P., and Kornberg, A. (1970). Biochemical studies of bacterial sporulation and germination. XXII. Energy metabolism in early stages of germination of *Bacillus megaterium* spores. J. Biol. Chem. *245*, 3637–3644.

Seto-Young, D.L., and Ellar, D.J. (1981). Studies on calcium transport during growth and sporulation. Microbios *30*, 191–208.

Singh, R.P., and Setlow, P. (1979). Purification and properties of phosphoglycerate phosphomutase from spores and cells of *Bacillus megaterium*. J. Bacteriol. *137*, 1024–1027.

Skomurski, J.F., Racine, F.M., and Vary, J.C. (1983). Steady-state fluorescence anisotropy changes of 1,6-diphenyl-1,3,5,-hexatriene in membranes from *Bacillus megaterium* spores. Biochimi. Biophys. Acta (BBA) – Biomembranes *731*, 428–436.

Slepecky, R., and Foster, J.W. (1959). Alterations in metal content of spores of *Bacillus megaterium* and the effect on some spore properties. J. Bacteriol. *78*, 117–123.

Smith, R.J. (1995). Calcium and bacteria. In Advances in Microbial Physiology, Poole, R.K., ed. (Academic Press, London), pp. 83–133.

Sonenshein, A.L., Hoch, J.A., and Losick, R. (2001). *Bacillus subtilis* and its closest relatives: from genes to cells (ASM Press, Washington, DC).

Southworth, T.W., Guffanti, A.A., Moir, A., and Krulwich, T.A. (2001). GerN, an endospore germination protein of *Bacillus cereus*, is an Na(+)/H(+)-K(+) antiporter. J. Bacteriol. *183*, 5896–5903.

Steichen, C., Chen, P., Kearney, J.F., and Turnbough, C.L., Jr. (2003). Identification of the immunodominant protein and other proteins of the *Bacillus anthracis* exosporium. J. Bacteriol. *185*, 1903–1910.

Stewart, B.T., and Halvorson, H.O. (1953). Studies on the spores of aerobic bacteria. I. The occurrence of alanine racemase. J. Bacteriol. *65*, 160–166.

Stewart, G.S., Eaton, M.W., Johnstone, K., Barrett, M.D., and Ellar, D.J. (1980). An investigation of membrane fluidity changes during sporulation and germination of *Bacillus megaterium* K.M. measured by electron spin and nuclear magnetic resonance spectroscopy. Biochim. Biophys. Acta *600*, 270–290.

Stewart, M., Somlyo, A.P., Somlyo, A.V., Shuman, H., Lindsay, J.A., and Murrell, W.G. (1980). Distribution of calcium and other elements in cryosectioned *Bacillus cereus* T spores, determined by high-resolution scanning electron probe x-ray microanalysis. J. Bacteriol. *143*, 481–491.

Stewart, M., Somlyo, A.P., Somlyo, A.V., Shuman, H., Lindsay, J.A., and Murrell, W.G. (1981). Scanning electron probe x-ray microanalysis of elemental distributions in freeze-dried cryosections of *Bacillus coagulans* spores. J. Bacteriol. *147*, 670–674.

Stiles, W. (1961). Trace Elements in Plants (Cambridge University Press, London).

Sugiyama, H. (1951). Studies on factors affecting the heat resistance of spores of *Clostridium botulinum*. J. Bacteriol. *62*, 81–96.

Sunde, E.P., Setlow, P., Hederstedt, L., and Halle, B. (2009). The physical state of water in bacterial spores. Proc. Natl. Acad. Sci. U.S.A. *106*, 19334–19339.

Swerdlow, R.D., and Setlow, P. (1984). Isolation and characterization of two distinct fractions from the inner membrane of dormant *Bacillus megaterium* spores. J. Bacteriol. *158*, 9–15.

Swerdlow, B.M., Setlow, B., and Setlow, P. (1981). Levels of H+ and other monovalent cations in dormant and germinating spores of *Bacillus megaterium*. J. Bacteriol. *148*, 20–29.

Takubo, Y., Atarashi, M., Nishihara, T., and Kondo, M. (1988). Isolation and characterization of outermost layer deficient mutant spores of *Bacillus megaterium*. Microbiol. Immunol. *32*, 973–979.

Tani, K., Watanabe, T., Matsuda, H., Nasu, M., and Kondo, M. (1996). Cloning and sequencing of the spore germination gene of *Bacillus megaterium* ATCC 12872: similarities to the NaH-antiporter gene of Enterococcus hirae. Microbiol. Immunol. *40*, 99–105.

Tauveron, G., Slomianny, C., Henry, C., and Faille, C. (2006). Variability among *Bacillus cereus* strains in spore surface properties and influence on their ability to contaminate food surface equipment. Int. J. Food Microbiol. *110*, 254–262.

Thackray, P.D., Behravan, J., Southworth, T.W., and Moir, A. (2001). GerN, an antiporter homologue important in germination of *Bacillus cereus* endospores. J. Bacteriol. *183*, 476–482.

Thomas, R.S. (1964). Ultrastructural localization of mineral matter in bacterial spores by micronincineration. J. Cell Biol. *23*, 113–133.

Tipper, D.J., and Gauthier, J.J. (1972). In Spores V, Halvorson, H.O., Hanson, R., and Campbell, L.L., eds. (ASM Press, Washington, DC), pp. 3–12.

Troger, R. (1959). Uber den Metallnachweis in Quecksilber- oder kupferbehandelten Bakterien. Arch. Mikrobiol. *33*, 186–190.

Ulanowski, Z., Ludlow, I.K., and Waites, W.M. (1987). Water content and size of spore components determined by laser diffractometry. FEMS Microbiol. Lett. *40*, 229–232.

Underwood, E.J. (1959). Mineral metabolism. Annu. Rev. Biochem. *28*, 499–526.

Vaca Chavez, F., Hellstrand, E., and Halle, B. (2006). Hydrogen exchange and hydration dynamics in gelatin gels. J. Phys. Chem. B. *110*, 21551–21559.

Vanounou, S., Parola, A.H., and Fishov, I. (2003). Phosphatidylethanolamine and phosphatidylglycerol are segregated into different domains in bacterial membrane. A study with pyrene-labelled phospholipids. Mol. Microbiol. *49*, 1067–1079.

Vinter, V. (1956). Sporulation of bacilli. Consumption of calcium by the cells and decrease in the proteolytic activity of the medium during sporulation of *Bacillus megaterium*. Folia Biol. *2*, 216–224.

Vinter, V. (1957). The effect of cysteine upon spore formation by *Bacillus megateriuim*. J. Appl. Bacteriol. *20*, 325–332.

Virtanen, A.I., and Pulkki, L. (1933). Biochemische Untersuchungen ulber Bakteriensporen. Arch. Mikrobiol. *4*, 99–122.

Wakisaka, Y., Masaki, E., and Nishimoto, Y. (1982). Formation of crystalline delta-endotoxin or poly-beta-hydroxybutyric acid granules by asporogenous mutants of *Bacillus thuringiensis*. Appl. Environ. Microbiol. *43*, 1473–1480.

Waldham, D.G., and Halvorson, H.O. (1954). Studies on the relationship between equilibrium vapor pressure and moisture content of bacterial endospores. Appl. Microbiol. *2*, 333–338.

Warth, A.D. (1978). Molecular structure of the bacterial spore. In Advances in Microbial Physiology, Rose, A.H., and Morris, J.G., ed. (Academic Press, London), pp. 1–45.

Watt, I.C. (1981). Water vapor adsorption by *Bacillus stearothermophilus* endospores. In Sporulation and Germination, Levinson, H.S., Sonenshein, A.L., and Tipper, D.J., eds. (ASM Press, Washington, DC), pp. 253–255.

Weinberg, E.D. (1964). Manganese requirement for sporulation and other secondary biosynthetic processes of *Bacillus*. Appl. Microbiol. *12*, 436–441.

Westphal, A.J., Price, P.B., Leighton, T.J., and Wheeler, K.E. (2003). Kinetics of size changes of individual *Bacillus thuringiensis* spores in response to changes in relative humidity. Proc. Natl. Acad. Sci. U.S.A. *100*, 3461–3466.

Williams, R.J.P. (1981). The Bakerian Lecture. Natural selection of the chemical elements. Proc. R. Soc. Lond. B. Biol. Sci. *213*, 361–397.

Windle, J.J., and Sacks, L.E. (1963). Electron paramagnetic resonance of managanese(II) and copper(II) in spores. Biochim. Biophys. Acta *66*, 173–179.

Young, S.B., and Setlow, P. (2003). Mechanisms of killing of *Bacillus subtilis* spores by hypochlorite and chlorine dioxide. J. Appl. Microbiol. *95*, 54–67.

Young, S.B., and Setlow, P. (2004a). Mechanisms of *Bacillus subtilis* spore resistance to and killing by aqueous ozone. J. Appl. Microbiol. *96*, 1133–1142.

Young, S.B., and Setlow, P. (2004b). Mechanisms of killing of *Bacillus subtilis* spores by Decon and Oxone, two general decontaminants for biological agents. J. Appl. Microbiol. *96*, 289–301.

Zaman, M.S., Goyal, A., Dubey, G.P., Gupta, P.K., Chandra, H., Das, T.K., Ganguli, M., and Singh, Y. (2005). Imaging and analysis of *Bacillus anthracis* spore germination. Microsc. Res. Tech. *66*, 307–311.

Zhang, P., Setlow, P., and Li, Y. (2009). Characterization of single heat-activated *Bacillus* spores using laser tweezers Raman spectroscopy. Opt. Express *17*, 16480–16491.

The Spore as an Infectious Agent: Anthrax Disease as a Paradigm

11

James M. Vergis, Christy L. Ventura, Louise D. Teel and Alison D. O'Brien

Abstract

The zoonotic disease anthrax is caused by the inhalation, ingestion, or cutaneous inoculation of *Bacillus anthracis* spores into a susceptible host. While inhalational anthrax is rare in humans, the 2001 distribution of *B. anthracis* spores through the US mail led to an acceleration of research on disease progression in animal models, vaccines and therapeutics. After inoculation, spores are rapidly phagocytosed by macrophages and dendritic cells where they germinate into vegetative bacilli that produce toxins and capsule, both of which facilitate dissemination in the host. The vegetative bacilli are susceptible to antibiotics, the only licensed post-exposure treatment against *B. anthracis*. Despite the fact that the spore is the infectious particle, the currently approved vaccines for humans in the US and UK are derived from the vegetative culture filtrate of a toxigenic, non-encapsulated *B. anthracis* strain; the vaccines primarily contain protective antigen (PA). Here we describe the role of the spore in anthrax disease progression. In addition, we discuss PA-based vaccines and therapeutics, as well as historical applications of spore-based vaccines. Finally, we present current research focused on spore antigen vaccines and germination modulators.

Introduction: spores and pathogenesis

Numerous highly pathogenic Gram-positive organisms produce infective spores that by nature are exceedingly difficult to rid from the environment. When the spores are introduced into a permissive environment within a susceptible host, they germinate into vegetative cells that are responsible for the clinical symptomatology that is associated with disease. A shared attribute of many spore-producing bacteria is the capacity to express toxins that, when elaborated as a result of vegetative outgrowth of the germinated spores, cause extensive damage to the host. Of particular medical concern are the anaerobic *Clostridium* and aerobic *Bacillus* species.

Disease caused by clostridial species

Once *Clostridium botulinum* spores germinate in food stuffs or in the intestines of susceptible hosts, particularly neonates, the vegetative cells elaborate lethal botulinum toxin that is responsible for flaccid paralysis (Quintaliani and Quintiliani, 2008). In contrast, *Clostridium tetani* spores that contaminate wounds germinate and release tetanus toxin that is responsible for spastic paralysis, or 'lockjaw' (Quintaliani and Quintiliani, 2008). *Clostridium difficile* spores are prevalent in hospitals and increasingly in the community because antibiotic use has selected for the proliferation of this ubiquitous soil and zoonotic agent (Quintaliani and Quintiliani, 2008). Once the gastrointestinal flora has been perturbed by antibiotics such that the environment becomes more favourable, *C. difficile* spores germinate and grow vegetatively in the gut. The vegetative clostridia produce cytotoxins A and B that damage the gut epithelium and lead to the development of potentially lethal pseudomembranous colitis (Quintaliani and Quintiliani, 2008). Spores shed from such patients are resistant to decontamination with typical cleaners and alcohol-based hand sanitizers; thus, the spores become firmly established as a nosocomial hazard in the hospital environment (reviewed in Dubberke *et al.*, 2011). In this review, we will focus on spores of *B. anthracis* and the pathogenesis and prevention of anthrax disease.

Anthrax disease

Anthrax is a zoonotic disease primarily of herbivorous mammals that ingest and/or inhale spores from the soil while grazing. Germination of spores in the pharynx, respiratory and/or gastrointestinal tract and elaboration of toxin during vegetative growth lead to a rapidly fatal septicaemia with few other predictive symptoms (Velimirovic, 2008). *B. anthracis* was responsible for huge die-offs of livestock prior to the implementation of vaccination with the live, attenuated, capsule-deficient *B. anthracis* Sterne strain in the middle of the last century; nonetheless, spores in the soil remain viable for decades and sporadic outbreaks among domesticated and wild animals occur worldwide, especially in warmer climates where temperatures favour the process of sporulation (Beyer and Turnbull, 2009).

Anthrax disease in humans, which are accidental hosts for *B. anthracis*, can be divided into two stages: early stage, which is defined by route of exposure, and late stage disease. Early stage disease results from introduction of spores into broken skin (cutaneous), the respiratory tract (inhalational), or the gastrointestinal tract (Velimirovic, 2008). The spores are taken up by resident phagocytes and undergo germination and replication at the site of infection and in local draining lymph nodes. Late stage disease results from systemic dissemination of the bacilli that continue to replicate and express lethal toxin (LT) and oedema toxin (ET). LT is composed of lethal factor (LF), a zinc metalloprotease that inactivates mitogen-activated protein kinase (MAPKK), and protective antigen (PA), the host cell binding domain (Mock and Fouet, 2001). ET is composed of oedema factor (EF), a calmodulin-dependent adenylate cyclase, and PA. PA binds to one of its host cell receptors, which are tumour endothelium marker-8 (TEM8 or ANTXR1) and capillary morphogenesis protein-2 (CMG2 or ANTXR2), and forms an oligomeric pore to which EF or LF can bind and be internalized and exert their effects. CMG2 appears to be the dominant PA receptor *in vivo* (Bradley *et al.*, 2001; Scobie *et al.*, 2003; Liu, S., *et al.*, 2009). In addition to the two anthrax toxins, a polyglutamic acid capsule that protects the organism from phagocytosis is produced by vegetative bacilli; the capsule is essential for disease in humans. *B. anthracis* contains two megaplasmids, pXO1 and pXO2, that are required for full virulence, as will be discussed below. The genes that encode the toxin components (*pag, lef, cya*) are present on pXO1 and the capsule operon (*capBCAD*) is on pXO2 (Mock and Fouet, 2001).

Cutaneous anthrax

The most common anthrax presentation (~95% of recognized cases in humans) is a localized, cutaneous infection acquired through the contact with animal hides, wool, or other animal products that are contaminated with viable spores (Tutrone *et al.*, 2002; Wright *et al.*, 2010). The characteristic lesion is an eschar, or black, flat-centred lesion with a red, oedematous raised border that is most often seen on exposed skin on the upper extremities. Approximately 80% of such lesions resolve spontaneously without medical intervention; however, antimicrobial therapy is indicated because of the 10–20% risk that cutaneous anthrax will progress with systemic toxaemia, septicaemia, meningoencephalitis and death. Cases are rare in the United States, but also occur, as recently documented in individuals who craft or play West African traditional drums made with goat or cow hide skins (CDC, 2008).

Inhalational anthrax

Inhalation of spores into the lungs is the cause of inhalational anthrax, the most lethal form of the disease with a 75–100% mortality rate (Velimirovic, 2008). The high mortality is in part due to late recognition of the disease because early symptoms such as mild fever, sore throat, and muscle aches are non-specific and resemble those associated with a cold or flu. After several days, the symptoms worsen and include high fever, difficulty breathing, and shock. Diagnosis of inhalational anthrax is made based on blood cultures that are positive for *B. anthracis* and a chest radiograph that displays mediastinal widening (Velimirovic, 2008). Unfortunately, diagnosis may be clear only when the patient is beyond the benefit of aggressive antimicrobial therapy as a result of the rapidly overwhelming effects of the toxins on the host.

Inhalational anthrax came into the spotlight with the intentional dissemination of anthrax spores through the United States Postal System in 2001. Nonetheless, decades prior to that episode, aerosol delivery of spores was developed for use as a biological weapon in both the former Soviet Union (Meselson *et al.*, 1994) and Iraq (Glassman, 1966). In the former cases, accidental dissemination of spores into the community of Sverdlovsk downwind of a military microbiological laboratory provided deadly evidence of their dispersal capacity (Abramova *et al.*, 1993; Meselson *et al.*, 1994; Grinberg *et al.*, 2001). Indeed, the relative ease with which *B. anthracis* spores can be produced, in conjunction with their capacity to cause high mortality upon inhalational exposure, caused the Centers for Disease Control and Prevention (CDC) to classify *B. anthracis* as a 'category A' bioterrorism agent (Rotz *et al.*, 2002). Accidental inhalational exposure in humans is often associated with handling of contaminated hides and fibres (woolsorter's disease). Cases of inhalational anthrax caused by closely related *Bacillus cereus* strains that produce anthrax toxins and a unique polysaccharide capsule have been documented in metal workers in the southern United States (Miller *et al.*, 1997; Hoffmaster *et al.*, 2004; Avashia *et al.*, 2007).

Gastrointestinal anthrax

Intestinal anthrax, which is the rarest form of anthrax, is lethal in 25–65% of cases and is acquired by consumption of food (generally meat) that is contaminated with spores. Early disease is characterized by gastroenteritis, low-grade fever and flu-like symptoms. As is the case with cutaneous and inhalational anthrax, late stage disease is characterized by rapid onset of septicaemia and toxaemia (Velimirovic, 2008).

Injectional anthrax

A new form of anthrax, termed 'injectional', was most recently reported following an outbreak of acute illness in intravenous heroin users in the UK (Booth *et al.*, 2010; Ramsay *et al.*, 2010; Powell *et al.*, 2011). An epidemiological investigation suggested that the heroin was contaminated with *B. anthracis* spores. While the disease presentation varied among infected individuals, none of the cases presented with the classic black eschar seen in cutaneous anthrax cases; patients developed localized lesions, cellulitis in or near the heroin injection site, and necrotizing fasciitis in some cases (Ramsay *et al.*, 2010). Roughly half of those individuals died as a result of the infection, which focused attention on the risk of both chemical and infectious adulterants in illicit drugs.

Animal models of *B. anthracis* disease

Studies of *B. anthracis* disease progression cannot be done in humans for ethical reasons. Therefore, models of human disease that use rodents, rabbits, and non-human primates have been developed (Goossens, 2009). The majority of animal models mimic inhalational or subcutaneous anthrax in humans. Inhalational models have been used most often, despite the relatively low frequency of inhalational anthrax (compared with the other routes of inoculation), because aerosolized spores have the potential to be used as bioweapons (Goossens, 2009). Like humans, non-human primates and rabbits are particularly susceptible to the effects of LT and ET, a characteristic that makes these animal models particularly useful for studies of late stage anthrax disease as well as vaccines and therapeutics. Mice and guinea pigs can be used to model both the infection (early) and toxaemia (late) stages of disease (Goossens, 2009). Mice have been used most extensively due to their low cost, ease of handling, and genetic diversity. However, use of the correct combination of *B. anthracis* strain and mouse strain is particularly important for elucidation of bacterial factors that are required for establishment of infection.

Non-human primate models

Rhesus macaques, cynomolgus macaques, and African green monkeys are the most commonly used non-human primates because inhalational anthrax disease in these species very closely mimics that of humans (Vasconcelos *et al.*, 2003; Twenhafel, 2010). Lever *et al.* (2008) recently demonstrated that disease progression in the common marmoset (*Callithrix jacchus*), a much smaller non-human primate that is more amenable to studies that require BSL-3 containment, is similar to that of humans as well. A single study in 1959 used chimpanzees as a model for human disease (Albrink and Goodlow, 1959). In that study, Albrink and Goodlow observed widespread oedema and haemorrhage that were most pronounced in the lungs, lymph nodes, and spleen in three of four chimpanzees that were inoculated via aerosol droplets. In addition, bacteraemia was observed in all animals within 2 days after inoculation; animals that succumbed to infection had high levels of bacteraemia. Subsequent studies have shown that anthrax disease that is clinically identical to human anthrax occurs in wild chimpanzees and gorillas in Africa (Leendertz *et al.*, 2004, 2006a,b; Klee *et al.*, 2006). Klee *et al.* used whole genome sequencing to characterize the aetiological agent of anthrax in these great apes as a variant of *B. cereus* that contains the *B. anthracis* virulence plasmids pXO1 and pXO2; they classified the strain as *B. cereus* variety (var.) anthracis (Klee *et al.*, 2010).

Rabbit model

Rabbits have been used extensively to assess the efficacy of vaccines against anthrax because, as noted above, they are particularly susceptible to the effects of LT and ET. Immunization with PA elicits an antibody response that is effective for prevention of anthrax disease (see below for a detailed discussion of anthrax vaccines). Disease progression after inhalational or subcutaneous (s.c.) inoculation of New Zealand White rabbits with fully virulent *B. anthracis* largely mimics the course of toxaemia (late stage disease) in humans (Zaucha *et al.*, 1998). One notable difference in response to *B. anthracis* infection between rabbits and humans is that rabbits display a less pronounced degree of pneumonia and meningitis (Zaucha *et al.*, 1998).

Guinea pig model

The immune system of guinea pigs is thought to be similar to that of humans. Thus, guinea pigs are useful for studies of the pathogenesis and prevention of *B. anthracis* disease. To study the early stages of inhalational anthrax, Ross inoculated guinea pigs with fully virulent *B. anthracis* via the intratracheal (i.t.) route and sacrificed animals at early time points post-challenge for histological examination of the organs (Ross, 1957). In this study, she found that i.t. administration of spores was more effective than aerosol inoculation of a 'spore cloud' for delivery of a lethal dose of spores to the lungs. Spores were rapidly taken up by alveolar phagocytes, which led to detachment of the phagocytes from the alveolar wall. Vegetative bacilli were present in the phagocytes, a finding that suggests that some germination occurred in the lungs of these guinea pigs. However, the majority of the spores germinated only after they were transported within alveolar phagocytes to the 'tracheobronchial glands', an event that occurred by 4 hours after inoculation. Germination and subsequent replication of the vegetative bacilli occurred in the tracheobronchial glands throughout the 24 hours of the studies. As with anthrax disease in humans and in other animal models, replication and dissemination of bacilli and toxins leads to a rapidly terminal septicaemia and toxaemia (Bloom *et al.*, 1947; Cromartie *et al.*, 1947; Hughes *et al.*, 1956).

Mouse models

Studies subsequent to Ross' work have primarily used mouse models of inhalational, intranasal (i.n.), or i.t. administration of spores. Inbred mouse strains differ in their susceptibility to *B. anthracis*. DBA/2J and A/J are highly susceptible to fully virulent *B. anthracis* as well as to *B. anthracis* Sterne (Welkos *et al.*, 1986), a toxigenic and non-encapsulated strain that is used as a vaccine for cattle (Sterne, 1939). In contrast, C57BL/6, CBA/J, and BALB/c mice, while susceptible to infection with fully virulent *B. anthracis*, are relatively resistant to *B. anthracis* Sterne and other non-encapsulated strains of

B. anthracis (Welkos *et al.*, 1986). In a series of studies that utilized the progeny of crosses of A/J mice (sensitive) and CBA/J mice (resistant), as well as passive transfer of serum from resistant mice to sensitive mice, Welkos and Friedlander determined that resistance to *B. anthracis* Sterne is associated with the ability to synthesize the C5 component of complement (Welkos and Friedlander, 1988b). C5a, a polypeptide released from complement component C5, acts as a chemoattractant for macrophages and neutrophils when a normally sterile site (i.e. lung) becomes compromised and an inflammatory response is initiated (Monk *et al.*, 2007). Neutrophils are recruited more slowly and fewer macrophages accumulate at the site of *B. anthracis* Sterne infection in A/J mice than in CBA/J mice (Welkos *et al.*, 1989). Thus, the antimicrobial activity that is associated with early neutrophil infiltration in response to infection is reduced or absent in A/J mice, which allows for more rapid anthrax spore germination, followed by replication and dissemination. To further assess the roles of macrophages and neutrophils in murine anthrax, Cote *et al.* depleted BALB/c mice of either cell type, then administrated *B. anthracis* Ames spores via intraperitoneal (i.p.) or aerosol routes. Macrophage-depleted mice harboured more vegetative bacilli in the peritoneal cavity than did untreated mice after i.p. inoculation, while bronchoalveolar spaces from macrophage-depleted mice contained mostly ungerminated spores after aerosol inoculation (Cote *et al.*, 2006). Neutropenic and untreated mice were equally affected by *B. anthracis* Ames administered by either route, except at a very high aerosol dose, when neutropenic mice were more susceptible than untreated mice (Cote *et al.*, 2006).

B. anthracis disease progression within macrophages

Ross' study in guinea pigs was the first to systematically describe deposition of spores into the alveoli and subsequent migration of spore-laden alveolar phagocytes to the tracheobronchial lymph glands (Ross, 1957). However, the cell type(s) and the location in which germination occurs after inoculation of the lungs with *B. anthracis* remain controversial. The dogma for many years suggested that spores administered to the lung were phagocytosed by alveolar macrophages and transported to the draining lymph nodes, where the spores germinated. The resulting bacilli were then thought to replicate intracellularly and express LT that killed the macrophages from within, thus releasing the bacteria into the lymph and eventually the bloodstream. Numerous *in vitro* studies have shown that macrophages can phagocytose spores and promote germination (Guidi-Rontani *et al.*, 1999; Dixon *et al.*, 2000; Welkos *et al.*, 2002; Ruthel *et al.*, 2004; Gutting *et al.*, 2005; Cleret *et al.*, 2007; Cote *et al.*, 2008; Oliva *et al.*, 2008). The role of macrophages in the progression of anthrax disease is three-fold. First, macrophages take up the spores and provide a permissive environment for germination. Second, macrophages have the capacity to kill spores and vegetative cells. Third, macrophages lysed by LT release cytokines that recruit neutrophils and other inflammatory mediators to the site of infection.

Phagocytosis of spores by macrophages

The integrin Mac-1, located on the phagocyte cell surface, binds to the exosporium glycoprotein *Bacillus* collagen-like protein of *anthracis* (BclA) to promote internalization of spores. This interaction is enhanced by the association of the rhamnose moieties on BclA with CD14, a known receptor for lipopolysaccharides and Gram-positive cell envelope components (Oliva *et al.*, 2008, 2009). This rhamnose–CD14 interaction then leads to activation of phosphoinositol-3-kinase (through Toll-like receptor-2 signalling), which, in turn, causes activation of Mac-1 and enhances the ability of Mac-1 to bind BclA. Opsonization is not required for phagocytic uptake of spores by macrophages, although anti-PA antibodies have been shown to enhance phagocytosis (Welkos *et al.*, 2002; Cote *et al.*, 2005). However, studies in our laboratory showed that antibody against BclA increased spore uptake 7-fold when compared with anti-PA, but, unlike anti-PA, anti-BclA reduced intracellular germination by 4-fold (Brahmbhatt *et al.*, 2007a).

Events that occur inside the macrophage after phagocytosis

Once inside the macrophage, the spore will either initiate germination or be killed. Whether the spores themselves are killed by macrophages or germination within macrophages is required prior to killing remains a matter of debate (Goossens, 2009). While the *in vivo* germinants for *B. anthracis* have yet to be identified, inhibitors of phago-endosomal acidification can promote germination and outgrowth of the bacilli because the intracellular environment is less hostile (Welkos *et al.*, 2002). Germination can also be enhanced by the presence of anti-PA antibodies, though anti-PA-mediated enhancement of spore germination can lead to increased killing by macrophages (Welkos *et al.*, 2002). *B. anthracis* Ames spores phagocytosed by RAW264.7 macrophages were shown either to die soon after uptake or to germinate (Welkos *et al.*, 2002). Further studies showed that the multiplicity of infection (MOI, ratio of spores to macrophages) dictated whether spores germinated or were killed within the cell. A lower MOI resulted in death of the spore and a higher MOI allowed for germination and outgrowth of vegetative bacilli (Kang *et al.*, 2005; Ruthel *et al.*, 2004). Kang *et al.* used a germination-deficient strain of *B. anthracis* to demonstrate that germination of spores is absolutely required before macrophage antibacterial activity is effective *in vitro* (Kang *et al.*, 2005). In addition, they showed that exosporium components confer protection against macrophage killing.

Protection from nitric oxide

Phagocytosis of spores causes activation of macrophages. This activation event results in the rapid upregulation of an inflammatory response that includes production of nitric oxide (Kang *et al.*, 2008). Nitric oxide promotes germination of spores but also has the capacity to kill spores after they germinate. The *B. anthracis* genome encodes four superoxide dismutases (SODs), and two of these SODs are spore-associated (Cybulski *et al.*, 2009a; Passalacqua *et al.*, 2006). A *B. anthracis* Sterne derivative in which all of the SOD genes were inactivated was attenuated for virulence in A/J mice and was significantly more sensitive to oxidative stress and macrophage killing (Cybulski *et al.*, 2009a). Of note, Shatalin *et al.* (2008) demonstrated that *B. anthracis* encodes a nitric oxide synthase (bNOS) that activates the bacterial catalase to reduce reactive oxygen species that would otherwise exert antibacterial activity on the bacilli (Shatalin *et al.*, 2008). Further, Weaver *et al.* (2007) showed that arginase activity present in the exosporium of *B. anthracis* also acts to decrease nitric oxide levels. Thus, *B. anthracis* retains multiple mechanisms to overcome nitric oxide stress within the macrophage.

Inflammatory response

In addition to rapid production of reactive oxygen species, activated macrophages also produce and secrete factors that are involved in recruitment of inflammatory mediators. Macrophages infected with spores *in vitro* secrete the pro-inflammatory cytokines IL-1β, IL-6, IL-10, IL-12, and TNF-α (Pickering and Merkel, 2004; Pickering *et al.*, 2004; Sabet *et al.*, 2006; Kang *et al.*, 2008). A/J mice challenged with an aerosol of *B. anthracis* 7702 ($pXO1^+$ $pXO2^-$) spores had higher levels of IL-6 and IFN-γ in their sera 3 and 5 days post inoculation (Pickering and Merkel, 2004). The pro-inflammatory cytokines secreted by macrophages serve many functions, one of which is to recruit neutrophils to the site of infection. In an *in vitro* system, human neutrophils were shown to both kill ungerminated spores and promote germination of *B. anthracis* spores (Mayer-Scholl *et al.*, 2005). In addition, neutrophils had the capacity to kill vegetative bacilli in a manner that was independent of reactive oxygen species. The neutrophil component required to kill the spores was determined to be the antimicrobial peptide α-defensin (Mayer-Scholl *et al.*, 2005). Similar studies have not been done with mouse neutrophils. However, Cote *et al.* used neutropenic mice to demonstrate that neutrophils play a negligible role in disease progression after i.p. or aerosol injection of spores, except in the case of high dose aerosol inoculation. Under the latter circumstances, all of the neutropenic mice died whereas only 40% of untreated mice died (Cote *et al.*, 2006).

Killing of macrophages and release of vegetative bacilli

B. anthracis spores that have germinated and survived the antimicrobial defences of the phagolysosome can escape to the cytosol and replicate as vegetative bacilli (Dixon *et al.*, 2000; Ruthel *et al.*, 2004). Expression of the genes that encode PA (*pag*), LF (*lef*), and EF (*cya*) and the toxin gene activator AtxA (*atxA*) is upregulated in macrophages infected *in vitro* with *B. anthracis* Sterne (Guidi-Rontani *et al.*, 1999). LT produced by the intracellular bacilli can kill the susceptible macrophages as a result of the downstream effects of inactivation of MAPKK that may lead to apoptosis of the infected cell (Bergman *et al.*, 2005; Cote *et al.*, 2008; Guidi-Rontani *et al.*, 2001), with subsequent release of the bacteria from the macrophage. Liu *et al.* showed that the interaction of elaborated PA with its receptor CMG2 on proximal neutrophils and macrophages to form an oligomeric pore that permits LF binding and activation can kill the local myeloid cells that would otherwise act to control the infection (Liu *et al.*, 2010). Susceptibility of macrophages to the activity of LT is dependent on a member of the NOD-like receptor family called Nlrp1b. These proteins act as sensors of intracellular danger signals for inflammasome complexes. Macrophages with the Nlrp1b$^{S/S}$ allele are sensitive to LT-induced lysis. LT causes activation of caspase-1 and induces release of IL-1β, a cytokine that is responsible for recruitment of neutrophils that limit dissemination of bacilli and control s.c. infection (Kang *et al.*, 2008; Liu *et al.*, 2010; Terra *et al.*, 2010). Spores can also induce IL-1β, caspase-1, and nitric oxide in an LT- and Nlrp1b-independent manner. The effect of spore-induced IL-1β is suppression of germination and macrophage lysis (Kang *et al.*, 2008). Thus, the downstream effects of IL-1β release by macrophages are dependent on whether spores or LT cause the cytokine induction.

Role of dendritic cells in disease progression

Recent studies suggest that resident dendritic cells of the lung play a greater role in spore uptake and transport to the lymph nodes than was previously thought. Cultured human dendritic cells were shown to take up spores, which caused maturation of the dendritic cells as evidenced by a decrease in tissue-specific chemokine receptors and an increase in lymph node-homing receptors (Brittingham *et al.*, 2005). Activation of the MAPK cascade and expression of pro-inflammatory genes occurred within 2 hours of spore uptake. By 6 hours, expression of the pro-inflammatory genes was downregulated and signalling molecules downstream of MAPK were cleaved; such MAPK cleavage suggested that the spores had germinated and begun to produce LT (Brittingham *et al.*, 2005). Cleret *et al.* (2007) subsequently demonstrated in mice inoculated i.n. that interstitial lung dendritic cells had the capacity to capture spores from the bronchoalveolar space via a transepithelial extension mechanism; the interstitial lung dendritic cells were observed in the thoracic lymph nodes within 30 minutes and up to 72 hours post inoculation. Shetron-Rama *et al.* (2010) extended these observations in transgenic mice depleted of CD11c^{+} dendritic cells. They demonstrated that CD11c^{+} dendritic cells are in fact required for efficient migration of spores to the lung-associated lymph nodes after i.t. inoculation.

Germination at the site of infection versus in the draining lymph nodes

As mentioned above, the dogma for many decades stated that the site of spore germination in the host is the draining lymph nodes proximal to the site of infection. Recent studies have applied, among other techniques, *in vivo* imaging technology in an attempt to identify the specific location(s) within the host in which germination occurs. In studies designed to assess the role of capsule in dissemination of *B. anthracis* in an immunocompetent host, BALB/c mice were inoculated i.t. with UT500 (pXO1^{+} pXO2^{+}) or a UT500 derivative that lacked *capBCAD* (Drysdale *et al.*, 2005). Spores from both strains germinated in the lungs of mice within 3 hours post inoculation, as determined by enumeration of heat-resistant spores and heat-sensitive bacilli in homogenized lung tissue. However, vegetative bacilli were not visualized in the lungs of mice infected with either strain until at least

72 hours post inoculation; by 72 hours, the wild-type bacilli had disseminated to the spleen and returned to the lung, while the *capBCAD* mutant had been cleared because it failed to disseminate (Drysdale *et al.*, 2005). Glomski *et al.* (2007) inoculated BALB/c mice via i.t., i.n., or aerosol routes with a derivative of a non-toxigenic encapsulated *B. anthracis* that produces a luminescent signal during vegetative growth. After i.t. administration, luminescent bacilli were found in both the mandibular lymph nodes and the larynx within 24 hours. Aerosol and i.n. inoculation led to vegetative outgrowth in the nasal cavity within 22–29 hours and in the mandibular lymph nodes 5–15 hours later. Spores were found in the lungs and mediastinal lymph nodes by enumeration of heat-resistant bacteria recovered from these tissues early after inoculation; however, the importance of these spores in the disease progression was unclear because luminescence was not observed in the lung until the bacilli had disseminated and returned to the lung. Regardless of the route of inoculation into the lung, germination occurred at the site of infection and bacilli disseminated throughout the body (Glomski *et al.*, 2007). A similar pattern of germination at the site of infection followed by systemic dissemination was observed after intragastric and s.c. administration of the luminescent *B. anthracis* strain as well (Glomski *et al.*, 2007).

Studies in our laboratory confirmed that *B. anthracis* can germinate in the lung. A/J mice were inoculated i.n. with spores of a *B. anthracis* Sterne derivative that emitted a luminescent signal upon germination, and *in vivo* imaging was used to assess germination and outgrowth of the bacilli (Sanz *et al.*, 2008). Luminescence was observed in lung tissue extracted from infected mice within 6 hours post-inoculation. Histological analysis of tissue sections from infected mice showed that spores were phagocytosed by macrophages within 30–60 minutes post challenge, and that neutrophils were present and contained spores within 6 hours. In addition, Glomski *et al.* extended their earlier observations of germination and outgrowth from the site of inoculation by aerosol or i.n. inoculation of A/J mice with luminescent toxigenic non-encapsulated *B. anthracis* spores (Glomski *et al.*, 2008). Aerosol inoculation of spores resulted in germination in the nasopharynx and was followed by dissemination to the kidneys, lungs and stomach. Spores administered i.n. germinated in the nasopharynx and/or the lungs (Glomski *et al.*, 2008). Taken together, the *in vivo* imaging studies suggest that germination probably occurs at the site of inoculation. None of the studies summarized here demonstrated a significant role for the mediastinal lymph nodes in spore germination or vegetative outgrowth of the bacilli.

Current vaccines and therapies against *B. anthracis*

The incidence of anthrax in the United States is very low, and sporadic cases are treated with antibiotics as described below. Nonetheless, 'Anthrax Vaccine Adsorbed' (AVA) is approved for human use, but administration of AVA is limited to individuals at elevated risk of exposure, such as biomedical researchers and military personnel. Here we discuss accepted therapies and vaccines, as well as potential limitations of and improvements to the current vaccine.

Antibiotics

The dilemma of treatment for *B. anthracis* with antibiotics is that such drugs only target spores that have initiated germination and vegetative cells. Because spores are metabolically dormant, they are resistant to antibiotics. While several antibiotics are effective against *B. anthracis* infection, host survival is highly correlated with the commencement of therapy relative to time of exposure. Delayed treatment, which often occurs because the initial clinical symptoms of the disease are non-specific, allows *B. anthracis* to replicate unchecked and to elaborate potent toxins that ultimately kill the host. Penicillin G is not recommended because some *B. anthracis* isolates encode the genes to express functional beta-lactamases (Chen *et al.*, 2003). Therefore, the CDC recommends treatment for 7–10 days with either oral ciprofloxacin or doxycycline for naturally acquired, uncomplicated cutaneous anthrax. For gastrointestinal and inhalational anthrax, intravenous ciprofloxacin is preferred because fluoroquinolones are bactericidal while tetracyclines are bacteristatic. One or more additional antibiotics are included in the treatment regimen due to the high mortality associated with these routes of infection; protein synthesis inhibitors, such as clindamycin, are highly recommended because they can reduce the level of toxin production (CDC, 2001; Stern *et al.*, 2008). A full 60-day course of antibiotic treatment after systemic disease is recommended because spores have been shown to cause inhalational anthrax up to 58 days after the initial exposure (Meselson *et al.*, 1994).

For prophylactic treatment after exposure to aerosolized *B. anthracis* spores, the CDC recommends that the 60-day antibiotic therapy be used in conjunction with a three dose vaccination regimen with AVA (administered at 0 weeks, 2 weeks and 4 weeks) (Wright *et al.*, 2010). The administration of AVA is not FDA-approved for post-exposure prophylaxis (PEP) but is available under an Emergency Use Authorization (EUA) or an Investigational New Drug (IND) protocol (CDC, 2010). The combined therapy is based on the observation that 100% of rhesus macaques survived inhalational exposure to a large dose of *B. anthracis* spores when given AVA and a shortened 14-day ciprofloxacin regimen, compared with 44% survival in an infected group treated with ciprofloxacin only (Vietri *et al.*, 2006). In theory, the combined administration of antibiotics and vaccine allows the immune system to develop protective antibodies against PA elaborated prior to eradication of vegetative cells. Vaccination may also address concerns of patient adherence to the 60-day antibiotic regimen raised after the 2001 bioterrorism-related release of spores when 10,000 people were given prophylactic antimicrobial treatment. Of the people who responded to surveys about PEP use, only 44% took the entire regimen as prescribed, possibly because of side-effects from long-term antibiotic use (Shepard *et al.*, 2002). While the rhesus macaque data suggest that a 14-day antibiotic regimen combined with vaccination is sufficient to confer full protection against anthrax, the FDA still recommends the full 60-day cycle until further data are available (CDC, 2010).

History of anthrax vaccines

Louis Pasteur is credited with development of the first animal vaccine against anthrax in 1881. His vaccine consisted of two doses of spores from *B. anthracis* vegetative cells grown at elevated temperature. This elevated temperature step was later reported to cure the cells of the toxin-encoding plasmid pXO1 and resulted in a non-toxigenic, encapsulated strain of *B. anthracis* (Mikesell *et al.*, 1983). The first dose given, the Pasteur I vaccine, was derived from cultures grown at 42–43°C for 15–20 days while the cells for the second dose, the Pasteur II vaccine, were obtained from cultures also grown at 42–43°C but harvested after only 10–12 days. The reduced incubation time resulted in increased virulence of the Pasteur II vaccine relative to Pasteur I (Hambleton *et al.*, 1984). Investigators have hypothesized that Pasteur's vaccines were actually mixtures of cultures that contained a small percentage of fully virulent bacteria as well as a much larger population of non-toxigenic bacteria since the presence of toxin in a vaccine is usually required to confer immunity (Ivins *et al.*, 1986; Scorpio *et al.*, 2006). To simplify vaccine production and administration, the Pasteur vaccine regimen was modified such that the Pasteur I culture was omitted and the Pasteur II vaccine was treated with saponin to reduce its virulence and enhance its immunogenicity (Hambleton *et al.*, 1984). Despite the effectiveness of the Pasteur vaccine, administration of the vaccine itself was shown to cause anthrax in a small subset of animals that received it (Artenstein, 2009). Furthermore, the Pasteur vaccine suffered from batch to batch variability in virulence, as well as a decline in effectiveness over time (Hambleton *et al.*, 1984; Baillie, 2001). As a result, the Pasteur vaccine is currently only used in Italy for vaccination of horses and goats (Turnbull, 2008). An additional attenuated strain of *B. anthracis* called Carbosap is used in Italy to vaccinate cattle and sheep. Even though the Carbosap strain is attenuated, it possesses both virulence plasmids and produces toxin and capsule (Fasanella *et al.*, 2001); despite rigorous analysis and comparison to a virulent *B. anthracis* strain, the reason that this strain is attenuated is as yet unknown (Adone *et al.*, 2002; La Rosa *et al.*, 2006).

Overview of types of vaccines in use today

As described above, the disease anthrax can occur in both animals and humans through the introduction of *B. anthracis* spores into the host by inhalation, ingestion, or skin puncture. Despite the fact that the spore is the infectious particle, in the United States and United Kingdom the only vaccines approved for human use are derived from the vegetative form of *B. anthracis* and mainly contain PA. However, spore-based human vaccines were used in the former Soviet Union, and are used in Russia and China now. In addition, spore-based animal vaccines are used worldwide. A summary of the current animal and human vaccines is provided in Table 11.1.

Vaccines for animals

The current animal vaccines used by most countries are derived from toxigenic, non-encapsulated strains of *B. anthracis* with a $pXO1^{+}/pXO2^{-}$ plasmid profile. *B. anthracis* Sterne 34F2 is the most widely used vaccine strain (Turnbull, 2008). *B. anthracis* Sterne 34F2 was isolated in the 1930s as a rough colony (i.e. non-encapsulated) that grew on a nutrient agar plate that contained 50% horse serum and was incubated under high CO_2 to promote capsule production; the plate was initially inoculated from a two-day culture of fully virulent *B. anthracis*. Initial tests of the Sterne vaccine showed it was safer and less reactogenic than the Pasteur vaccine, and, importantly, it maintained the same efficacy (Artenstein, 2009). The most common formulation of the Sterne vaccine consists of *B. anthracis* Sterne 34F2 spores suspended in saponin and glycerol (Turnbull, 2008). One dose is typically given to animals and provides immunity for at least one year, after which annual boosters are recommended (Hambleton *et al.*, 1984; Wright *et al.*, 2010). Even though the Sterne vaccine is less reactogenic than the Pasteur vaccine, it does retain some reactogenicity due to the expression of toxins encoded on pXO1. New veterinary strains of *B. anthracis* derived from *B. anthracis* Sterne have been engineered to overcome this issue but are not currently in use. In one strain designated RPL686, an engineered His686Ala mutation in the catalytic site of LF results in an inactive form of the protein. Guinea pigs vaccinated with either *B. anthracis* Sterne or *B. anthracis* RPL686 spores were protected 100% from challenge with a lethal dose of *B. anthracis* 17JB. Serum IgG anti-PA and IgG anti-LF titres were similar between the vaccinated groups (Brossier *et al.*, 1999).

Table 11.1 Anthrax vaccines currently in use

B. anthracis strain	Plasmid content	Source	Country
Human vaccines			
V770-NP1-R[1]	pXO1	Adsorbed culture filtrate	United States
Sterne 34F22	pXO1	Precipitated culture filtrate	United Kingdom
STI-1	pXO1	Live spores	Russian Federation
A16R	pXO1	Live spores	China
Animal vaccines			
Sterne 34F[2]	pXO1	Live spores	Worldwide except countries below
Carbosap	pXO1/pXO2	Live spores	Italy
Pasteur I	pXO2	Live spores	Italy
1190R-Stamatin	pXO1	Live spores	Romania
55-VNIIVVIM	pXO1	Live spores	Russian Federation

Adapted from Turnbull (1998), Fasanella *et al.* (2001) and Kuznetsovskii *et al.* (2009).
[1]Vaccine known as 'Anthrax Vaccine Adsorbed' (AVA) or BioThrax®.
[2]Vaccine known as 'Anthrax Vaccine Precipitated' (AVP).

Vaccines for use in humans

The vaccines currently used in humans originate from toxigenic, non-encapsulated strains of *B. anthracis* (pXO1$^+$/pXO2$^-$). AVA, also known as Biothrax®, is a non-viable, toxin-based vaccine derived from a vegetative culture filtrate of attenuated *B. anthracis* V770-NP1-R grown under anaerobic conditions. The vaccine is produced by adsorption of the culture filtrate to aluminium hydroxide gel, which results in a vaccine that primarily contains PA as well as small amounts of LF and EF. The analogous human vaccine used in the United Kingdom, 'Anthrax Vaccine Precipitated' (AVP), is prepared in a similar manner but is derived from *B. anthracis* Sterne 34F2 and contains smaller amounts of PA and higher concentrations of LF and EF (Turnbull *et al.*, 1986; Turnbull, 1991; Baillie *et al.*, 2003; Whiting *et al.*, 2004). The higher level of LF in the AVP vaccine may provide enhanced protection since the presence of LF has been shown in animal models to increase the immune response to PA (Price *et al.*, 2001; Baillie *et al.*, 2004). The presence of LF and EF as well as PA in the vaccine provides all of the components needed for holotoxin formation after vaccine administration. Thus, small amounts of holotoxins may account for the minor reactogenicity (i.e. soreness, redness, rash, swelling and lumps) at the injection site that can be observed after vaccination (Marano *et al.*, 2008; Niu and Ball, 2009). However, these side-effects are commonly observed after administration of any vaccine and may not be specifically attributed to AVA or AVP. The vaccination schedules are five doses of AVA over 18 months or four doses of AVP over 6 months, and both vaccines require annual boosters (Turnbull, 1998; Friedlander and Little, 2009; Wright *et al.*, 2010).

In contrast to the US and UK vaccines, the vaccines used in the former Soviet Union and China are live spores of the avirulent *B. anthracis* strains STI-1 and A16R, respectively. Both strains are toxigenic and non-encapsulated (i.e. pXO1$^+$/pXO2$^-$) like *B. anthracis* Sterne. Concerns about the safety of a live spore vaccine due to the residual virulence observed in the spore-based animal vaccine have prevented their use in humans in the United States, despite reports that the STI-1 spore vaccine causes no adverse reactions (Shlyakhov and Rubinstein, 1994; Turnbull, 2000). The Russian STI-1 vaccine requires two doses 21 days apart and annual boosters, while the Chinese A16R vaccine is a single dose vaccine (Turnbull, 1998; Wright *et al.*, 2010).

Problems/shortcomings with current vaccines

An optimal vaccine is effective after only a few doses and provides long term immunity against disease. In addition, a vaccine should have a defined composition and minimal reactogenicity. Lastly, and especially in the context of mass vaccination after a bioterrorist event, an ideal vaccine should be self-administered, have a long shelf life, be stable at room temperature, and be amenable to rapid, large scale production. As mentioned above, neither AVA nor AVP satisfies these criteria, especially with regard to the number of doses required to confer protection against anthrax disease. In fact, until recently, the AVA vaccination regimen included six doses rather than the currently approved five doses (Wright *et al.*, 2010). Studies to evaluate the efficacy of a further reduction of the vaccination regimen are under way. Production of AVA and AVP is further complicated by lot to lot variability in the culture supernatants from which these vaccines are derived; the concentrations of PA, EF, LF, and other bacterial products can vary (Ivins *et al.*, 1998). The lack of a 'predefined' vaccine composition can explain the variable immunologic responses seen in humans and in animal models.

The spore-based vaccines used in humans and animals elicit much better protection than do their culture supernatant-derived counterparts and require only one or two doses. All of the human vaccines, except the Chinese A16R vaccine, require annual boosters to provide long term protection (Turnbull, 1998). The effectiveness of spore-based vaccines varies in the different animal models that have been tested. Administration of live spore vaccines to guinea pigs elicited greater protection against vaccine-resistant *B. anthracis* isolates than did immunization with PA-based vaccines (Little and Knudson, 1986; Turnbull *et al.*, 1986). In mice, live-spore vaccines were not as effective as PA-based vaccines when mice were subsequently challenged with virulent *B. anthracis*.

This observation was especially pronounced for A/J mice, a strain of mice that is particularly susceptible to *B. anthracis* Sterne infection, as the live-spore vaccines were completely ineffective in these animals (Welkos and Friedlander, 1988a). However, in both guinea pigs and mice, immunization with a combination of formaldehyde-inactivated spores (FIS) and PA provided protection against challenge with two virulent strains of *B. anthracis* (Brossier *et al.*, 2002). Immunocompetent mice immunized with live *B. anthracis* Sterne spores were completely (CBA/J) or partially (BALB/c) protected from subsequent challenge with fully virulent *B. anthracis* spores. Interestingly, CBA/J mice immunized with a *B. subtilis* strain that produces PA were protected against infection with *B. anthracis*. In contrast, immunization with PA or passive transfer of anti-PA antiserum did not protect mice from challenge with virulent *B. anthracis* spores (Brossier *et al.*, 2002). Thus, components of the *Bacillus* spore itself, and not only PA that is produced after the spores germinate, appear to play a role in spore-vaccine-based immunity against anthrax disease.

Recombinant PA-based vaccines in development

As mentioned above, the current therapy to treat anthrax in humans is limited to antimicrobial drugs and co-administration of anthrax vaccine. However, strains of *B. anthracis* can be engineered to be resistant to current antibiotics or vaccines. Scientists routinely insert antibiotic resistance markers into bacteria to facilitate genetic studies, and bacteria can naturally acquire mutations or genes that confer antibiotic resistance. Indeed, Pomerantsev *et al.* generated a derivative of the virulent *B. anthracis* H-7 strain that contains the *B. cereus* VKM-B164 cereolysin AB genes that encode a phospholipase C and a sphingomyelinase. Hamsters immunized with the Russian STI-1 live spore vaccine strain, which confers protection against *B. anthracis* H-7, were not protected against challenge with the *B. anthracis* H-7 cereolysin-expressing derivative. Protection against the engineered strain was restored only when hamsters were immunized with an STI-1 derivative that expressed the cereolysin AB genes (Pomerantsev *et al.*, 1997). Protection afforded by AVA was not tested against this engineered strain; however, AVA is efficacious against challenge with a variety of *B. anthracis* strains in both rabbits and non-human primates (Fellows *et al.*, 2001). A US government panel believes that the likelihood that an AVA-resistant *B. anthracis* strain will be engineered is remote because PA, the main component of AVA, is highly conserved among *B. anthracis* strains and mutant forms of PA inactivate PA oligomers such that they cannot deliver EF and LF into the cell (Joellenbeck *et al.*, 2002). Nevertheless, immunization with AVA does not confer 100% protection against all *B. anthracis* strains that have been tested, a finding that suggests the generation of AVA-resistant *B. anthracis* strains is within the realm of possibility (Fellows *et al.*, 2001).

Improvement of AVA through use of a different adjuvant

Efficacy and safety of AVA have been well established and research to modify the current aluminium-based adjuvant is under way. Aluminium (i.e. aluminium hydroxide, aluminium phosphate, alum) is a very common adjuvant in licensed vaccines and has been in use in the US for over 60 years. Aluminium-based adjuvants have been proven safe, though they can elicit local reactions (Baylor *et al.*, 2002). A new formulation of AVA that contains the adjuvant CpG 7909, designated AV7909 (Table 11.2), is currently in Phase I clinical trials (Clinical Trial Identifier NCT01263691). The CpG 7909 adjuvant is an unmethylated synthetic oligonucleotide that is 24 bases in length and contains three copies of the GTCGTT motif; this motif has been reported to agonize human Toll-like receptor nine (TLR-9) and stimulate B cells. This stimulation of cells causes a more rapid and increased immune response (Krieg *et al.*, 1995; Cooper *et al.*, 2004). Indeed, BALB/c mice immunized with a hepatitis B vaccine with CpG as the adjuvant generated a 10-fold higher response than those immunized with the alum-based vaccine (Weeratna *et al.*, 2003). Additionally, immunization with 10-fold less antigen in the presence of both alum and CpG as adjuvants elicited a 10-fold higher antibody titre when compared with alum as an adjuvant

Table 11.2 Recent anthrax vaccine and passive therapy US clinical trials

Clinical trial identifier	Name	Company	Description	Phase
Vaccines				
NCT01263691	AV7909	Emergent BioSolutions	Currently licensed AVA vaccine with CPG7909 adjuvant	I
NCT00103467	PreviThrax™(rPA102)	Emergent BioSolutions	rPA-based vaccine	I
NCT00100724				II
NCT00170456	SparVax™	PharmAthene	rPA-based vaccine	II
NCT00170469				II
Passive therapies				
NCT00639678	Raxibacumab™(PamAb)	Human Genome Sciences	Recombinant human IgG anti-PA antibody	III
NCT01202695	Thravixa™(AVP-21D9)	Emergent BioSolutions	Human anti-PA mAb	I
NCT00845650	Anthrivig™(AIGIV)	Emergent BioSolutions	Human IgG anti-PA isolated from the plasma of healthy AVA-vaccinated human donors	I
NCT01204866	Valortim® (MDX-1303)	PharmAthene	Human anti-PA mAb	I
NCT00829582	Anthim™ (ETI-204)	Elusys Therapeutics	Humanized anti-PA mAb	I

Non-commercial biological name is provided in parenthesis where available.
Data compiled from http://clinicaltrials.gov.

alone (Weeratna *et al.*, 2003). The Phase I clinical trial of AV7909 was designed to address the following:

1 Is the formulation safe?
2 Is an increase in anti-PA antibody observed in people immunized with the combined alum and CpG adjuvants when compared with those immunized with either adjuvant alone?
3 Are the anti-PA antibody responses elicited by different AV7909 formulations (which contain lower amounts of AVA) comparable to the responses to the licensed AVA vaccine?

A reduction in number of immunizations and possible elimination of annual boosters, so long as efficacy is equal or increased, would be marked improvements to the current AVA.

Recombinant protective antigen vaccine

As mentioned previously, the main component of both the AVA and AVP vaccines is PA. Two major second generation anthrax vaccines that contain purified recombinant PA (rPA) and use aluminium-based adjuvants are currently in development (Table 11.2). Benefits of a rPA-based vaccine include a defined vaccine composition and the capacity for large-scale production. PreviThrax™, a variant of the rPA102 vaccine originally developed at the US Army Medical Research Institute of Infectious Diseases, is under development at Emergent BioSolutions Inc. (Baillie, 2009). The rPA is overexpressed from a multicopy plasmid in a sporulation-deficient *B. anthracis* Sterne derivative (Baillie, 2009). The other rPA-based vaccine, SparVax™, was originally developed at the Defence Science Technology Laboratory in the UK and is currently in development by PharmAthene Inc. (Baillie, 2009). The SparVax™ rPA is produced from a codon-optimized DNA construct in *Escherichia coli* (Baillie, 2009; Williamson *et al.*, 2005). Both of these rPA-vaccine candidates have completed Phase II clinical trials (Clinical Trial Identifiers NCT00103467, NCT00100724, NCT00170456, NCT00170469) in which their safety and immunogenicity in healthy test subjects was confirmed. When compared with two doses of AVA given at 0 and 4 weeks, rPA102 (the PreviThrax™ predecessor) elicited less local reactogenicity but more systemic reactogenicity that typically consisted of a headache after the first inoculation. In addition, two doses of rPA102 produced similar anti-PA antibody responses when compared with two doses of AVA, while a third rPA102 dose given at week 8 further increased the antibody response; however, a third dose of AVA was not given for comparison (Gorse *et al.*, 2006). The clinical trials with SparVax™ also suggest that a three-dose regimen given over a two-month period is sufficient to provide protective immunity; in preclinical studies SparVax™ was shown to protect against a lethal *B. anthracis* Ames aerosol challenge in both rabbits and non-human primates (Baillie, 2009).

The two rPA-based vaccines show promise as second generation anthrax vaccines in part because they satisfy most of the criteria for an optimized vaccine: reduced number of doses, defined composition, minimal reactogenicity, and capacity for large-scale production. As the current clinical trial data suggest, only three doses of vaccine given within two months should be sufficient to provide protection when compared with AVA and AVP. Both of these vaccines currently use aluminium-based adjuvants that, as described earlier, may not be optimal. Inclusion of an oligonucleotide adjuvant similar to CpG7909 that is used in the new AVA formulation AV7909 may induce a more robust anti-PA response that could further reduce the number of rPA-vaccine doses and/or the amount of antigen required. Because these vaccines are derived from recombinant proteins, they would most probably need to be stored at 4°C and would have a predefined shelf life, characteristics that require continual replenishment of expired lots in stockpiles (Baillie, 2009). The next generation of SparVax™ is as effective as the current SparVax™ but requires only two injections, uses a different adjuvant, and is lyophilized. The latter feature increases the stability of the vaccine to three years at room temperature (Potera, 2009).

Another benefit to an rPA-based vaccine is that mutations can be introduced into *pag*, the gene that encodes PA, to generate a dominant negative inhibitory (DNI) form of PA. The DNI form of PA would retain the capacity to oligomerize and bind LF and EF but not the capacity to translocate the holotoxins across the cell membrane (Sellman *et al.*, 2001a,b). A successful DNI mutant rPA-based vaccine elicited a robust neutralizing anti-PA response in mice (Aulinger *et al.*, 2005; Yan *et al.*, 2008). Mutant forms of PA that blocked oligomerization generated high levels of anti-PA antibodies in mice but had significantly lower neutralizing capacity than did wild-type PA (Yan *et al.*, 2008). While DNI forms of rPA are not in clinical trials, their 'toxoid status' could allow them to be used both as vaccines and as PEP (Cybulski Jr., *et al.*, 2009b).

Alternative vaccine delivery methods

Administration of the currently licensed vaccines still requires the use of needles (and injections) which makes mass vaccination strategies difficult to implement. Several alternative routes of vaccination are under investigation and include oral, intranasal, transcutaneous, and microneedle intradermal inoculation.

Oral administration of vaccines requires the antigen to be protected from the harsh environment of the stomach and typically uses attenuated bacteria as the delivery vehicle. Attenuated strains of *Bacillus*, *Salmonella*, and *Lactobacillus* have been explored as oral vehicles for *B. anthracis* vaccination with variable success rates (Coulson *et al.*, 1994; Rengel and Bohnel, 1994; Zegers *et al.*, 1999; Aloni-Grinstein *et al.*, 2005; Duc le *et al.*, 2007; Shakya *et al.*, 2007; Stokes *et al.*, 2007; Baillie *et al.*, 2008; Mohamadzadeh *et al.*, 2009, 2010; Osorio *et al.*, 2009). All of the bacterial strains used as vaccine candidates express some form of PA and rely on this protein to elicit protective immunity. Goats immunized orally by consumption of live *B. anthracis* Sterne 34F2 vaccine spores in their feed produced significantly higher anti-PA titres than did goats that had the vaccine sprayed directly onto their abraded gums; however, the levels of anti-PA antibody in both groups were typically low and variable compared with those obtained after the s.c. vaccination (Shakya *et al.*, 2007). These animals were not subsequently challenged with virulent *B. anthracis*, so the protective efficacy of this strategy is unknown.

The MASC-13 vaccine strain of *B. anthracis* is

a non-toxigenic, non-encapsulated strain that constitutively expresses a DNI mutant form rPA. Live MASC-13 spores administered orally to guinea pigs provided partial protection against lethal challenge with virulent *B. anthracis* Vollum; the partial protection is most probably due to the variable anti-PA antibody responses observed in each group (Aloni-Grinstein *et al.*, 2005). PA expressed in *Salmonella enterica* serovar *Typhi* strain Ty21a, a licensed oral vaccine strain, has been shown to elicit strong anti-PA responses in mice and to protect mice from lethal i.n. challenge with *B. anthracis* 7702 and *B. anthracis* STI-1 (Stokes *et al.*, 2007; Osorio *et al.*, 2009). PA produced by orally administered *Lactobacillus casei* was shown to elicit an immune response in BALB/c mice (Zegers *et al.*, 1999). Furthermore, oral administration of *L. acidophilus* that expressed a PA/dendritic cell-targeting peptide fusion protein resulted in greater protection of mice from *B. anthracis* Sterne challenge compared with *L. acidophilus* that expressed PA alone; in addition, levels of total anti-PA and neutralizing anti-PA antibodies were similar to those observed after s.c. immunization with rPA in aluminium hydroxide adjuvant (Mohamadzadeh *et al.*, 2009). The *L. acidophilus* vaccine was subsequently improved with a change in delivery vehicle to *L. gasseri* and the use of a high copy, stable vector to express the PA/dendritic cell-targeting peptide fusion protein. Complete protection from *B. anthracis* Sterne challenge was provided by the *L. gasseri* vaccine (Mohamadzadeh *et al.*, 2010). Oral vaccination is an optimal delivery method for any vaccine since it is non-invasive and can elicit both mucosal and systemic immune responses. However, in the case of *B. anthracis* vaccination, this route may not be as effective as parenteral routes because anti-PA responses are highly variable after oral immunization. The possible exception to this conclusion is the *Lactobacillus*-based candidates.

Intranasal administration may be the most appropriate route for immunization, since the most probable means of infection from a biological attack would be inhalation of aerosolized spores. Delivery of a vaccine to the same sites that would be exposed during infection would increase the likelihood that secretory anti-PA antibodies are already present to reduce or prevent disease (Cybulski *et al.*, 2009b). i.n. administration of rPA in mice elicited high IgA anti-PA and IgG anti-PA titres; s.c. immunization with rPA also produced high IgG anti-PA titres but did not elicit an IgA anti-PA response (Gaur *et al.*, 2002). Complete protection of A/J mice against *B. anthracis* Sterne challenge was afforded by two 15 μl i.n. doses of AVA given 3 weeks apart. Like rPA, AVA elicited both IgG anti-PA and IgA anti-PA antibody responses (Zeng *et al.*, 2007). i.n. vaccination of BALB/c mice and guinea pigs with rPA in a soybean oil and water-based adjuvant produced higher IgA anti-PA and IgG anti-PA responses than did rPA in CpG and alum-based adjuvants. Guinea pigs subsequently challenged intradermally (i.d.) with *B. anthracis* spores were protected, while guinea pigs challenged i.n. were partially protected (Bielinska *et al.*, 2007). In the late 1950s, researchers in the former Soviet Union demonstrated that administration of dry spores to humans via aerosol, s.c., or scarification routes induced similar Anthraxin skin test (used to assess post-vaccine immunity) results up to one year after vaccination. While their method of simultaneous vaccination of groups of individuals by aerosol exposure to high doses of spores for 5–15 minutes was very effective, it was reserved for emergency situations (Shlyakhov and Rubinstein, 1994). A powdered form of rPA delivered i.n. to New Zealand white rabbits in three doses 3 weeks apart was shown to completely protect against *B. anthracis* Ames aerosol challenge, while a liquid form of rPA also delivered i.n. conferred only partial protection (Mikszta *et al.*, 2005). This same phenomenon was observed by another group who used only two doses of rPA either i.n. or i.m. prior to *B. anthracis* Ames challenge. Increased protection was provided by the powdered rPA despite lower anti-PA and neutralizing anti-PA antibody titres in the i.n. immunized groups than in the i.m. immunized group (Huang *et al.*, 2007). Even one dose of powdered-rPA administered i.n. conferred protection to 80% of the rabbits challenged 9 weeks after vaccination (Klas *et al.*, 2008). While the i.n. administration of liquid rPA was somewhat effective for prevention of disease after spore challenge, i.n. administration of powdered vaccines was more effective. This finding may be due to better particulate dispersion in the lungs after powdered vaccine administration. A powdered rPA vaccine would allow for needle-less mass vaccinations and would probably have an increased shelf life when compared with liquid forms of rPA vaccine.

Transcutaneous vaccination involves skin patches that contain the antigen in addition to a strong skin adjuvant such as the bacterial ADP-ribosylating toxins cholera toxin or *E. coli* heat-labile toxin (Glenn *et al.*, 2003). The inclusion of the labile toxin adjuvant with the protein antigen on the skin patch was required to elicit an immune response in humans (Guerena-Burgueno *et al.*, 2002). Skin patch immunization of A/J mice with rPA delivered with or without labile toxin adjuvant afforded 100% protection against a lethal s.c. challenge of *B. anthracis* Sterne (Matyas *et al.*, 2004). A similar aerosol challenge in A/J mice with *B. anthracis* STI-1 demonstrated that the labile toxin adjuvant was necessary to elicit IgG and IgA responses in the lungs (Kenney *et al.*, 2004). Skin patch immunization of CBA/J mice also conferred partial protection against an i.n. challenge with fully virulent *B. anthracis* Ames (Peachman *et al.*, 2006).

The apparatus used in microneedle vaccinations can consist of either solid protrusions less than 1 mm in length or hollow protrusions 1–1.5 mm in length. The solid protrusions are typically arrayed and used to pierce or scrape the skin before administration of the antigen, coated with a dry vaccine, or constructed from vaccine-encapsulated polymers that degrade or dissolve in the skin. Hollow microneedles are miniaturized versions of traditional hypodermic needles that are used to inject a liquid vaccine i.d. with a specialized syringe or pump (Prausnitz *et al.*, 2009). A study in New Zealand White rabbits demonstrated no protection against *B. anthracis* Ames aerosol challenge when rPA was delivered via microabrasion; however, when rPA was delivered via i.d. injection with 1 mm microneedles, rabbits were protected 100% from *B. anthracis* Ames aerosol challenge (Mikszta *et al.*, 2005). Microneedle injections could be done by non-medical personnel or could be self-administered, since the length of the needle itself dictates the depth of the injection; thus, vaccination via microneedle injection could be rapidly accomplished in the case of a mass exposure incident.

Anti-capsule vaccines

As mentioned above, fully virulent vegetative *B. anthracis* produces an antiphagocytic poly-γ-D-glutamic acid (PGA) capsule (Ezzell and Welkos, 1999). Since encapsulated strains of *B. anthracis* are protected from the host immune system, antibodies directed to the PGA capsule may be of therapeutic value. In fact, anti-PGA antibodies bind to and kill encapsulated bacilli *in vitro* via opsonophagocytosis (Rhie *et al.*, 2003; Schneerson *et al.*, 2003). The PGA capsule itself is not very immunogenic despite the identification of anti-PGA antibodies in convalescent sera from patients naturally infected by *B. anthracis* (Goodman and Nitecki, 1966; Sirisanthana *et al.*, 1988; Harrison *et al.*, 1989). However, the PGA capsule is immunogenic when conjugated to a protein carrier, though the elicited antibodies are not always protective. s.c. vaccination of mice with purified capsule crosslinked to BSA elicited a high IgG anti-capsule titre that was similar in magnitude to the response to capsule conjugated to PA; however, only the PA–PGA conjugate was protective for mice challenged with *B. anthracis*. In addition, while the anti-capsule titres were enhanced by conjugation of PGA to PA, the IgG anti-PA titres were similar in the groups immunized s.c. with PA, PGA + PA, and the PGA–PA conjugate, observations which suggest that inclusion of PGA with PA do not enhance (or hinder) the anti-PA response (Chabot *et al.*, 2004). Immunization of guinea pigs with purified capsule or 10mer, 15mer, or 20mer peptides of PGA conjugated to PA elicited IgG anti-capsule antibodies; the PA–20mer conjugate induced the greatest response. All formulations conferred 100% protection against *B. anthracis* H9401 spore challenge (Lee *et al.*, 2009). Dry powder vaccine formulations that contained free PA + PA-PGA 10mer peptide conjugate or free PA + free PGA peptide administered i.n. to rabbits conferred 100% protection against *B. anthracis* Ames challenge, though only the PA–PGA peptide conjugate group produced anti-capsule antibody and displayed no clinical signs post challenge (Wimer-Mackin *et al.*, 2006). Since no anti-capsule antibody was generated by the non-conjugated vaccine group, the protection was most probably due to the free PA alone; the benefit of the peptide conjugate to the formulation is only suggested by the lack of clinical signs post challenge. A vaccine group immunized with just the conjugated PA–PGA peptide is necessary to show the contribution of the conjugate alone; however, this group was not included in the study. Purified capsule has also been conjugated to other proteins such as the outer membrane protein complex of *Neisseria meningitidis* serotype B (Joyce *et al.*, 2006), *Pseudomonas aeruginosa* exotoxin A, and tetanus toxoid (Kubler-Kielb *et al.*, 2006); all elicited anti-capsule antibody responses (Joyce *et al.*, 2006; Kubler-Kielb *et al.*, 2006). Since PA is highly immunogenic on its own and is an effective vaccine against *B. anthracis*, a capsule–PA conjugate would probably improve an anthrax vaccine because it would target both the vegetative bacilli and the toxins associated with the disease.

DNA vaccines

The use of DNA-based vaccines removes the need to introduce a purified protein or a delivery vehicle since the host generates the antigen *in vivo*. The antigen is encoded on a vector under the control of a mammalian or viral promoter. To date there are no FDA-approved human DNA-based vaccines, but according to www.clinicaltrials.gov, numerous clinical trials are under way or have been completed. The potential for human DNA vaccines is demonstrated by the approval of animal DNA vaccines against West Nile virus in horses, haematopoietic necrosis virus in salmon, and melanoma in dogs (Moss, 2009). *B. anthracis* DNA vaccine development is in its infancy, but numerous studies with PA-based DNA vaccines have indicated that an anti-PA antibody response can be elicited in various models (Galloway *et al.*, 2004; Hermanson *et al.*, 2004; Hashimoto *et al.*, 2005; Hahn *et al.*, 2006a; McConnell *et al.*, 2007; Luxembourg *et al.*, 2008; Keitel *et al.*, 2009; Liu, *et al.*, 2009; Livingston *et al.*, 2010). While DNA vaccines do show promise in some animal models, they are not as efficacious as AVA in non-human primate models. The use of different adjuvants and delivery methods may improve the efficacy of these vaccines in the future.

Passive immunization

As mentioned above, upon exposure to *B. anthracis*, the current treatment for an unvaccinated individual involves a 60-day antibiotic regimen and possibly vaccination. If the infection is discovered and treated relatively quickly after exposure, the patient's prognosis is good since the bacilli have not had the chance to replicate unchecked in the host and produce LT and ET. One therapeutic strategy is to neutralize the anthrax toxins with antibodies and inhibit vegetative growth with antibiotics; this strategy has been shown to be effective in DBA/2 mice in which ciprofloxacin and IgG anti-PA administered together protected 90–100% of mice from *B. anthracis* Sterne challenge, while ciprofloxacin alone only protected 50% of the mice (Karginov *et al.*, 2004). Several neutralizing antibody therapeutics are currently in development (Table 11.2) (Little *et al.*, 1988; Sawada-Hirai *et al.*, 2004; Wang *et al.*, 2004; Mohamed *et al.*, 2005; Peterson *et al.*, 2006, 2007; Vitale *et al.*, 2006; Migone *et al.*, 2009). All of the passive antibody therapies in development show promise, particularly for a post-exposure setting. In the case of the monoclonal anti-PA antibodies, it may be possible to administer a cocktail of therapeutics with different modes of action (as long as they do not interfere with each other). Co-administration of one or more of these immunotherapies in conjunction with an antibiotic regimen should provide a robust and effective treatment against *B. anthracis*.

Anti-spore vaccination strategies

The proven success of the Russian and Chinese spore-based human vaccinations and worldwide veterinary vaccinations with live, attenuated spores demonstrates the efficacy of spores as a vaccine. Nonetheless, protection from intoxication is an important component of a *B. anthracis* vaccine as well, as has been demonstrated by the failure of spores from non-toxigenic, non-encapsulated strains to protect guinea pigs from *B. anthracis* Ames challenge. However, when the bacteria were complemented with a plasmid that encodes the PA gene, the spores generated from those vegetative cells provided some protection (Barnard and Friedlander, 1999). Furthermore,

NMRI mice vaccinated with rPA and live Sterne spores were protected from *B. anthracis* Ames challenge (Hahn *et al.*, 2006b). Detoxified strains of *B. anthracis* in which the genes that encode the toxin components have been mutated to render the toxins inactive have provided comparable protective immunity to *B. anthracis* Sterne in animals (Brossier *et al.*, 1999, 2000). Nevertheless, concerns of residual toxicity, virulence, and safety associated with the use of any human live-spore vaccine would preclude its use in the US and Europe.

Inactivated spore vaccine

The use of spore particles as an immunogen requires permanent inactivation of the spores, and one strategy to accomplish this spore inactivation is treatment with formalin. Immunization with formalin inactivated spores (FIS) of a non-toxigenic *B. anthracis* strain has been shown to provide no protection against s.c. challenge with virulent *B. anthracis* 9602 in mice and minimal protection (33%) in guinea pigs. However, when PA was added to the FIS vaccination, 100% protection was observed in both animal models; immunization with PA alone afforded similar protection to FIS alone (Brossier *et al.*, 2002). In a later study, mice vaccinated with any combination of PA and FIS s.c. or i.n. were not protected against i.n. challenge with *B. anthracis* 9602. In guinea pigs challenged i.n. or by aerosol with *B. anthracis* 9602, only FIS and PA administered together provided 100% or 80% protection, respectively, and no protection was afforded by immunization with PA alone (Gauthier *et al.*, 2009). Guinea pigs are a better animal model than mice to study the predicted efficacy of a potential *B. anthracis* vaccine or therapeutic in humans because the guinea pig immune system is more similar to the human immune system. The high level of protection attained after immunization with FIS and PA is very promising and highlights the need to develop vaccines that target both the infectious particle and the toxins. In addition, the capacity of inactive whole spores to contribute significantly to protection against disease suggests that individual components of the spores themselves may confer protection.

Spore proteins as vaccine candidates

The exosporium is the outermost, 'balloon-like' layer of the *B. anthracis* spore and comprises mostly lipids, proteins, and carbohydrates (Matz *et al.*, 1970; Beaman *et al.*, 1971). As determined from electron microscopy and X-ray diffraction studies of the *B. cereus* exosporium, the exosporium itself consists of two layers, an inner basal layer that displays crystal-like hexagonal properties and an outer hairy nap layer (Gerhardt and Ribi, 1964). More recent electron microscopy experiments have shown arrays of crown-like structures, each of which surrounds a cavity, and the presence of 23–34Å pores that permeate the basal layer. This configuration indicates that the exosporium is a semi–permeable barrier that allows the passage of germinants and other small molecules through this outermost layer of the spore (Ball *et al.*, 2008). Since the exosporium makes up the outermost layer of a *B. anthracis* spore and, as such, is the part of the spore that interacts with the host environment, the proteins associated with and embedded in the exosporium should be a primary focus for possible vaccine candidates. To date, almost 40 exosporium proteins or exosporium-associated proteins have been identified in the literature (Table 11.3). The majority of these proteins are localized to the basal layer of the exosporium while the nap layer consists of the glycoproteins BclA (Sylvestre *et al.*, 2002; Steichen *et al.*, 2003) and possibly BclB (Waller *et al.*, 2005; Thompson *et al.*, 2007; Thompson and Stewart, 2008).

Antibodies raised against *B. anthracis* Sterne spores or purified exosporium react predominately with BclA, specifically to the protein rather than to the carbohydrate constituents (Steichen *et al.*, 2003). The possible vaccine benefits of BclA were first demonstrated with a combination DNA vaccine that contained plasmids that encoded the genes for PA and BclA; vaccination of NMRI mice with either plasmid separately conferred significant protection (Hahn *et al.*, 2006b). Immunization of A/J mice with recombinant non-glycosylated BclA (rBclA) increased the mean time to death (MTD) after s.c. challenge with *B. anthracis* Sterne (Brahmbhatt *et al.*, 2007a).

Table 11.3 Predicted exosporium proteins

Gene ID	Name	Reference
BA0108	Translation elongation factor Tu	Liu *et al.* (2004), Redmond *et al.* (2004)
BA0252	Alanine racemase	Steichen *et al.* (2003), Liu *et al.* (2004), Redmond *et al.* (2004)
BA0267	GroEL	Charlton *et al.* (1999), Redmond *et al.* (2004)
BA0355	CotB homologue	Redmond *et al.* (2004)
BA0803	CotJC	Lai *et al.* (2003), Liu *et al.* (2004)
BA0804	CotJB	Liu *et al.* (2004)
BA0805	CotJA	Liu *et al.* (2004)
BA0887	S-layer protein EA1	Redmond *et al.* (2004)
BA1222	BclA	Sylvestre *et al.* (2002),Steichen *et al.* (2003)
BA1234	CotZ1	Liu *et al.* (2004), Redmond *et al.* (2004)
BA1237	BxpB/ExsFA	Liu *et al.* (2004), Redmond *et al.* (2004)
BA1238	CotY/CotZ2	Liu *et al.* (2004), Redmond *et al.* (2004)
BA1359	ExsB	Todd *et al.* (2003)
BA1489	Fe-Mn superoxide dismutase (SOD15)	Liu *et al.* (2004)
BA1629	Cold-shock protein (CspB-1)	Redmond *et al.* (2004)
BA1786	ExsE	Todd *et al.* (2003)
BA2150	ExsG	Todd *et al.* (2003)
BA2162	BxpA	Steichen *et al.* (2003), Liu *et al.* (2004)
BA2292	Hypothetical protein (p2138)	Liu *et al.* (2004)
BA2332	BxpC	Boydston *et al.* (2006)
BA2450	BclB	Waller *et al.* (2005), Thompson *et al.* (2007), Thompson and Stewart (2008)
BA2554	ExsK/hypothetical protein (p2377)	Liu *et al.* (2004), Redmond *et al.* (2004)
BA2617	ExsD	Todd *et al.* (2003)
BA2894	ExsC	Todd *et al.* (2003)
BA3211	Hypothetical protein (p2986)	Lai *et al.* (2003)
BA3668	Glycosyl hydrolase, family 18	Liu *et al.* (2004)
BA3906	CotE	Todd *et al.* (2003)
BA4181	Dihydrolipoamide dehydrogenase (PdhD)	Redmond *et al.* (2004)
BA4266	Hypothetical protein (p3957)	Liu *et al.* (2004)
BA4387	Leucine dehydrogenase	Redmond *et al.* (2004)
BA4499	Mn superoxide dismutase (SODA1)	Liu *et al.* (2004)
BA4722	ThiJ/PfpI family protein	Liu *et al.* (2004)
BA4898	Small, acid-soluble spore protein B	Liu *et al.* (2004)
BA5640	Cell wall hydrolase	Liu *et al.* (2004)
BA5641	YwdL	Liu *et al.* (2004)
BA5699	Hypothetical protein (p5303)	Liu *et al.* (2004)

Furthermore, mice immunized with sub-optimal amounts of rPA and boosted with rBclA were fully protected from *B. anthracis* Sterne challenge. In that study, anti-BclA antibodies both increased opsonophagocytosis of spores and inhibited their germination within macrophages (Brahmbhatt *et al.*, 2007a). One possible drawback to the use of BclA as a vaccine candidate is that BclA appears to be unnecessary for full virulence of *B. anthracis*; the *B. anthracis* Sterne Δ*bclA* strain germinated more quickly *in vitro* and caused death of A/J mice more rapidly than the wild-type *B. anthracis* Sterne despite similar LD_{50} values for both strains (Sylvestre *et al.*, 2002; Brahmbhatt *et al.*, 2007b; Oliva *et al.*, 2008); a *B. anthracis* Ames Δ*bclA* derivative was shown to be as virulent as *B. anthracis* Ames as well (Bozue *et al.*, 2007).

In addition to BclA, numerous uncharacterized potential antigens and vaccine candidates are located on the *B. anthracis* exosporium. Of five probable exposporium candidates tested, immunization of A/J mice with BxpB and p5303 separately or in combination was shown to confer protection against s.c. *B. anthracis* Sterne challenge. The vaccine regimen included a suboptimal PA prime immunization followed by immunization with either or both recombinant proteins 15 days later (Cybulski *et al.*, 2008). BxpB is a structural component of the exosporium and, like the *B. anthracis* Δ*bclA* mutant, a Δ*bxpB* mutant is devoid of the characteristic hair-like nap that surrounds the spore (Steichen *et al.*, 2005; Giorno *et al.*, 2009; Severson *et al.*, 2009). Unlike the exosporium of the Δ*bclA* mutant spores, which retain a similar stability to those of wild-type spores, a *B. anthracis* Δ*bxpB* mutant forms a fragile, unstable exosporium around the spore (Sylvestre *et al.*, 2002; McPherson *et al.*, 2010). The p5303 protein has no known function but is highly conserved among *Bacillus* species. There are still many other possible exosporium antigen candidates left to consider and we believe that further exploration is warranted.

Spore germination effectors

Once spores are inhaled, they can remain dormant in the lungs for months after exposure (Glassman, 1966). Before spores undergo germination, they are metabolically inactive and, as such, resistant to antibiotic treatment. One possible anti-spore therapeutic strategy involves inhibition of the germination step. Inhibition of germination would prevent the spore from replication as a vegetative cell and, thus, would halt the disease. Two known classes of germinants in *B. anthracis* include amino acids and nucleotides (Fisher and Hanna, 2005). It is logical then to search for germination inhibitors derived from these two classes of chemicals. *In vitro* experiments have identified the unnatural D-isomers of alanine and histidine as spore germination inhibitors (Hu *et al.*, 2006, 2007). In fact, *B. anthracis* spores have an alanine racemase in their exosporium that allows them to prevent premature germination by conversion of the germinant L-alanine into the germination inhibitor D-alanine when conditions are not optimal for vegetative growth (McKevitt *et al.*, 2007). Both D-alanine and D-histidine have been shown at millimolar concentrations to inhibit germination of spores in macrophages; 70% inhibition was observed with D-alanine, 25–30% inhibition was observed with D-histidine, and greater than 90% inhibition occurred when with both D-amino acids were present (Hu *et al.*, 2006, 2007). D-alanine alone was calculated to have a 50% inhibitory concentration (IC_{50}) value of 20 µM in macrophages (Alvarez *et al.*, 2010). Several purine analogues have also been identified as germination inhibitors *in vitro*. Interestingly, the capacity of a molecule to inhibit germination *in vitro* did not translate to inhibition of spore germination in macrophages; in fact, some analogues that were not inhibitory *in vitro* actually protected the macrophages from *B. anthracis* spore-mediated killing (Akoachere *et al.*, 2007, 2010). The best inhibitor in macrophages was 6-thioguanosine (6-TG), with an IC_{50} value of 4 µM. When 6-TG was combined with D-alanine (both at suboptimal inhibitory concentrations), these two compounds displayed a synergistic inhibitory effect within macrophages (Alvarez *et al.*, 2010).

In addition to small molecule inhibitors of germination, antibodies have been shown to inhibit germination both *in vitro* and in macrophages. Sera from animals vaccinated with a live *B. anthracis* Sterne-like strain were shown to inhibit germination and stimulate phagocytosis by macrophages (Stepanov *et al.*, 1996). Later studies using anti-rPA, anti-AVA sera, and purified anti-PA antibodies suggested that the anti-PA response was the source of the anti-spore activities; the authors hypothesized that there are transient but detectable amounts of PA expressed on the spore surface (Welkos *et al.*, 2001, 2002; Cote *et al.*,

2005). At the very least, PA mRNA was detected by RT-PCR only 15 minutes after the onset of germination, and proteomics studies suggested that PA was expressed very early in the germination process (Cote *et al.*, 2005). While somewhat controversial, the apparent anti-spore properties of anti-PA antibodies suggest that the current or next-generation *B. anthracis* vaccines that contain PA may have the capacity to engage an immune response that will prevent or reduce infection at the germination stage as well as during the toxigenic phase of the disease.

Controversy exists whether alteration of spore germination, either through inhibition or induction, is a sound therapeutic strategy. Mouse experiments have demonstrated that alteration of spore germination directly within the lungs by either inhibition or induction provided no benefit and did not prevent infection; in addition, the authors hypothesized that inhibition or induction of germination could inadvertently increase the severity of the disease (Cote *et al.*, 2009). Additional studies to include a broader screen of germination effectors *in vitro* and *in vivo* are necessary before the use of germination modulation is deemed a non-viable therapeutic strategy.

Summary

The capacity to sporulate provides *B. anthracis* and related species with unique survival advantages that include protection from the environment, longevity of the infective stage, and access to susceptible hosts. As we have summarized, the interaction of spores with host phagocytes in the course of infection also highlights the special adaption that sporulation provides in persistence, host immune cell modulation, and resistance to inactivation until conditions are suitable for vegetative proliferation. Elaboration of toxins by vegetative bacilli is key to lysis of target cells and subsequent death of the host, and PA is the primary target for immunologic protection. However, the spore itself also provides us with an attractive target for protection against anthrax.

Acknowledgements

Preparation of this chapter was supported by funding from the Biological Defense Research Directorate to ADO. The views expressed are those of the authors and do not reflect official policy or position of the Department of the Navy, Department of Defence, or the US government.

References

Abramova, F.A., Grinberg, L.M., Yampolskaya, O.V., and Walker, D.H. (1993). Pathology of inhalational anthrax in 42 cases from the Sverdlovsk outbreak of 1979. Proc. Natl. Acad. Sci. U.S.A. *90*, 2291–2294.

Adone, R., Pasquali, P., La Rosa, G., Marianelli, C., Muscillo, M., Fasanella, A., Francia, M., and Ciuchini, F. (2002). Sequence analysis of the genes encoding for the major virulence factors of *Bacillus anthracis* vaccine strain 'Carbosap'. J. Appl. Microbiol. 93, 117–121.

Akoachere, M., Squires, R.C., Nour, A.M., Angelov, L., Brojatsch, J., and Abel-Santos, E. (2007). Identification of an *in vivo* inhibitor of *Bacillus anthracis* spore germination. J. Biol. Chem. *282*, 12112–12118.

Albrink, W.S., and Goodlow, R.J. (1959). Experimental inhalation anthrax in the chimpanzee. Am. J. Pathol. 35, 1055–1065.

Aloni-Grinstein, R., Gat, O., Altboum, Z., Velan, B., Cohen, S., and Shafferman, A. (2005). Oral spore vaccine based on live attenuated nontoxinogenic *Bacillus anthracis* expressing recombinant mutant Protective Antigen. Infect. Immun. 73, 4043–4053.

Alvarez, Z., Lee, K., and Abel-Santos, E. (2010). Testing nucleoside analogues as inhibitors of *Bacillus anthracis* spore germination *in vitro* and in macrophage cell culture. Antimicrob. Agents Chemother. *54*, 5329–5336.

Artenstein, A.W. (2009). Vaccine: A Biography, 1st edn (Springer, New York).

Aulinger, B.A., Roehrl, M.H., Mekalanos, J.J., Collier, R.J., and Wang, J.Y. (2005). Combining anthrax vaccine and therapy: a dominant-negative inhibitor of anthrax toxin is also a potent and safe immunogen for vaccines. Infect. Immun. 73, 3408–3414.

Avashia, S.B., Riggins, W.S., Lindley, C., Hoffmaster, A., Drumgoole, R., Nekomoto, T., Jackson, P.J., Hill, K.K., Williams, K., Lehman, L., *et al.* (2007). Fatal pneumonia among metalworkers due to inhalation exposure to *Bacillus cereus* containing *Bacillus anthracis* toxin genes. Clin. Infect. Dis. *44*, 414–416.

Baillie, L. (2001). The development of new vaccines against *Bacillus anthracis* J. Appl. Microbiol. *91*, 609–613.

Baillie, L., Hebdon, R., Flick-Smith, H., and Williamson, D. (2003). Characterisation of the immune response to the UK human anthrax vaccine. FEMS Immunol. Med. Microbiol. *36*, 83–86.

Baillie, L.W. (2009). Is new always better than old?: The development of human vaccines for anthrax. Hum. Vaccin. *5*, 806–816.

Baillie, L.W., Rodriguez, A.L., Moore, S., Atkins, H.S., Feng, C., Nataro, J.P., and Pasetti, M.F. (2008). Towards a human oral vaccine for anthrax: the utility of a *Salmonella typhi* Ty21a-based prime-boost immunization strategy. Vaccine *26*, 6083–6091.

Baillie, L., Townend, T., Walker, N., Eriksson, U., and Williamson, D. (2004). Characterization of the human immune response to the UK anthrax vaccine. FEMS Immunol. Med. Microbiol. *42*, 267–270.

Ball, D.A., Taylor, R., Todd, S.J., Redmond, C., Couture-Tosi, E., Sylvestre, P., Moir, A., and Bullough, P.A. (2008). Structure of the exosporium and sublayers of spores of the *Bacillus cereus* family revealed by electron crystallography. Mol. Microbiol. *68*, 947–958.

Barnard, J.P., and Friedlander, A.M. (1999). Vaccination against anthrax with attenuated recombinant strains of *Bacillus anthracis* that produce protective antigen. Infect. Immun. *67*, 562–567.

Baylor, N.W., Egan, W., and Richman, P. (2002). Aluminum salts in vaccines – US perspective. Vaccine *20 Suppl 3*, S18–23.

Beaman, T.C., Pankratz, H.S., and Gerhardt, P. (1971). Paracrystalline sheets reaggregated from solubilized exosporium of *Bacillus cereus* J. Bacteriol. *107*, 320–324.

Bergman, N.H., Passalacqua, K.D., Gaspard, R., Shetron-Rama, L.M., Quackenbush, J., and Hanna, P.C. (2005). Murine macrophage transcriptional responses to *Bacillus anthracis* infection and intoxication. Infect. Immun. 73, 1069–1080.

Beyer, W., and Turnbull, P.C. (2009). Anthrax in animals. Mol. Aspects Med. *30*, 481–489.

Bielinska, A.U., Janczak, K.W., Landers, J.J., Makidon, P., Sower, L.E., Peterson, J.W., and Baker, J.R., Jr. (2007). Mucosal immunization with a novel nanoemulsion-based recombinant anthrax protective antigen vaccine protects against *Bacillus anthracis* spore challenge. Infect. Immun. 75, 4020–4029.

Bloom, W.L., Watson, D.W., Cromartie, W.J., and Freed, M. (1947). Studies on infection with *Bacillus anthracis*; preparation and characterization of an anthracidal substance from various animal tissues. J. Infect. Dis. *80*, 41–52.

Booth, M.G., Hood, J., Brooks, T.J., and Hart, A. (2010). Anthrax infection in drug users. Lancet 375, 1345–1346.

Boydston, J.A., Yue, L., Kearney, J.F., and Turnbough, C.L., Jr. (2006). The ExsY protein is required for complete formation of the exosporium of *Bacillus anthracis*. J. Bacteriol. *188*, 7440–7448.

Bozue, J., Cote, C.K., Moody, K.L., and Welkos, S.L. (2007). Fully virulent *Bacillus anthracis* does not require the immunodominant protein BclA for pathogenesis. Infect. Immun. 75, 508–511.

Bradley, K.A., Mogridge, J., Mourez, M., Collier, R.J., and Young, J.A. (2001). Identification of the cellular receptor for anthrax toxin. Nature *414*, 225–229.

Brahmbhatt, T.N., Darnell, S.C., Carvalho, H.M., Sanz, P., Kang, T.J., Bull, R.L., Rasmussen, S.B., Cross, A.S., and O'Brien, A.D. (2007a). Recombinant exosporium protein BclA of *Bacillus anthracis* is effective as a booster for mice primed with suboptimal amounts of protective antigen. Infect. Immun. 75, 5240–5247.

Brahmbhatt, T.N., Janes, B.K., Stibitz, E.S., Darnell, S.C., Sanz, P., Rasmussen, S.B., and O'Brien, A.D. (2007b). *Bacillus anthracis* exosporium protein BclA affects spore germination, interaction with extracellular matrix proteins, and hydrophobicity. Infect. Immun. 75, 5233–5239.

Brittingham, K.C., Ruthel, G., Panchal, R.G., Fuller, C.L., Ribot, W.J., Hoover, T.A., Young, H.A., Anderson, A.O., and Bavari, S. (2005). Dendritic cells endocytose *Bacillus anthracis* spores: implications for anthrax pathogenesis. J. Immunol. *174*, 5545–5552.

Brossier, F., Mock, M., and Sirard, J.C. (1999). Antigen delivery by attenuated *Bacillus anthracis*: new prospects in veterinary vaccines. J. Appl. Microbiol. *87*, 298–302.

Brossier, F., Weber-Levy, M., Mock, M., and Sirard, J.C. (2000). Role of toxin functional domains in anthrax pathogenesis. Infect. Immun. *68*, 1781–1786.

Brossier, F., Levy, M., and Mock, M. (2002). Anthrax spores make an essential contribution to vaccine efficacy. Infect. Immun. *70*, 661–664.

CDC. (2001). Update: investigation of bioterrorism-related anthrax and interim guidelines for exposure management and antimicrobial therapy, October 2001. MMWR *50*, 909–919.

CDC. (2008). Cutaneous anthrax associated with drum making using goat hides from West Africa – Connecticut, 2007. MMWR *57*, 628–631.

CDC. (2010). Use of anthrax vaccine in the United States – recommendations of the advisory committee on immunization practices (ACIP), 2009. MMWR *59*, 1–25.

Chabot, D.J., Scorpio, A., Tobery, S.A., Little, S.F., Norris, S.L., and Friedlander, A.M. (2004). Anthrax capsule vaccine protects against experimental infection. Vaccine *23*, 43–47.

Charlton, S., Moir, A.J., Baillie, L., and Moir, A. (1999). Characterization of the exosporium of *Bacillus cereus* J. Appl. Microbiol. *87*, 241–245.

Chen, Y., Succi, J., Tenover, F.C., and Koehler, T.M. (2003). Beta-lactamase genes of the penicillin-susceptible *Bacillus anthracis* Sterne strain. J. Bacteriol. *185*, 823–830.

Cleret, A., Quesnel-Hellmann, A., Vallon-Eberhard, A., Verrier, B., Jung, S., Vidal, D., Mathieu, J., and Tournier, J.N. (2007). Lung dendritic cells rapidly mediate anthrax spore entry through the pulmonary route. J. Immunol. *178*, 7994–8001.

Cooper, C.L., Davis, H.L., Morris, M.L., Efler, S.M., Krieg, A.M., Li, Y., Laframboise, C., Al Adhami, M.J., Khaliq, Y., Seguin, I., *et al.* (2004). Safety and immunogenicity of CPG 7909 injection as an adjuvant to Fluarix influenza vaccine. Vaccine *22*, 3136–3143.

Cote, C.K., Rossi, C.A., Kang, A.S., Morrow, P.R., Lee, J.S., and Welkos, S.L. (2005). The detection of protective antigen (PA) associated with spores of *Bacillus anthracis* and the effects of anti-PA antibodies on spore germination and macrophage interactions. Microb. Pathog. *38*, 209–225.

Cote, C.K., Van Rooijen, N., and Welkos, S.L. (2006). Roles of macrophages and neutrophils in the early host response to *Bacillus anthracis* spores in a mouse model of infection. Infect. Immun. *74*, 469–480.

Cote, C.K., DiMezzo, T.L., Banks, D.J., France, B., Bradley, K.A., and Welkos, S.L. (2008). Early interactions between fully virulent *Bacillus anthracis* and macrophages that influence the balance between spore clearance and development of a lethal infection. Microbes Infect. *10*, 613–619.

Cote, C.K., Bozue, J., Twenhafel, N., and Welkos, S.L. (2009). Effects of altering the germination potential of *Bacillus anthracis* spores by exogenous means in a mouse model. J. Med. Microbiol. *58*, 816–825.

Coulson, N.M., Fulop, M., and Titball, R.W. (1994). *Bacillus anthracis* protective antigen, expressed in *Salmonella typhimurium* SL 3261, affords protection against anthrax spore challenge. Vaccine *12*, 1395–1401.

Cromartie, W.J., Watson, D.W., Bloom, W.L., and Heckly, R.J. (1947). Studies on infection with *Bacillus anthracis*. II. The immunological and tissue damaging properties of extracts prepared from lesions of *B. anthracis* infection. J. Infect. Dis. *80*, 14–27.

Cybulski, R.J., Jr., Sanz, P., McDaniel, D., Darnell, S., Bull, R.L., and O'Brien, A.D. (2008). Recombinant *Bacillus anthracis* spore proteins enhance protection of mice primed with suboptimal amounts of protective antigen. Vaccine *26*, 4927–4939.

Cybulski, R.J., Jr., Sanz, P., Alem, F., Stibitz, S., Bull, R.L., and O'Brien, A.D. (2009a). Four superoxide dismutases contribute to *Bacillus anthracis* virulence and provide spores with redundant protection from oxidative stress. Infect. Immun. *77*, 274–285.

Cybulski, R.J., Jr., Sanz, P., and O'Brien, A.D. (2009b). Anthrax vaccination strategies. Mol. Aspects Med. *30*, 490–502.

Dixon, T.C., Fadl, A.A., Koehler, T.M., Swanson, J.A., and Hanna, P.C. (2000). Early *Bacillus anthracis*–macrophage interactions: intracellular survival survival and escape. Cell Microbiol. *2*, 453–463.

Drysdale, M., Heninger, S., Hutt, J., Chen, Y., Lyons, C.R., and Koehler, T.M. (2005). Capsule synthesis by *Bacillus anthracis* is required for dissemination in murine inhalation anthrax. EMBO J. *24*, 221–227.

Dubberke, E.R., Haslam, D.B., Lanzas, C., Bobo, L.D., Burnham, C.A., Grohn, Y.T., and Tarr, P.I. (2011). The ecology and pathobiology of *Clostridium difficile* infections: an interdisciplinary challenge. Zoonoses Public Health *58*, 4–20.

Duc le, H., Hong, H.A., Atkins, H.S., Flick-Smith, H.C., Durrani, Z., Rijpkema, S., Titball, R.W., and Cutting, S.M. (2007). Immunization against anthrax using *Bacillus subtilis* spores expressing the anthrax protective antigen. Vaccine *25*, 346–355.

Ezzell, J.W., and Welkos, S.L. (1999). The capsule of *Bacillus anthracis*, a review. J. Appl. Microbiol. *87*, 250.

Fasanella, A., Losito, S., Trotta, T., Adone, R., Massa, S., Ciuchini, F., and Chiocco, D. (2001). Detection of anthrax vaccine virulence factors by polymerase chain reaction. Vaccine *19*, 4214–4218.

Fellows, P.F., Linscott, M.K., Ivins, B.E., Pitt, M.L., Rossi, C.A., Gibbs, P.H., and Friedlander, A.M. (2001). Efficacy of a human anthrax vaccine in guinea pigs, rabbits, and rhesus macaques against challenge by *Bacillus anthracis* isolates of diverse geographical origin. Vaccine *19*, 3241–3247.

Fisher, N., and Hanna, P. (2005). Characterization of *Bacillus anthracis* germinant receptors *in vitro*. J. Bacteriol. *187*, 8055–8062.

Friedlander, A.M., and Little, S.F. (2009). Advances in the development of next-generation anthrax vaccines. Vaccine *27*, D28–D32.

Galloway, D., Liner, A., Legutki, J., Mateczun, A., Barnewall, R., and Estep, J. (2004). Genetic immunization against anthrax. Vaccine *22*, 1604–1608.

Gaur, R., Gupta, P.K., Banerjea, A.C., and Singh, Y. (2002). Effect of nasal immunization with protective antigen of *Bacillus anthracis* on protective immune response against anthrax toxin. Vaccine *20*, 2836–2839.

Gauthier, Y.P., Tournier, J.-N., Paucod, J.-C., Corre, J.-P., Mock, M., Goossens, P.L., and Vidal, D.R. (2009). Efficacy of a vaccine based on protective antigen and killed spores against experimental inhalational anthrax. Infect. Immun. *77*, 1197–1207.

Gerhardt, P., and Ribi, E. (1964). Ultrastructure of the exosporium enveloping spores of *Bacillus cereus*. J. Bacteriol. *88*, 1774–1789.

Giorno, R., Mallozzi, M., Bozue, J., Moody, K.S., Slack, A., Qiu, D., Wang, R., Friedlander, A., Welkos, S., and Driks, A. (2009). Localization and assembly of proteins comprising the outer structures of the *Bacillus anthracis* spore. Microbiology *155*, 1133–1145.

Glassman, H.N. (1966). Discussion. Bacteriol. Rev. *30*, 657–659.

Glenn, G.M., Kenney, R.T., Ellingsworth, L.R., Frech, S.A., Hammond, S.A., and Zoeteweij, J.P. (2003). Transcutaneous immunization and immunostimulant strategies: capitalizing on the immunocompetence of the skin. Expert Rev. Vaccines 2, 253–267.

Glomski, I.J., Piris-Gimenez, A., Huerre, M., Mock, M., and Goossens, P.L. (2007). Primary involvement of pharynx and peyer's patch in inhalational and intestinal anthrax. PLoS Pathog. *3*, e76.

Glomski, I.J., Dumetz, F., Jouvion, G., Huerre, M.R., Mock, M., and Goossens, P.L. (2008). Inhaled non-capsulated *Bacillus anthracis* in A/J mice: nasopharynx and alveolar space as dual portals of entry, delayed dissemination, and specific organ targeting. Microbes Infect. *10*, 1398–1404.

Goodman, J.W., and Nitecki, D.E. (1966). Immunochemical studies on the poly-gamma-D-glutamyl capsule of *Bacillus anthracis*. I. Characterization of the polypeptide and of the specificity of its reaction with rabbit antisera. Biochemistry *5*, 657–665.

Goosens, P.L. (2009). Animal models of human anthrax: The question for the Holy Grail. Mol. Aspects. Med. *30*, 467–480.

Gorse, G.J., Keitel, W., Keyserling, H., Taylor, D.N., Lock, M., Alves, K., Kenner, J., Deans, L., and Gurwith, M. (2006). Immunogenicity and tolerance of ascending doses of a recombinant protective antigen (rPA102) anthrax vaccine: a randomized, double-blinded, controlled, multicenter trial. Vaccine *24*, 5950–5959.

Grinberg, L.M., Abramova, F.A., Yampolskaya, O.V., Walker, D.H., and Smith, J.H. (2001). Quantitative pathology of inhalational anthrax I: quantitative microscopic findings. Mod. Pathol. *14*, 482–495.

Guerena-Burgueno, F., Hall, E.R., Taylor, D.N., Cassels, F.J., Scott, D.A., Wolf, M.K., Roberts, Z.J., Nesterova, G.V., Alving, C.R., and Glenn, G.M. (2002). Safety and immunogenicity of a prototype enterotoxigenic *Escherichia coli* vaccine administered transcutaneously. Infect. Immun. *70*, 1874–1880.

Guidi-Rontani, C., Weber-Levy, M., Labruyere, E., and Mock, M. (1999). Germination of *Bacillus anthracis* spores within alveolar macrophages. Mol. Microbiol. *31*, 9–17.

Guidi-Rontani, C., Levy, M., Ohayon, H., and Mock, M. (2001). Fate of germinated *Bacillus anthracis* spores in primary murine macrophages. Mol. Microbiol. *42*, 931–938.

Gutting, B.W., Gaske, K.S., Schilling, A.S., Slaterbeck, A.F., Sobota, L., Mackie, R.S., and Buhr, T.L. (2005). Differential susceptibility of macrophage cell lines to *Bacillus anthracis*-Vollum 1B. Toxicol. in Vitro *19*, 221–229.

Hahn, U.K., Aichler, M., Boehm, R., and Beyer, W. (2006a). Comparison of the immunological memory after DNA vaccination and protein vaccination against anthrax in sheep. Vaccine *24*, 4595–4597.

Hahn, U.K., Boehm, R., and Beyer, W. (2006b). DNA vaccination against anthrax in mice-combination of anti-spore and anti-toxin components. Vaccine *24*, 4569–4571.

Hambleton, P., Carman, J.A., and Melling, J. (1984). Anthrax: the disease in relation to vaccines. Vaccine *2*, 125–132.

Harrison, L.H., Ezzell, J.W., Abshire, T.G., Kidd, S., and Kaufmann, A.F. (1989). Evaluation of serologic tests for diagnosis of anthrax after an outbreak of cutaneous anthrax in Paraguay. J. Infect. Dis. *160*, 706–710.

Hashimoto, M., Boyer, J.L., Hackett, N.R., Wilson, J.M., and Crystal, R.G. (2005). Induction of protective immunity to anthrax lethal toxin with a nonhuman primate adenovirus-based vaccine in the presence of preexisting anti-human adenovirus immunity. Infect. Immun. *73*, 6885–6891.

Hermanson, G., Whitlow, V., Parker, S., Tonsky, K., Rusalov, D., Ferrari, M., Lalor, P., Komai, M., Mere, R., Bell, M., *et al.* (2004). A cationic lipid-formulated plasmid DNA vaccine confers sustained antibody-mediated protection against aerosolized anthrax spores. Proc. Natl. Acad. Sci. U.S.A. *101*, 13601–13606.

Hoffmaster, A.R., Ravel, J., Rasko, D.A., Chapman, G.D., Chute, M.D., Marston, C.K., De, B.K., Sacchi, C.T., Fitzgerald, C., Mayer, L.W., *et al.* (2004). Identification of anthrax toxin genes in a *Bacillus cereus* associated with an illness resembling inhalation anthrax. Proc. Natl. Acad. Sci. U.S.A. *101*, 8449–8454.

Hu, H., Sa, Q., Koehler, T.M., Aronson, A.I., and Zhou, D. (2006). Inactivation of *Bacillus anthracis* spores in murine primary macrophages. Cell Microbiol. *8*, 1634–1642.

Hu, H., Emerson, J., and Aronson, A.I. (2007). Factors involved in the germination and inactivation of *Bacillus anthracis* spores in murine primary macrophages. FEMS Microbiol. Lett. *272*, 245–250.

Huang, J., Mikszta, J.A., Ferriter, M.S., Jiang, G., Harvey, N.G., Dyas, B., Roy, C.J., Ulrich, R.G., and Sullivan, V.J. (2007). Intranasal administration of dry powder anthrax vaccine provides protection against lethal aerosol spore challenge. Hum. Vaccin. *3*, 90–93.

Hughes, R., May, A.J., and Widdicombe, J.G. (1956). The role of the lymphatic system in the pathogenesis of anthrax. Br. J. Exp. Pathol. *37*, 343–349.

Ivins, B.E., Ezzell, J.W., Jr., Jemski, J., Hedlund, K.W., Ristroph, J.D., and Leppla, S.H. (1986). Immunization studies with attenuated strains of *Bacillus anthracis*. Infect. Immun. *52*, 454–458.

Ivins, B.E., Pitt, M.L., Fellows, P.F., Farchaus, J.W., Benner, G.E., Waag, D.M., Little, S.F., Anderson, G.W., Jr., Gibbs, P.H., and Friedlander, A.M. (1998). Comparative efficacy of experimental anthrax vaccine candidates against inhalation anthrax in rhesus macaques. Vaccine *16*, 1141–1148.

Jernigan, D.B., Raghunathan, P.L., Bell, B.P., Brechner, R., Bresnitz, E.A., Butler, J.C., Cetron, M., Cohen, M., Doyle, T., Fischer, M., *et al.* (2002). Investigation of bioterrorism-related anthrax, United States, 2001: epidemiologic findings. Emerg. Infect. Dis. *8*, 1019–1028.

Joellenbeck, L.M., Zwanziger, L.L., Durch, J.S., and Strom, B.L., eds. (2002). The anthrax vaccine: is it safe? does it work? (National Academy Press, Washington D.C.).

Joyce, J., Cook, J., Chabot, D., Hepler, R., Shoop, W., Xu, Q., Stambaugh, T., Aste-Amezaga, M., Wang, S., Indrawati, L., *et al.* (2006). Immunogenicity and protective efficacy of *Bacillus anthracis* poly-gamma-D-glutamic acid capsule covalently coupled to a protein carrier using a novel triazine-based conjugation strategy. J. Biol. Chem. *281*, 4831–4843.

Kang, T.J., Basu, S., Zhang, L., Thomas, K.E., Vogel, S.N., Baillie, L., and Cross, A.S. (2008). *Bacillus anthracis* spores and lethal toxin induce IL-1beta via functionally distinct signaling pathways. Eur. J. Immunol. *38*, 1574–1584.

Kang, T.J., Fenton, M.J., Weiner, M.A., Hibbs, S., Basu, S., Baillie, L., and Cross, A.S. (2005). Murine macrophages kill the vegetative form of *Bacillus anthracis*. Infect. Immun. *73*, 7495–7501.

Karginov, V.A., Robinson, T.M., Riemenschneider, J., Golding, B., Kennedy, M., Shiloach, J., and Alibek, K. (2004). Treatment of anthrax infection with combination of ciprofloxacin and antibodies to protective antigen of *Bacillus anthracis*. FEMS Immunol. Med. Microbiol. *40*, 71–74.

Keitel, W.A., Treanor, J.J., El Sahly, H.M., Evans, T.G., Kopper, S., Whitlow, V., Selinsky, C., Kaslow, D.C., Rolland, A., Smith, L.R., *et al.* (2009). Evaluation of a plasmid DNA-based anthrax vaccine in rabbits, nonhuman primates and healthy adults. Hum. Vaccine *5*, 536–544.

Kenney, R.T., Yu, J., Guebre-Xabier, M., Frech, S.A., Lambert, A., Heller, B.A., Ellingsworth, L.R., Eyles, J.E., Williamson, E.D., and Glenn, G.M. (2004). Induction of protective immunity against lethal anthrax challenge with a patch. J. Infect. Dis. *190*, 774–782.

Klas, S.D., Petrie, C.R., Warwood, S.J., Williams, M.S., Olds, C.L., Stenz, J.P., Cheff, A.M., Hinchcliffe, M., Richardson, C., and Wimer, S. (2008). A single immunization with a dry powder anthrax vaccine protects rabbits against lethal aerosol challenge. Vaccine *26*, 5494–5502.

Klee, S.R., Ozel, M., Appel, B., Boesch, C., Ellerbrok, H., Jacob, D., Holland, G., Leendertz, F.H., Pauli, G., Grunow, R., *et al.* (2006). Characterization of *Bacillus anthracis*-like bacteria isolated from wild great apes from Cote d'Ivoire and Cameroon. J. Bacteriol. *188*, 5333–5344.

Klee, S.R., Brzuszkiewicz, E.B., Nattermann, H., Brüggemann, H., Dupke, S., Wollherr, A., Franz, T., Pauli, G., Appel, B., Liebl, W., *et al.* (2010). The genome of a *Bacillus* isolate causing anthrax in chimpanzees combines chromosomal properties of *B. cereus* with *B. anthracis* virulence plasmids. PLoS One *5*, e10986.

Krieg, A.M., Yi, A.-K., Matson, S., Waldschmidt, T.J., Bishop, G.A., Teasdale, R., Koretzky, G.A., and Klinman, D.M. (1995). CpG motifs in bacterial DNA trigger direct B cell activation. Nature *374*, 546–549.

Kubler-Kielb, J., Liu, T.Y., Mocca, C., Majadly, F., Robbins, J.B., and Schneerson, R. (2006). Additional conjugation methods and immunogenicity of *Bacillus anthracis* poly-gamma-D-glutamic acid–protein conjugates. Infect. Immun. *74*, 4744–4749.

Kuznetsovskii, A., Migova, L., Fedorov, Y., Kibirev, Y., Vorob'ev, A., Darmov, I., Borisevich, I., Savinykh, A., and Gorin, O. (2009). Genotyping of *Bacillus anthracis* vaccine strains by multilocus VNTR analysis. Mol. Gen. Microbiol. Virol. *24*, 177–182.

La Rosa, G., Muscillo, M., Sali, M., De Carolis, E., Marianelli, C., Ciuchini, F., Fasanella, A., and Adone, R. (2006). Molecular study of genes involved in virulence regulatory pathways in *Bacillus anthracis* vaccine strain 'Carbosap'. New Microbiol. *29*, 307–310.

Lai, E.M., Phadke, N.D., Kachman, M.T., Giorno, R., Vazquez, S., Vazquez, J.A., Maddock, J.R., and Driks, A. (2003). Proteomic analysis of the spore coats of *Bacillus subtilis* and *Bacillus anthracis*. J. Bacteriol. *185*, 1443–1454.

Lee, D.Y., Chun, J.H., Ha, H.J., Park, J., Kim, B.S., Oh, H.B., and Rhie, G.E. (2009). Poly-gamma-d-glutamic acid and protective antigen conjugate vaccines induce functional antibodies against the protective antigen and capsule of *Bacillus anthracis* in guinea-pigs and rabbits. FEMS Immunol. Med. Microbiol. *57*, 165–172.

Leendertz, F.H., Ellerbrok, H., Boesch, C., Couacy-Hymann, E., Matz-Rensing, K., Hakenbeck, R., Bergmann, C., Abaza, P., Junglen, S., Moebius, Y., *et al.* (2004). Anthrax kills wild chimpanzees in a tropical rainforest. Nature *430*, 451–452.

Leendertz, F.H., Lankester, F., Guislain, P., Neel, C., Drori, O., Dupain, J., Speede, S., Reed, P., Wolfe, N., Loul, S., *et al.* (2006a). Anthrax in Western and Central African great apes. Am. J. Primatol. *68*, 928–933.

Leendertz, F.H., Yumlu, S., Pauli, G., Boesch, C., Couacy-Hymann, E., Vigilant, L., Junglen, S., Schenk, S., and Ellerbrok, H. (2006b). A new *Bacillus anthracis* found in wild chimpanzees and a gorilla from West and Central Africa. PLoS Pathog. *2*, e8.

Lever, M.S., Stagg, A.J., Nelson, M., Pearce, P., Stevens, D.J., Scott, E.A., Simpson, A.J., and Fulop, M.J. (2008). Experimental respiratory anthrax infection in the common marmoset (*Callithrix jacchus*). Int. J. Exp. Pathol. *89*, 171–179.

Little, S.F., and Knudson, G.B. (1986). Comparative efficacy of *Bacillus anthracis* live spore vaccine and protective antigen vaccine against anthrax in the guinea pig. Infect. Immun. *52*, 509–512.

Little, S.F., Leppla, S.H., and Cora, E. (1988). Production and characterization of monoclonal antibodies to the protective antigen component of *Bacillus anthracis* toxin. Infect. Immun. *56*, 1807–1813.

Liu, H., Bergman, N.H., Thomason, B., Shallom, S., Hazen, A., Crossno, J., Rasko, D.A., Ravel, J., Read, T.D., Peterson, S.N., *et al.* (2004). Formation and composition of the *Bacillus anthracis* Endospore. J. Bacteriol. *186*, 164–178.

Liu, S., Crown, D., Miller-Randolph, S., Moayeri, M., Wang, H., Hu, H., Morley, T., and Leppla, S.H. (2009). Capillary morphogenesis protein-2 is the major receptor mediating lethality of anthrax toxin *in vivo*. Proc. Natl. Acad. Sci. U.S.A. *106*, 12424–12429.

Liu, S., Miller-Randolph, S., Crown, D., Moayeri, M., Sastalla, I., Okugawa, S., and Leppla, S.H. (2010). Anthrax toxin targeting of myeloid cells through the CMG2 receptor is essential for establishment of *Bacillus anthracis* infections in mice. Cell Host Microbe *8*, 455–462.

Liu, T.H., Oscherwitz, J., Schnepp, B., Jacobs, J., Yu, F., Cease, K.B., and Johnson, P.R. (2009). Genetic vaccines for anthrax based on recombinant adeno-associated virus vectors. Mol. Ther. *17*, 373–379.

Livingston, B.D., Little, S.F., Luxembourg, A., Ellefsen, B., and Hannaman, D. (2010). Comparative performance of a licensed anthrax vaccine versus electroporation based delivery of a PA encoding DNA vaccine in rhesus macaques. Vaccine *28*, 1056–1061.

Luxembourg, A., Hannaman, D., Nolan, E., Ellefsen, B., Nakamura, G., Chau, L., Tellez, O., Little, S., and Bernard, R. (2008). Potentiation of an anthrax DNA vaccine with electroporation. Vaccine *26*, 5216–5222.

McConnell, M.J., Hanna, P.C., and Imperiale, M.J. (2007). Adenovirus-based prime-boost immunization for rapid vaccination against anthrax. Mol. Ther. *15*, 203–210.

McKevitt, M.T., Bryant, K.M., Shakir, S.M., Larabee, J.L., Blanke, S.R., Lovchik, J., Lyons, C.R., and Ballard, J.D. (2007). Effects of endogenous D-alanine synthesis and autoinhibition of *Bacillus anthracis* germination on *in vitro* and *in vivo* infections. Infect. Immun. *75*, 5726–5734.

McPherson, S.A., Li, M., Kearney, J.F., and Turnbough, C.L., Jr. (2010). ExsB, an unusually highly phosphorylated protein required for the stable attachment of the exosporium of *Bacillus anthracis*. Mol. Microbiol.*76*, 1527–1538.

Marano, N., Plikaytis, B.D., Martin, S.W., Rose, C., Semenova, V.A., Martin, S.K., Freeman, A.E., Li,

H., Mulligan, M.J., Parker, S.D., *et al.* (2008). Effects of a reduced dose schedule and intramuscular administration of anthrax vaccine adsorbed on immunogenicity and safety at 7 months: a randomized trial. JAMA *300*, 1532–1543.

Matyas, G.R., Friedlander, A.M., Glenn, G.M., Little, S., Yu, J., and Alving, C.R. (2004). Needle-free skin patch vaccination method for anthrax. Infect. Immun. *72*, 1181–1183.

Matz, L.L., Beaman, T.C., and Gerhardt, P. (1970). Chemical composition of exosporium from spores of *Bacillus cereus*. J. Bacteriol. *101*, 196–201.

Mayer-Scholl, A., Hurwitz, R., Brinkmann, V., Schmid, M., Jungblut, P., Weinrauch, Y., and Zychlinsky, A. (2005). Human neutrophils kill *Bacillus anthracis*. PLoS Pathog. *1*, e23.

Meselson, M., Guillemin, J., Hugh-Jones, M., Langmuir, A., Popova, I., Shelokov, A., and Yampolskaya, O. (1994). The Sverdlovsk anthrax outbreak of 1979. Science *266*, 1202–1208.

Migone, T.S., Subramanian, G.M., Zhong, J., Healey, L.M., Corey, A., Devalaraja, M., Lo, L., Ullrich, S., Zimmerman, J., Chen, A., *et al.* (2009). Raxibacumab for the treatment of inhalational anthrax. N. Engl. J. Med. *361*, 135–144.

Mikesell, P., Ivins, B.E., Ristroph, J.D., and Dreier, T.M. (1983). Evidence for plasmid-mediated toxin production in *Bacillus anthracis*. Infect. Immun. *39*, 371–376.

Mikszta, J.A., Sullivan, V.J., Dean, C., Waterston, A.M., Alarcon, J.B., Dekker, J.P., Brittingham, J.M., Huang, J., Hwang, C.R., Ferriter, M., *et al.* (2005). Protective immunization against inhalational anthrax: a comparison of minimally invasive delivery platforms. J. Infect. Dis. *191*, 278–288.

Miller, J., Hair, J., Hebert, M., Hebert, L., Roberts, F., Jr, and Weyant, R. (1997). Fulminating bacteremia and pneumonia due to *Bacillus cereus* [published erratum appears in J. Clin. Microbiol. 1997; *35*, 1294]. J. Clin. Microbiol. 35, 504–507.

Mock, M., and Fouet, A. (2001). Anthrax. Ann. Rev. Microbiol. 55, 647–671.

Mohamadzadeh, M., Duong, T., Sandwick, S.J., Hoover, T., and Klaenhammer, T.R. (2009). Dendritic cell targeting of *Bacillus anthracis* protective antigen expressed by *Lactobacillus acidophilus* protects mice from lethal challenge. Proc. Natl. Acad. Sci. U.S.A. *106*, 4331–4336.

Mohamadzadeh, M., Durmaz, E., Zadeh, M., Pakanati, K.C., Gramarossa, M., Cohran, V., and Klaenhammer, T.R. (2010). Targeted expression of anthrax protective antigen by *Lactobacillus gasseri* as an anthrax vaccine. Future Microbiol. 5, 1289–1296.

Mohamed, N., Clagett, M., Li, J., Jones, S., Pincus, S., D'Alia, G., Nardone, L., Babin, M., Spitalny, G., and Casey, L. (2005). A high-affinity monoclonal antibody to anthrax protective antigen passively protects rabbits before and after aerosolized *Bacillus anthracis* spore challenge. Infect. Immun. 73, 795–802.

Monk, P.N., Scola, A.M., Madala, P., and Fairlie, D.P. (2007). Function, structure and therapeutic potential of complement C5a receptors. Br. J. Pharmacol. *152*, 429–448.

Moss, R.B. (2009). Prospects for control of emerging infectious diseases with plasmid DNA vaccines. J. Immune. Based Ther. Vaccines 7, 3.

Niu, M., and Ball, R. (2009). Adverse events after anthrax vaccination reported to the Vaccine Adverse Event Reporting System (VAERS), 1990–2007 [Vaccine 2009; *27*, 290–297]. Vaccine 27, 6654–6655.

Oliva, C.R., Swiecki, M.K., Griguer, C.E., Lisanby, M.W., Bullard, D.C., Turnbough, C.L., Jr., and Kearney, J.F. (2008). The integrin Mac-1 (CR3) mediates internalization and directs *Bacillus anthracis* spores into professional phagocytes. Proc. Natl. Acad. Sci. U.S.A. *105*, 1261–1266.

Oliva, C., Turnbough, C.L., Jr., and Kearney, J.F. (2009). CD14–Mac-1 interactions in *Bacillus anthracis* spore internalization by macrophages. Proc. Natl. Acad. Sci. U.S.A. *106*, 13957–13962.

Osorio, M., Wu, Y., Singh, S., Merkel, T.J., Bhattacharyya, S., Blake, M.S., and Kopecko, D.J. (2009). Anthrax protective antigen delivered by *Salmonella enterica* serovar *Typhi* Ty21a protects mice from a lethal anthrax spore challenge. Infect. Immun. *77*, 1475–1482.

Passalacqua, K.D., Bergman, N.H., Herring-Palmer, A., and Hanna, P. (2006). The superoxide dismutases of *Bacillus anthracis* do not cooperatively protect against endogenous superoxide stress. J. Bacteriol. *188*, 3837–3848.

Peachman, K.K., Rao, M., Alving, C.R., Burge, R., Leppla, S.H., Rao, V.B., and Matyas, G.R. (2006). Correlation between lethal toxin-neutralizing antibody titers and protection from intranasal challenge with *Bacillus anthracis* Ames strain spores in mice after transcutaneous immunization with recombinant anthrax protective antigen. Infect. Immun. *74*, 794–797.

Peterson, J.W., Comer, J.E., Noffsinger, D.M., Wenglikowski, A., Walberg, K.G., Chatuev, B.M., Chopra, A.K., Stanberry, L.R., Kang, A.S., Scholz, W.W., *et al.* (2006). Human monoclonal anti-protective antigen antibody completely protects rabbits and is synergistic with ciprofloxacin in protecting mice and guinea pigs against inhalation anthrax. Infect. Immun. *74*, 1016–1024.

Peterson, J.W., Comer, J.E., Baze, W.B., Noffsinger, D.M., Wenglikowski, A., Walberg, K.G., Hardcastle, J., Pawlik, J., Bush, K., Taormina, J., *et al.* (2007). Human monoclonal antibody AVP-21D9 to protective antigen reduces dissemination of the *Bacillus anthracis* Ames strain from the lungs in a rabbit model. Infect. Immun. *75*, 3414–3424.

Pickering, A.K., and Merkel, T.J. (2004). Macrophages release tumor necrosis factor alpha and interleukin-12 in response to intracellular *Bacillus anthracis* spores. Infect. Immun. *72*, 3069–3072.

Pickering, A.K., Osorio, M., Lee, G.M., Grippe, V.K., Bray, M., and Merkel, T.J. (2004). Cytokine response to infection with *Bacillus anthracis* spores. Infect. Immun. *72*, 6382–6389.

Pomerantsev, A.P., Staritsin, N.A., Mockov, Y.V., and Marinin, L.I. (1997). Expression of cereolysin AB genes in *Bacillus anthracis* vaccine strain ensures protection against experimental hemolytic anthrax infection. Vaccine *15*, 1846–1850.

Potera, C. (2009). PharmAthene secures biodefense position – company stays solely focused on developing bioterrorism countermeasure products in genetic engineering & biotechnology news. Available at: http://www.genengnews.com/keywordsandtools/print/1/12808/.

Powell, A.G., Crozier, J.E., Hodgson, H., and Galloway, D.J. (2011). A case of septicaemic anthrax in an intravenous drug user. BMC Infect. Dis. *11*, 21.

Prausnitz, M.R., Mikszta, J.A., Cormier, M., and Andrianov, A.K. (2009). Microneedle-based vaccines. Curr. Top. Microbiol. Immunol. *333*, 369–393.

Price, B.M., Liner, A.L., Park, S., Leppla, S.H., Mateczun, A., and Galloway, D.R. (2001). Protection against anthrax lethal toxin challenge by genetic immunization with a plasmid encoding the lethal factor protein. Infect. Immun. *69*, 4509–4515.

Quintiliani, R., Jr., and Quintiliani, Sr., R. (2008). Clostridia. In Clinical Infectious Disease, Schlossberg, D., ed. (Cambridge University Press, New York), pp. 929–935.

Ramsay, C.N., Stirling, A., Smith, J., Hawkins, G., Brooks, T., Hood, J., Penrice, G., Browning, L.M., and Ahmed, S. (2010). An outbreak of infection with *Bacillus anthracis* in injecting drug users in Scotland. Euro. Surveill. *15*, 2–4.

Redmond, C., Baillie, L.W., Hibbs, S., Moir, A.J., and Moir, A. (2004). Identification of proteins in the exosporium of *Bacillus anthracis*. Microbiology *150*, 355–363.

Rengel, J., and Bohnel, H. (1994). Berl Munch Tierarztl Wochenschr *107*, 145–149.

Rhie, G.E., Roehrl, M.H., Mourez, M., Collier, R.J., Mekalanos, J.J., and Wang, J.Y. (2003). A dually active anthrax vaccine that confers protection against both bacilli and toxins. Proc. Natl. Acad. Sci. U.S.A. *100*, 10925–10930.

Ross, J.M. (1957). The pathogenesis of anthrax following the administration of spores by the respiratory route. J Pathol. Bacteriol. 73, 485–494.

Rotz, L.D., Khan, A.S., Lillibridge, S.R., Ostroff, S.M., and Hughes, J.M. (2002). Public health assessment of potential biological terrorism agents. Emerg. Infect. Dis. *8*, 225–230.

Ruthel, G., Ribot, W.J., Bavari, S., and Hoover, T.A. (2004). Time-lapse confocal imaging of development of *Bacillus anthracis* in macrophages. J. Infect. Dis. *189*, 1313–1316.

Sabet, M., Cottam, H.B., and Guiney, D.G. (2006). Modulation of cytokine production and enhancement of cell viability by TLR7 and TLR9 ligands during anthrax infection of macrophages. FEMS Immunol. Med. Microbiol. *47*, 369–379.

Sanz, P., Teel, L.D., Alem, F., Carvalho, H.M., Darnell, S.C., and O'Brien, A.D. (2008). Detection of *Bacillus anthracis* spore germination *in vivo* by bioluminescence imaging. Infect. Immun. *76*, 1036–1047.

Sawada-Hirai, R., Jiang, I., Wang, F., Sun, S.M., Nedellec, R., Ruther, P., Alvarez, A., Millis, D., Morrow, P.R., and Kang, A.S. (2004). Human anti-anthrax protective antigen neutralizing monoclonal antibodies derived from donors vaccinated with anthrax vaccine adsorbed. J. Immune. Based Ther. Vaccines *2*, 5.

Schneerson, R., Kubler-Kielb, J., Liu, T.Y., Dai, Z.D., Leppla, S.H., Yergey, A., Backlund, P., Shiloach, J., Majadly, F., and Robbins, J.B. (2003). Poly(gamma-D-glutamic acid) protein conjugates induce IgG antibodies in mice to the capsule of *Bacillus anthracis*: a potential addition to the anthrax vaccine. Proc. Natl. Acad. Sci. U.S.A. *100*, 8945–8950.

Scobie, H.M., Rainey, G.J., Bradley, K.A., and Young, J.A. (2003). Human capillary morphogenesis protein 2 functions as an anthrax toxin receptor. Proc. Natl. Acad. Sci. U.S.A. *100*, 5170–5174.

Scorpio, A., Blank, T.E., Day, W.A., and Chabot, D.J. (2006). Anthrax vaccines: Pasteur to the present. Cell Mol. Life Sci. *63*, 2237–2248.

Sellman, B.R., Mourez, M., and Collier, R.J. (2001a). Dominant negative mutants of a toxin subunit: an approach to therapy of anthrax. Science *292*, 695–697.

Sellman, B.R., Nassi, S., and Collier, R.J. (2001b). Point mutations in anthrax protective antigen that block translocation. J. Biol. Chem. *276*, 8371–8376.

Severson, K.M., Mallozzi, M., Bozue, J., Welkos, S.L., Cote, C.K., Knight, K.L., and Driks, A. (2009). Roles of the *Bacillus anthracis* spore protein ExsK in exosporium maturation and germination. J. Bacteriol. *191*, 7587–7596.

Shakya, K.P., Hugh-Jones, M.E., and Elzer, P.H. (2007). Evaluation of immune response to orally administered Sterne strain 34F2 anthrax vaccine. Vaccine *25*, 5374–5377.

Shatalin, K., Gusarov, I., Avetissova, E., Shatalina, Y., McQuade, L.E., Lippard, S.J., and Nudler, E. (2008). *Bacillus anthracis*-derived nitric oxide is essential for pathogen virulence and survival in macrophages. Proc. Natl. Acad. Sci. U.S.A. *105*, 1009–1013.

Shepard, C.W., Soriano-Gabarro, M., Zell, E.R., Hayslett, J., Lukacs, S., Goldstein, S., Factor, S., Jones, J., Ridzon, R., Williams, I., *et al.* (2002). Antimicrobial postexposure prophylaxis for anthrax: adverse events and adherence. Emerg. Infect. Dis. *8*, 1124–1132.

Shetron-Rama, L.M., Herring-Palmer, A.C., Huffnagle, G.B., and Hanna, P. (2010). Transport of *Bacillus anthracis* from the lungs to the draining lymph nodes is a rapid process facilitated by CD11c+ cells. Microb. Pathog. *49*, 38–46.

Shlyakhov, E.N., and Rubinstein, E. (1994). Human live anthrax vaccine in the former USSR. Vaccine *12*, 727–730.

Sirisanthana, T., Nelson, K.E., Ezzell, J.W., and Abshire, T.G. (1988). Serological studies of patients with cutaneous and oral-oropharyngeal anthrax from northern Thailand. Am. J. Trop. Med. Hyg. *39*, 575–581.

Steichen, C., Chen, P., Kearney, J.F., and Turnbough, C.L., Jr. (2003). Identification of the immunodominant protein and other proteins of the *Bacillus anthracis* exosporium. J. Bacteriol. *185*, 1903–1910.

Steichen, C.T., Kearney, J.F., and Turnbough, C.L., Jr. (2005). Characterization of the exosporium basal layer protein BxpB of *Bacillus anthracis*. J. Bacteriol. *187*, 5868–5876.

Stepanov, A.V., Marinin, L.I., Pomerantsev, A.P., and Staritsin, N.A. (1996). Development of novel vaccines against anthrax in man. J. Biotechnol. *44*, 155–160.

Stern, E.J., Uhde, K.B., Shadomy, S.V., and Messonnier, N. (2008). Conference report on public health and clinical guidelines for anthrax. Emerg. Infect. Dis. *14*.

Sterne, M. (1939). The use of anthrax vaccines prepared from avirulent (unencapsulated) variants of *Bacillus anthracis*. Onderstepoort J. Vet Sci. Animal Ind. *13*, 307–312.

Stokes, M.G., Titball, R.W., Neeson, B.N., Galen, J.E., Walker, N.J., Stagg, A.J., Jenner, D.C., Thwaite, J.E., Nataro, J.P., Baillie, L.W., *et al.* (2007). Oral administration of a *Salmonella enterica*-based vaccine expressing *Bacillus anthracis* protective antigen confers protection against aerosolized *B. anthracis*. Infect. Immun. *75*, 1827–1834.

Sylvestre, P., Couture-Tosi, E., and Mock, M. (2002). A collagen-like surface glycoprotein is a structural component of the *Bacillus anthracis* exosporium. Mol. Microbiol.*45*, 169–178.

Terra, J.K., Cote, C.K., France, B., Jenkins, A.L., Bozue, J.A., Welkos, S.L., LeVine, S.M., and Bradley, K.A. (2010). Cutting edge: resistance to *Bacillus anthracis* infection mediated by a lethal toxin sensitive allele of Nalp1b/Nlrp1b. J. Immunol. *184*, 17–20.

Thompson, B.M., and Stewart, G.C. (2008). Targeting of the BclA and BclB proteins to the *Bacillus anthracis* spore surface. Mol. Microbiol. *70*, 421–434.

Thompson, B.M., Waller, L.N., Fox, K.F., Fox, A., and Stewart, G.C. (2007). The BclB glycoprotein of *Bacillus anthracis* is involved in exosporium integrity. J. Bacteriol. *189*, 6704–6713.

Todd, S.J., Moir, A.J., Johnson, M.J., and Moir, A. (2003). Genes of *Bacillus cereus* and *Bacillus anthracis* encoding proteins of the exosporium. J. Bacteriol. *185*, 3373–3378.

Turnbull, P.C. (1991). Anthrax vaccines: past, present and future. Vaccine *9*, 533–539.

Turnbull, P.C. (1998). Guidelines for the Surveillance and Control of Anthrax in Humans and Animals (World Health Organization Division of Emerging & Other Communicable Diseases, Geneva, Switzerland).

Turnbull, P.C. (2000). Current status of immunization against anthrax: old vaccines may be here to stay for a while. Curr. Opin. Infect. Dis. *13*, 113–120.

Turnbull, P.C. (2008). Anthrax in humans and animals, 4th edn (WHO Press, Geneva, Switzerland).

Turnbull, P.C., Broster, M.G., Carman, J.A., Manchee, R.J., and Melling, J. (1986). Development of antibodies to protective antigen and lethal factor components of anthrax toxin in humans and guinea pigs and their relevance to protective immunity. Infect. Immun. *52*, 356–363.

Tutrone, W.D., Scheinfeld, N.S., and Weinberg, J.M. (2002). Cutaneous anthrax: a concise review. Cutis *69*, 27–33.

Twenhafel, N.A. (2010). Pathology of inhalational anthrax animal models. Vet. Pathol. *47*, 819–830.

Vasconcelos, D., Barnewall, R., Babin, M., Hunt, R., Estep, J., Nielsen, C., Carnes, R., and Carney, J. (2003). Pathology of inhalation anthrax in cynomolgus monkeys (*Macaca fascicularis*). Lab. Invest. *83*, 1201–1209.

Velimirovic. (2008). Anthrax and other *Bacillus* species. In Clinical Infectious Disease, Schlossberg, D., ed. (Cambridge University Press, New York), pp. 897–902.

Vietri, N.J., Purcell, B.K., Lawler, J.V., Leffel, E.K., Rico, P., Gamble, C.S., Twenhafel, N.A., Ivins, B.E., Heine, H.S., Sheeler, R., *et al.* (2006). Short-course postexposure antibiotic prophylaxis combined with vaccination protects against experimental inhalational anthrax. Proc. Natl. Acad. Sci. U.S.A. *103*, 7813–7816.

Vitale, L., Blanset, D., Lowy, I., O'Neill, T., Goldstein, J., Little, S.F., Andrews, G.P., Dorough, G., Taylor, R.K., and Keler, T. (2006). Prophylaxis and therapy of inhalational anthrax by a novel monoclonal antibody to protective antigen that mimics vaccine-induced immunity. Infect. Immun. *74*, 5840–5847.

Waller, L.N., Stump, M.J., Fox, K.F., Harley, W.M., Fox, A., Stewart, G.C., and Shahgholi, M. (2005). Identification of a second collagen-like glycoprotein produced by *Bacillus anthracis* and demonstration of associated spore-specific sugars. J. Bacteriol. *187*, 4592–4597.

Wang, F., Ruther, P., Jiang, I., Sawada-Hirai, R., Sun, S.M., Nedellec, R., Morrow, P.R., and Kang, A.S. (2004). Human monoclonal antibodies that neutralize anthrax toxin by inhibiting heptamer assembly. Hum. Antibodies *13*, 105–110.

Weaver, J., Kang, T.J., Raines, K.W., Cao, G.L., Hibbs, S., Tsai, P., Baillie, L., Rosen, G.M., and Cross, A.S. (2007). Protective role of *Bacillus anthracis* exosporium in macrophage-mediated killing by nitric oxide. Infect. Immun. *75*, 3894–3901.

Weeratna, R., Comanita, L., and Davis, H.L. (2003). CPG ODN allows lower dose of antigen against hepatitis B surface antigen in BALB/c mice. Immunol. Cell Biol. *81*, 59–62.

Welkos, S.L., and Friedlander, A.M. (1988a). Comparative safety and efficacy against *Bacillus anthracis* of protective antigen and live vaccines in mice. Microbial Pathogenesis *5*, 127–139.

Welkos, S.L., and Friedlander, A.M. (1988b). Pathogenesis and genetic control of resistance to the Sterne strain of *Bacillus anthracis*. Microb Pathog *4*, 53–69.

Welkos, S.L., Keener, T.J., and Gibbs, P.H. (1986). Differences in susceptibility of inbred mice to *Bacillus anthracis*. Infect. Immun. *51*, 795–800.

Welkos, S.L., Trotter, R.W., Becker, D.M., and Nelson, G.O. (1989). Resistance to the Sterne strain of *B. anthracis*: phagocytic cell responses of resistant and susceptible mice. Microb. Pathog. *7*, 15–35.

Welkos, S., Little, S., Friedlander, A., Fritz, D., and Fellows, P. (2001). The role of antibodies to *Bacillus anthracis* and anthrax toxin components in inhibiting the early stages of infection by anthrax spores. Microbiology *147*, 1677–1685.

Welkos, S., Friedlander, A., Weeks, S., Little, S., and Mendelson, I. (2002). *In vitro* characterisation of the phagocytosis and fate of anthrax spores in macrophages and the effects of anti-PA antibody. J. Med. Microbiol. *51*, 821–831.

Whiting, G.C., Rijpkema, S., Adams, T., and Corbel, M.J. (2004). Characterisation of adsorbed anthrax vaccine by two-dimensional gel electrophoresis. Vaccine *22*, 4245–4251.

Williamson, E.D., Hodgson, I., Walker, N.J., Topping, A.W., Duchars, M.G., Mott, J.M., Estep, J., LeButt, C., Flick-Smith, H.C., Jones, H.E., *et al.* (2005). Immunogenicity of recombinant protective antigen and efficacy against aerosol challenge with anthrax. Infect. Immun. *73*, 5978–5987.

Wimer-Mackin, S., Hinchcliffe, M., Petrie, C.R., Warwood, S.J., Tino, W.T., Williams, M.S., Stenz, J.P., Cheff, A., and Richardson, C. (2006). An intranasal vaccine targeting both the *Bacillus anthracis* toxin and bacterium provides protection against aerosol spore challenge in rabbits. Vaccine *24*, 3953–3963.

Wright, J.G., Quinn, C.P., Shadomy, S., Messonnier, N., Centers for Disease, C., and Prevention (2010). Use of anthrax vaccine in the United States: recommendations of the Advisory Committee on Immunization Practices (ACIP), 2009. MMWR *59*, 1–30.

Yan, M., Roehrl, M.H., Basar, E., and Wang, J.Y. (2008). Selection and evaluation of the immunogenicity of protective antigen mutants as anthrax vaccine candidates. Vaccine *26*, 947–955.

Zaucha, G.M., Pitt, L.M., Estep, J., Ivins, B.E., and Friedlander, A.M. (1998). The pathology of experimental anthrax in rabbits exposed by inhalation and subcutaneous inoculation. Arch. Pathol. Lab. Med. *122*, 982–992.

Zegers, N.D., Kluter, E., van Der Stap, H., van Dura, E., van Dalen, P., Shaw, M., and Baillie, L. (1999). Expression of the protective antigen of *Bacillus anthracis* by *Lactobacillus casei*: towards the development of an oral vaccine against anthrax. J. Appl. Microbiol. *87*, 309–314.

Zeng, M., Xu, Q., and Pichichero, M.E. (2007). Protection against anthrax by needle-free mucosal immunization with human anthrax vaccine. Vaccine *25*, 3588–3594.

Heterogeneity in Bacterial Spore Populations

12

Peter Setlow, Jintao Liu and James R. Faeder

Abstract

Genetically identical populations of spores of *Bacillus* and *Clostridium* species invariably exhibit significant heterogeneity, including wide variation in individual spores' resistance to wet heat, speed of germination, and requirements for heat activation before germination. This heterogeneity appears due to both stochasticity in gene expression and variations in culture conditions during sporulation, and presumably facilitates the survival of spore populations. However, small percentages of individuals in spore populations that germinate extremely slowly (superdormant spores) or are much more resistant to environmental insults greatly complicates spore eradication in the food and health care industries. Recently a number of methods have been developed to simultaneously analyse the behaviour and biochemical properties of multiple individual spores in populations, in particular their (i) response to wet heat; (ii) germination; (iii) levels of specific components including dipicolinic acid and individual proteins; and (iv) protein status, whether native or denatured. Results using these methods have led to a much greater appreciation of the scope of spore heterogeneity. In addition, analysis of the germination of multiple individual spores has facilitated the development of a quantitative model for spore germination, and such a model suggests experiments to further elucidate the mechanism(s) of spore germination.

Introduction

Spores of various Gram-positive bacteria of *Bacillus* and *Clostridium* species and their close relatives are metabolically dormant and extremely resistant to a host of environmental stresses (Setlow, 2006; Setlow and Johnson, 2007). These spore formers are found in diverse habitats, and thus are constantly interacting with both humans and our foodstuffs. The ability of spore-forming species to cause disease and the importance of spores in food spoilage have made spores of major concern to the food and health industries. The emergence of spores of *Bacillus anthracis* as a potential weapon of biological terrorism has further heightened the awareness of bacterial spores and their potential to cause harm.

Given the potential deleterious effects of spores of these various species, there has been an enormous amount of work on appropriate methods for spore elimination. Indeed, many of the regulations for sterilization of foods and medical products have been developed to ensure the killing of spores, in particular the killing of spores that are (i) among the most resistant; and (ii) potentially the most harmful – for example, spores of *Clostridium botulinum* in the food industry (Setlow and Johnson, 2007). Work over the past years has also led to great progress in the understanding of spore structure, chemical composition, resistance, killing and germination. One striking feature that has become clear from this work is that populations of spores prepared from a single isolate are heterogeneous even though they are genetically identical. This heterogeneity has been most evident in studies of spore resistance and spore germination (Russell, 1982; Gould, 2006; Indest *et al.*, 2009). Thus killing curves of spores during a treatment such as high temperature often show tails of apparently more resistant spores. There can be a number of trivial explanations for such 'tails', for example the presence of a few large clumps of spores in populations. However, even in the absence of spore clumps, it is not uncommon for a small percentage of spores in populations to have much higher heat resistance than the majority of the population. Similarly, in studies of spore germination, it is common to find that while the great majority of spores in a population germinate relatively rapidly, some small percentage germinates extremely slowly ('superdormant spores') (Gould, 2006; Setlow and Johnson, 2007; Indest *et al.*, 2009). The presence of these slow-germinating superdormant spores in spore populations has frustrated attempts to eradicate spores by first stimulating spore germination and then applying a relatively mild treatment to kill the now more sensitive germinated spores. One additional example of the insidious nature of superdormant spores in spore populations concerns decisions on the length of time that individuals exposed to *B. anthracis* spores need to be treated with antibiotics (Brookmeyer *et al.*, 2003; Heine *et al.*, 2007). When *B. anthracis* spores germinate in the human body, they can cause fatal disease. However, germinated spores are readily destroyed by a number of antibiotics, in contrast to dormant spores. Consequently, dormant spores remaining in the body are capable of causing disease after antibiotic treatment is discontinued. One factor in decisions on the length of time that antibiotics need to be taken after a *B. anthracis* spore exposure must therefore be knowledge of the normal heterogeneity in the germination times of such spores.

The heterogeneity in otherwise genetically identical spore populations has been recognized for many years (Vary and Halvorson, 1965; Woese *et al.*, 1968; Hashimoto *et al.*, 1969a,b; Dean and Douthit, 1974; Russell, 1982). However, information on factors involved in such heterogeneity was slow to accumulate, in large part because of the necessity for analysing properties of individual spores in populations. A few studies of single spore's behaviour were carried out more than 40 years ago (Vary and Halvorson, 1965; Hashimoto *et al.*, 1969a,b). These studies showed that the germination of individual spores in populations was very heterogeneous. The number of such studies has increased dramatically in recent years as new methods have been developed. In this review, we will first discuss methods to assess the properties and behaviour of individual spores. We will then focus on the actual heterogeneity in spore populations in terms of chemical and physiological properties. Finally, we will discuss spore heterogeneity with emphasis on the possible advantages to spore populations of this population heterogeneity. We will also present a mathematical model that relates heterogeneity in the physical properties of spores, such as the expression level of germinant receptors, with heterogeneity in spore behavioural properties, such as germination kinetics. We will also discuss how such models may aid in the identification of the key mechanisms underlying the heterogeneity of spore behaviour.

Methods for analysis of spore population heterogeneity

In order to analyse the causes of spore population heterogeneity, methods to examine the properties of individual spores, preferably multiple individual spores are essential. In older work on spore germination heterogeneity, the germination of one individual spore at a time was followed by phase contrast microscopy (Vary and Halvorson, 1965; Hashimoto *et al.*, 1969a,b; Billon *et al.*, 1997; Leuschner and Lillford, 1999). These works showed that spore germination heterogeneity was due to a variable lag period, T_{lag}, between the time of mixing of nutrient germinants with spores and the time for initiation of dipicolinic acid (DPA) release. However, the necessity for examining one spore at a time made this method cumbersome. In recent years a number of methods have been developed that use phase contrast or differential interference contrast (DIC) microscopy to simultaneously follow the germination of thousands of individual spores over many hours (Zhang *et al.*, 2010a,b,d; Kong *et al.*, 2011). Other microscopy techniques have also been adapted to analyse the germination of multiple individual spores. Thus fluorescence microscopy and flow cytometry have been used to follow spore germination in the presence of nucleic acid dyes that only stain germinated spores (Coote *et al.*, 1995; Tracy *et al.*, 2008; Kong *et al.*, 2010a,b). The potential is also there to use fluorescence-activated cell sorting to isolate remaining dormant spores in spore populations that have germinated ≥95%.

CaDPA release, a major event in spore germination, has also been measured in individual spores, by following the burst of fluorescence when CaDPA is released in the presence of Tb^{3+}, forming highly fluorescent Tb-DPA (Yung *et al.*, 2007). CaDPA release during germination or wet heat treatment has also been followed by Raman spectroscopy of individual spores trapped by laser tweezers [laser tweezers Raman spectroscopy (LTRS)] (Chen *et al.*, 2006; Kong *et al.*, 2011). Consequently, LTRS has provided detailed information on the germination and wet heat-killing of individual spores in populations. LTRS originally analysed only one spore at a time, but can now analyse multiple individual trapped spores simultaneously (Zhang *et al.*, 2010c). Fluorescence spectroscopy and elastic light scattering have also been used to analyse the germination of individual spores trapped by laser tweezers (Peng *et al.*, 2009; Kong *et al.*, 2010a,b, 2011; Zhang *et al.*, 2010d). The germination of individual spores has also been analysed by microbiological methods, in particular using spores of *Clostridium* species (Barker *et al.*, 2005; Stringer *et al.*, 2005; Webb *et al.*, 2007). In these analyses individual wells of microlitre plates are inoculated with ≤1 spore and spore germination, outgrowth and eventual vegetative growth in the wells is examined by microscopic and turbidometric analyses.

The methods described above to follow the germination of individual spores have provided a wealth of information on spore germination heterogeneity. However it is essential that many additional properties of individual dormant spores also be examined. This analysis may well facilitate the elucidation of specific causes of spore germination heterogeneity. The ultimate goal of such efforts is to create a predictive model that relates heterogeneity in spore properties to heterogeneity in spore germination. Such a model would have obvious practical applications to food safety and would also provide a mechanistic understanding of the germination process.

A number of methods have been developed or adapted for analysis of the physical and chemical properties of individual dormant spores as follows.

1. Electron microscopy (EM) can be used to determine individual spore size and dimensions (Carrera *et al.*, 2007). While EM is technically demanding and has the potential for artefacts, with care this can provide valuable information on the distribution of overall sizes of various spore layers in individual spores in populations.
2. Raman spectroscopy can be used to assess individual spores' CaDPA content (Huang *et al.*, 2007), although this normally varies very little in individual spores in populations. Raman spectroscopy can also provide information about the average conformation of protein in spores, and this has been helpful in assessing extents of protein denaturation during spore killing by wet heat (Coleman *et al.*, 2007, 2010).
3. Nano small ion mass spectroscopy (NanoSIMS) can be used to assess the distribution of ions in spores and potentially the permeability properties of individual spores (Ghosal *et al.*, 2008, 2010; Weber *et al.*, 2010).
4. Fourier-transform infrared spectroscopy can be used to determine levels and states of individual spore components (Subramanian *et al.*, 2006, 2007). This technique may also be useful in assessing the state of protein, whether native or denatured, in spores.
5. Phase contrast microscopy and buoyant density gradient centrifugation can be used to determine individual spore core's refractive indices and/or their core water content (the major factor in spore wet heat resistance) (Dean and Douthit, 1974; Gerhardt and Marquis, 1989). Phase contrast microscopy has also been adapted to examine large

numbers of individual spores simultaneously, and in real time (Kong *et al.*, 2011c).

6 Finally, fluorescence microscopy has been used to determine the permeability of individual spore's outer layers (Stecchini *et al.*, 2009). Perhaps even more importantly, fluorescence microscopy can also be used to quantitatively determine individual spore's levels of crucial proteins, when such proteins are expressed as functional fusions to fluorescent reporter proteins. This latter type of analysis is currently being carried out to assess the heterogeneity in levels of various germination proteins in individual *Bacillus subtilis* spores (Griffiths *et al.*, 2011).

The various techniques listed above can provide significant information about variations in properties between individual spores in populations. However, it will be essential to correlate structural and biochemical differences between individual spore with differences in individual spores' physiological behaviour, such as in germination or during high-temperature treatment.

Heterogeneity in spore physical and chemical properties

Determination of the heterogeneity in chemical and physical properties of spores in populations requires that multiple individual spores be analysed to allow comparisons of properties of individual spores. A number of properties of individual spores have been examined in attempts to quantify spore heterogeneity.

Spore size and dimensions

A number of studies have examined the size and dimensions of spores of a variety of species, (Carrera *et al.*, 2007; Stecchini *et al.*, 2009). While there is certainly the potential for artefacts in using electron microscopy to assess overall spore size as well as the dimensions of various spore structural layers, such studies have invariably concluded that there is significant heterogeneity in these properties of individual spores in populations, with variations of ±25% in overall spore size, as well as significant variation in the sizes of spore layers such as the core, cortex and coat.

Spore staining

Examination of spores stained with several fluorescent dyes has indicated that spore populations exhibit rather heterogeneous staining, with some individuals staining well and some only poorly (Magge at al., 2009; Stecchini *et al.*, 2009). The reasons for such differences are not known nor what this might imply about these spores' physiological properties. In addition, it is also not clear whether differences in this spore staining reflect differences due to damage to spores well after spore formation, such as loss of bits of the exosporium in those spores that have this structure, or damage to the spore coat.

Spore CaDPA content

CaDPA constitutes 15–25% of core dry weight, although this value varies between species and between spores of the same strain prepared under different conditions (Gerhardt and Marquis, 1989). Given the latter observation, it is not surprising that the CaDPA content of individual spores in a population also varies as determined by Raman spectroscopy of individual spores (Huang *et al.*, 2007). Some of this variation is probably due to variations in spore size, but variability in CaDPA levels may also contribute to variations in spore core water content (Gerhardt and Marquis, 1989).

Spore core water content

It is well established that the water content in the dormant spore core is invariably much lower than for growing cells or germinated spores (Gerhardt and Marquis, 1989). This value is 25–50% of core wet weight in spores of different species and this low core water content is the major factor in dormant spore resistance to wet heat, and probably also in long-term survival at lower temperatures (Gerhardt and Marquis, 1989). The low core water content is also probably the major factor in spore dormancy, as otherwise soluble proteins appear to be immobile in the spore core (Cowan *et al.*, 2003; Sunde *et al.*, 2009). As with spore CaDPA content, core water content can also vary depending on sporulation conditions (Russell, 1982; Gerhardt and Marquis, 1989; Melly *et al.*, 2002). Consequently it would be surprising if core water content did not vary significantly between individual spores in a population, and there are some data indicating that such variation exists. For example, a small percentage of spores in populations of several *Bacillus* species have significant lower core water content, higher wet heat resistance and germinate more slowly than the population as a whole (Ghosh *et al.*, 2009).

Currently, the physiological effects of the factors noted above are not known. However, It should be possible with current methodology to correlate core water content by phase contrast microscopy, core CaDPA levels by Raman spectroscopy and either wet heat inactivation or spore germination rates using individual spores. Such data would be invaluable in allowing definitive causal correlations between heterogeneity in chemical properties of individual spores and their physiological properties such as wet heat resistance and germination.

Heterogeneity in spores' physiological properties

Given the heterogeneity in spore's physical and chemical properties, it is not surprising that there is also significant heterogeneity in spores' physiological properties. The spore properties that have been best studied in this regard are heat activation prior to spore germination, wet heat resistance and spore germination. In some cases, large amounts of work have been carried out analysing the behaviour of multiple individual spores. In other cases, heterogeneity has been studied by separating spore fractions from initial populations and demonstrating differences in the properties of the initial population and the fractionated spores (Ghosh and Setlow, 2009, 2010; Ghosh *et al.*, 2009).

Heat activation

It is well known that germination of spores can be greatly stimulated by an activation treatment (Keynan and Evenchick, 1969). The most common method for spore activation is short incubation in water at high but sublethal temperatures. The mechanism of this heat activation is not known but the process is reversible and may involve changes in protein structure (Keynan and Evenchick, 1969; Zhang *et al.*, 2009). The requirement for heat activation varies significantly between individual spores in a population. Some spores germinate well without any heat activation, while others exhibit a relatively strict requirement for heat activation. Recent work has identified a dramatic example of this heterogeneity in spores' requirement for heat activation., When spores of a number of *Bacillus* species that were very slow to germinate (e.g. superdormant spores) were isolated, these spores were found to exhibit 8–15°C higher temperature requirements for optimal heat activation than did the initial spore population (Ghosh *et al.*, 2009). This difference was probably not genetic (Ghosh and Setlow, 2009; Ghosh *et al.*, 2009) and was associated with a significantly lower water content in these superdormant spores. However, the reason for this difference in the heat activation requirement of different individual spores has not been determined.

Wet heat resistance

Of the physiological properties known to exhibit significant heterogeneity in individual members of spore populations, the one that is most notable is the level of spore resistance to wet heat. It thus seems appropriate to briefly review the factors that determine the levels of spore heat resistance (Gerhardt and Marquis, 1989; Setlow, 2006). A number of factors determine the level of spore wet heat resistance including: (i) sporulation temperature; (ii) the optimum growth temperature of the strain/species; (iii) the spore's levels of CaDPA and its associated divalent cations; iv) the saturation of spore DNA with the spore-specific DNA binding proteins, the α/β-type small, acid-soluble spore proteins (SASP); and most importantly v) the level of water in the spore core. Of these variables, obviously sporulation temperature and an organism's optimal growth temperature are constants for all individuals in a population of sporulating cells. Levels of α/β-type SASP are also not known to vary significantly between individuals in wild-type spore populations, although certainly could. However, as noted above, levels of spore CaDPA and its associated divalent metal ions can vary significantly depending on the sporulation media, and core CaDPA levels can affect levels of spore core water. The spore core's water content can also be influenced by sporulation conditions such as ionic strength and by the specific structure of the spore's peptidoglycan cortex. Consequently, it appears probably that core water content can vary significantly between spores in a population formed in a single sporulation process. Given that there is generally a strong direct correlation between the core water content of spores in populations and their wet heat resistance, it seems probable that variations in core water content between individual spores will: (i) be present in spore populations; and (ii) result in differences in wet heat resistance of individual spores in populations.

That spore populations do indeed exhibit significant heterogeneity in resistance properties has been known for many years (Russell, 1982; Periago *et al.*, 2004; Kort *et al.*, 2005; Cronin and Wilkinson, 2008; Smelt *et al.*, 2008; Hornstra *et al.*, 2009; Xu *et al.*, 2009; Corradini *et al.*, 2010), but it is only relatively recently that this heterogeneity has been studied at the single spore level. Unfortunately, there is still no perfect way to simultaneously assess the killing of multiple individual spores over extended periods. However, events such as CaDPA release that are known to indicate that a spore is truly dead have been monitored during spore wet heat treatment, although CaDPA release actually takes place after spore death (Coleman *et al.*, 2007, 2010). In one study, individual spores of several *Bacillus* species that were captured by laser tweezers were incubated in water at elevated temperatures and Raman spectroscopy was used to follow the spores' CaDPA content and overall protein conformation (Zhang *et al.*, 2010b). This work found that the release of ≥ 90% of CaDPA and the great majority of protein denaturation took place in a period of only 1–2 minutes. However, the time when these drastic changes began varied greatly between individual spores in a population. These results suggest that individual spores in populations have significantly different wet heat resistance. However, no correlations have yet been made between the precise time at which CaDPA is released from spores incubated at high temperatures in water and some other variable spore property such as core water content. It is also clear that CaDPA release is not the cause of spore killing by wet heat, since work with spore populations has shown that spore killing by wet heat precedes CaDPA release (Coleman *et al.*, 2007, 2010).

Heterogeneity in spore germination

The major variables that determine the rate of germination of spore populations are the efficacy of heat activation, the level of nutrient germinants, and the number of nutrient germinant receptors (nGRs) that recognize the particular germinant in question (Vary and Halvorson, 1965; Hashimoto *et al.*, 1969a,b; Billon *et al.*, 1997; Leuschner and Lillford, 1997; Setlow, 2003; Zhang *et al.*, 2010a). These nGRs are specific for one nutrient germinant or a mixture of nutrient germinants. Although physiological factors that might affect nGR gene expression are not known, average levels of nGR gene expression can vary 3–4-fold depending on the sporulation medium, and spores with higher levels of appropriate nGR expression germinate faster with their cognate nutrient germinant (Cabrera-Martinez *et al.*, 2003; Hornstra *et al.*, 2006). Consequently, increasing the levels of nGR gene expression by placing the genes under stronger forespore-specific promoters: (i) increases appropriate nGR levels markedly; and (ii) results in greatly increased rates of germination with the cognate nutrient germinants. These as well as other data indicate that a major variable that affects spores' germination rates is the number of nGRs per spore (Zhang *et al.*, 2010a). This is especially so since it appears probable that average levels of nGRs per spore are in the tens of molecules (Setlow 2003). In addition, the number of nGRs/spore varies significantly between individual spores in the same population (Setlow 2003; Griffiths *et al.*, 2011).

It is well known that individual spores in populations are quite heterogeneous in their germination, with some germinating rapidly and some germinating only very slowly (Indest *et al.*, 2009). Indeed, some spores in populations germinate only after many hours or even days and have been termed superdormant spores (see 'Superdormant spores' section). Studies carried out over 40 years ago using phase contrast microscopy to study the inosine germination of *Bacillus cereus* spores showed that the major heterogeneity in the germination of these individual spores with this germinant was a highly variable T_{lag} period following germinant addition, during which there was little if any change in phase contrast intensity (Vary and Halvorson, 1965; Hashimoto *et al.*, 1969a,b). This variable T_{lag} period was followed by a rapid loss in ~70% of phase contrast intensity due to CaDPA release. The rate of this rapid loss in phase contrast intensity was relatively constant and independent of the length of T_{lag}. The length of the T_{lag} period was decreased by spore heat activation prior to germinant addition, as well as increasing germinant concentrations. More recent work using phase contrast microscopy, DIC microscopy and LTRS has confirmed and extended these early observations (Zhang *et al.*, 2010a). This recent work has studied the nutrient germination of thousands of individual spores of *B. cereus*, *Bacillus megaterium* and *B. subtilis*, and has shown that times for the rapid release of > 85% of spore CaDPA are relatively constant at 0.5–3 minutes at 37°C and are unaffected by heat activation, germinant concentrations, and levels of nGRs per spore. However, loss of the cortex lytic enzyme CwlJ in either *B. subtilis* or *B. megaterium* spores slows this period of rapid CaDPA release 5- to 15-fold. This period of rapid CaDPA release during spore germination is then followed by a slow further decrease in the spore's refractility due to hydrolysis of the peptidoglycan cortex. The time for this cortex hydrolysis seems to vary somewhat for individual spores, and this is shown most dramatically by the heterogeneity in the release of CaDPA in germination of spores of *Bacillus* species with exogenous CaDPA, since the latter agent triggers germination by activating the cortex-lytic enzyme CwlJ. However, the contribution of heterogeneity in rates of cortex hydrolysis to the overall heterogeneity in spore germination is not clear.

What is clear, however, is that the major contributor to the heterogeneity in rates of spore germination is the large variability in T_{lag} following addition of nutrient germinants and prior to initiation of rapid CaDPA release, since T_{lag} values can vary tremendously for individual spores (Zhang *et al.*, 2010a). As found many years ago and confirmed recently, the value of T_{lag} is decreased by heat activation and also decreases as germinant concentrations increase, at least up to saturating germinant concentrations. T_{lag} values are also decreased markedly by increases in the average numbers of nGRs per spore, and this is true for at least two different nGRs. Overall, these results suggest that the major factor determining the length of the T_{lag} period in the germination of an individual spore is the level of the appropriate 'activated' nGR in that particular spore. The word 'activated' is in quotation marks here, because we do not know exactly what a nGR does or how an 'activated' nGR differs from a non-activated one.

If the major factor in determining the T_{lag} value of individual spores is indeed the level of activated nGRs, then for heat-activated spores germinating with saturating germinant levels, there should be a good correlation between a spore's level of the appropriate nGR and its T_{lag} value. Indeed, preliminary work has indicated that for *B. subtilis* spores there is a fair correlation between the T_{lag} value of individual spores germinating in the nutrient combination of L-asparagine, D-glucose, D-fructose and K^+ (AGFK) and the level of the GerK nGR essential for AGFK germination (J. Zhang, K.K. Griffiths, A.E. Cowan, P. Setlow and J. Yu, unpublished). This correlation was made by examining the fluorescence intensity of a functional GerK nGR subunit fused to a fluorescent protein in individual spores, as well as the times for the germination of these spores with AGFK.

Given that the length of the T_{lag} period appears to be the major variable causing the heterogeneity in spore germination, an obvious question is what is happening in T_{lag} and how do levels of activated nGRs affect what is happening during this period? At present, the only event in T_{lag} that has been reasonably well studied is commitment, whereby a spore treated with a nutrient germinant will continue through germination even after the germinant is removed and further germinant binding is blocked (Yi and Setlow, 2010). Thus spores of *B. cereus* or *B. subtilis* germinating with L-alanine or L-valine exhibit no germination events if given an excess of D-alanine at the same time as the L-alanine is added. However, if D-alanine addition is delayed for a few minutes when only a few per cent of spores have initiated CaDPA release, committed spores will continue through germination and release all CaDPA after D-alanine addition, while uncommitted spores remain dormant. The precise mechanism of commitment is not known, but recent work has indicated that the factors that influence the timing of commitment to germinate ($T_{commitment}$) of spores of *B. cereus* and *B. subtilis* are identical to those that influence values of T_{lag}. Thus, $T_{commitment}$ is decreased by heat activation, elevated germinant concentrations and increased average levels of nGRs per spore. Consequently, the variability in $T_{commitment}$ appears to be a major factor in the variability in spore germination rates, although there also appears to be variability in the time between commitment and CaDPA release for individual spores. However, the factors that influence this latter variability are unclear. Consequently, while precisely how levels of activated nGRs per spore determine $T_{commitment}$ and T_{lag} values is not known, it appears most probable that the variability in levels of activated nGRs per spore is the major factor determining the heterogeneity in spore germination. We will return to this question in discussing a model for spore germination in 'Quantitative model for spore germination,' below.

Superdormant spores

The most extreme example of the heterogeneity in the germination of spores in populations is the presence of superdormant spores, those that germinate many hours after the great majority of spores have germinated (Gould, 2006; Indest *et al.*, 2009). While the existence of such superdormant spores has long been known, it has only been recently that such spores were isolated by repeated germination of spore populations for 1–2 hours and removal of all spores that germinated (Ghosh and Setlow, 2009, 2010; Ghosh *et al.*, 2009). The resulting superdormant spores of *B. cereus*, *B. megaterium* and *B. subtilis* germinated extremely poorly with the germinants used to initially germinate them, although they did germinate fairly well with mixtures of several nutrient germinants and germinated normally with non-nutrient germinants that do not act via the spore's nGRs (Ghosh and Setlow, 2009, 2010; Ghosh *et al.*, 2009; Wei *et al.*, 2010). Analysis of factors that influence the amount of superdormant spores isolated under various conditions indicated that yields of superdormant spores in populations are decreased by: (i) heat activation; (ii) increased levels of nutrient germinants targeting one nGR; (iii) mixtures of nutrient germinants that target several nGR-dependent pathways; and (iv) increased average levels of nGRs per spore. Strikingly, these factors are the same as those that decrease values of T_{lag} and $T_{commitment}$, suggesting that superdormant spores are generally those with the lowest levels of nGRs per spore.

Heterogeneity in spore outgrowth

In addition to significant heterogeneity in the germination process, there is also significant heterogeneity in spore outgrowth, as has been well documented with spores of *Clostridium* species (Barker *et al.*, 2005; Webb *et al.*, 2007; Stringer *et al.*, 2008). However, the causes of this heterogeneity are not known.

Stages of spore germination

Based on our current understanding, germination of a single spore can be divided into four stages (Fig. 12.1) based on two well-characterized events: commitment to germination and CaDPA release. The four stages in order are: (i) pre-commitment;

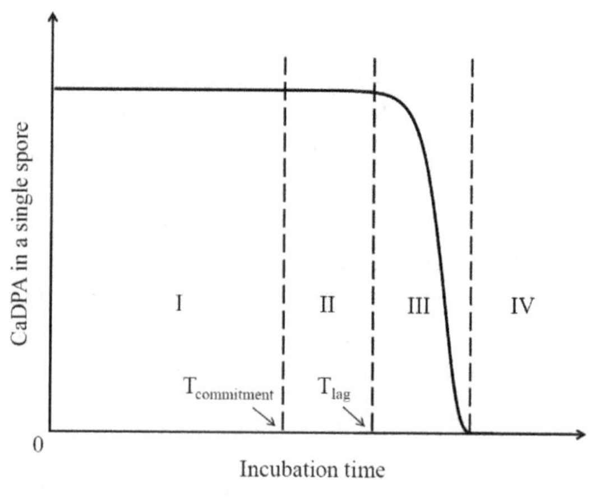

Figure 12.1 Four stages of spore germination with nutrient germinants: (I) commitment to germination, (II) pre-CaDPA release, (III) CaDPA release, and (IV) post-CaDPA release. Note that durations of the various stages are not drawn to scale, and that there can be a small amount (~10%) of slow CaDPA release prior to T_{lag} starting at the beginning of stage I with spores of some species (Chen *et al.*, 2006; Peng *et al.*, 2009; Zhang *et al.*, 2010a; Kong *et al.*, 2011).

(ii) pre-CaDPA release; (iii) CaDPA release; and (iv) post CaDPA release.

Stage I starts from the addition of germinant and ends the moment that the spore commits to germination. Nutrient–nGR interaction plays a major role in Stage I and a large set of experiments have been performed, in particular with *B. subtilis* spores since: (i) the nGRs are already known; (ii) many germinants and inhibitors for each nGR are known; (iii) levels of nGRs can be controlled and probably even measured directly in the near future; and (iv) there are spores with nGR variants that exhibit modified nutrient–nGR interaction. The next question is what other molecules and processes are involved in Stage I, and this will be key to understanding how nutrient–nGR interaction triggers germination, why $T_{\text{commitment}}$ varies between spores and is decreased by heat activation and increasing the nGR level per spore.

Stage II starts from the moment of commitment and ends at the initiation of rapid CaDPA release. Little is known about Stage II, in large part because there is as yet no method for direct measurement of the length and heterogeneity of its duration in individual spores. The length of this stage was suggested to correlate positively with the $T_{\text{commitment}}$ value (Yi and Setlow, 2010), but a different analysis showed that a model assuming no correlation is also possible (J. Liu and J.R. Faeder, unpublished). The molecular changes involved in Stage II are unknown, except that the release of monovalent ions should begin during one of the first two stages, since this event is prior to the release of CaDPA (Swerdlow *et al.*, 1981). While the initial nutrient germinant is no longer required at this stage, it is unclear whether germinant–nGR interaction is still required.

Stage III starts from the initiation of rapid CaDPA release and ends at the completion of CaDPA release. This stage has been well characterized by a series of recent experiments (Chen *et al.*, 2006; Peng *et al.*, 2009; Kong *et al.*, 2010a, 2011). As was mentioned previously, the duration of Stage III is relatively constant and unaffected by heat activation, germinant concentration and levels of nGRs per spore. In addition, experiments measuring the uptake of a nucleic acid dye by single germinating *Bacillus* spores showed that the uptake starts at the beginning of Stage III (Kong *et al.*, 2010b). This finding implies that drastic changes inside the spore core take place in Stage III, although what these changes are is not clear. One remaining question is what determines the duration of CaDPA release, which lasts 3 minutes in wild-type *B. subtilis* spores at 37°C (Peng *et al.*, 2009) but only takes 30 seconds in wild-type *B. cereus* spores (Kong *et al.*, 2010a). The protein CwlJ is certainly one of the players in this stage, since its deletion results in significantly slower CaDPA release.

The final stage in germination, Stage IV, starts from the completion of CaDPA release and covers the remainder of the germination process, including the progression into spore outgrowth. As mentioned, significant heterogeneity has been observed in this stage but little is know about the contributing factors or mechanisms. Since we are only concerned with the mechanism that initiates spore germination, we will not discuss Stage IV further.

Quantitative model for spore germination

The observations about heterogeneity in spore germination presented above can be used to assemble a quantitative model of germination that recapitulates the major experimental observations and suggests several avenues for future experimentation. In the following, we present a quantitative model that focuses on Stage I, which is where we have the most knowledge and can perform the most experiments. The model is a generalization of a model originally proposed by Woese *et al.* (1968). It is based on the fact that the nGR (corresponding to the hypothetical 'germination enzyme' proposed by Woese *et al.*, 1968) level per spore is heterogeneous and on the assumption that the variability in levels of activated nGRs per spore determines the heterogeneity in germination. Fig. 12.2 shows the key elements of the predictive model for single spore germination kinetics (Fig. 12.2D) based on three key elements: (1) the distribution of nGR receptor numbers in a spore population (Fig. 12.2A); (2) the concentration dependence of nutrient germinant binding and activation of the particular nGR (Fig. 12.2B); and (3) the dependence of $T_{\text{commitment}}$ for a spore on the number of bound nGRs

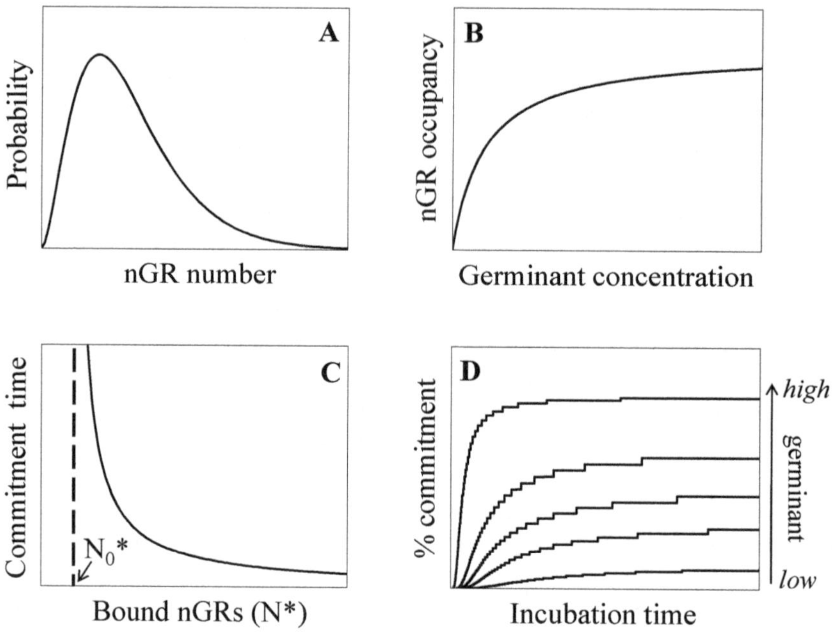

Figure 12.2 Model for spore germination with nutrient germinants. (A–C) Components of the model. (A) Distribution of number of nGRs of a given type per spore in a spore population, which can be determined from experiments using fluorescently tagged nGRs (K.K. Griffiths, J. Zhang, A.E. Cowan, J. Yu and P. Setlow, unpublished). (B) Fraction of nGRs bound to germinant as a function of germinant concentration. (C) Time required for commitment to germinate ($T_{commitment}$) as a function of number of bound nGRs (N^*). No germination happens if N^* is smaller than the threshold N_0^*. (D) Predicted germination kinetics for different germinant levels. The speed and extent of germination increase as germinant concentration is increased.

(Fig. 12.2C). Based on our four-stage division of the germination process presented in 'Stages of spore germination' above, it is clear that the model is directly applicable to prediction of $T_{commitment}$ rather than T_{lag}, since some as yet unknown processes are involved between commitment and the initiation of CaDPA release, which do not require the continued presence of the nutrient germinant.

Of the three basic components of the model, only the nGR distribution (Fig. 12.2A) can be directly determined from experiments. The other components, the nGR activation and the commitment time curves must be determined by fitting data for germination kinetics as a function of germinant concentration. The nGRs are present at low levels, with 24–40 molecules per spore for the GerB receptor (Paidhungat and Setlow, 2001) and other nGRs are expected to be at similar levels. In addition, recent experiments with spores expressing fluorescently labelled nGRs have provided detailed quantitative information about receptor distributions in spores (Griffiths *et al.*, 2011). Preliminary analysis of these data (J. Liu and J.R. Faeder, unpublished) shows that, instead of the Poisson distribution assumed by Woese *et al.* (1968), the nGR number on a single spore follows a gamma distribution, as was recently shown to be the case for most of the proteins in the *E. coli* proteome (Taniguchi *et al.*, 2010).

We assume nGRs are activated by germinant binding (Fig. 12.1B). In addition, binding and activation are assumed to be fast in comparison to the time required for germination. These assumptions appear to be reasonable given that germinants are typically small molecules, so binding equilibration times should be on the order of milliseconds and the associated nGR conformational changes are expected to be even faster. This step was only minimally studied by Woese *et al.* (1968), probably due to the unclear nature of the 'germination enzyme' and the large amount of labour needed to measure spore germination at many different conditions using the methods for analysing single spores available at that time. A recent preliminary analysis of germination kinetics of *B. subtilis* spores showed that activation of either the GerA or GerB* nGRs (GerB* is a variant nGR that can trigger germination with L-asparagine alone; Paidhungat and Setlow, 1999; Setlow, 2003) was consistent with non-cooperative binding of germinants to each of these nGRs (J. Liu, X. Yi, P. Setlow and J.R. Faeder, unpublished). However, it has not been ruled out that the nGRs may organize and function as higher order complexes, especially given that they have been shown to be localized into clusters on the inner membrane of the spore (Griffiths *et al.*, 2011).

The number of bound nGRs in a spore is the product of the number of nGRs in the spore and the average nGR occupancy at the given germinant concentration. Therefore, increased germinant concentrations would increase levels of bound nGRs per spore, as would higher average levels of total nGRs per spore. We assume $T_{commitment}$ depends only on the number of bound nGRs in a spore (Fig. 12.1C). This assumption is a simplification because other spore properties, such as the expression level of other components of the germination pathway or spore water content are also likely to affect the germination rate. The importance of such factors is possibly reflected by the imperfect correlation of individual spore's germination times with nGR levels (Griffiths *et al.*, 2011). We also assume that below a threshold number of bound nGRs, N^*, germination does not occur, but above the threshold $T_{commitment}$ decreases with increasing numbers of bound nGRs N^* (note that the value of N^* may be different for different types of nGRs). The assumption of an activation threshold is required to capture the phenomenon that the percentage of spores that germinate decreases as the nutrient concentration is decreased. Woese *et al.* (1968) postulated a particular mechanism for the dependence of T_{lag} on the number of activated 'germination enzymes' [although work on the commitment event (Halmann and Keynan, 1962), then called the trigger reaction, probably had not drawn attention at that time], where germination was initiated by the accumulation of a hypothetical substance above a threshold amount, and the rate of the accumulation was proportional to the number of active 'germination enzymes'. This results in a T_{lag} curve that is qualitatively similar to Fig. 12.2C, but where the functional form of the curve is determined by as yet unknown rate constants. In our general model, no form of the curve is assumed and the dependence of $T_{commitment}$ (instead of T_{lag}) on the number of active nGR's is fit to data. The existence of a threshold number of active 'germination enzymes' for germination to occur arises in the Woese *et al.* (1968) model from the assumption that the hypothetical substance was degraded at certain rate, with the result that some spores could never reach the threshold level required for germination to occur.

The model provides a simple explanation for superdormant spores, which is that the superdormant fraction comes from the low-expression end of the nGR distribution (Fig. 12.2A). This is supported by the experimental observation that superdormant spores germinate normally with CaDPA or dodecylamine (Ghosh and Setlow, 2009, 2010), which do not involve the activation of the nGRs, but remain dormant with respect to the initial nutrients used to isolate them. What is less clear is why superdormant spores are also less responsive to other nutrients that stimulate different nGRs yet germinate relatively normally with nutrient mixtures that stimulate multiple nGRs (Ghosh and Setlow, 2009). The model suggests that the levels of different nGRs on single spores could be correlated. However, only weak correlations have been observed (Griffiths *et al.*, 2011), even though the expression of the genes encoding different nGR subunits are regulated by the same transcription factors during sporulation, including the same RNA polymerase sigma factor, σ^G and the DNA binding protein SpoVT (Wang *et al.*, 2006). One additional possibility is that the diminished responses of superdormant spores to single nutrients combined with normal responses to nutrient combinations is a reflection of cooperative interactions between different nGRs.

The model also provides the basis for including the effects of other factors, such as heat activation or pH, on the kinetics or extent of germination, and for predicting the effects of germinant mixtures that interact with multiple types of nGRs simultaneously (X. Yi, J. Liu, J.R. Faeder and P.

Setlow, unpublished). In the case of heat activation, it is not known what this process actually does to increase spore germination rates, but the model of Fig. 12.2B and C suggests two possibilities. First, heat activation could increase the affinity of nGRs for their cognate germinants, increasing the number of bound nGRs in a spore for a given germinant concentration. A second possibility is that heat activation increases the efficacy of each bound nGR, shortening $T_{commitment}$ for a given number of bound nGRs and/or decreasing the threshold N^* for germination. The latter seems to be favoured by the fact that non-heat-activated spores give much higher yields of superdormant spores than do optimally activated populations (Ghosh and Setlow, 2009). However, the two possibilities may coexist, and other factors not yet considered in the model may also play some role.

Reasons for heterogeneity between individual spores in populations

Clearly, there is significant heterogeneity between individual spores in populations. The question then is what causes this heterogeneity. There appear to be two main possible causes: (i) differences in the environment to which sporulating cells in populations are exposed or extrinsic factors; and (ii) stochastic effects or intrinsic factors (Elowitz *et al.*, 2002).

Differences in environment of individual sporulating cells in populations

Spores prepared under different conditions, whether these be differences in temperature, growth medium, pH, aeration, solid or liquid medium, etc., exhibit markedly different properties (Gerhardt and Marquis, 1989; Melly *et al.*, 2002; Rose *et al.*, 2007; Stecchini *et al.*, 2009). While individuals in populations of sporulating cells, at least in the laboratory, are all exposed to the same temperature, this is not the case for the medium in which the individual cells are sporulating. This is because sporulation in a population is not perfectly synchronous, such that the environment of the first cohort of cells in the population that initiate and complete sporulation will be different than the environment to which the last cohort of cells to sporulate are exposed, since the sporulation process will alter the growth medium (Hornstra *et al.*, 2009). It is not known what these changes are and how they influence spore properties. However, it should be possible to isolate the first spores formed in populations and compare their properties with the total spore population formed, and it would be surprising if there were no differences. Indeed, given that individuals in spore populations formed in nature are probably exposed to even more different conditions during their sporulation than in the laboratory, one would expect even more heterogeneity in spore populations formed in natural environments.

Stochastic reasons

A second probable reason for heterogeneity in spore properties appears to be stochastic variation in levels of key spore proteins, in particular the spore's nGRs. Recent years have seen a number of studies indicating that stochastic variation in chemical reactions and/or the amount of transcription and/or translation of key gene products can play an important role in population heterogeneity in bacteria, including in sporulating cells (Maamar *et al.*, 2007; Naoki *et al.*, 2008; Veening *et al.*, 2009; Choi *et al.*, 2010; de Jong *et al.*, 2010). Indeed, such stochastic effects can be particularly important for low abundance gene products, for which a decrease of one or two mRNAs transcribed from a given gene, or slight decreases in mRNA translation could have drastic effects on gene products' levels. This stochastic variation could thus have drastic effects on levels of nGRs in individual spores, as for the one nGR (GerB of *B. subtilis* spores) for which data are available, there are only ~25 molecules of the GerBA protein/spore. In addition, the subunit composition of the nGRs is not known with surety, but each GerB nGR seems likely to contain more than one molecule of the GerBA protein, giving an average number of ≤ 12 GerB nGRs per spore. Analysis of spores with nGR subunits fused to fluorescent reporter proteins has indicated that levels of fluorescence in these spores vary over an ≥ 20-fold range, with some spores having minimal if any fluorescence, consistent with a wide range in nGRs numbers in individual spores (Griffiths *et al.*, 2011). These variations in nGR numbers may not be due only to stochastic effects, but also to effects of different conditions over the course of sporulation of populations, since different sporulation conditions give different levels of expression of genes encoding nGRs, with corresponding effects on rates of germination of the resultant spores (Hornstra *et al.*, 2006).

Advantages of spore population heterogeneity

As noted above, it is clear that there is significant heterogeneity in bacterial spore populations, in both their chemical composition as well as their physiological properties One obvious question is what is the potential advantage to a genetically homogeneous population to have such heterogeneity – other than to frustrate the food microbiologist. In the case of a property such as heat resistance as well as other resistance properties, the advantages are not immediately obvious. In some respects it might be most advantageous for all spores in a population to exhibit maximum resistance and thus ensure maximum spore survival no matter what conditions the spores encounter during dormancy. However, it may be that achievement of maximum spore resistance comes with penalties in such areas as speed of germination. Thus individuals in a population may adopt a heterogeneous distribution of resistance to hedge the population's bet that the spore's environment will never be so stressful that some spores will not survive yet survivors cannot still rapidly germinate and grow. This concept makes the prediction that the most resistant spores in a population are also those that germinate the slowest, and while there is at least one report that is consistent with this idea (Ghosh *et al.*, 2009), it has by no means been proven.

The second major physiological property of spores in which there is significant heterogeneity is germination, as some spores germinate fast and some extremely slowly, perhaps even after days. Here the advantage to the population as a whole of such heterogeneity is more obvious. If all spores in a population germinated rapidly and then the environment was in fact not conducive to growth, or conditions such as temperature or water content rapidly changed to deleterious levels, then the fate of whole populations would be jeopardized. Similarly, if no spores in the population ever germinate, the spore population is maintained but will not expand, and will lose out to those populations whose spores do germinate and go on to grow. Consequently it would seem advantageous for spore populations to be a mixture of individuals that exhibit a wide variation in germination rates, such that the population as a whole has the best chance of both surviving and multiplying in the many different environments in which spore populations exist. In this model superdormant spores would simply be members of spore populations that are held in reserve so to speak, in case all the advance scouts (those that germinate rapidly) are destroyed.

Conclusions

Certainly the major conclusion to be taken from this chapter is that populations of spores are extremely heterogeneous in their properties, in particular their resistance and germination such that individuals in spore population die or germinate at often quite different rates. As a consequence, spore heterogeneity has major applied implications in many areas, and consequently understanding the causes of spore heterogeneity is of significant importance. The work summarized in this chapter has led to a number of advances in our understanding of spore heterogeneity. Major conclusions from this work have included the following

1 A variety of new techniques have been developed for determining the composition of individual dormant spores, including the CaDPA content, core water content and numbers of specific proteins per spore. These methodologies have begun to provide data that quantifies spore heterogeneity.
2 A number of methods have also been developed to examine the behaviour of individual spores during processes such as wet heat killing or germination.
3 Correlation of data from analysis of the constituents of individual spores with the

behaviour of these individual spores in processes such as wet heat treatment and germination has begun to give an understanding of the molecular causes of spore heterogeneity in these processes.

4 Combination of data on heterogeneity in specific spore constituents and physiological behaviour, for example between variations in GR numbers per spore with germination heterogeneity, may well lead to mathematical models for the behaviour of spore populations that have predictive value, not only for spore germination but ultimately also for spore resistance and killing. Such models may have significant applied applications and provide new insight into mechanisms of spore resistance, killing and germination.

Acknowledgements

Work in the authors' laboratories have been supported over the years by grants from the National Institutes of Health (PS, JRF), the Army Research Office (PS), the NSF (JRF) and a Multi University Research Initiative (MURI) award from the US Department of Defence (JRF, PS).

References

Bagyan, I., Hobot, J., and Cutting, S. (1996). A compartmentalized regulator of developmental gene expression in *Bacillus subtilis*. J. Bacteriol. *178*, 4500–4507.

Barker, G.C., Malakar, P.K., and Peck, M.W. (2005). Germination and growth from spores: variability and uncertainty in the assessment of food borne hazards. Int. J. Food Microbiol. *100*, 67–76.

Billon, C.M., McKirgan, C.J., McClure, P.J., and Adair, C. (1997). The effect of temperature on the germination of single spores of *Clostridium botulinum* 62A. J. Appl. Microbiol. *82*, 48–56.

Brookmeyer, R.E., Johnson, E., and Bollinger, R. (2003). Modeling the optimum duration of antibiotic prophylaxis in an anthrax outbreak. Proc. Natl. Acad. Sci. U.S.A. *100*, 10129–10132.

Cabrera-Martinez, R.-M., Tovar-Rojo, F., Vepachedu, V.R., and Setlow, P. (2003). Effects of overexpression of nutrient receptors on germination of spores of *Bacillus subtilis*. J. Bacteriol. *185*, 2457–2464.

Carrera, M., Zandomeni, R.O., Fitzgibbon, J., and Sagripanti, J.-L. (2007). Difference between the spore sizes of *Bacillus anthracis* and other *Bacillus* species. J. Appl. Microbiol. *102*, 303–312.

Chen, D., Huang, S.-S., and Li, Y.-Q. (2006). Real-time detection of kinetic germination and heterogeneity of single *Bacillus* spores by laser tweezers Raman spectroscopy. Anal. Chem. *78*, 6936–6941.

Choi, P.J., Xie, X.S., and Shakhnovich, E.I. (2010). Stochastic switching in gene networks can occur by a single-molecule event or many molecular steps. J. Mol. Biol. *396*, 230–244.

Coleman, W.H., Chen, D., Li, Y.Q., Cowan, A.E., and Setlow, P. (2007). How moist heat kills spores of *Bacillus subtilis*. J. Bacteriol. *189*, 8458–8466.

Coleman, W.H., Zhang, P., Li, Y.Q., and Setlow, P. (2010). Mechanism of killing of spores of *Bacillus cereus* and *Bacillus megaterium* by wet heat. Lett. Appl. Microbiol. *50*, 507–514.

Coote, P.J., Billon, C.M.-P., Pennell, S., McClure, P.J., Ferdinando, D.P., and Cole, M.B. (1995). The use of confocal scanning laser microscopy (CSLM) to study the germination of individual spores of *Bacillus cereus*. J. Microbiol. Meth. *21*, 193–208.

Corradini, M.G., Normand, M.D., Eisenberg, M., and Peleg, M. (2010). Evaluation of a stochastic inactivation model of heat-activated *Bacilli* spores. Appl. Environ. Microbiol. *76*, 4402–4412.

Cowan, A.E., Koppel, D.E., Setlow, P., and Setlow, P. (2003). A soluble protein is immobile in dormant spores of *Bacillus subtilis* but is mobile in germinated spores: implications for spore dormancy. Proc. Natl. Acad. Sci. U.S.A. *100*, 4209–4214.

Cronin, U.P., and Wilkinson, M.G., (2008). *Bacillus cereus* endospores exhibit a heterogeneous response to heat treatment and low-temperature storage. Food Microbiol. *25*, 235–243.

Dean, D.H., and Douthit, H.A. (1974). Buoyant density heterogeneity in spores of *Bacillus subtilis*: biochemical and physiological basis. J. Bacteriol. *117*, 601–610.

Elowitz, M.B., Levine, A.J., Siggia, E.D., and Swain, P.S. (2002). Stochastic gene expression in a single cell. Science *297*, 1183–1186.

Gerhardt, P., and Marquis, R.E. (1989). Spore thermoresistance mechanisms. In Regulation of Prokaryotic Development, Smith, I., Slepecky, R.A., and Setlow, P., eds. (American Society for Microbiology, Washington D.C.), pp. 43–63.

Ghosal, S., Fallon, S.J., Leighton, T.J., Wheeler, K.E., Kristo, M.J., Hutcheon, I.D., and Weber, P.K. (2008). Imaging and 3D elemental characterization of intact bacterial spores by high-resolution secondary ion mass spectrometry. Anal. Chem. *80*, 5986–5992.

Ghosal, S., Leighton, T.J., Wheeler, K.E., Hutcheon, I.D., and Weber, P.K. (2010). Spatially resolved characterization of water and ion incorporation in *Bacillus* spores. Appl. Environ. Microbiol. *76*, 3275–3282.

Ghosh, S., and Setlow, P. (2009). Isolation and characterization of superdormant spores of *Bacillus* species. J. Bacteriol. *191*, 1787–1797.

Ghosh, S., and Setlow, P. (2010). The preparation, germination properties and stability of superdormant spores of *Bacillus cereus*. J. Appl. Microbiol. *108*, 582–590.

Ghosh, S., Zhang, P., Li, Y.-q., and Setlow, P. (2009). Superdormant spores of *Bacillus* species have elevated wet-heat resistance and temperature requirements for heat activation. J. Bacteriol. *191*, 5584–5591.

Gould, G.W. (2006). History of science – spores. J. Appl. Microbiol. *101*, 507–513.

Griffiths, K.K., Zhang, J., Cowan, A.E., Yu, J., and Setlow, P. (2011). Germination proteins in the inner membrane of dormant *Bacillus subtilis* spores colocalize in a discrete cluster. Mol. Microbiol. *81*, 1061–1077.

Halmann, M., and Keynan, A. (1962). Stages in germination of spores of *Bacillus licheniformis*. J. Bacteriol. *84*, 1187–1193.

Hashimoto, T., Frieben, W.R., and Conti, S.F. (1969a). Microgermination of *Bacillus cereus* spores. J. Bacteriol. *100*, 1385–1392.

Hashimoto, T., Frieben, W.R., and Conti, S.F. (1969b). Germination of single bacterial spores. J. Bacteriol. *98*, 1011–1020.

Heine, H.S., Bassett, J., Miller, L., Hartings, J.M., Irvine, H.E., Pitt, M.L., Fritz, D., Norris, S.L., and Byrne, W.R. (2007). Determination of antibiotic efficacy against *Bacillus anthracis* in a mouse aerosol challenge model. Antimicrob. Agents Chemother. *51*, 1373–1379.

Hornstra, L.M., de Vries, Y.P., de Vos, W.M., and Abee, T. (2006). Influence of sporulation medium composition on transcription of *ger* operons and the germination response of spores of *Bacillus cereus* ATCC 14579. Appl. Environ. Microbiol. *72*, 3746–3749.

Hornstra, L.M., Ter Beek, A., Smelt, J.P., Kallemeijn, W.W., and Brul, S. (2009). On the origin of heterogeneity in (preservation) resistance of *Bacillus* spores: input for a 'systems' analysis approach of bacterial spore outgrowth. Int. J. Food Microbiol. *134*, 9–15.

Huang, S.-S., Chen, D., Pelczar, P.L., Vepachedu, V.R., Setlow, P., and Li, Y.-Q. (2007). Levels of Ca^{2+}-dipicolinic acid in individual *Bacillus* spores determined using microfluidic Raman tweezers. J. Bacteriol. *13*, 4681–4687.

Indest, K.J., Buchholz, W.G., Faeder, J.R., and Setlow, P. (2009). Workshop report: modeling the molecular mechanism of bacterial spore germination and elucidating reasons for germination heterogeneity. J. Food Sci. *74*, R73–R78.

de Jong, I.G., Veening, J.-W., and Kuipers, O.P. (2010). Heterochronic phosphorelay gene expression as a source of heterogeneity in *Bacillus subtilis* spore formation. J. Bacteriol. *192*, 2053–2067.

Keynan, A., and Evenchick, Z. (1969). Activation. In The Bacterial Spore. Gould, G.W., and Hurst, H., eds. (Academic Press, New York), pp. 359–396.

Kong, L., Zhang, P., Setlow, P., and Li, Y.-Q. (2010a). Characterization of bacterial spore germination using integrated phase contrast microscopy, Raman spectroscopy, and optical tweezers. Anal. Chem. *82*, 3840–3847.

Kong, L., Zhang, P., Yu, J., Setlow, P., and Li, Y.-Q. (2010b) Monitoring the kinetics of uptake of a nucleic acid dye during the germination of single spores of *Bacillus* species. Anal. Chem. *82*, 8717–8724.

Kong, L., Zhang, P., Wang, G., Yu, J., Setlow, P., and Li, Y.-Q. (2011). Phase contrast microscopy, Raman spectroscopy, fluorescence spectroscopy and optical tweezers to characterize bacterial spore germination. Nat. Protocols *6*, 625–639.

Kort, R., O'Brien, A.C., van Stokkum, I.H.M., Oomes, S.J.C.M., Crielaard, W., Hellingwerf, K.J., and Brul, S. (2005). Assessment of heat resistance of bacterial spores from food product isolates by fluorescence monitoring of dipicolinic acid release. Appl. Environ. Microbiol. *71*, 3556–3564.

Leuschner, R.G.K., and Lillford, P.J. (1999). Effects of temperature and heat activation on germination of individual spores of *Bacillus subtilis*. Lett. Appl. Microbiol. *29*, 228–232.

Maamar, H., Raj, A., and Dubnau, D. (2007). Noise in gene expression determines cell fate in *Bacillus subtilis*. Science *317*, 526–529.

Magge, A., Setlow, B., Cowan, A.E., and Setlow, P. (2009). Analysis of dye binding by and membrane potential in spores of *Bacillus* species. J. Appl. Microbiol. *106*, 814–824.

Melly, E., Genest, P.C., Gilmore, M.E., Little, S., Popham, D.L., Driks, A., and Setlow, P. (2002). Analysis of the properties of spores of *Bacillus subtilis* prepared at different temperatures. J. Appl. Microbiol. *92*, 1105–1115.

Naoki, H., Sakumura, Y., and Ishii, S. (2008). Stochastic control of spontaneous signal generation for gradient sensing in chemotaxis. J. Theor. Biol. *255*, 259–266.

Paidhungat, M., and Setlow, P. (1999). Isolation and characterization of mutations in *Bacillus subtilis* that allow spore germination in the novel germinant D-alanine. J. Bacteriol. *181*, 3341–3350.

Paidhungat, M., and Setlow, P. (2001). Localization of a germinant receptor protein (GerBA) to the inner membrane of *Bacillus subtilis* spores. J. Bacteriol. *183*, 3982–3990.

Peng, L., Chen, D., Setlow, P., and Li, Y.-Q. (2009). Elastic and inelastic light scattering from single bacterial spores in an optical trap allows the monitoring of spore germination dynamics. Anal. Chem. *81*, 4035–4042.

Periago, P.M., van Zuijlen, A., Fernandez, P.S., Klapwijk, P.M., ter Steeg, P.F., Corradini, M.G., and Peleg, M. (2004). Estimation of the non-isothermal inactivation patterns of *Bacillus sporothermodurans* IC4 spores in soups from their isothermal survival data. Int. J. Food Microbiol. *95*, 205–218.

Rose, R., Setlow, B., Monroe, A., Mallozzi, M., Driks, A., and Setlow, P. (2007). Comparison of the properties of *Bacillus subtilis* spores made in liquid or on plates. J. Appl. Microbiol. *103*, 691–699.

Russell, A.D. (1982). The Destruction of Bacterial Spores (Academic Press, New York).

Setlow, P. (2003). Spore germination. Curr. Opinion Microbiol. *6*, 550–556.

Setlow, P. (2006). Spores of *Bacillus subtilis*: their resistance to and killing by radiation, heat and chemicals. J. Appl. Microbiol. *101*, 514–525.

Setlow, P., and Johnson, E.A. (2007). Spores and their significance. In Food Microbiology, Fundamentals and Frontiers, 3rd edition, Doyle, M.P., and Beuchat, L.R. eds. (ASM Press, Washington, DC), pp. 35–67.

Smelt, J.P.P.M., Bos, A.P., Kort, R., and Brul, S. (2008). Modelling the effect of sub(lethal) heat treatment of *Bacillus subtilis* spores on germination rate and outgrowth to exponentially growing vegetative cells. Int. J. Food Microbiol. *128*, 34–40.

Stecchini, M.L., Spaziani, M., Del Torre, M., and Pacor, S. (2009). *Bacillus cereus* cell and spore properties as influenced by the micro-structure of the medium. J. Appl. Microbiol. *106*, 1838–1848.

Stringer, S.C., Webb, M.D., George, S.M., Pin, C., and Peck, M.W. (2005). Heterogeneity of times required for germination and outgrowth from single spores of nonproteolytic *Clostridium botulinum*. Appl. Environ. Microbiol. *71*, 4998–5003.

Subramanian, S.A., Ahn, J., Balasubramaniam, V.M., and Rodriguez-Saona, L. (2006). Determination of spore inactivation during thermal and pressure-assisted thermal processing using FT-IR spectroscopy. J. Agric. Food Chem. *54*, 10300–10306.

Subramanian, S.A., Ahn, J., Balasubramaniam, V.M., and Rodriguez-Saona, L. (2007). Monitoring biochemical changes in bacterial spore during thermal and pressure-assisted thermal processing using FT-IR spectroscopy. J. Agric. Food Chem. *55*, 9311–9317.

Sunde, E.P., Setlow, P., Hederstedt, L., and Halle, B. (2009). The physical state of water in bacterial spores. Proc. Natl. Acad. Sci. U.S.A. *106*, 19334–19339.

Swerdlow, B.M., Setlow, B., and Setlow, P. (1981). Levels of H^+ and other monovalent cations in dormant and germinating spores of *Bacillus megaterium*. J. Bacteriol. *148*, 20–29.

Taniguchi, Y., Choi, P.J., Li, G.W., Chen, H., Babu, M., Hearn, J., Emili, A., and Xie, X.S. (2010). Quantifying *E. coli* proteome and transcriptome with single-molecule sensitivity in single cells. Science *329*, 533–538.

Tracy, B.P., Gaida, S.M., and Papoutsakis, E.T. (2008). Development and application of flow-cytometric techniques for analyzing and sorting endospore-forming Clostridia. Appl. Environ. Microbiol. *74*, 7497–7506.

Vary, J.C., and Halvorson, H.O. (1965). Kinetics of germination of bacterial spores. J. Bacteriol. *89*, 1340–1347.

Veening, J.-W., Klaas Smits, W., and Kuipers, O.P. (2008). Bistability, epigenetics, and bet-hedging in bacteria. Annu. Rev. Microbiol. *62*, 193–210.

Wang, S.T., Setlow, B., Conlon, E.M., Lyon, J.L., Imamura, D., Sato, T., Setlow, P., Losick, R., and Eichenberger, P. (2006). The forespore line of gene expression in *Bacillus subtilis*. J. Mol. Biol. *358*, 16–37.

Webb, M.D., Pin, C., Peck, M.W., and Stringer, S.C. (2007). Historical and contemporary NaCl concentrations affect the duration and distribution of lag times from individual spores of nonproteolytic *Clostridium botulinum*. Appl. Environ. Microbiol. 73, 2118–2127.

Weber, P.K., Graham, G.A., Teslich, N.E., Chan, W.M., Ghosal, S., Leighton, T.J., and Wheeler, K.E. (2010). NanoSIMS of *Bacillus* spores sectioned by focused ion beam. J. Microsc. *238*, 189–199.

Wei, J., Shah, I., Ghosh, S., Dworkin, J., Hoover, D.G., and Setlow, P. (2010). Superdormant spores of *Bacillus* species germinate normally with high pressure, peptidoglycan fragments and bryostatin. J. Bacteriol. *192*, 1455–1458.

Woese, C.R., Vary, J.C., and Halvorson, H.O. (1968). A kinetic model for bacterial spore germination. Proc. Natl. Acad. Sci. U.S.A. *59*, 869–875.

Xu, H., He, X., Gou, J., Lee, Y.-Y., and Ahn, J. (2009). Kinetic evaluation of physiological heterogeneity in bacterial spores during thermal inactivation. J. Gen. Appl. Microbiol. *55*, 295–299.

Yi, X., and Setlow, P. (2010). Studies of the commitment step in the germination of spores of *Bacillus* species. J. Bacteriol. *192*, 3424–3433.

Yung, P.T., Shafaat, H.S., Connon, S.A., and Ponce, A. (2007). Quantification of viable endospores from a Greenland ice core. FEMS Microbiol. Ecol. *59*, 300–306.

Zhang, P., Garner, W., Yi, X., Yu, J., Li, Y.-Q., and Setlow, P. (2010a). Factors affecting variability in time between addition of nutrient germinants and rapid dipicolinic acid release during germination of spores of *Bacillus* species. J. Bacteriol. *192*, 3608–3619.

Zhang, P., Kong, L., Setlow, P., and Li, Y.-Q. (2010b). Characterization of wet-heat inactivation of single spores of *Bacillus* species by dual-trap Raman spectroscopy and elastic light scattering. Appl. Environ. Microbiol. *76*, 1796–1805.

Zhang, P., Kong, L., Setlow, P., and Li, Y.-Q. (2010c). Multiple-trap laser tweezers Raman spectroscopy for simultaneous monitoring of the biological dynamics of multiple individual cells. Opt. Express. 35, 3221–3223.

Zhang, P., Kong, L., Wang, G., Setlow, P., and Li, Y.-q. (2010d). Combination of Raman tweezers and quantitative differential interference microscopy for measurement of dynamics and heterogeneity during the germination of individual bacterial spores. J. Biomed. Opt. *15*, 056010-1–056010-9.

Zhang, P., Setlow, P., and Li, Y. (2009). Characterization of single heat-activated *Bacillus* spores using laser tweezers Raman spectroscopy. Opt. Express *17*, 16480–16491.

Detection of Bacterial Spores: Prospects and Challenges

13

Sainath Rao Shilpakala, Chintamani D. Atreya and Krishna Mohan V. Ketha

Abstract

Bacteria of several genera are able to form endospores when subjected to certain starvation conditions. The endospores are dormant forms that are structurally and biochemically different from the corresponding growing or vegetative cells. These bacterial endospores resist antibiotics, desiccation, and ordinary boiling than the vegetative cells. The detection of bacterial endospores can be important for a wide variety of purposes. In the sanitation and hygiene fields, detection of bacterial spores can be critical to monitor indoor environments, water quality and food quality. Similarly, in the public health perspective detection of bacterial spores is very important in ensuring regulation of safer transfusion and other therapeutic products, administered either orally or intravenously. Monitoring of bacterial spores is gaining importance even in other areas such as agriculture wherein soil or plant samples are periodically monitored for bacterial spores to ensure high levels of the bacterial population/toxin to be effective against insect pests. More recently, with the possibility of bacterial spores being used as bio-threat agents there has been an augmented effort in developing much more sensitive, specific and rapid detection systems for bacterial spores. The review discusses about current methods of bacterial spore detection and the challenges involved.

Introduction

Bacteria of several genera form endospores when subjected to certain environmental conditions. These include nutrient starvation or limitation, toxic chemical concentrations, changes in temperature or pressure. Sporulation involves transcriptional and morphological changes resulting in the formation of endospore. These endospores can remain in their dormant state for years. They withstand high temperatures, desiccation, radiation and toxic chemicals. These extraordinary properties make spores resistant to antimicrobials such as lysozyme (Gilbert *et al.*, 1990). Numerous signals stimulate the spore's exit from dormancy, of which the presence of nutrients is the key signal. Bacterial cells can also exit dormancy in response to signals released from other bacteria (Votyakova *et al.*, 1994; Kaprelyants and Kell, 1996).

Spore properties of various sporulating bacteria

The unique properties of a bacterial spore can be attributed to its cellular structure. Many bacterial spores have characteristic proteins that differentiate them from their respective vegetative forms. Spore-specific proteins form the basis for many diagnostic/detection assays. Hence, in depth analysis of the spore cellular structures is critical. The outer coat surrounding the spore protects it from toxic chemicals and enzymes (Henriques and Moran, 2007). Beneath this outer coat is a layer of peptidoglycan referred to as cortex. A germ cell wall resides under the cortex and this layer of peptidoglycan eventually becomes the cell wall of the bacterium once the endospore germinates. The inner membrane, under the germ cell wall, acts as a barrier against several chemicals.

The spore core at the centre of the endospore, is metabolically inactive (Driks, 2009) and houses the cell's DNA, ribosomes and large amounts of calcium–dipicolinic acid complexes (Henriques and Moran, 2000, 2007). Dipicolinic acid is an endospore-specific chemical and can constitute up to 10% of the spore's dry weight and appears to play a role in maintaining spore dormancy. In addition, Small acid-soluble proteins (SASPs) are endospore-specific proteins that tightly bind to DNA and protect spore DNA from thermal denaturation and other harsh environmental conditions (Henriques and Moran, 2000). Other endospore species-specific structures and chemicals include stalks, toxin crystals, or an additional outer glycoprotein layer called the exosporium (Errington, 2003; Piggot and Hilbert, 2004). It is known that at least some of the above mentioned spore-differentiating cellular features have been exploited in developing a sensitive and rapid detection method for detecting bacterial spores. A physical feature, the position of endospore differs among bacterial species and is useful in identification by simple microscopy (Melly *et al.*, 2002).

Endospores are produced by several genus *Bacillus*, *Clostridium*, *Desulfosporomusa*, *Sporosarcina*, *Sporolactobacillus*, *Oscillospira*, and *Thermoactinomyces*. The genus *Myxococcus* forms myxospores or fruiting bodies in response to nutrient deprivation (Wireman and Dworkin, 1975), whereas the medically important *Streptomyces* sp produce spores called conidia on filaments called mycelia or sporangiospores in specialized sacs called sporangia. It was recently reported that mycobacteria can also form spore-like structures (Singh *et al.*, 2010; Traag *et al.*, 2010). For classical identification of these bacterial spores it is essential to know how they are formed. Bacteria belonging to the genus *Azotobacter* form bacterial cysts in response to unfavourable environmental conditions. Cysts differ from endospores in the way they are formed and are resistant to heat and desiccation but to a lesser degree than are endospores (Layne and Johnson, 1964).

Significance of spores in pathology and as biothreat agents

Spores of some bacterial species are highly pathogenic (Table 13.1). Dormant spores remain viable for long periods of time in the environment (Cano and Borucki, 1995; Vreeland *et al.*, 2000; Setlow, 2006). Infection may result from ingestion, inhalation or intradermal inoculation of spores (Friedlander *et al.*, 1993; Zaucha *et al.*, 1998; Rupnik *et al.*, 2009). For example, ingested spores of *Clostridium difficle* survive the harsh enzymes and low pH of the stomach, facilitating their entry into the lower gastrointestinal tract where they germinate and initiate infection (Fekety *et al.*, 1981; Huang *et al.*, 2006; Gerding *et al.*, 2008). Spores of *B. anthracis* survive attacks by the host's phagocytic cells and using them as a means of transport to the lymph nodes (Ruthel *et al.*, 2004; Weaver *et al.*, 2007).

Many spore-forming species such as *C. botulinum* are pathogens that can cause a specific type of disease (botulism), while others such as *B. anthracis* can cause pulmonary, cutaneous and gastrointestinal disease manifestations (Table 13.1). In some cases more than one species can cause similar disease (gas gangrene is caused by *C. perfringens* and *C. sordelli*).

Adherence of the spores to the host cell also play a key role in the initiation of infection (Andersson *et al.*, 1998). For example, Spores of *Bacillus anthracis* adhere to the host epithelial cells allowing the bacteria to cross the epithelial barrier and persist until the conditions are favourable for infection (Bozue *et al.*, 2007; Oliva *et al.*, 2008, 2009; Russell *et al.*, 2008).

Laboratory studies have shown that spores of *B. subtilis* germinate in response to nutrients such as sugars, amino acids such as L-alanine, and a mixture of nutrients like glucose, fructose, asparagine and K^+. Nutrients when added, binds to the *gerA* receptors present on the spore's membrane. This activates the release of ions and dipicolinic acid. This process results in the hydrolysis of spore's peptidoglycan spore cortex and expansion of germ cell wall resulting in spore germination (Setlow, 2006; Dworkin and Shah, 2010). In addition to nutrients, *in vitro* spore germination is initiated by several physical and chemical agents that include Ca^{2+}- DPA, high pressure, salts etc. (Moir, 2006). The importance of germination receptors and cortex-lytic enzymes to virulence has been demonstrated in many species (Moir, 2006; Burns *et al.*, 2010; Carr *et al.*, 2010).

Table 13.1 Examples of pathogenic spore-forming bacteria and the disease they cause

Strain	Disease caused
B. anthracis	Pulmonary anthrax, cutaneous anthrax, gastrointestinal anthrax
B. cereus	Food poisoning, endocarditis, gas gangrene, neonatal infections
B. subtilis	Hepatotoxicity
B. thuringiensis	Periodontitis, wound infections, ocular infections, bacteraemia, gastroenteritis
B. megaterium	Ocular keratitis
C. botulinum	Botulism
C. perfringens	Diarrhoea, gas gangrene, food poisoning, necrotic enteritis
C. tetani	Tetanus
C. difficile	Antibiotic-associated diarrhoea, pseudomembranous colitis
C. sordellii	Gas gangrene, bacteraemia, toxic shock syndrome

In some cases heat-stable toxins produced in contaminated food (mostly by *B. cereus*) can cause food poisoning without a concurrent infection (Taylor *et al.*, 2005; Stenfors Arnesen *et al.*, 2007, 2008).

Challenges in detection of bacterial spores

Despite the development of numerous detection assays, it is still a problem to differentiate between spores of different *Bacillus* species (Schnepf *et al.*, 1998). In order for them to be detected, germination of the resistant spore is often a prerequisite and hence culturing of the resulting vegetative bacteria becomes essential. Additionally, DNA- and RNA-based identifications require complex sample preparations (Lim *et al.*, 2005; Henriques and Moran, 2007).

The major difficulty with *B. anthracis* detection is its high similarity with other strains in the genera, which leads to false-positive identifications. Even though *B. anthracis* forms a highly monomorphic lineage within the *B. cereus* group, but some strains of *B. thuringiensis* and *B. cereus* are closely related to the *B. anthracis* posing a challenge to differentiate between these strains (Kolsto *et al.*, 2009). The main feature of *B. anthracis* is the presence of two virulence plasmids. However the Sterne and Pasteur vaccine strains of *B. anthracis* lack either the capsule plasmid pXO2 or the toxin plasmid pXO1, respectively, whereas the Carbosap strain includes both plasmids (Fasanella *et al.*, 2001). Loss of pXO1 in environmental samples is rare, while loss of pXO2 is more common (Mock and Fouet, 2001). *B. thuringiensis* is indistinguishable from *B. cereus* except for the insecticidal genes (Helgason *et al.*, 2000). Hence, nucleic acid analysis of these strains could be an important tool for identification. There is some argument among the scientists that the three members of the *B. cereus* group belong to the same species, with varying plasmids due to horizontal genetic exchange (Helgason *et al.*, 2000). Furthermore, some striking differences exist between *B. anthracis* and other closely related species such as *B. cereus* and *B. thuringiensis* wherein it became evident that *B. anthracis* while causing disease in humans produces a polypeptide capsule (under appropriate growth conditions).

Recent findings report that the presence of pXO1 and pXO2 no longer principally separates *B. anthracis* from other *Bacilli* (Hoffmaster *et al.*, 2004, 2006a). In 1994, a *B. cereus* strain G9241 isolated from a welder from Louisiana with severe pneumonia reportedly contained a complete pXO1 plasmid. This strain also had a capsule, but the capsule was not composed of D-glutamyl polypeptides and was not encoded by *B. anthracis cap* genes located on the *B. anthracis* pXO2 plasmid. Few years later in October and November 2003 two more strains of *B. cereus* causing pneumonia were isolated from workers in Texas (Hoffmaster *et al.*, 2006b). Recently, two *B. cereus* group strains, CI and CA, carrying two plasmids with close identity to *B. anthracis* pXO1 and pXO2 were isolated from great apes that had died due to anthrax-like disease. The pXO1 and pXO2 plasmids in these strains carried the anthrax toxin genes and the polyglutamate capsule biosynthesis genes, respectively along with the associated regulatory genes (*atxA*, *acpA*, and *acpB*) (Klee *et al.*, 2006; Kolsto *et al.*, 2009)

Bacillus: spore characteristics and differences between spores and vegetative forms

Difference between spores of various *Bacillus* strains

Spore development has been extensively studied in the model organism, *B. subtilis*. However the sequence of steps during sporulation is almost same in all endospore-forming organisms that have been examined. It is critical that the characteristic proteins in each spore species not only aid in differentiating between the vegetative and sporulating forms but also be able to differentiate pathogenic from non-pathogenic strains within the same genera.

Electron microscopic studies of thin sections of spores reveal that spores of different *Bacillus* species differ in their exterior layers whereas the spore interior is similar in all species (Holt and Leadbetter, 1969; Waller *et al.*, 2004). The outermost spore structure of *B. subtilis* and other endospores is a protein shell called the coat. However the spores of pathogenic organisms are encased in an additional layer called the exosporium formed by a basal layer and a hair-like glycoprotein nap (Henriques and Moran, 2007; Driks, 2009).

There are differences in the thickness and appearance of the coat layer among spores of various species or strains. For example, the coat layer in *B. cereus*, *B. anthracis*, *B. pumilus*, *B. megaterium*, and *B. clausii* is compact (Chada *et al.*, 2003), whereas the spores of *B. thuringiensis* show an inner coat with comb-like layer consisting of a patchwork of striated packets and an outer coat that is granular and loosely fitting with irregular thickness (Gerhardt *et al.*, 1976).

B. subtilis spores stained with ruthenium red stain followed by electron microscopy expose a glycoprotein layer that is not seen by transmission electron microscopy (TEM) or atomic force microscopy (AFM) (Gerhardt *et al.*, 1976; Waller *et al.*, 2004). AFM studies also reveal that the spore surface of *B. cereus*, *B. anthracis* and *B. subtilis* contain a series of bumps ranging from 7 to 40 nm in diameter depending on the species, permitting species-specific identification (Chada *et al.*, 2003; Waller *et al.*, 2004). *B. subtilis* spore surface consists of smaller circular bumps 7–20 nm in diameter and contain pores, around 24 nm in size or smaller, present in the spaces between the bumps (Chada *et al.*, 2003; Waller *et al.*, 2004). *B. anthracis* and *B. cereus* strains without exosporium contain no bumps or ridges but in strains with exosporium, ridges and bumps are often present. The ridges in *B. cereus* have a zigzag pattern and form bifurcations that are morphologically different from those seen in *B. anthracis* and *B. subtilis* (Chada *et al.*, 2003; Waller *et al.*, 2004).

Proteomic studies suggest that around 50 coat proteins and morphogenetic factors of *B. subtilis* have been encoded in the genomes of *B. cereus* and *B. anthracis* indicating that the composition of these coat layers may be conserved to some extent (Lai *et al.*, 2003; Waller *et al.*, 2004). However, the coat layers could be exploited in differentiating between spores of various *Bacillus* sp as they differ in their appearance and thickness among *Bacillus* spores. The coats of *B. cereus* and *B. anthracis* are thinner and more compact whereas the coats of *B. pumilus*, *B. megaterium* and *B. clausii* are thinner than the coat of *B. subtilis* and *B. anthracis* (Waller *et al.*, 2004; Giorno *et al.*, 2007). Most of the *B. subtilis* coat proteins are species specific or have orthologues (similar, but could be differentiated) in spores of other species or strains (Sylvestre *et al.*, 2002; Henriques and Moran, 2007).

The structure of exosporium in *Bacillus* species is strain specific (Henriques and Moran, 2007). The exosporium layer of *B. thuringiensis* and *B. cereus* spores are 25–40 nm thick and consists of a repetitive pattern of 12–14 nm subunits and the edge of the exosporium show protrusion from which hair-like filaments protrude (Waller *et al.*, 2004; Plomp *et al.*, 2005). In *B. pumilus* and *B. clausii*, the exosporium is closer to the outer coat and is formed by closely packed material with soft porous appearance and short-hair like protrusions. The length of hair-like nap in the exosporium varies between strains, which is due to polymorphisms in the gene encoding for the nap-forming protein (Castanha *et al.*, 2006).

In *B. anthracis*, the filaments of hair-like nap is formed by a single collagen-like protein BclA (Sylvestre *et al.*, 2002) with multiple copies of a pentasaccharide side chain. This oligosaccharide possesses an unusual terminal sugar called anthrose, followed by three rhamnose residues and a protein-bound *N*-acetylgalactosamine. Though BclA is present in *B. cereus* and *B. thuringiensis*, the monosaccharide anthrose is not present in these organisms and is considered a potential biomarker for the detection of anthrax (Sylvestre *et al.*, 2002; Daubenspeck *et al.*, 2004). In *B. cereus* several glycoproteins may contribute to nap formation in addition to BclA and further in-depth analysis of these proteins could reveal potential diagnostic applications that could further aid in differentiating between the various strains of *Bacillus*.

Proteins in the spore core can also be used to distinguish between different spores. For example, small acid soluble proteins (SASPs) found in *B. anthracis* spores are used as protein markers to differentiate between the various *Bacillus* species (Pribil *et al.*, 2005). SASP β is used in discriminating *B. anthracis* from *B. cereus* (Pribil *et al.*, 2005; Castanha *et al.*, 2006, 2007; Callahan *et al.*, 2008). However *B. cereus is not distinguishable from B. thuringiensis* as they exhibit a highly similar SASP profile. Further analysis of SASP α and β profiles aid in differentiating *B. cereus* from the *B. mycoides* group 2 and *B. weihenstephanensis* as they demonstrate a distinct SASP profile. Additionally the SASP α-β analyses could also be utilized to differentiate within *B. mycoides* and *B. weihenstephanensis* spores (Callahan *et al.*, 2009).

Detection platforms for *Bacillus* spores

The *Bacillus cereus* group includes three closely related species, *B. anthracis*, *B. cereus* and *B. thuringiensis*, as well as the more distantly related species *B. mycoides*, *B. pseudomycoides* and *B. weihenstephanensis*. These Gram-positive, spore-forming bacteria form a highly homogeneous subdivision of the genus *Bacillus*, which also contains several other organisms belonging to the *B. subtilis* group. The importance and public awareness of *B. cereus* group organisms are associated with their distinct phenotypes and pathological manifestations. *B. anthracis* is the causative agent of anthrax, a disease that affects humans and animals worldwide and has also been developed as a biological warfare agent (Inglesby *et al.*, 1999). *B. cereus* is an opportunistic human pathogen which is responsible mainly for gastrointestinal illnesses resulting from food contamination (Drobniewski, 1993), whereas *B. thuringiensis* is an insect pathogen whose toxin is a biological pesticide widely used globally in agriculture. The identification of the members of the *B. cereus* group poses significant challenges due to very high level of chromosomal and protein identity (Rasko *et al.*, 2005). Intense efforts have focused on overcoming these challenges, and there has been a particular focus on developing methods for specific detection of *B. anthracis* and for differentiating among strains of these closely related organisms.

Bacillus anthracis

Recent use of *B. anthracis* as a potential biowarfare agent, has led to heightened interest in the organism. The genetic similarity between *B. anthracis* and various *Bacillus* species, and its existence in both spore and vegetative form requires complex laboratory techniques for the detection and specific identification of the organism. *B. anthracis* is highly monomorphic, i.e. it shows little genetic variation and the homogeneity of this group of bacilli causes problems in the design of highly specific detection systems.

Endospores can be visualized microscopically by staining with Malachite green wherein the spores appear green and vegetative cells pink. Based on the morphology of the spore and the sacs that contain the spores, called sporangium, they can be divided into three groups.

- Group 1: Gram-positive that produces central or, terminal ellipsoidal or, cylindrical spores that do not distend the sporangium.Organisms: *B. anthracis, B. cereus, B. megaterium, B. mycoides, B. thuringiensis.*
- Group 2: Gram-variable (a mix of pink and purple cells) with ellipsoidal spores and swollen sporangia. Organisms: *B. alvei, B. circulans, B. brevis, B. coagulans, B. licheniformis, B. macerans, B. subtilis, B. pumilus.*

- Group 3: Gram-variable swollen sporangia with terminal or subterminal spores.

Organisms: B. *sphaericus*

The significant drawback of the Malachite Green method to differentiate spores is that any debris/artefacts on the slide could also potentially take up and retain the green stain resulting in false-positive identifications. Differentiation based purely on morphology requires expertise as endospores are small and typically oval while the large or irregular globs of green on the slide invariably are artefacts. Another method of staining developed recently involves the usage of the ruthenium red stain followed by electron microscopy to visualize acid polysaccharides on the outer surface of cells. This method demonstrated the presence of an external glycoprotein layer surrounding the spore of both *Bacillus anthracis* and *Bacillus subtilis* (Waller *et al.*, 2004). Additional information about these techniques have been discussed earlier in two reviews on laboratory diagnosis of *B. anthracis* (Edwards *et al.*, 2006; Rao *et al.*, 2010). While the review by Edwards *et al.* (2006) provides a background about the toxicology and epidemiology of *B. anthracis* focusing on challenges associated with the detection of *B. anthracis* with greater emphasis to immunological and nucleic-acid-based detection methods. The review by Rao *et al.* discusses current challenges and various improvements associated specifically with detection of *B. anthracis* spores and vegetative cells (Edwards *et al.*, 2006; Rao *et al.*, 2010).

Among all the diagnostic assays immunoassays are the most commonly used method for the detection of *B. anthracis* spores and vegetative cells (Iqbal *et al.*, 2000). A majority of these immunoassays exploit the distinctive surface antigens for rapid detection of these spores (Table 13.2). Spores of *B. anthracis* can be detected by immunofluorescence (IF) and immuno-radiometric (IRMA) assays (Phillips and Martin, 1983). The first differentiating spore surface protein BclA (*Bacillus* collagen-like protein of *B. anthracis*) was discovered in *B. anthracis* (Sylvestre *et al.*, 2003). BclA contains multiple collagen-like XXG repeats in its central region. The number of XXG repeats in BclA varies among different strains that aid in differentiating between the spores (Sylvestre *et al.*, 2003; Castanha *et al.*, 2006). Monoclonal and polyclonal antibodies generated against BclA have been successfully used for the detection of *B. anthracis* spores (Kuehn *et al.*, 2009).

Monoclonal antibodies raised against anti-tetrasaccharide and anti-anthrose-rhamnose disaccharide of the protein BclA for detecting spores of *B. anthracis* using a luminex assay has been partially successful. Owing to the high sensitivity of luminex reagent the method inherently exhibits cross-reactivity against two *B. cereus* strains. (Henriques and Moran, 2000; Castanha *et al.*, 2006; Tamborrini *et al.*, 2010). Carbohydrates and glycoprotein structures are also used as important targets for detection of *B. anthracis*. BclA is glycosylated with two O-linked carbohydrates, a 715-Da tetrasaccharide and a 324-Da disaccharide. The tetrasaccharide contains three mannose residues and a non-reducing terminal sugar called anthrose. Anthrose may be useful for species-specific identification of *B. anthracis* (Daubenspeck *et al.*, 2004).

B. anthracis cells are surrounded by the Sap-S layer proteins during exponential growth phase. This Sap S-layer protein is replaced by the EA1 S-layer protein when the cells enter the stationary phase (Mesnage *et al.*, 1997). The protein is also present in spores of *B. anthracis*. Monoclonal antibodies against EA1 protein bind to both *B. anthracis* spores and intact vegetative cells with high species-specificity and affinity with a detection limit of 10^7 CFU/ml. However, at lower concentrations, some cross-reactivity with other strains does occur (Wang *et al.*, 2009b).

The field of immunofluorescence diagnostics underwent a revolutionary improvement in enhancing sensitivity with the introduction of Quantum dots (Q-dots) as fluorochromes. Q-dots are fluorescent nanocrystals that exhibit a wide absorption spectra and a narrow emission peak that is ideal for multiplexing. The Q-dots are superior to fluorescent dyes by being very photostable and highly sensitive. Flow cytometry using Q–dot conjugates have been used for the detection of *Bacillus* spores; however, the drawback with this technique is the low sensitivity with a limit of detection of 10^3 spores (Zahavy *et al.*, 2010). High sensitivity is obtained by fluorescent

Table 13.2 Summary of various detection platforms for bacterial spores

S. no.	Analytical method used	Target organism	Limit of detection	Reference
1	ELISA	*B. anthracis*	104	Kuehn *et al.* (2009)
2	Bead-based immunoassay	*B. anthracis*	103	Farrell *et al.* (2005)
3	IRMA	*B. anthracis*	103–104	Phillips and Martin (1983)
4	Immunofluorescence assay	*B. anthracis* *B. cereus* *M. xanthus*	105 105 –	Phillips and Martin (1983) Ikanovic *et al.* (2007) Shepherd and Dailey (2005)
5	FRET (immuno-FRET)	*B. cereus*	105	Bruno *et al.* (2001)
6	Flow cytometry	*B. anthracis*	103	Makino *et al.* (2001)
		C. tryobutyricum	103 sfu/100 ml	Lavilla *et al.* (2010)
7	PCR	*B. cereus*	32.6 sfu	Ankolekar *et al.* (2009)
		C. tryobutyricum	3–30 sfu/100 ml	Herman *et al.* (1995)
8	Real-time PCR	*B. anthracis*	1 sfu	Makino *et al.* (2001)
		B. cereus	60 spores/ml	Martinez-Blanch *et al.* (2010)
		B. thuringiensis	103 spores/g	Guidi *et al.* (2010)
		C. tryobutyricum	1 spore/ml	Lopez-Enriquez *et al.* (2007)
9	Multiplex PCR	*B. subtilis*	104	Jorgensen and Leser (2007)
		C. perfringens	104 spu/g	Kanakaraj *et al.* (1998)
		C. botulinum	~100 spore/kg	Fach *et al.* (2002)
10	Nested-PCR	*B. anthracis*	10 sfu/100 g	Beyer *et al.* (1995)
11	PCR-ELISA	*C. botulinum*	10 spores/25 g	Fach *et al.* (2002)
12	PCR-based hybridization	*B. anthracis*	2 sfu	Reif *et al.* (1994)
13	PCR filter-based assay	*B. anthracis*	5 spu	Lampel *et al.* (2004)
14	Lateral immunoassays			
	BTA test	*B. anthracis*	106	King *et al.* (2003)
	BADD test	*B. anthracis*	105	
	Anthrax SMART II	*B. anthracis*	105	
		B. thuringiensis	106	
15	Membrane-based optometry	*B. anthracis*	500 spores	Floriano *et al.* (2005)
16	Membrane-based biosensor	*B. anthracis*	10 spores	Baeumner *et al.* (2004)
17	MALDI-TOF	*B. anthracis*	–	DelVecchio *et al.* (2006)
		B. cereus		
		B. thuringiensis		
18	MS and FTIR	*B. cereus*	–	Goodacre *et al.* (2000)
		B. licheniformis		
		B. subtilis		
19	Raman spectroscopy	*B. anthracis*	104	Pestov *et al.* (2008)
		B. cereus	–	Pai *et al.* (2005)
20	Colony hybridization	*C. tryobutyricum*	6 × 10–5 cfu/g	Heikinheimo *et al.* (2004)
21	Phage display peptide ligands	*B. anthracis*	–103	Brigati *et al.* (2004)
		C. tryobutyricum		Lavilla *et al.* (2009)

BTA, biothreat alert; BADD, biowarfare agent detection device; ELISA, enzyme-linked Immunosorbent assay; IRMA, indirect immunoradiometric assay; RT-PCR-F, real-time polymerase chain reaction and fluorescence; FRET, fluorescence resonance energy transfer; MS, mass spectrometry; FTIR, Fourier transform infrared spectroscopy; sfu, spore-forming unit; – denotes not done or not available.

microscopy wherein even a single bait would be visualized using Q-dots (Sainathrao *et al.*, 2009). Though antigen and antibody-based detection methods provide potential for rapid bio-detection, they lack sensitivity with detection limits of around 10^5 CFU which is much lower than nucleic acid-based assays. Many immunoassays have a sensitivity threshold of 100- to 1000-fold above the estimated infectious dose. An infectious dose is the minimum number of the infectious organism required to cause an infection. Hence, by failing to detect the critical infectious dose these immunoassays fall short. To provide a scenario, the infectious dose (ID_{50}) of inhalational anthrax is estimated to be around 8000–50,000 spores, though minimum infective dose may be less (Franz *et al.*, 2001). These immunoassays are less specific and results need confirmation using other methods. Most of the methods described above need complicated protocols such as cell disruption and protein purification, and cannot support convenient, rapid and real-time detection of *B. anthracis*. Therefore, direct on-site point-of-care detection of *B. anthracis* is still a challenge especially with regards to time, sensitivity and specificity of the assay.

Nucleic-acid approaches are currently being used for the rapid detection and identification of *B. anthracis* spores as these systems are more sensitive than antigen/antibody-based detection systems. These techniques rely on the selection of nucleic acid probes that are specific for *B. anthracis* and do not cross react with other species. Primers are chosen to amplify targets such as *lef, pag or cap* that reside on either or on both of the virulence plasmids pXO1 and pXO2 (Makino *et al.*, 2001; Tamborrini *et al.*, 2010). A PCR-based assay using dual-probe sandwich hybridization targeting the *cap* genes has been reported to be highly sensitive with detection limit of 2 spore-forming units (sfu/ml) (Reif *et al.*, 1994) and using the same PCR-based assays with primers for *capA* gene the detection levels could be as few as 1 sfu (Makino *et al.*, 1993). Similarly, nested-PCR utilizing primers targeting the capsule gene seem to reveal a sensitivity of less than 10 spores per 100 g of soil (Beyer *et al.*, 1995). Recent studies show that using DNA–DNA hybridization and PCR assays for *cap C, pag* and *lef* genes, 10^5–10^6 *B. anthracis* spores/g of soil could be detected in untreated soil and the detection limit could be different in different soil conditions (Gulledge *et al.*, 2010). However, the disadvantages of hybridization is that the technique is highly laborious and the inherent impossibility of establishing a central database for hybridization profiles (Cho and Tiedje, 2001). Likewise, PCR for the detection of DNA from *B. anthracis* spores from powder samples was reported to be rapid and could detect ≤ 10 *B. anthracis* spores per gram of the sample (Luna *et al.*, 2003). Rapid strides have been made in the field of sample extraction and processing as well. The GeneXpert Q-PCR DNA detection system (Cepheid, Sunnyvale, CA, USA) for example combines the Q-PCR features of smart-cycler system with the automated disposable testing cartridges thus eliminating the need to extract DNA from the samples prior to analysis (Belgrader *et al.*, 2000). Similarly, the commercially available LightCycler *B. anthracis* detection kit (Roche Applied Science) consists of a disposable glass capillary system into which extracted DNA is placed and analysed (Uhl *et al.*, 2002). Real-time PCR using primers and probes for regions on pXO1, pXO2, and a segment of the *B. anthracis* chromosome are known to detect 1 pg DNA corresponding to around 5–10 spores (Hoffmaster *et al.*, 2002). Nucleic acid sequence-based amplification (NASBA) in combination with lateral flow assay has been successfully demonstrated. This assay is rapid and could be completed in 15 minutes and visualized by handheld reflectometer. For this assay, two mRNA sequences *atxA* and *pag*, were chosen for the detection of low numbers of spores. Using *atxA*, detection as low as one spore could be detected after 8 hours of sporulation and culture followed by NASBA amplification and detection. In the case of *pag* the detection time was 4 hours and the detection limit was 10 spores. In both cases no cross-reactivity with any other *Bacillus* species was noticed (Baeumner *et al.*, 2004).

Nucleic acid-based assays can detect fewer microorganisms in a short period of time (Ibekwe and Grieve, 2003). However the drawback of this system is the efficiency of extraction and purification of nucleic acid from spores which in turn influences the sensitivity and accuracy of the nucleic acid-based detection. Substances in sample matrices can inhibit the enzymatic and hybridization reactions, degrade the nucleic acid and reduce the efficiency of spore lysis (Lim *et al.*, 2005). Although there are a number of nucleic acid-based assays for detection however no one system satisfies all of the criteria, such as sensitivity, specificity, and being able to be performed at the site.

Limitations in immunoassays and nucleic acid-based detection assays have constantly resulted in attempts at discovering other novel ways of detection tools. One such recent discovery is the application of biosensors as diagnostic tools. There are various types of biosensors. Evanescent wave fibre-optic biosensors are analytical devices that measures the interaction between biomolecules. Specific ligands are immobilized to a sensor surface and a solution of receptor or antibody in injected over the top. Binding is measured by recording the changes in the refractive index caused by the molecules interacting near the sensor surface within the evanescent field. The strength of the biosensor depends on two factors. One is the sensitivity of the transducer and the other is the immobilized effect of bioactive material. The RAPTOR biosensor developed by the Naval Research Laboratory and Analyte 2000 (Research International) has been developed for the rapid detection of *B. anthracis* spores from powders. These sensors have a polystyrene fibre optic waveguide onto which the capture antibodies are coated. When the sample is passed over the waveguides, the target analytes are captured by the antibodies (Tims and Lim, 2004).

Cantilever and acoustic wave biosensor exploit change in mass to detect target agents. Capture molecules are adsorbed onto the surface substrate. As the target is captured, the added mass results in a frequency shift from normal oscillation of the cantilever. An additional shift in resonance frequency upon capture of the target is observed and measured (Campbell and Mutharasan, 2006a,b, 2007). Biosensors based on acoustic wave technology have been tested successfully using spores of *B. thuringiensis* (as a surrogate for *B. anthracis* spores). A monoclonal antibody immobilized on the surface captures the target resulting in acoustic measurements (Branch and Brozik, 2004). Laser Doppler vibrometer (LDV) biosensors are resonant mass sensors for the detection of *B. anthracis* Sterne spores in air and liquid. These sensors measure the decrease in resonant frequency driven by thermally induced oscillations when spores bind to the antibody immobilized on the surface of the sensor. Sensitivity of this technique is as few as 50 spores on the cantilever detected in water (Campbell and Mutharasan, 2007; Davila *et al.*, 2007; Pal and Alocilja, 2009).

Magnetoelastic biosensors are made up of amorphous ferromagnetic ribbons. It is based on the principle of change in the frequency of magnetoelastic materials when the material binds to the target. Within this specialized field, a phage-based magnetoelastic wireless biosensor for the detection of *B. anthracis* spores has recently been reported (Bruno and Kiel, 1999; Cable *et al.*, 2007; Petrenko, 2008; Huang *et al.*, 2009). This biosensor utilizes a filamentous phage-derived biomolecular recognition probe to capture target agents. The use of phage as a biomolecular recognition element improves the sensitivity and detection limit as it is robust and has strong resistance against heat (up to 80°C), organic solvents, urea (up to 6 M), acid, alkali and other chemicals. The detection limit of the biosensor is in the range of 10^3 spores/ml. The sensors can be stored at temperatures of 25°C, 45°C, and 65°C for period of three months with about 51%, 60%, and 75% loss in binding affinity, respectively.

Electrically active polyaniline coated magnetic nanoparticles (EAPM) have also been used for the detection of *B. anthracis* spores in food samples. The detection mechanism of the biosensor depends on the capillary flow of the captured spores on the biosensor surface with direct-charge transfer across the EAPM nanoparticles. Sensitivity of this biosensor is reported to be in the range of $4.2 \times 10^{2-3}$ spores/ml. The capture efficiency of the EAPM nanoparticles is better than that of commercially available immunomagnetic beads which can capture at a lower limit of 10^4 CFU/ml (Davila *et al.*, 2007; Pal and Alocilja, 2009).

An attenuated total reflection Fourier transform infrared spectroscopy (FTIR-ATR) in combination with electric field was used to detect spores. In this method, the negatively charged spores migrate and are concentrated on the surface of a ZnSe internal reflection element (IRE) under the influence of an electric field. Detection limit by this method was reported to be 2.5×10^7 spores/cm^3 (Li and Tripp, 2008; Li *et al.*, 2010).

Piezoelectric-excited millimetre-sized cantilever (PEMC) sensors consists of piezoelectric and a borosilicate glass layer. PEMC sensors are immobilized with antibodies to *B. anthracis* and exposed to spores. The resonant frequency decreases at a rate proportional to the spore concentration, depending on the number of spores bound to the sensor surface. The detection limit of *B. anthracis* spores using this method was found to be 300 spores/ml. The method is highly selective by detecting only *B. anthracis* spores when a sample containing mixture of *B. thuringiensis* spores and *B. anthracis* spores were used (Campbell and Mutharasan, 2007; Pal and Alocilja, 2009). In the same line, Quartz crystal microbalance (QCM) Biosensors consist of a thin quartz disc with electrodes plated on it. The target is captured by the antibodies or small molecules, immobilized on the surface of the sensor thereby causing a change in the resonance frequency. Polyclonal antibody-based quartz crystal microbalance sensors developed to detect *B. anthracis* simulates are one of the more recent techniques wherein the sensor is highly specific and recognizes *B. anthracis* spores and vegetative cells with a limit of detection of 10^3 CFU or 10^3 spores/ml in less than 30 minutes (Hao *et al.*, 2009; Pal and Alocilja, 2009). Microchip technology using a membrane-based optical analysis system has been reported for the detection of anthrax spores (Floriano *et al.*, 2005). Experiments using *B. atrophaeus*, a common substitute for anthrax demonstrated that the system could detect less than 500 spores within 5 minutes.

Surface Plasma Resonance (SPR) biosensors have become a promising tool for studying the biomolecular interactions. These sensors detect target analytes by measuring the refractive index changes that occur when the target binds to the surface of the sensor. They are highly sensitive, label-free, convenient, and work in real time. SPR sensor for the detection of whole *B. anthracis* spores is also sensitive with a detection limit of 10^4 CFU/ml (Table 13.2). This method seem to be efficient as it was rapid with a detection time is within 40 min, and it also differentiated other related *Bacillus* spores from *B. anthracis* spores (Wang *et al.*, 2009a). The limitations of SPR biosensors are that the fluid forces that acts on the cells has to be overcome before the cells can reach the sensor surface where they are captured as the force of the laminar flow may wash away the unbound cells. Another limiting factor is that the refractive changes occur only within the 300 nm distance within the surface and when cells as large as bacterial cells interact, only the portion of the cell that is in close contact with sensor surface may contribute to the change in refractive index. (Skottrup *et al.*, 2008)

Phage-display libraries have improved the development of affinity probes by a process called 'biopanning' where the library is exposed to a target and the best binding peptide is selected against a specific antigen. Probes of this sort have been developed to selectively bind *B. anthracis* spores (Williams *et al.*, 2003; Brigati *et al.*, 2004; Petrenko, 2008). However this assay is laborious and needs several rounds of panning in order to identify the best binding peptide. The peptide needs to be tagged in order for it to be applied in detection platforms.

Detection of other *Bacillus* spp.

B. cereus is a rod-shaped bacterium that is widely distributed in the environment (Drobniewski, 1993). *B. cereus* is mainly associated with food poisoning and is also responsible for various opportunistic and fatal non-gastrointestinal infections.

B. cereus produces two types of toxin responsible for food poisoning, one is the emetic toxin (cereulide, which is a small, heat-stable, non-ribosomally synthesized dodecadepsipeptide) and the enterotoxins (Lund *et al.*, 2000; Rajkovic *et al.*, 2008). *B. cereus* also produces other virulence factors like phospholipases, haemolysins, proteases, and other degradative enzymes.

Enumeration of *B. cereus* is usually done by the conventional culture-based plate count technique on selective media such as mannitol egg yolk polymyxin (MYP) and polymyxin egg yolk mannitol bromothymol blue agar (PEMBA), individual biochemical confirmation tests, and other specific approaches (e.g. spore morphology and the presence of the crystal protein) for species identification. These procedures are laborious and time-consuming and usually lead to misidentification due to the high level of phenotypic similarity among the species in the *B. cereus* group.

Molecular techniques offer the possibility of rapid and accurate identification of *B. cereus* isolates. Among these techniques, the quantitative real-time PCR (qPCR) assays for nucleic acid detection provide high speed, sensitivity and reproducibility while minimizing contamination (Table 13.2). Real-time PCRs targeting the haemolysin gene (Wang *et al.*, 1997), the phospholipase C gene (Martinez-Blanch *et al.*, 2009, 2010), the 16S rRNA gene (Reekmans *et al.*, 2009), the non-haemolytic enterotoxin genes (Yang *et al.*, 2007), the haemolysin BL genes (Gore *et al.*, 2003) and the emetic toxin genes (Fricker *et al.*, 2007) have already been attempted with good success rate in detection. More recently, a total of 180 spore-forming food contaminants representing 38 species and eight genera were tested by using ready-to-use PCR assay in artificially contaminated food samples. The assay demonstrated a detection limit of one spore of *B. cereus* spore in 25 g of food sample (Postollec *et al.*, 2010).

Another method quantified *B. cereus* spores in the gastrointestinal environment using qPCR and culture plating for monitoring viability and physiology of the *B. cereus* spores in various gastrointestinal compartments (Ceuppens *et al.*, 2010). The draw back of these molecular methods is that the spores are more resistant to DNA extraction than are vegetative cells and therefore can remain undetected when PCR approaches are used. To overcome this drawback, different methods have been adopted to release DNA from spores of the *B. cereus* group, for example sonication, mechanical lysis, commercial kits, and germination with subsequent lysis of the vegetative cells.

B. thuringiensis is an insect pathogen; however, it can act as opportunistic pathogen in humans and animals by causing tissue necrosis, pulmonary infection, or food poisoning (Schnepf *et al.*, 1998). *B. thuringiensis* is characterized by the production of insecticidal proteinaceous crystals during sporulation (Schnepf *et al.*, 1998). The toxins exhibit different specificities towards various classes of insects. *B. thuringiensis* is the most commonly used biopesticide worldwide. The genes encoding Cry and Cyt toxins are commonly located on large plasmids, which when lost, make the bacterium indistinguishable from *B. cereus.* On the other hand *B. thuringiensis* can be distinguished from *B. cereus* and *B. anthracis* by microbiological and biochemical tests. Other conventional methods include culturing the organism in selective growth media, H-flagellar serotyping, etc. (Shisa *et al.*, 2002). In recent years real-time PCR-based methods have become the most common technique for the sensitive detection and quantification of *B. thuringiensis* directly from environmental samples (Guidi *et al.*, 2010).

Clostridium: spore characteristics and differences

Clostridia are anaerobic, Gram-positive endospore-forming bacteria. The genus includes important pathogens *C. perfringens, C. tetani, C. botulinum, C. difficile* and strains of economic importance to humans like *C. acetobutylicum, C. beijerinckii, C. phytofermentans, C. thermocellum* and others.

Clostridia species, like *Bacillus^species,* produce endospores under nutritionally deprived conditions. However, the mechanism of sporulation in *Clostridia* has not been extensively studied as it has been in *B. subtilis. Bacillus* and *Clostridium* species sporulate by a wide range of environmental and physiological signals, which arise from nutrient starvation, cell density and Krebs cycle (Stragier and Losick, 1996). Recent studies identified inorganic phosphate (Pi) as a signal for inducing sporulation of *C. perfringens* (Philippe *et al.*, 2006).

The sequence of morphological events during the formation of *Clostridium* spores is almost identical to the events in *Bacillus* species sporulation (Paredes-Sabja *et al.*, 2008). During the first 3 hours, the original rod-shaped cells become ellipsoid to ovoid in shape; by 5–6 hours subterminal spores develop within these enlarged cells. A unique feature is the development of a convoluted, membranous exosporium which adheres to the outer surface of the two coats and has an unusual fine structure resembling a rectangular array of subunits (Hoeniger *et al.*, 1968; Paredes-Sabja *et al.*, 2008).

The classical phosphorylation pathway required to phosphorylate SpoA in *Bacillus* species is absent in *Clostridium* species indicating that Clostridial SpoA can directly phosphorylate by sensor histidine kinases (Henriques and Moran, 2000). Furthermore, expression of SpoA is also essential for the production of *C. perfringens* enterotoxin (CPE) (Henriques and Moran, 2000). Expression of key mother cell transcription factor (SpoIIID) depends on σ^E-associated RNA polymerase in *B. subtilis* but not in *C. perfringens* (Harry *et al.*, 2009).

In solvent-forming *C. acetobutylicum*, sporulation is a stationary phase event which also includes solvent formation and autolysis. However in some pathogenic *Clostridia*, such as *C. perfringens*, toxin production occurs during stationary phase and the enterotoxin CPE is produced during sporulation (Varga *et al.*, 2004). Sporulation in *C. acetobutylicum* seems to require carbon source and sufficient ATP with some form of stress in contrast to the simple sporulation in *B. subtilis* (Bahl *et al.*, 1995). In the case of *C. difficile*, large *Clostridial toxins* A and B are produced during stationary phase (Paredes *et al.*, 2005).

Heat-resistance of the spores of *C. perfringens* is influenced by environmental and genetic factors. Heat-resistance properties of *C. perfringens* spores is mainly influenced by the medium in which they are heated. Spores of some *C. perfringens* can survive boiling temperatures for an hour or longer in a protective medium. Food poisoning isolates of *C. perfringens* possess greater heat resistance properties than normal *C. perfringens*. It has been reported that incomplete cooking or inadequate heating may not only fail to kill *C. perfringens* spores, but can actually favour development of *C. perfringens* food poisoning by inducing spore germination (Paredes-Sabja *et al.*, 2008; Harry *et al.*, 2009). *C. perfringens* spores can also survive in lower temperatures. Food poisoning due to *C. perfringens* spores can result if viable spores present in frozen foods germinate when food is warmed for consumption (Li and McClane, 2006).

Difference between spores of various *Clostridium* strains

The position of endospores differs among bacterial species which can be used to differentiate spores of various Clostridial strains (Knaysi, 1948). These differences can be visually seen under a light microscope by first staining the endospores with the Schaeffer-Fulton stain. Spores may be located in the middle of the bacterium (central), at the end of the bacterium (terminal) or near the end of the bacterium (subterminal). *Clostridium tetani* has terminal and spherical endospore. *C. tertium* contains oval and terminal spore whereas *C. welchii* and *C. sporogenes* contains central or subterminal spores.

Detection methods for *Clostridium* spores

C. perfringens

Clostridium perfringens causes a variety of diseases in humans, including gas gangrene (Clostridial myonecrosis), enteritis necroticans (pigbel), and acute food poisoning. The symptoms are induced by the production of *C. perfringens* enterotoxin (CPE), a 35-kDa protein that is heavily expressed during sporulation (Collie and McClane, 1998).

The conventional approach to confirm a *C. perfringens* food-borne illness involves detection of $> 10^5$ bacterial cells or spores per gram of faeces or detection of CPE after inducing sporulation of an isolate. Flow cytometric method using fluorescent labelled anti-CPE antibody detects CPE exposed on the surface of the *C. perfringens* spores and is useful for diagnosing contamination (Kusunoki *et al.*, 1998). Several PCR methods for the detection of cpe-positive *C. perfringens* from faecal samples have been standardized so far with some success (Kanakaraj *et al.*, 1998). Hydrophobic grid membrane filter–colony hybridization (HGMF-CH) method has also been employed for the enumeration and isolation of cpe-positive *C. perfringens* spores from faecal samples. This method consists of HGMF filters through which the samples are filtered and these filters are placed on tryptose-sulphite-cycloserine (TSC) agar plates. Bacterial colonies grown on these agar plates upon incubation are replicated on a nylon membrane disc.

Later the replicas are transferred on to a blotting paper followed by hybridization and chromogenic detection (Heikinheimo *et al.*, 2004) (Table 13.2).

C. botulinum

C. botulinum produces several neurotoxins known as botulinum neurotoxins types A to G, which cause flaccid muscular paralysis (Zhang *et al.*, 2010). Routine laboratory diagnosis of botulism is based on the detection of botulinum neurotoxins. Food materials are often contaminated with spores of non-proteolytic *C. botulinum* and/or proteolytic *C. botulinum*. Fish and shellfish (e.g. crustaceans, mollusks) are among the food materials most frequently and heavily contaminated with spores of non-proteolytic *C. botulinum* and/or proteolytic *C. botulinum*. A method that combines a selective enrichment culture with multiplex PCR has been developed and validated to enumerate specifically the spores of non-proteolytic *C. botulinum* (Peck *et al.*, 2010). A method using PCR-enzyme-linked immunosorbent assay (ELISA) was used to detect *C. botulinum* spores in environmental samples. The method was found to be more sensitive than the mouse bioassay, wherein the samples are injected into mice to detect the presence of toxins (Fach *et al.*, 2002).

C. difficile

C. difficile causes diarrhoea and more serious intestinal conditions such as colitis. It can be found in the intestinal tract of humans and animals (Pirs *et al.*, 2008). Laboratory diagnosis of *C. difficile* depends on isolation of spores by broth enrichment cultures that could detect few spores as reported in a recent study (Cloud and Kelly, 2007). However the study lacks information about the number of organisms in the sample. Recent studies were performed that could evaluate the qualitative and quantitative analysis of *C. difficile* spores. This method showed that as low as 10 spores/g of meat could be isolated using enrichment cultures (Kamiya *et al.*, 1987). Similarly, another study reported that by using enrichment cultures, ≤10 spores/g could be detected. However, the detection threshold of quantitative culture was higher with a detection limit of 20 spores/g (Weese *et al.*, 2009).

Myxococcus: spore characteristics and differences

Myxobacteria belong to the class Deltaproteobacteria and order Myxococcales. Several species have been identified and when the nutrients are scarce, individual cells aggregate to form fruiting bodies, unlike the sporulation of other bacteria which occurs in a single cell (Kroos, 2007). These fruiting bodies can take different shape and colour depending on the species (Kroos, 2007). The cells begin as rod-shaped vegetative cells, within the fruiting bodies and develop into rounded myxospores with thick cell walls. These myxospores, analogous to spores in other organisms, survive until nutrients are abundant (Velicer and Vos, 2009). The size of myxobacteria fruiting bodies varies between 10–1000 μm or greater depending on the species (Kroos, 2007).

Fruiting body formation is triggered by amino acid depletion (Kaplan and Plamann, 1996; Shimkets, 1999). The cells aggregate to form mounds. The mounds increases in size containing up to 10^5 cells and then forms into macroscopic fruiting bodies (Sliusarenko *et al.*, 2007). The rod shaped cells inside the fruiting bodies form spherical spores that are metabolically inactive and resistant to temperature and desiccation. When nutrients become available, the spores germinate and complete the life cycle (Shimkets, 1999; Zusman *et al.*, 2007).

Fruiting body morphology differs among species and is regulated by complex genetic networks and multiple-stage specific intercellular signals (Shimkets, 1999). Detailed research of myxobacteria at the molecular and genetic level has been carried out mostly in the model species *Myxococcus xanthus* which forms relatively simple fruiting bodies (Shimkets, 1999).

The only practical application myxobacteria have found so far is the production of secondary metabolites (Krug *et al.*, 2008). Around 650 chemical structures and its variants have been obtained from myxobacteria (Reichenbach and Hofle, 1993). Some of the compounds produced by myxobacteria are stigmetellin, myxothiazol, myxalamids, aurachins, phenoxan, thiangazol (Reichenbach and Hofle, 1993). These metabolites are used as antibiotics, electron transport inhibitors, inhibitors of eubacterial RNA polymerase etc.which once characterized may be exploited as a biomarker for detection (Rosenberg *et al.*, 1973; Vaks *et al.*, 1974)

Different types of myxospores of the genus *Myxococcus*

Many different fruiting body shapes occur among the 40 species of myxobacteria demonstrating the versatility of cell-contact signalling (Sliusarenko *et al.*, 2007). Fruiting bodies of *M. xanthus*, *M. virescens* and *M. fulvus* consist of a spherical sporangium resting on a short stalk (Shimkets, 1999). The sporangium of *M. stipitatis* is elevated on a tall stalk. *Stigmatella aurantiaca* and *Chondromyces* spp. create thick stalks topped with several sporangioles (spherical spores not enclosed in a sporangium) (Kaiser, 2004). Soft slimy fruiting body is seen in *M. kofleria*, whereas a hard and cartilaginous fruiting body is seen in the species *Corallococcus* and *Archangium* (Kaiser, 2004).

Based on the number of sporangioles, the species of Myxobacteria can be differentiated, where single sporangiole on top of a stalk is seen in *Mellitangium*, *Stigmetella erecta*, whereas a group of sporangioles on or in the substrate is seen in *Cystobacter*, *Pyxicoccus*, *Polyangium*, *Sorangium*, *Nannocystis*, *Haliangium*, and *Byssophaga* (Shimkets, 1999). Group of small spherical sporangioles that are often empty and appearing glassy transparent, arranged in extended dense sheets or often in chains is seen in the species *Hylangium* (Shimkets, 1999). Cluster of sporangioles on top of an unbranched stalk is seen in *Stigmetella auraantiaca*, and in most of *Chondromyces* species. A cluster of sporangioles on a branched stalk is seen in *Chondromyces crocatus* (Shimkets, 1999).

Detection of *Myxococcus*

Myxospores tend to stick together and hence single colonies may not represent equal number of cells from environmental samples (Kroos, 2007). This hinders the development of cultural methods for estimating populations of myxobacteria in environmental samples

A strategy that combines PCR and denaturing gradient gel electrophoresis by screening the eubacterial 16S rDNA clone libraries followed by sequencing was reported for *M. xanthus* (Pinar *et al.*, 2010).Only the DNA of *M. xanthus* was amplified by this method indicating the specificity of the primers. A single round PCR assay for the detection of *M. cerebralis* based on the amplification of a segment of the gene of 18S ribosomal subunit was reported. The sensitivity and specificity of this method was 100% with a lower detection limit of around 50 myxospores per reaction (Baldwin and Myklebust, 2002).

Streptomyces: spore characteristics and differences

Streptomycetes are Gram-positive bacteria with a complex life cycle. *Streptomyces* form spores on aerial hyphae, and sporulation involves alterations in growth, morphogenesis and cell cycle processes like cell division and chromosome segregation (Chater *et al.*, 2010). Unlike the endospores of other Gram-positive bacteria, such as *Bacillus* and *Clostridium*, *Streptomyces* exospores are not resistant to extreme heat or pH and are less dormant; however, they are fairly resistant to desiccation (Staley and Gosink, 1999). Streptomycetes are highly abundant in nature and are found primarily in soil, where they can stay as dormant spores for long periods until conditions become favourable for their growth (Haiser *et al.*, 2009). When growth of vegetative mycelium has given rise to a colony, nutrient limitation triggers formation of aerial mycelium (Flardh and Buttner, 2009). The aerial hyphae represent reproductive structures and are transformed into pigmented spore chains that mature and eventually release spores. (Staley and Gosink, 1999)

Using atomic force microscopy, the cell surface changes that accompany the complex life cycle of *S. coelicolor* has been characterized (Del *et al.*, 2007). Vegetative hyphae possess a relatively smooth surface and granulated extracellular matrix which is absent in aerial hyphae and spores, while the surface of the mature spores on the other hand contain tightly organized fibrous mosaic layer that is absent in vegetative hyphae (Del *et al.*, 2007).

Additionally, *Streptomyces* spores differ in their spore morphology and colour from species to species that is useful for differentiating between species (Anderson and Wellington, 2001). About 80 per cent of the species studied have smooth spores (*S. albidoflavus, S. albus, S. acedomyceticus, S. alboniger, S. albosporeus,* and *S. anulatus*) (Anderson and Wellington, 2001). Some *Streptomyces* strains are hairy-spored (*S. albogriseolus, S. calvus,* and *Streptomyces sp*), and some warty-spored (*S. diastaticus, S. griseoplanus,* and *S. olivaceus*), The amount or length of hair on the spores of may differ from strain to strain that again may contribute in proper identification of the organism (Kumar and Goodfellow, 2010).

Detection of *Streptomyces* spores

Streptomycetes are well known for their production of antibiotics and secondary metabolites (Behal, 2000). Though these organisms may be isolated from clinical specimens, they generally turn out to be contaminants or colonizers (Carey *et al.*, 2001). *Streptomyces* species can sometimes cause human disease especially in immunocompromised host. One of the most common presentation of *Streptomyces* infection is Mycetoma, a chronic suppurative infection of the skin and underlying soft tissue (Dunne *et al.*, 1998). Some species like *S. albus, S. olivaceus, S. parvulus, S. humidus,* and *S. herbaricolor* are known to be isolated from lung nodules of patients with acute myelogenous leukaemia and from central venous catheter related blood stream infection from patients with metastatic breast cancer and metastatic cholangiocarcinoma (Mossad *et al.*, 1995; Carey *et al.*, 2001).

Presumptive diagnosis of *Streptomyces* sp includes Gram-positive staining, where they appear as long filamentous bacteria (Kapadia *et al.*, 2007). *Streptomyces* sp are negative for partial acid-fast staining. This feature distinguishes *Streptomyces* from other species (Kapadia *et al.*, 2007). Colony morphology shows powdery to waxy colonies and off-white in colour. With time, most colonies become pigmented with colours ranging from white to grey, buff, orange, red or purple (Anderson and Wellington, 2001). Most cultures have a strong 'earthy' aroma and by exploiting the characteristic aromas of each organism the different strains of *Streptomyces* can be identified. Certain *Streptomyces* sp are also known to produce geosmin, a secondary metabolite, responsible for the muddy and musty odour in potable water (Canhoto and Magan, 2003). A recent technique, E-nose consisting of an array of 14 conducting polymer sensors has been successfully used to detect *Streptomyces* spores (Canhoto and Magan, 2003). This method is also used to discriminate between different species based on their volatile odour production patterns (Canhoto and Magan, 2003).

Conclusions

Thus, it is evident that development and evaluation of high-sensitive, specific and rapid assays for the detection of various bacterial spores is gaining importance due to the fact that the easily spreading trait of these spores (both naturally and due to human involvement) has created an immediate need for the spore detection. Consequently, timely detection paves way for prevention of some of the highly transmissible and fatal diseases caused by the spore-forming organisms. While it is ideal to directly detect spores in a sample (without the need to induce them into vegetative forms) however, current knowledge on the various spore detection platforms described above indicate that there still is a need to improve on sensitivity, specificity and rapidity of the spore-specific detection assays. A vast majority of these assays are still in the developmental phases and efforts are on-going to develop assays that could detect these spores with minimal sample processing and that too in a short span of time. Some of these rapid assays involve analysis of samples on the site using portable devices and demonstrate appreciable sensitivity, specificity and are less-time consuming.. Interestingly, while the nucleic-acid-based techniques are highly sensitive than antibody-based detection systems, however these systems require clean samples for detection that may involve time-consuming sample processing and are expensive than the conventional immunoassays. Moreover, as an additional step the nucleic-acid-based assays require confirmation with lab-based confirmatory tests for the final diagnosis. The introduction of Q-dots and other novel biosensor assays has certainly enhanced the levels of detection and are in the process of further refinement.

Detection platforms should be specific enough to detect the target analytes to minimize false-positive results. Another important factor is the reproducibility. Factors such as stability, consistency and assay conditions may affect the reproducibility of the results. Surprisingly, some of the basic and simple detection techniques like microscopy are very useful in rapid identification of spores. However, such direct methods require technical expertise. Hence, one of the major emphasis towards development of spore detection assays is that the assays require minimal end-user expertise and are consistently reliable.

Acknowledgements

SS is a recipient of a postdoctoral fellowship at the Center for Biologics Evaluation and Research (CBER) administered by the Oak Ridge Institute for Science and Education (ORISE) through an intra agency agreement between the U. Department of Energy and the US Food and Drug Administration. Authors wish to thank Dr Xuan Chi, DH, OBRR, CBER and Dr Anita Verma, DBPAP, OVRR, CBER for their review of the manuscript.

Disclaimer

The findings and conclusions in this article have not been formally disseminated by the Food and Drug Administration and should not be construed to represent any Agency determination policy.

References

Anderson, A.S., and Wellington, E.M. (2001). The taxonomy of *Streptomyces* and related genera. Int. J. Syst. Evol. Microbiol. *51*, 797–814.

Andersson, A., Granum, P.E., and Ronner, U. (1998). The adhesion of *Bacillus cereus* spores to epithelial cells might be an additional virulence mechanism. Int. J. Food Microbiol. *39*, 93–99.

Baeumner, A.J., Leonard, B., McElwee, J., and Montagna, R.A. (2004). A rapid biosensor for viable *B. anthracis* spores. Anal. Bioanal. Chem. *380*, 15–23.

Bahl, H., Muller, H., Behrens, S., Joseph, H., and Narberhaus, F. (1995). Expression of heat shock genes in *Clostridium acetobutylicum*. FEMS Microbiol. Rev. *17*, 341–348.

Baldwin, T.J., and Myklebust, K.A. (2002). Validation of a single round polymerase chain reaction assay for identification of *Myxobolus cerebralis* myxospores. Dis. Aquat. Organ *49*, 185–190.

Behal, V. (2000). Bioactive products from *Streptomyces*. Adv. Appl. Microbiol. *47*, 113–156.

Belgrader, P., Okuzumi, M., Pourahmadi, F., Borkholder, D.A., and Northrup, M.A. (2000). A microfluidic cartridge to prepare spores for PCR analysis. Biosens. Bioelectron. *14*, 849–852.

Beyer, W., Glockner, P., Otto, J., and Bohm, R. (1995). A nested PCR method for the detection of *Bacillus anthracis* in environmental samples collected from former tannery sites. Microbiol. Res. *150*, 179–186.

Bozue, J., Moody, K.L., Cote, C.K., Stiles, B.G., Friedlander, A.M., Welkos, S.L., and Hale, M.L. (2007). *Bacillus anthracis* spores of the bclA mutant exhibit increased adherence to epithelial cells, fibroblasts, and endothelial cells but not to macrophages. Infect. Immun. *75*, 4498–4505.

Branch, D.W., and Brozik, S.M. (2004). Low-level detection of a bacillus anthracis simulant using Love-wave biosensors on 36 degrees YX LiTaO3. Biosens. Bioelectron. *19*, 849–859.

Brigati, J., Williams, D.D., Sorokulova, I.B., Nanduri, V., Chen, I.H., Turnbough, C.L., Jr., and Petrenko, V.A. (2004). Diagnostic probes for *Bacillus anthracis* spores selected from a landscape phage library. Clin. Chem. *50*, 1899–1906.

Bruno, J.G., and Kiel, J.L. (1999). *In vitro* selection of DNA aptamers to anthrax spores with electrochemiluminescence detection. Biosens. Bioelectron. *14*, 457–464.

Burns, D.A., Heap, J.T., and Minton, N.P. (2010). *Clostridium difficile* spore germination: an update. Res. Microbiol. *161*, 730–734.

Cable, M.L., Kirby, J.P., Sorasaenee, K., Gray, H.B., and Ponce, A. (2007). Bacterial spore detection by [Tb3+(macrocycle)(dipicolinate)] luminescence. J. Am. Chem. Soc. *129*, 1474–1475.

Callahan, C., Castanha, E.R., Fox, K.F., and Fox, A. (2008). The *Bacillus cereus* containing sub-branch most closely related to *Bacillus anthracis*, have single amino acid substitutions in small acid-soluble proteins, while remaining sub-branches are more variable. Mol. Cell Probes *22*, 207–211.

Callahan, C., Fox, K., and Fox, A. (2009). The small acid soluble proteins (SASP alpha and SASP beta) of *Bacillus weihenstephanensis* and *Bacillus mycoides* group 2 are the most distinct among the *Bacillus cereus* group. Mol. Cell Probes *23*, 291–297.

Campbell, G.A., and Mutharasan, R. (2006a). Detection of *Bacillus anthracis* spores and a model protein using PEMC sensors in a flow cell at 1 mL/min. Biosens. Bioelectron. *22*, 78–85.

Campbell, G.A., and Mutharasan, R. (2006b). Piezoelectric-excited millimeter-sized cantilever (PEMC) sensors detect *Bacillus anthracis* at 300 spores/mL. Biosens. Bioelectron. *21*, 1684–1692.

Campbell, G.A., and Mutharasan, R. (2007). Method of measuring *Bacillus anthracis* spores in the presence of copious amounts of *Bacillus thuringiensis* and *Bacillus cereus*. Anal. Chem. *79*, 1145–1152.

Canhoto, O.F., and Magan, N. (2003). Potential for detection of microorganisms and heavy metals in potable water using electronic nose technology. Biosens. Bioelectron. *18*, 751–754.

Cano, R.J., and Borucki, M.K. (1995). Revival and identification of bacterial spores in 25- to

40-million-year-old Dominican amber. Science *268*, 1060–1064.

Carey, J., Motyl, M., and Perlman, D.C. (2001). Catheter-related bacteremia due to *Streptomyces* in a patient receiving holistic infusions. Emerg. Infect. Dis. *7*, 1043–1045.

Carr, K.A., Lybarger, S.R., Anderson, E.C., Janes, B.K., and Hanna, P.C. (2010). The role of *Bacillus anthracis* germinant receptors in germination and virulence. Mol. Microbiol. *75*, 365–375.

Castanha, E.R., Swiger, R.R., Senior, B., Fox, A., Waller, L.N., and Fox, K.F. (2006). Strain discrimination among *B. anthracis* and related organisms by characterization of bclA polymorphisms using PCR coupled with agarose gel or microchannel fluidics electrophoresis. J. Microbiol. Methods *64*, 27–45.

Castanha, E.R., Vestal, M., Hattan, S., Fox, A., Fox, K.F., and Dickinson, D. (2007). *Bacillus cereus* strains fall into two clusters (one closely and one more distantly related) to *Bacillus anthracis* according to amino acid substitutions in small acid-soluble proteins as determined by tandem mass spectrometry. Mol. Cell Probes *21*, 190–201.

Ceuppens, S., Boon, N., Rajkovic, A., Heyndrickx, M., Van de Wiele, T., and Uyttendaele, M. (2010). Quantification methods for *Bacillus cereus* vegetative cells and spores in the gastrointestinal environment. J. Microbiol. Methods *83*, 202–210.

Chada, V.G., Sanstad, E.A., Wang, R., and Driks, A. (2003). Morphogenesis of bacillus spore surfaces. J. Bacteriol. *185*, 6255–6261.

Chater, K.F., Biro, S., Lee, K.J., Palmer, T., and Schrempf, H. (2010). The complex extracellular biology of *Streptomyces*. FEMS Microbiol. Rev. *34*, 171–198.

Cho, J.C., and Tiedje, J.M. (2001). Bacterial species determination from DNA–DNA hybridization by using genome fragments and DNA microarrays. Appl. Environ. Microbiol. *67*, 3677–3682.

Cloud, J., and Kelly, C.P. (2007). Update on *Clostridium difficile* associated disease. Curr. Opin. Gastroenterol. *23*, 4–9.

Collie, R.E., and McClane, B.A. (1998). Evidence that the enterotoxin gene can be episomal in *Clostridium perfringens* isolates associated with non-food-borne human gastrointestinal diseases. J. Clin. Microbiol. *36*, 30–36.

Daubenspeck, J.M., Zeng, H., Chen, P., Dong, S., Steichen, C.T., Krishna, N.R., Pritchard, D.G., and Turnbough, C.L., Jr. (2004). Novel oligosaccharide side chains of the collagen-like region of BclA, the major glycoprotein of the *Bacillus anthracis* exosporium. J. Biol. Chem. *279*, 30945–30953.

Davila, A.P., Jang, J., Gupta, A.K., Walter, T., Aronson, A., and Bashir, R. (2007). Microresonator mass sensors for detection of *Bacillus anthracis* Sterne spores in air and water. Biosens. Bioelectron. *22*, 3028–3035.

Del, S.R., Armstrong, I., Wright, C., and Dyson, P. (2007). Characterization of changes to the cell surface during the life cycle of *Streptomyces coelicolor*: atomic force microscopy of living cells. J. Bacteriol. *189*, 2219–2225.

Driks, A. (2009). The *Bacillus anthracis* spore. Mol. Aspects Med. *30*, 368–373.

Drobniewski, F.A. (1993). *Bacillus cereus* and related species. Clin. Microbiol. Rev. *6*, 324–338.

Dunne, E.F., Burman, W.J., and Wilson, M.L. (1998). *Streptomyces pneumonia* in a patient with human immunodeficiency virus infection: case report and review of the literature on invasive streptomyces infections. Clin. Infect. Dis. *27*, 93–96.

Dworkin, J., and Shah, I.M. (2010). Exit from dormancy in microbial organisms. Nat. Rev. Microbiol. *8*, 890–896.

Edwards, K.A., Clancy, H.A., and Baeumner, A.J. (2006). *Bacillus anthracis*: toxicology, epidemiology and current rapid-detection methods. Anal. Bioanal. Chem. *384*, 73–84.

Errington, J. (2003). Regulation of endospore formation in *Bacillus subtilis*. Nat. Rev. Microbiol. *1*, 117–126.

Fach, P., Perelle, S., Dilasser, F., Grout, J., Dargaignaratz, C., Botella, L., Gourreau, J.M., Carlin, F., Popoff, M.R., and Broussolle, V. (2002). Detection by PCR-enzyme-linked immunosorbent assay of *Clostridium botulinum* in fish and environmental samples from a coastal area in northern France. Appl. Environ. Microbiol. *68*, 5870–5876.

Fasanella, A., Losito, S., Trotta, T., Adone, R., Massa, S., Ciuchini, F., and Chiocco, D. (2001). Detection of anthrax vaccine virulence factors by polymerase chain reaction. Vaccine *19*, 4214–4218.

Fekety, R., Kim, K.H., Brown, D., Batts, D.H., Cudmore, M., and Silva, J., Jr. (1981). Epidemiology of antibiotic-associated colitis; isolation of *Clostridium difficile* from the hospital environment. Am. J. Med. *70*, 906–908.

Flardh, K., and Buttner, M.J. (2009). *Streptomyces* morphogenetics: dissecting differentiation in a filamentous bacterium. Nat. Rev. Microbiol. *7*, 36–49.

Floriano, P.N., Christodoulides, N., Romanovicz, D., Bernard, B., Simmons, G.W., Cavell, M., and McDevitt, J.T. (2005). Membrane-based on-line optical analysis system for rapid detection of bacteria and spores. Biosens. Bioelectron. *20*, 2079–2088.

Franz, D.R., Jahrling, P.B., McClain, D.J., Hoover, D.L., Byrne, W.R., Pavlin, J.A., Christopher, G.W., Cieslak, T.J., Friedlander, A.M., and Eitzen, E.M., Jr. (2001). Clinical recognition and management of patients exposed to biological warfare agents. Clin. Lab Med. *21*, 435–473.

Fricker, M., Messelhausser, U., Busch, U., Scherer, S., and Ehling-Schulz, M. (2007). Diagnostic real-time PCR assays for the detection of emetic *Bacillus cereus* strains in foods and recent food-borne outbreaks. Appl. Environ. Microbiol. *73*, 1892–1898.

Friedlander, A.M., Welkos, S.L., Pitt, M.L., Ezzell, J.W., Worsham, P.L., Rose, K.J., Ivins, B.E., Lowe, J.R., Howe, G.B., Mikesell, P., *et al.* (1993). Postexposure prophylaxis against experimental inhalation anthrax. J. Infect. Dis. *167*, 1239–1243.

Gerding, D.N., Muto, C.A., and Owens, R.C., Jr. (2008). Treatment of *Clostridium difficile* infection. Clin. Infect. Dis. *46 Suppl 1*, S32–S42.

Gerhardt, P., Pankratz, H.S., and Scherrer, R. (1976). Fine structure of the *Bacillus thuringiensis* spore. Appl. Environ. Microbiol. *32*, 438–440.

Gilbert, P., Collier, P.J., and Brown, M.R. (1990). Influence of growth rate on susceptibility to antimicrobial agents: biofilms, cell cycle, dormancy, and stringent response. Antimicrob. Agents Chemother. *34*, 1865–1868.

Giorno, R., Bozue, J., Cote, C., Wenzel, T., Moody, K.S., Mallozzi, M., Ryan, M., Wang, R., Zielke, R., Maddock, A., *et al.* (2007). Morphogenesis of the *Bacillus anthracis* spore. J. Bacteriol. *189*, 691–705.

Gore, H.M., Wakeman, C.A., Hull, R.M., and McKillip, J.L. (2003). Real-time molecular beacon NASBA reveals hblC expression from *Bacillus* spp. in milk. Biochem. Biophys. Res. Commun. *311*, 386–390.

Guidi, V., De, R.S., Benagli, C., Luthy, P., and Tonolla, M. (2010). A real-time PCR method to quantify spores carrying the *Bacillus thuringiensis* var. israelensis cry4Aa and cry4Ba genes in soil. J. Appl. Microbiol. *109*, 1209–1217.

Gulledge, J.S., Luna, V.A., Luna, A.J., Zartman, R., and Cannons, A.C. (2010). Detection of low numbers of *Bacillus anthracis* spores in three soils using five commercial DNA extraction methods with and without an enrichment step. J. Appl. Microbiol. *109*, 1509–1520.

Haiser, H.J., Yousef, M.R., and Elliot, M.A. (2009). Cell wall hydrolases affect germination, vegetative growth, and sporulation in *Streptomyces coelicolor*. J. Bacteriol. *191*, 6501–6512.

Hao, R., Wang, D., Zhang, X., Zuo, G., Wei, H., Yang, R., Zhang, Z., Cheng, Z., Guo, Y., Cui, Z., *et al.* (2009). Rapid detection of *Bacillus anthracis* using monoclonal antibody functionalized QCM sensor. Biosens. Bioelectron. *24*, 1330–1335.

Harry, K.H., Zhou, R., Kroos, L., and Melville, S.B. (2009). Sporulation and enterotoxin (CPE) synthesis are controlled by the sporulation-specific sigma factors SigE and SigK in *Clostridium perfringens*. J. Bacteriol. *191*, 2728–2742.

Heikinheimo, A., Lindstrom, M., and Korkeala, H. (2004). Enumeration and isolation of cpe-positive *Clostridium perfringens* spores from feces. J. Clin. Microbiol. *42*, 3992–3997.

Helgason, E., Okstad, O.A., Caugant, D.A., Johansen, H.A., Fouet, A., Mock, M., Hegna, I., and Kolsto, A.B. (2000). *Bacillus anthracis, Bacillus cereus,* and *Bacillus thuringiensis* – one species on the basis of genetic evidence. Appl. Environ. Microbiol. *66*, 2627–2630.

Henriques, A.O., and Moran, C.P., Jr. (2000). Structure and assembly of the bacterial endospore coat. Methods *20*, 95–110.

Henriques, A.O., and Moran, C.P., Jr. (2007). Structure, assembly, and function of the spore surface layers. Annu. Rev. Microbiol. *61*, 555–588.

Hoeniger, J.F., Stuart, P.F., and Holt, S.C. (1968). Cytology of spore formation in *Clostridium perfringens*. J. Bacteriol. *96*, 1818–1834.

Hoffmaster, A.R., Meyer, R.F., Bowen, M.D., Marston, C.K., Weyant, R.S., Thurman, K., Messenger, S.L., Minor, E.E., Winchell, J.M., Rassmussen, M.V., *et al.* (2002). Evaluation and validation of a real-time polymerase chain reaction assay for rapid identification of *Bacillus anthracis*. Emerg. Infect. Dis. *8*, 1178–1182.

Hoffmaster, A.R., Ravel, J., Rasko, D.A., Chapman, G.D., Chute, M.D., Marston, C.K., De, B.K., Sacchi, C.T., Fitzgerald, C., Mayer, L.W., *et al.* (2004). Identification of anthrax toxin genes in a *Bacillus cereus* associated with an illness resembling inhalation anthrax. Proc. Natl. Acad. Sci. U.S.A *101*, 8449–8454.

Hoffmaster, A.R., Hill, K.K., Gee, J.E., Marston, C.K., De, B.K., Popovic, T., Sue, D., Wilkins, P.P., Avashia, S.B., Drumgoole, R., *et al.* (2006a). Characterization of *Bacillus cereus* isolates associated with fatal pneumonias: strains are closely related to *Bacillus anthracis* and harbor *B. anthracis* virulence genes. J. Clin. Microbiol. *44*, 3352–3360.

Hoffmaster, A.R., Hill, K.K., Gee, J.E., Marston, C.K., De, B.K., Popovic, T., Sue, D., Wilkins, P.P., Avashia, S.B., Drumgoole, R., *et al.* (2006b). Characterization of *Bacillus cereus* isolates associated with fatal pneumonias: strains are closely related to *Bacillus anthracis* and harbor *B. anthracis* virulence genes. J. Clin. Microbiol. *44*, 3352–3360.

Holt, S.C., and Leadbetter, E.R. (1969). Comparative ultrastructure of selected aerobic spore-forming bacteria: a freeze-etching study. Bacteriol. Rev. *33*, 346–378.

Huang, S., Yang, H., Lakshmanan, R.S., Johnson, M.L., Wan, J., Chen, I.H., Wikle, H.C., III, Petrenko, V.A., Barbaree, J.M., and Chin, B.A. (2009). Sequential detection of Salmonella typhimurium and *Bacillus anthracis* spores using magnetoelastic biosensors. Biosens. Bioelectron. *24*, 1730–1736.

Huang, S.S., Datta, R., and Platt, R. (2006). Risk of acquiring antibiotic-resistant bacteria from prior room occupants. Arch. Intern. Med. *166*, 1945–1951.

Ibekwe, A.M., and Grieve, C.M. (2003). Detection and quantification of *Escherichia coli* O157:H7 in environmental samples by real-time PCR. J. Appl. Microbiol. *94*, 421–431.

Inglesby, T.V., Henderson, D.A., Bartlett, J.G., Ascher, M.S., Eitzen, E., Friedlander, A.M., Hauer, J., McDade, J., Osterholm, M.T., O'Toole, T., *et al.* (1999). Anthrax as a biological weapon: medical and public health management. Working Group on Civilian Biodefense. JAMA *281*, 1735–1745.

Iqbal, S.S., Mayo, M.W., Bruno, J.G., Bronk, B.V., Batt, C.A., and Chambers, J.P. (2000). A review of molecular recognition technologies for detection of biological threat agents. Biosens. Bioelectron. *15*, 549–578.

Kaiser, D. (2004). Signaling in myxobacteria. Annu. Rev. Microbiol. *58*, 75–98.

Kamiya, S., Yamakawa, K., Ogura, H., and Nakamura, S. (1987). Effect of various sodium taurocholate preparations on the recovery of *Clostridium difficile* spores. Microbiol. Immunol. *31*, 1117–1120.

Kanakaraj, R., Harris, D.L., Songer, J.G., and Bosworth, B. (1998). Multiplex PCR assay for detection of *Clostridium perfringens* in feces and intestinal contents of pigs and in swine feed. Vet. Microbiol. *63*, 29–38.

Kapadia, M., Rolston, K.V., and Han, X.Y. (2007). Invasive *Streptomyces* infections: six cases and literature review. Am. J. Clin. Pathol. *127*, 619–624.

Kaplan, H.B., and Plamann, L. (1996). A *Myxococcus xanthus* cell density-sensing system required for multicellular development. FEMS Microbiol. Lett. *139*, 89–95.

Kaprelyants, A.S., and Kell, D.B. (1996). Do bacteria need to communicate with each other for growth? Trends Microbiol. *4*, 237–242.

Klee, S.R., Nattermann, H., Becker, S., Urban-Schriefer, M., Franz, T., Jacob, D., and Appel, B. (2006). Evaluation of different methods to discriminate *Bacillus anthracis* from other bacteria of the *Bacillus cereus* group. J. Appl. Microbiol. *100*, 673–681.

Knaysi, G. (1948). The endospore of bacteria. Bacteriol. Rev. *12*, 19–77.

Kolsto, A.B., Tourasse, N.J., and Okstad, O.A. (2009). What sets *Bacillus anthracis* apart from other *Bacillus* species? Annu. Rev. Microbiol. *63*, 451–476.

Kroos, L. (2007). The *Bacillus* and *Myxococcus* developmental networks and their transcriptional regulators. Annu. Rev. Genet. *41*, 13–39.

Krug, D., Zurek, G., Revermann, O., Vos, M., Velicer, G.J., and Muller, R. (2008). Discovering the hidden secondary metabolome of *Myxococcus xanthus*: a study of intraspecific diversity. Appl. Environ. Microbiol. *74*, 3058–3068.

Kuehn, A., Kovac, P., Saksena, R., Bannert, N., Klee, S.R., Ranisch, H., and Grunow, R. (2009). Development of antibodies against anthrose tetrasaccharide for specific detection of *Bacillus anthracis* spores. Clin. Vaccine Immunol. *16*, 1728–1737.

Kumar, Y., and Goodfellow, M. (2010). Reclassification of *Streptomyces hygroscopicus* strains as *Streptomyces aldersoniae* sp. nov., *Streptomyces angustmyceticus* sp. nov., comb. nov., *Streptomyces ascomycinicus* sp. nov., *Streptomyces decoyicus* sp. nov., comb. nov., *Streptomyces milbemycinicus* sp. nov. and *Streptomyces wellingtoniae* sp. nov. Int. J. Syst. Evol. Microbiol. *60*, 769–775.

Kusunoki, H., Hu, D., Piyankarage, R.H., Sugii, S., and Uemura, T. (1998). Flow cytometric analysis for enterotoxin exposed on *Clostridium perfringens* spores. J. Vet. Med. Sci. *60*, 1357–1359.

Lai, E.M., Phadke, N.D., Kachman, M.T., Giorno, R., Vazquez, S., Vazquez, J.A., Maddock, J.R., and Driks, A. (2003). Proteomic analysis of the spore coats of *Bacillus subtilis* and *Bacillus anthracis*. J. Bacteriol. *185*, 1443–1454.

Layne, J.S., and Johnson, E.J. (1964). Resistant properties of *Azotobacter* cysts induced in response to mineral deficiencies. J. Bacteriol. *88*, 956–959.

Li, H., and Tripp, C.P. (2008). Detection of *Bacillus globigii* spores using a Fourier transform infrared-attenuated total reflection method. Appl. Spectrosc. *62*, 963–967.

Li, H., Doucette, L.D., Bousfield, D., and Tripp, C.P. (2010). Detection of spores using electric field-assisted FTIR-ATR. Anal. Chem. *82*, 5053–5059.

Li, J., and McClane, B.A. (2006). Further comparison of temperature effects on growth and survival of *Clostridium perfringens* type A isolates carrying a chromosomal or plasmid-borne enterotoxin gene. Appl. Environ. Microbiol. 72, 4561–4568.

Lim, D.V., Simpson, J.M., Kearns, E.A., and Kramer, M.F. (2005). Current and developing technologies for monitoring agents of bioterrorism and biowarfare. Clin. Microbiol. Rev. *18*, 583–607.

Luna, V.A., King, D., Davis, C., Rycerz, T., Ewert, M., Cannons, A., Amuso, P., and Cattani, J. (2003). Novel sample preparation method for safe and rapid detection of *Bacillus anthracis* spores in environmental powders and nasal swabs. J. Clin. Microbiol. *41*, 1252–1255.

Lund, T., De Buyser, M.L., and Granum, P.E. (2000). A new cytotoxin from *Bacillus cereus* that may cause necrotic enteritis. Mol. Microbiol. *38*, 254–261.

Makino, S.I., Iinuma-Okada, Y., Maruyama, T., Ezaki, T., Sasakawa, C., and Yoshikawa, M. (1993). Direct detection of *Bacillus anthracis* DNA in animals by polymerase chain reaction. J. Clin. Microbiol. *31*, 547–551.

Makino, S.I., Cheun, H.I., Watarai, M., Uchida, I., and Takeshi, K. (2001). Detection of anthrax spores from the air by real-time PCR. Lett. Appl. Microbiol. *33*, 237–240.

Martinez-Blanch, J.F., Sanchez, G., Garay, E., and Aznar, R. (2009). Development of a real-time PCR assay for detection and quantification of enterotoxigenic members of *Bacillus cereus* group in food samples. Int. J. Food Microbiol. *135*, 15–21.

Martinez-Blanch, J.F., Sanchez, G., Garay, E., and Aznar, R. (2010). Evaluation of a real-time PCR assay for the detection and quantification of *Bacillus cereus* group spores in food. J. Food Prot. 73, 1480–1485.

Melly, E., Genest, P.C., Gilmore, M.E., Little, S., Popham, D.L., Driks, A., and Setlow, P. (2002). Analysis of the properties of spores of *Bacillus subtilis* prepared at different temperatures. J. Appl. Microbiol. *92*, 1105–1115.

Mesnage, S., Tosi-Couture, E., Mock, M., Gounon, P., and Fouet, A. (1997). Molecular characterization of the *Bacillus anthracis* main S-layer component: evidence that it is the major cell-associated antigen. Mol. Microbiol. 23, 1147–1155.

Mock, M., and Fouet, A. (2001). Anthrax. Annu. Rev. Microbiol. 55, 647–671.

Moir, A. (2006). How do spores germinate? J. Appl. Microbiol. *101*, 526–530.

Mossad, S.B., Tomford, J.W., Stewart, R., Ratliff, N.B., and Hall, G.S. (1995). Case report of *Streptomyces endocarditis* of a prosthetic aortic valve. J. Clin. Microbiol. 33, 3335–3337.

Oliva, C., Turnbough, C.L., Jr., and Kearney, J.F. (2009). CD14–Mac-1 interactions in *Bacillus anthracis* spore internalization by macrophages. Proc. Natl. Acad. Sci. U.S.A *106*, 13957–13962.

Oliva, C.R., Swiecki, M.K., Griguer, C.E., Lisanby, M.W., Bullard, D.C., Turnbough, C.L., Jr., and Kearney, J.F. (2008). The integrin Mac-1 (CR3) mediates internalization and directs *Bacillus anthracis* spores into professional phagocytes. Proc. Natl. Acad. Sci. U.S.A. *105*, 1261–1266.

Pal, S., and Alocilja, E.C. (2009). Electrically active polyaniline coated magnetic (EAPM) nanoparticle as novel transducer in biosensor for detection of *Bacillus anthracis* spores in food samples. Biosens. Bioelectron. *24*, 1437–1444.

Paredes, C.J., Alsaker, K.V., and Papoutsakis, E.T. (2005). A comparative genomic view of clostridial sporulation and physiology. Nat. Rev. Microbiol. *3*, 969–978.

Paredes-Sabja, D., Torres, J.A., Setlow, P., and Sarker, M.R. (2008). *Clostridium perfringens* spore germination: characterization of germinants and their receptors. J. Bacteriol. *190*, 1190–1201.

Peck, M.W., Plowman, J., Aldus, C.F., Wyatt, G.M., Izurieta, W.P., Stringer, S.C., and Barker, G.C. (2010). Development and application of a new method for specific and sensitive enumeration of spores of nonproteolytic *Clostridium botulinum* types B, E, and F in foods and food materials. Appl. Environ. Microbiol. *76*, 6607–6614.

Petrenko, V.A. (2008). Landscape phage as a molecular recognition interface for detection devices. Microelectronics. J. *39*, 202–207.

Philippe, V.A., Mendez, M.B., Huang, I.H., Orsaria, L.M., Sarker, M.R., and Grau, R.R. (2006). Inorganic phosphate induces spore morphogenesis and enterotoxin production in the intestinal pathogen *Clostridium perfringens*. Infect. Immun. *74*, 3651–3656.

Phillips, A.P., and Martin, K.L. (1983). Comparison of direct and indirect immunoradiometric assays (IRMA) for *Bacillus anthracis* spores immobilised on multispot microscope slides. J. Appl. Bacteriol. *55*, 315–324.

Piggot, P.J., and Hilbert, D.W. (2004). Sporulation of *Bacillus subtilis*. Curr. Opin. Microbiol. *7*, 579–586.

Pinar, G., Jimenez-Lopez, C., Sterflinger, K., Ettenauer, J., Jroundi, F., Fernandez-Vivas, A., and Gonzalez-Munoz, M.T. (2010). Bacterial community dynamics during the application of a *Myxococcus xanthus*-inoculated culture medium used for consolidation of ornamental limestone. Microb. Ecol. *60*, 15–28.

Pirs, T., Ocepek, M., and Rupnik, M. (2008). Isolation of *Clostridium difficile* from food animals in Slovenia. J. Med. Microbiol. *57*, 790–792.

Plomp, M., Leighton, T.J., Wheeler, K.E., and Malkin, A.J. (2005). The high-resolution architecture and structural dynamics of *Bacillus* spores. Biophys. J. *88*, 603–608.

Postollec, F., Bonilla, S., Baron, F., Jan, S., Gautier, M., Mathot, A.G., Hallier-Soulier, S., Pavan, S., and Sohier, D. (2010). A multiparametric PCR-based tool for fast detection and identification of spore-forming bacteria in food. Int. J. Food Microbiol. *142*, 78–88.

Pribil, P.A., Patton, E., Black, G., Doroshenko, V., and Fenselau, C. (2005). Rapid characterization of *Bacillus* spores targeting species-unique peptides produced with an atmospheric pressure matrix-assisted laser desorption/ionization source. J. Mass Spectrom. *40*, 464–474.

Rajkovic, A., Uyttendaele, M., Vermeulen, A., Andjelkovic, M., Fitz-James, I., in 't Veld, P., Denon, Q., Verhe, R., and Debevere, J. (2008). Heat resistance of *Bacillus cereus* emetic toxin, cereulide. Lett. Appl. Microbiol. *46*, 536–541.

Rao, S.S., Mohan, K.V., and Atreya, C.D. (2010). Detection technologies for *Bacillus anthracis*: prospects and challenges. J. Microbiol. Methods *82*, 1–10.

Rasko, D.A., Altherr, M.R., Han, C.S., and Ravel, J. (2005). Genomics of the *Bacillus cereus* group of organisms. FEMS Microbiol. Rev. *29*, 303–329.

Reekmans, R., Stevens, P., Vervust, T., and De, V.P. (2009). An alternative real-time PCR method to detect the *Bacillus cereus* group in naturally contaminated food gelatine: a comparison study. Lett. Appl. Microbiol. *48*, 97–104.

Reichenbach, H., and Hofle, G. (1993). Biologically active secondary metabolites from myxobacteria. Biotechnol. Adv. *11*, 219–277.

Reif, T.C., Johns, M., Pillai, S.D., and Carl, M. (1994). Identification of capsule-forming *Bacillus anthracis* spores with the PCR and a novel dual-probe hybridization format. Appl. Environ. Microbiol. *60*, 1622–1625.

Rosenberg, E., Vaks, B., and Zuckerberg, A. (1973). Bactericidal action of an antibiotic produced by *Myxococcus xanthus*. Antimicrob. Agents Chemother. *4*, 507–513.

Rupnik, M., Wilcox, M.H., and Gerding, D.N. (2009). *Clostridium difficile* infection: new developments in epidemiology and pathogenesis. Nat. Rev. Microbiol. *7*, 526–536.

Russell, B.H., Vasan, R., Keene, D.R., Koehler, T.M., and Xu, Y. (2008). Potential dissemination of *Bacillus anthracis* utilizing human lung epithelial cells. Cell Microbiol. *10*, 945–957.

Ruthel, G., Ribot, W.J., Bavari, S., and Hoover, T.A. (2004). Time-lapse confocal imaging of development of *Bacillus anthracis* in macrophages. J. Infect. Dis. *189*, 1313–1316.

Sainathrao, S., Mohan, K.V., and Atreya, C. (2009). Gamma-phage lysin PlyG sequence-based synthetic peptides coupled with Qdot-nanocrystals are useful for developing detection methods for *Bacillus anthracis* by using its surrogates, *B. anthracis*-Sterne and *B. cereus*-4342. BMC. Biotechnol. *9*, 67.

Schnepf, E., Crickmore, N., Van, R.J., Lereclus, D., Baum, J., Feitelson, J., Zeigler, D.R., and Dean, D.H. (1998). *Bacillus thuringiensis* and its pesticidal crystal proteins. Microbiol. Mol. Biol. Rev. *62*, 775–806.

Setlow, P. (2006). Spores of *Bacillus subtilis*: their resistance to and killing by radiation, heat and chemicals. J. Appl. Microbiol. *101*, 514–525.

Shimkets, L.J. (1999). Intercellular signaling during fruiting-body development of *Myxococcus xanthus*. Annu. Rev. Microbiol. *53*, 525–549.

Shisa, N., Wasano, N., Ohgushi, A., Lee, D.H., and Ohba, M. (2002). Extremely high frequency of common flagellar antigens between *Bacillus thuringiensis* and *Bacillus cereus*. FEMS Microbiol. Lett. *213*, 93–96.

Singh, B., Ghosh, J., Islam, N.M., Dasgupta, S., and Kirsebom, L.A. (2010). Growth, cell division and sporulation in mycobacteria. Antonie Van Leeuwenhoek *98*, 165–177.

Skottrup, P.D., Nicolaisen, M., and Justesen, A.F. (2008). Towards on-site pathogen detection using antibody-based sensors. Biosens. Bioelectron. *24*, 339–348.

Sliusarenko, O., Zusman, D.R., and Oster, G. (2007). Aggregation during fruiting body formation in *Myxococcus xanthus* is driven by reducing cell movement. J. Bacteriol. *189*, 611–619.

Staley, J.T., and Gosink, J.J. (1999). Poles apart: biodiversity and biogeography of sea ice bacteria. Annu. Rev. Microbiol. *53*, 189–215.

Stenfors Arnesen, L.P., Fagerlund, A., and Granum, P.E. (2008). From soil to gut: *Bacillus cereus* and its food poisoning toxins. FEMS Microbiol. Rev. *32*, 579–606.

Stenfors Arnesen, L.P., O'sullivan, K., and Granum, P.E. (2007). Food poisoning potential of *Bacillus cereus* strains from Norwegian dairies. Int. J. Food Microbiol. *116*, 292–296.

Stragier, P., and Losick, R. (1996). Molecular genetics of sporulation in *Bacillus subtilis*. Annu. Rev. Genet. *30*, 297–241.

Sylvestre, P., Couture-Tosi, E., and Mock, M. (2002). A collagen-like surface glycoprotein is a structural component of the *Bacillus anthracis* exosporium. Mol. Microbiol. *45*, 169–178.

Sylvestre, P., Couture-Tosi, E., and Mock, M. (2003). Polymorphism in the collagen-like region of the *Bacillus anthracis* BclA protein leads to variation in exosporium filament length. J. Bacteriol. *185*, 1555–1563.

Tamborrini, M., Holzer, M., Seeberger, P.H., Schurch, N., and Pluschke, G. (2010). Anthrax spore detection by a luminex assay based on monoclonal antibodies that recognize anthrose-containing oligosaccharides. Clin. Vaccine Immunol. *17*, 1446–1451.

Taylor, J.M., Sutherland, A.D., Aidoo, K.E., and Logan, N.A. (2005). Heat-stable toxin production by strains of *Bacillus cereus, Bacillus firmus, Bacillus megaterium, Bacillus simplex* and *Bacillus licheniformis*. FEMS Microbiol. Lett. *242*, 313–317.

Tims, T.B., and Lim, D.V. (2004). Rapid detection of *Bacillus anthracis* spores directly from powders with an evanescent wave fiber-optic biosensor. J. Microbiol. Methods *59*, 127–130.

Traag, B.A., Driks, A., Stragier, P., Bitter, W., Broussard, G., Hatfull, G., Chu, F., Adams, K.N., Ramakrishnan, L., and Losick, R. (2010). Do mycobacteria produce endospores? Proc. Natl. Acad. Sci. U.S.A *107*, 878–881.

Uhl, J.R., Bell, C.A., Sloan, L.M., Espy, M.J., Smith, T.F., Rosenblatt, J.E., and Cockerill, F.R., III (2002). Application of rapid-cycle real-time polymerase chain reaction for the detection of microbial pathogens: the Mayo-Roche Rapid Anthrax Test. Mayo Clin. Proc. *77*, 673–680.

Vaks, B., Zuckerberg, A., and Rosenberg, E. (1974). Purification and partial characterization of an antibiotic produced by *Myxococcus xanthus*. Can. J. Microbiol. *20*, 155–161.

Varga, J., Stirewalt, V.L., and Melville, S.B. (2004). The CcpA protein is necessary for efficient sporulation and enterotoxin gene (cpe) regulation in *Clostridium perfringens*. J. Bacteriol. *186*, 5221–5229.

Velicer, G.J., and Vos, M. (2009). Sociobiology of the myxobacteria. Annu. Rev. Microbiol. *63*, 599–623.

Votyakova, T.V., Kaprelyants, A.S., and Kell, D.B. (1994). Influence of viable cells on the resuscitation of dormant cells in *Micrococcus luteus* cultures held in an extended stationary phase: the population effect. Appl. Environ. Microbiol. *60*, 3284–3291.

Vreeland, R.H., Rosenzweig, W.D., and Powers, D.W. (2000). Isolation of a 250 million-year-old halotolerant bacterium from a primary salt crystal. Nature *407*, 897–900.

Waller, L.N., Fox, N., Fox, K.F., Fox, A., and Price, R.L. (2004). Ruthenium red staining for ultrastructural visualization of a glycoprotein layer surrounding the spore of *Bacillus anthracis* and *Bacillus subtilis*. J. Microbiol. Methods *58*, 23–30.

Wang, D.B., Bi, L.J., Zhang, Z.P., Chen, Y.Y., Yang, R.F., Wei, H.P., Zhou, Y.F., and Zhang, X.E. (2009a). Label-free detection of *B. anthracis* spores using a surface plasmon resonance biosensor. Analyst *134*, 738–742.

Wang, D.B., Yang, R., Zhang, Z.P., Bi, L.J., You, X.Y., Wei, H.P., Zhou, Y.F., Yu, Z., and Zhang, X.E. (2009b). Detection of *B. anthracis* spores and vegetative cells with the same monoclonal antibodies. PLoS. One. *4*, e7810.

Wang, R.F., Cao, W.W., and Cerniglia, C.E. (1997). A universal protocol for PCR detection of 13 species of food-borne pathogens in foods. J. Appl. Microbiol. *83*, 727–736.

Weaver, J., Kang, T.J., Raines, K.W., Cao, G.L., Hibbs, S., Tsai, P., Baillie, L., Rosen, G.M., and Cross, A.S. (2007). Protective role of *Bacillus anthracis* exosporium in macrophage-mediated killing by nitric oxide. Infect. Immun. *75*, 3894–3901.

Weese, J.S., Avery, B.P., Rousseau, J., and Reid-Smith, R.J. (2009). Detection and enumeration of *Clostridium difficile* spores in retail beef and pork. Appl. Environ. Microbiol. *75*, 5009–5011.

Williams, D.D., Benedek, O., and Turnbough, C.L., Jr. (2003). Species-specific peptide ligands for the detection of *Bacillus anthracis* spores. Appl. Environ. Microbiol. *69*, 6288–6293.

Wireman, J.W., and Dworkin, M. (1975). Morphogenesis and developmental interactions in myxobacteria. Science *189*, 516–523.

Yang, I.C., Shih, D.Y., Wang, J.Y., and Pani, T.M. (2007). Development of rapid real-time PCR and most-probable-number real-time PCR assays to quantify enterotoxigenic strains of the species in the *Bacillus cereus* group. J. Food Prot. *70*, 2774–2781.

Zahavy, E., Heleg-Shabtai, V., Zafrani, Y., Marciano, D., and Yitzhaki, S. (2010). Application of fluorescent nanocrystals (q-dots) for the detection of pathogenic bacteria by flow-cytometry. J. Fluoresc. *20*, 389–399.

Zaucha, G.M., Pitt, L.M., Estep, J., Ivins, B.E., and Friedlander, A.M. (1998). The pathology of experimental anthrax in rabbits exposed by inhalation and subcutaneous inoculation. Arch. Pathol. Lab. Med. *122*, 982–992.

Zhang, J.C., Sun, L., and Nie, Q.H. (2010). Botulism, where are we now? Clin. Toxicol. *48*, 867–879.

Zusman, D.R., Scott, A.E., Yang, Z., and Kirby, J.R. (2007). Chemosensory pathways, motility and development in *Myxococcus xanthus*. Nat. Rev. Microbiol. *5*, 862–872.

Properties and Detection Methods of *Bacilli* Spores in Food and in Medical Settings

14

Olga Tarasenko, Pierre Alusta, Sergey Kazakov and Kalle Levon

Abstract

Numerous members of *Bacillus and Clostridium* genera are widely distributed in nature. Their control is considerably important since these species are responsible for a variety of food spoilage, foodborne illnesses, and food-poisoning problems, in addition to their potential use as biological warfare agents. The present study concentrates on the taxonomic *Bacillus cereus* group covering studies of genetics and morphology, including that of harmful bacteria. Moreover, several classical issues in spore resistance biology are being considered. Examining a number of reports will provide us with morphological properties of spores as well as their biochemical composition. We will then extrapolate on some of the findings, thus enabling us to determine the role of how to identify proteins in spore immunology and outline a possible issue in this regard. This paper presents an overview of different physicochemical techniques meaningful for spore detection and identification as a basis for biosensor construction and/or other detection techniques. Special attention is given to different methods of spore detection and identification including but not limited to microbiological, immunological, genetic, and radiometric methods, automation of assay procedures, biosensors as well as new approaches. Future directions for development and problems related to biosensors are discussed in the final part of this review.

Detection versus identification

In general, detection is merely the extraction of meaningful information from a larger stream of information without specifically interacting with it. Detection can be used in biology to determine an organism (e.g. wolf, human, bacteria, or cocci).

In biology, the term 'identification', however, refers to the process of assigning a pre-existing individual or class name to an individual organism. 'Identification' of organisms may be based on (i) individual natural features and (ii) specific molecular markers. The more common form of identification is that of organisms (e.g. used in microbiology) ranging from identifying genus (e.g. '*Mycobacterium*' or '*Bacillus*') to species (e.g. '*Mycobacterium tuberculosis*' or '*Bacillus anthracis*'). The features may be either morphological, anatomical, physiological, behavioural, or molecular. Specific molecular markers will only occur in closely related species. Identification may either rely on manual or computerized methods. Identification may involve the use of identification keys, browsing through field guidebooks that contain (often illustrated) species accounts, or comparing individual characteristics of genus and species.

Both, detection and identification are very critical in the early determination of both genus and species of a pathogen caused a specific infection, disease outbreak, and water or food contamination, or biological attacks.

Pathogens

Bacterial pathogens are ubiquitous in soil, marine and estuarine waters, the intestinal tract of animals, or in water contaminated with faecal matter. An average adult carries over 150 different types of bacteria that exist both inside and outside the body (Madigan *et al.*, 1997). The majority of microorganisms carry out essential activities, many of which are closely associated with plants or animals in symbiosis. However, certain potentially harmful microorganisms can have profound effects on mammals and may be the cause of various infectious diseases.

Every year millions of food-borne illnesses cases are reported. The latter are steadily rising due to the presence of bacteria contaminating food products (Raymond *et al.*, 2010). In addition, recent *B. anthracis* attacks in the U.S.A. have demonstrated the vulnerability of air, drinking water, food supply, and energy sources. These sources are also the most difficult ones to protect due to their wide accessibility to the public, making constant monitoring very challenging.

Bacillus and *Clostridium* are responsible for a variety of food spoilage, food-borne illnesses, and food-poisoning problems, in addition to its potential use as biological warfare agents (Doyle *et al.*, 1997; Jensen *et al.*, 2003; Shiota *et al.*, 2010).

Food safety is a major focus of food microbiology. Pathogenic bacteria, viruses and toxins produced by pathogens are all possible contaminants of food. Water and food can transmit many potential biological agents, their toxins, and or dormant cells (spores or cysts), leading to disease. *B. anthracis* is a main concern since it causes anthrax, a highly infectious disease of hoofed animals that can be readily transmitted to humans. There is evidence that *Bacillus* and *Clostridium* species can be weaponized for aerosol application, food contamination in biological warfare (Marshall, 1997; Zilinskas, 1997), or even transmitted through water or water supplies.

Various members of *Bacillus* and *Clostridium* may cause changes in the character of food during handling, preparation, transportation, or storage. Microbial growth leads to food deterioration and spoilage by decay, and food poisoning, mainly caused by different and less widespread bacteria. As microorganisms grow, they release their own enzymes into the liquid surrounding them, and absorb the products of external digestion. Pathogens may also produce waste products which act as poisons or toxins, thus causing the renowned ill-effects.

Food poisoning is defined as an illness caused by the consumption of food or water contaminated with bacteria and/or their toxins, or with parasites, viruses, or chemicals. The symptoms, varying in degree and combination, include abdominal pain, vomiting, diarrhoea, and headache; more serious cases can result in life-threatening neurological, hepatic and renal syndromes leading to permanent disability or death.

Bacillus cereus has been recognized as an agent of food poisoning since 1955. Since then, 52 outbreaks of food-borne disease associated with *B. cereus* were reported to the CDC (in 2003, there were two). *B. cereus* causes two types of food-borne illnesses. One type is characterized by nausea and vomiting and abdominal cramps and has an incubation period of 1–6 hours. This is the 'short-incubation' or emetic form of the disease. The second type is manifested primarily by abdominal cramps and diarrhoea following an incubation period of 8 to 16 hours. Diarrhoea may be a small volume or profuse and watery. This type is referred to as the 'long-incubation' or diarrheal form of the disease, and it resembles food poisoning caused by *Clostridium perfringens.*

C. perfringens food poisoning results when person eat improperly cooked and stored foods because this pathogen release toxins. These toxins can make a person ill if enough is taken into the body though (i) swallowing a large amount of the bacteria, which then multiply and produce toxin in the intestine; or (ii) consumption of contaminated food with the toxin already in it.

The importance of *Bacilli* and *Clostridium* species should not be underestimated since many of these species produce highly resistant spores (Gerhardt and Marquis, 1989; Setlow, 1994; Carlin *et al.*, 2000; Carr *et al.*, 2010) (Fig. 14.1). Spores can withstand a variety of harsh conditions. Research has shown that spores are indeed capable of surviving cold temperatures (Carlin et. al, 2000), radiation, heat, ultraviolet light, and food preservation or preparation (Gerhardt and Marquis, 1989; Setlow, 1994; Carr *et al.*, 2010).

B. anthracis has been taxonomically aligned with *B. cereus, B. thuringiensis, B. megaterium, and B. mycoides* based on morphological and DNA similarities (Drobniewski, 1993). All five species have been classified into the *Bacillus* subgroup one

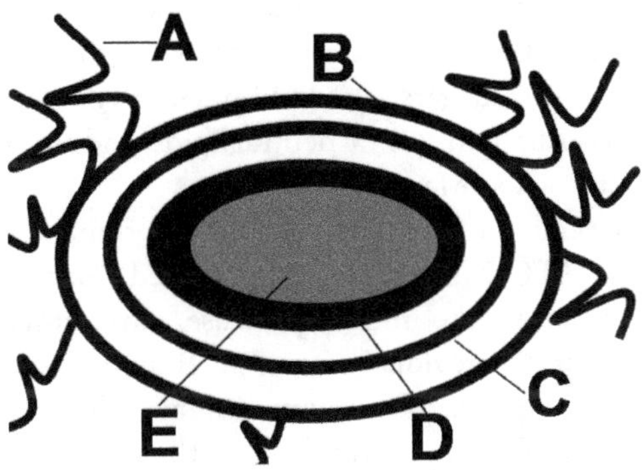

Figure 14.1 Generalized diagram of a bacterial spore: appendages (A), exosporium (B), outer coat (C), inner coat (D) and cortex core (E).

based on their large cell width (> 1mm) and spores that do not distend the sporangia (Gordon *et al.*, 1973). A comparison of the rRNA gene sequences is in concordance with this taxonomic arrangement (Ash *et al.*, 1991, 1992). DNA studies have further united *B. anthracis, B. cereus, B. thuringiensis, and B. mycoides,* all of which have remarkably AT-rich genomes (MacDonald *et al.*, 1963) and are highly homologous (Seki *et al.*, 1978). It was suggested that these four species should be reorganized into a single species with four subspecies (Turnbull *et al.*, 1991; Jackson *et al.*, 2001). These species are exhibiting morphological variation of vegetative cells that can undergo differentiation into spores. Theoretically, non-harmful bacteria can be used a vehicle to deliver genes, and proteins of harmful *B. anthracis*, for instance.

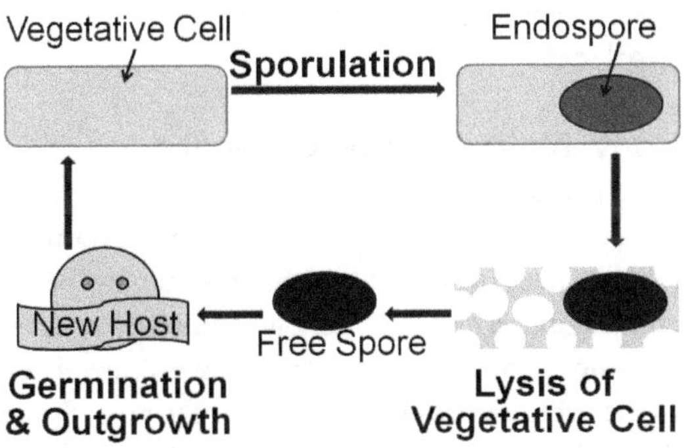

Figure 14.2 Sporulation process scheme.

B. cereus is one of the leading bacteria that contaminates food products and leads to severe food-borne illnesses and poisonings (Doyle *et al.*, 1997; Jensen *et al.*, 2003; Shiota *et al.*, 2010; Bottone, 2010). Outbreaks caused by *B. cereus* have occurred through contaminated rice, meat, mashed potatoes (Taylor *et al.*, 1975; Drobniewski, 1993), dairy products (Becker *et al.*, 1994), macaroni-and-cheese servings (Holmes *et al.*, 1981) and powdered milk products (Schmitt *et al.*, 1976). There, *B. cereus*-like organisms were recognized in outbreaks caused by *B. cereus* contracted through milk products. Positive identification of the isolates, however, was not performed (Schmitt *et al.*, 1976). Concerns over rapid identification of *B. cereus* or other spore-forming agents in dairy products and powdered milk products rose sharply, since milk is a base ingredient in many food products. *B. cereus* causes two types of food-borne intoxications, namely emetic and diarrheal. In addition, *B. cereus* spores can survive antiseptic treatments, making disease control ineffective. Terrorists could use *Bacillus* spores as biological weapons in imported food products (Barnaby, 1997; Stratis, 2007).

Spore structure and resistance

When challenged by stress factors such as lack of nutrients, *Bacillus* and *Clostridium* species produce dormant cell types, known as spores (Fig. 14.1) through sporulation process (Fig. 14.2) (Van Ness 1971; Gerhardt *et al.*, 1989; Setlow, 1994).

The bacterial spore structure is rather of sophisticated nature and exhibits the following parts; namely appendages, exosporium, outer coat, inner coat, cortex, and core (Fig. 14.1) (Gerhardt and Ribi, 1964; Hachizuka *et al.*, 1966, 1984; Moberly *et al.*, 1966; Kramer and Roth, 1968; Holt *et al.*, 1969; Beaman *et al.*, 1972; Aronson *et al.*, 1976; Souse *et al.*, 1976, 1978; Kozuka *et al.*, 1985; Hultgren *et al.*, 1993; Driks, 1999). *Bacillus* species were classified into four groups based on the presence of exosporium and appendages (Hashisuka *et al.*, 1984; Kozuka, 1984).

The application of solubilizing agents such as urea, mercaptoethanol, or sodium hydroxide causes exosporium layer to be pulled away from the outer coat (Sousa *et al.*, 1978). Weaponized spores lack appendages and exosporium.

Dormant spores are indeed resistant (Gerhardt and Marquis, 1989; Lindeque *et al.*, 1994; Setlow, 1994), and can survive for long periods in contaminated soils, silk or bones (De Vos, 1990). *B. anthracis* spores are heat or chlorination resistant (Burrows and Renner, 1999).

Low water content and high level of minerals in the spore core, decreased permeability, and saturation of the spore chromosome with a group of small acid-soluble proteins (SASP) α/β type are appear to be responsible for increased spore resistancy (Gerhardt *et al.*, 1989; Setlow, 1994). Van Ness (1971) postulated that alkaline pH, calcareous soils, and the presence of organic matter might buffer spore supply of calcium in addition to maintaining viable spores in a certain environment for longer periods of time. The presence of calcium only favours spore survival.

Biochemical composition is directly related to resistancy and infectivity of spores. *B. cereus* appendages of have shown a high concentration of protein, but less lipids and carbohydrates. Isolated exosporium fractions from *B. cereus* (Marz *et al.*, 1970) and *C. botulinum type* A (Takumi *et al.*, 1979) mainly consist of proteins and lipids. Exosporium is exhibiting strong hydrophobicity may be due to hydrophobicprotein and lipid components (Koshikawa *et al.*, 1989). The spore's coat consists largely of hydrophobic proteins (Warth *et al.*, 1963; Aronson and Pandey, 1978; Kozuka and Tochikubo, 1985; Koshikawa *et al.*, 1989). Hydrophobic chromatography of *B. anthracis* spores underlined the strong hydrophobicity (Doyle *et al.*, 1984). The exosporium of *B. subtilis* (Warth *et al.*, 1963; Kornberg A *et al.*, 1968; Hiragi, 1972), *B. cereus* (Martz *et al.*, 1970), *B. thuringiensis* (Scherer *et al.*, 1977), and *C. botulinum type A* (Takumi *et al.*, 1979) were isolated from spores and examined in detail regarding their chemical composition. The coat of *B. subtilis* is largely composed of proteins rich in amino acids, namely tyrosine and cysteine, minor amounts of carbohydrates and lipids (Warth *et al.*, 1963; Aroson and Fitz-James, 1968; Kornberg *et al.*, 1968; Hiragi, 1972; Jenkinson *et al.*, 1981). Spore exosporium of *B. cereus* consists mainly of proteins (52%), polysaccharides (20%), lipids (18%), and ash (4%). As many as 17 common amino acids were detected. Glucose and rhamnose were the principal neutral sugars. The lipid fraction contained 5.5% cardiolipin, 12.5% neutral lipids, and at least 19 fatty acids (Martz *et al.*, 1970).

Several proteins from the exosporium of *B. cereus* ATCC 10876 were recovered, amongst these were zinc metalloprotease, an immune inhibitor A, previously found in *B. thuringiensis*. The presence of this immune inhibitor has been shown in an active 73-kDa form on the exosporium of *B. cereus*. Other proteins associated with the exosporium include the molecular chaperone GroEL and a homologue of RocA [1-pyrroline-5-carboxylate dehydrogenase (EC 1.5.1.12) of *B. subtilis* (Charlton *et al.*, 1999)].

It was shown that low water in spore' core due to the increase in pyridine-2,6-dicarboxylic acid (dipicolinic acid) and calcium ions which causes the spore to become dehydrated, thus preventing DNA damage (Paidhungat and Setlow, 2001). The dehydrated state of spores causes them to be further resistant to heat (Carr *et al.*, 2010).

Overall, resistancy is directly related to infectious dose of the pathogen. It was shown, that an oral challenge of as few as *B. anthracis* 108 spores led to an anthrax infection in animal models (guinea pigs, rabbits, and rhesus monkeys) (Beatty *et al.*, 2003). For inhalation anthrax, the infectious dose is estimated to be 6000 spores (Watson and Keir, 1994). This corresponds to a drinking water concentration of 57 spores/l consumed at a rate of 15 l/day for 7 days, or 171 spores/l consumed at rate of 5 l/day (Burrows and Renner, 1999).

Bacterial spore detection methods

Conventional microbiological methods

Effective testing for the presence of spores requires reliable analytical methods that meet a number of challenging criteria. Timely and precise analyses are among the most important constraints related to the usefulness of microbiological testing. Spore detection methods are expected to be rapid and extremely sensitive, since the presence of even a single pathogenic organism may be that of an infectious dose. A remarkably selective detection methodology is required for a low number of pathogenic spores, often in the presence of a natural background of other non-pathogenic organisms. Conventional identification methods usually include a morphological evaluation of the microorganism as well as tests for the organism's ability to grow in various media under a variety of conditions. Although standard microbiological techniques allow the detection of a single spore, amplification of the signal is required through growth of a single cell into a colony. This process is relatively time-consuming. Since the development of a colony containing 10^6 organisms will take between 18 and 24 hours. Generally, no single test provides a definitive identification of an unknown bacterium.

Traditional microbiological methods for bacteria detection involve the following basic steps: pre-enrichment, selective enrichment, direct culture, qualitative and quantitative analysis, biochemical screening and serological confirmation (Helrich 1990; Kaspar *et al.*, 1990; Tietjen *et al.*, 1995; Hobson *et al.*, 1996; Rodriguez-Palacios *et al.*, 2007, 2009; Pirs *et al.*, 2008; Weese *et al.*, 2009).

While the infectious dose for sporeformers is not known, it presumably varies between healthy individuals and those with underlying risk factors (i.e. age, antibiotics or other drugs use, co-infections, immunosuppressed individuals, intravenous drug abusers, and neonates). It is thought that a low number of spores is less important than a larger number.

The enrichment method for the detection of sporeformers is more sensitive than that involving a direct culture. An enrichment culture is a medium with specific and known qualities that favours the growth of a particular microorganism. The enrichment medium supports the growth of a selected microorganism, while inhibiting the growth of others. Direct culture consists of a liquid or gelatinous substance containing nutrients in which microorganisms or tissues are cultivated for scientific purposes.

The enrichment method was shown to be able detect < 10 spores/g in meat (Weese *et al.*, 2009), while the detection of the quantitative procedure is somewhat higher (20 spores/g). The highest level of contamination was reported to be 240 spores/g (Rodriguez-Palacios *et al.*, 2007, 2009; Weese *et al.*, 2009). These findings were lower than those reported in a different study (Songer *et al.*, 2009). Presumably, regional variations in meat contamination by spores were due to the application of different methods and materials.

Hence, a complex series of tests is often required before any identification can be confirmed with certainty. The results of such tests are often difficult to interpret and not available on the time scale desired in a clinical laboratory. Standardization of sampling and culture methodology is required to evaluate (Weese *et al.*, 2009).

Microscopy methods

Optical microscopy (Fig. 14.3A), scanning electron microscopy (SEM) (Fig. 14.3B and C) and atomic force microscopy (AFM) (Fig. 14.3D) have been employed to examine morphology of bacterial spores of different species, including that of *B. atrophaeus* (a in Fig. 14.3A–D) and *B. cereus* spores (b in Fig. 14.3A–D) (Tarasenko *et al.*, 2003, 2004). These modes turned out to be supplementary, but valuable, techniques to differentiate spore types based on spores' morphology (Tarasenko *et al.*, 2003, 2004).

Genetic methods

Some new technologies are satisfyingly sensitive. However, analysis time is lengthy. For instance, a polymerase chain reaction (PCR) can be employed to amplify small quantities of genetic material to determine the presence of bacteria. The PCR method is extremely sensitive, but requires pure samples, hours of processing and expertise in molecular biology (Meng *et al.*, 1996; Sperveslage *et al.*, 1996).

A real-time PCR assay was developed for the quantitative detection of *Clostridium tyrobutyricum,* identified as the major causal agent of late blowing in cheese. This assay was optimized by a detergent treatment and enzymatic pretreatment of the sample prior to centrifugation and nucleic acid extraction. Assay detects from 25 to 300 *C. tyrobutyricum* sp. in 25 ml of artificially contaminated raw or ultrahigh-temperature treated whole milk (López-Enríquez *et al.*, 2007).

Preparing DNA for PCR from *B. anthracis*

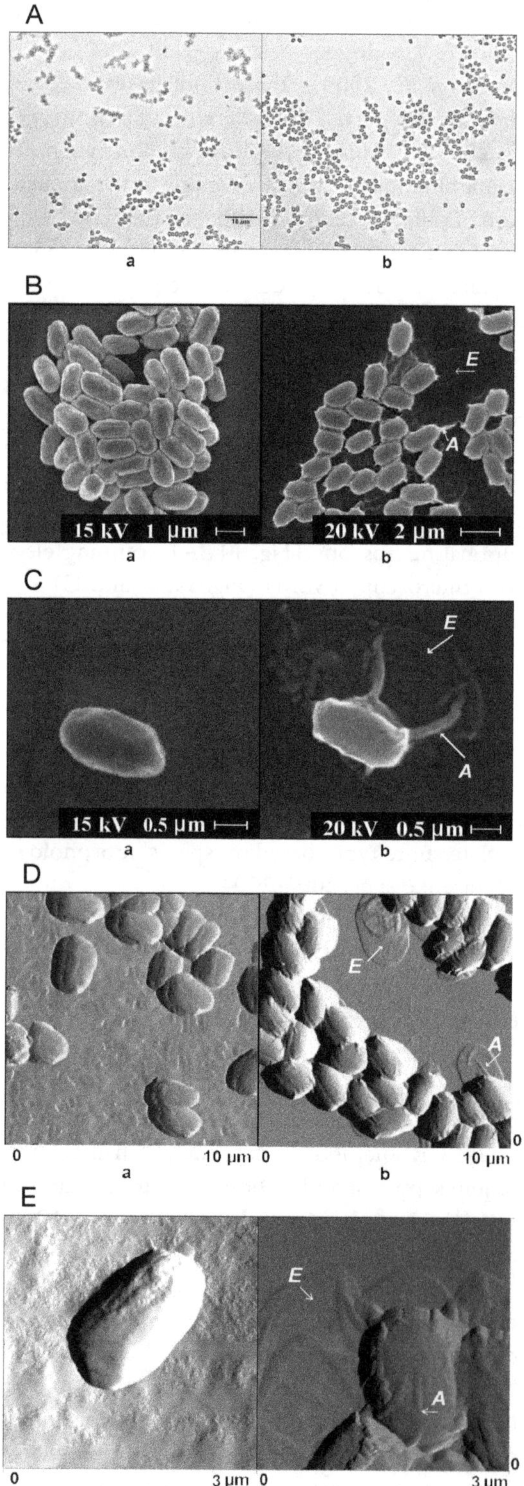

Figure 14.3 Optical (A), SEM (B, C), and AFM (D, F) images of *B. atrophaeus* (a) and *B. cereus* (b) spores. High (A, C, D) and low (B) magnification micrographs are shown: 'A' stands for appendages; 'E' stands for exosporium (in white).

spores is particularly challenging (Coyne *et al.*, 2001). PCR detection system for *B. anthracis* using real time PCR technology from colonies grown on selective agar plates was established (Makino *et al.*, 2001). This method allowed to detect one *B. anthracis* spore in 100 l of air sample. Similar method was use to isolate *B. anthracis* vegetative cells from meat and tissue (Makino *et al.*, 2001).

Using techniques developed by Redmond *et al.* (2004), it was proven possible to extract and characterize highly virulent Ames strain of *B. anthracis* based on sequence of spore's exosporium. The protein profiles were separated by SDS-PAGE-separated exosporium, sequence of bands was determined by either nanospray-MS/MS or N-terminal sequencing using automated Edman degradation in an Applied Biosystems gas-phase sequencer. Sequences and chromosomal open reading frames (ORFs) were compared with available sequences at Institute for Genomic Research (TIGR) (Redmond *et al.*, 2004).

B. anthracis is found in soil and in the same ecological niche as is *B. cereus*. All of these strains belong to the *B. cereus* group, which also includes *B. thuringiensis*. The distinction between these species is rather intricate. The simplest markers for distinguishing these strains are in fact attributable to their plasmids. A randomly amplified polymorphic DNA marker has been described, but it is uninformative about the plasmid content and consequently the strain virulence (Daffonchio *et al.*, 1999). Repetitive element polymorphism-PCR with BOX-A1R primer was further used to discriminate 52 strains of all six species of the *B. cereus* group (Cherif *et al.*, 2002). Signature sequences specific to *B. anthracis* were identified (Cherif *et al.*, 2002).

Ribotyping and more or less random polymerase chain reaction techniques have also been compared, and none one of these methods have been developed into a practical analytical tool (Shangkuan *et al.*, 2000). A *B. anthracis*–specific chromosomal DNA marker has been isolated, and a multiplex polymerase chain reaction analysis has been established based on this chromosomal DNA and plasmid markers (Patra *et al.*, 1996; Ramisse *et al.*, 1996). The combination is perfectly specific for virulent *B. anthracis* and also indicates its virulence state. With this tool, avirulent *B. anthracis* cannot be unambiguously distinguished from other *B. cereus* strains (Patra *et al.*, 1998). DNA–DNA hybridization experiments suggest that *B. anthracis* and *B. cereus* were closer related than the taxonomic classification initially implied (Kaneko *et al.*, 1978).

According to molecular analyses of the ORF, *B. anthracis* is extremely monomorphic and thus constitutes of only a single lineage of *B. cereus*. Both show little diversity (0.125% and 0.21% diversity per nucleotide, respectively) (Price *et al.*, 1999; Schupp *et al.*, 2000). The monomorphism was further demonstrated by the amplified fragment length polymorphism (AFLP) method (Keim *et al.*, 1997). Thus, AFLP is an extremely powerful tool for bacterial identification. Indeed, *B. anthracis* may be the most genetically uniform bacterial species known. Nevertheless, two distinct *B. anthracis* genetic subgroups (termed A and B) were identified.

Another molecular typing system was, therefore, developed based on variable-number tandem repeat (VNTR) (Keim *et al.*, 2000). Multiple-locus VNTR analysis (MLVA) using eight VNTR regions of the *B. anthracis* genome, including one for each plasmid, were tested using 426 strains isolated in different parts of the world. Eighty-nine distinct genotypes clustering into six major genetic groups were defined. They fall into two major phylogenetic groups identical to the A and B lineages defined by AFLP analysis. Group A is geographically widely distributed, whereas group B seems to be limited to a few regions. The association between *B. anthracis* endemicity and soil characteristics such as calcium content and pH has been established (Dragon *et al.*, 1995). Specifically, high levels of calcium in the soil promote spore vitality for prolonged periods, thereby increasing the chance of spores encountering and infecting a new host. Cycles of runoff and evaporation promote dispersion of spores from previous epidemics into storage areas, thereby concentrating them. Uptake of large doses of viable spores from storage areas by susceptible animals, via altered feeding or breeding behaviour, may then allow the bacterium to establish infection and cause a new epidemic (Dragon *et al.*, 1995).MLVA of *B. anthracis* in Southern Africa correlated to soil analysis, indicates that group A is associated with a broader range of soils than is group B (Smith *et al.*, 2000). The differences in geographic diversity may be a consequence of ecological factors affecting persistence differentially. MLVA appears to be a good tool for exploring the epidemiology and ecology of anthrax (Smith *et al.*, 2000).

Typically, a culture method combined with a standard mouse bioassay (SMB) used for a definitive detection of Botulinum neurotoxin (BoNT) producing clostridia in clinical, food, environmental samples. During SMB, mice are injected with studied sample (toxin and food extract) and the time taken for the animal to die is recorded. Highly toxic extracts shall be diluted to ensure that mortality occurs within 5–15 minutes. The toxicity of the sample is then calculated with reference to dose response curves established with standards and expressed in mouse units (MU). Although, the SMB is highly sensitive and specific, it is costly, lengthy, and there are ethical concerns due to the usage of laboratory animals. The multiplex PCR was developed to detect spores and to distinguish BoNT-producing clostridia possessing type A, B, E, and/or F botulinum toxin gene(s). This method demonstrated high levels of specificity (100%) and a relatively high accuracy (99.2%) (De Medici *et al.*, 2009) in comparison with a SMB method. The multiplex PCR method provides a rapid alternative for initial screening for BoNT-producing clostridia in clinical, food and environmental samples (De Medici *et al.*, 2009).

A direct detection method for *Clostridium tyrobutyricum* spores and vegetative cells was developed. DNA was liberated from spores by microwave treatment. A two-step PCR method with two nested pairs of primers derived from the 16S-23S rRNA spacer region was applied. The specificity of each of primer for *C. tyrobutyricum* was demonstrated. The detection demonstrated between 3 and 30 *C. tyrobutyricum* spores in 100 ml of raw milk (Herman *et al.*, 1995).

The aforementioned methods therefore provide a promising alternative to traditional microbiological methods for the rapid, sensitive detection of spores in clinical, food and environmental samples.

Radiometric techniques

Classical microbiology methods used for bacterial detection require a period of about one day up to weeks to certify the presence or absence of bacteria in canned foods. Detection of bacteria in blood or urine samples can be achieved within a matter of a few hours. Detection of aerobic and anaerobic species in foods involves radiometric measurement of $^{14}CO_2$ evolved from the metabolism of ^{14}C-glucose (Previte, 1972). The vials with tryptic soy broth ^{14}C-glucose were inoculated with anaerobic organisms. After addition of a spore inoculum, each vial was rendered anaerobic by sparging the broth with sterile N_2 for 2 minutes through a sterile cotton-plugged syringe and 3-inch (7.6-cm) 22-gauge needle at a pressure of 10 psi. The excess gas was vented through a second syringe. The effluent $^{14}CO_2$ was trapped in a series of five vaccine-capped Erlenmeyer flasks. A reading of 20% of full scale was considered as positive evidence of metabolism in the culture vial. Detection time ranges from 10 minutes to 3 hours for vegetative cells and from 3–4 hours for spores. The results demonstrate the potential of applying a radiometric technique for sterility testing of foods. Although, the radiometric methods are highly sensitive and specific, they are costly and required highly trained.

Immunological techniques

There is a need to be able to characterize structural differences between closely related *Bacillus* species. Unfortunately, there are limited immunological reagents available to characterize such differences. There is, without a doubt, a need for rapid immunodiagnostic techniques to detect and identify *B. anthracis*, and to distinguish spores of related species (Turnbull *et al.*, 1992).

The immunization of rabbits using spores of *B. anthracis* demonstrates that the surface of these spores is well immunogenic (Phillips *et al.*, 1988). The sensitivity and specificity of polyclonal and monoclonal antibodies raised against *B. anthracis* spore preparations has been assessed by Western blotting. None of the antibodies studied were completely specific in recognizing *B. anthracis* spore surface. A polyclonal serum was reported to recognize a wide range of spore surface epitopes and demonstrated limited cross-reaction with the near-neighbour species *B. cereus* spore surface (Longchamp *et al.* 1999).

Immunobloting analysis on extracts of whole spores and exosporium fragments showed that several antibodies raised against *B. anthracis* by immunization of mice were directed against an epitope present on an antigen larger than 250 kDa. This antigen was shown to be a glycoprotein and is located in the outer regions of exosporium (Sylvestre *et al.*, 2001). Two monoclonal antibodies to *B. cereus* spores were isotyped B183 and B48. These were found to be IgG1 and IgG2. Western blots of *B. cereus* spores extract showed that B183 antibody reacts with an antigen with a molecular mass of 250 kDa, whereas B48 reacted with a 76-kDa protein located on the exosporium. Both of these proteins may be glycoproteins (Quinlan *et al.*, 1997). The presence of glycoproteins is controversial, as it has been demonstrated that some covalently attached sugars on *Bacillus* proteins are the results of non-enzymatic glycosylation (Bhattacharya *et al.*, 1993). non-enzymatic glycosylation is a variation of the classical Maillard reaction between reducing sugars and amino acids. The sugars attach to free, non-ionized amino groups such as the N terminus and lysine side chains. The initial step is the formation of a Schiff base between an amino group and the open-chain form of the sugar. At this stage, the reaction is still fully reversible. However, when the Schiff base (an aldimine) undergoes the Amadori rearrangement to a ketamine, the attachment becomes effectively irreversible. Because of the requirement for a non-ionized amino group, the reaction is pH dependent. It proceeds rapidly at pHs of.8 and slowly at pHs of < 6 (Day *et al.*, 1979; Dolhoffer *et al.*, 1979). It is yet not determined whether the presence of sugars on the antigens arises from enzymatic glycosylation or is a result of environmental factors. The glycosylation on antigens does not affect the ability of antibodies to detect antigens (Quinlan *et al.*, 1997).

Antigen capture dipstick assays have been developed for a series of infectious agents including an assay for *B. anthracis* protective antigen as well as one for *B. anthracis* spores. These immunochromatographic assays utilize colloidal gold to visualize the reaction and take approximately 15 minutes to complete (Long *et al.*, 1999). It was shown that the created single chain of Fv (scFv) antibody fragments, recognizes inactivated spores as well (Frith *et al.*, 2001).

The ELISA for detecting anti-PA antibodies in human serum is relatively sensitive and rapid (Johnson-Winegar, 1984) compared with the indirect haemagglutination assay (Buchanan, 1971). A similar ELISA is still used today to monitor anti-PA antibody titres in humans, in a routine serosurvey of anthrax vaccines, and in laboratory animals (Turnbull *et al.*, 1986; Little *et al.*, 1990; Ivins *et al.*, 1992). One drawback in both of these assays is that false-positive results can occur. A second drawback of these assays is the requirement for purified PA. Purifying PA, however, is labour intensive and requires biological laboratory containment. In addition, purified PA is not always readily available. Lastly, these assays do not allow detection of spores and can only be used after spore germination, a point in time when the infection process is in progress and difficult to stop.

An antigen capture ELISA which does not require purified PA for detection of human antibodies to *B. anthracis* PA and which eliminates false-positive results was developed (Iacono-Connors *et al.*, 1994). Lysates of Spodoptera frugiperda (Sf-9) cells infected with recombinant baculovirus containing the PA gene were used as the source of PA to develop the ELISA. Recombinant PA from crude Sf-9 cell lysates or PA purified from *B. anthracis* Sterne strain was captured by an anti-PA monoclonal antibody coated onto microlitre plates. This capture ELISA can accurately detect anti-PA antibody in serum from vaccinated humans regardless of the source of PA. Thus, the antigen capture ELISA with crude preparations of baculovirus-expressed PA is reliable, safe, and inexpensive for determining anti-PA antibody levels in human sera. In addition, this antigen capture ELISA eliminates false-positive results because it allowed to detect PA (Iacono-Connors *et al.*, 1994).

Foodstuffs have traditionally been challenging matrices for conducting immunoassays.

Proteins, carbohydrates, and other macromolecules present in food may interfere with both immunoassays and PCR-based assays. Removal of particulate matter is another challenge for sample preparation used for either immunoassays or PCR-based tests. Although culture methods for bacterial identification remain the gold standard in food research, faster alternative methods are in demand.

Enzyme-linked immunosorbent assays (ELISA) and PCR are time-consuming. In addition, both methods can detect either only a single target (or pathogen) (ELISA) or a microbial analyte (pathogen and its sequence) (ELISA, PCR). Microsphere-based flow cytometric assays provide sensitive, specific recognition. Microsphere-based flow cytometric assays can be multiplexed to a high level for the detection of food contamination by bacterial species including sporeformers. Compared with culture methods, microsphere Luminex assays are faster (a few hours), require fewer reagents, and demand less technical expertise. Fluorescence-coded MagPlex microspheres coated with antibodies are used to capture these bacteria and toxins from spiked foodstuffs. Apple juice, green pepper, tomato, ground beef, alfalfa sprouts, milk, lettuce, spinach, and chicken washes were used as spiked foodstuffs. Samples were evaluated by the Luminex flow. Although MagPlex microspheres facilitate the recovery of microspheres and targets from the complex matrices, assay sensitivity is sometimes inhibited by up to one to three orders of magnitude; e.g. the detection limits of *E. coli* spiked in apple juice or milk, increased 100-fold, from 1000 to 100,000 CFU/ml. Thus, while the magnetic and fluorescent properties of the Luminex MagPlex microspheres allow for rapid, multiplexed testing for bacterial contamination in food matrices, analysis have shown that achieving desired limits of detection remains challenging.

Paramagnetic Dynal® bead-based sandwich immunoassay coupled with fluorescence detection was used to detect *B. globigii* spores as a simulant of *B. anthracis* in water (Farrell *et al.*, 2005). Paramagnetic Dynal® beads were used as a solid support. Primary antibody was attached to the beads by streptavidin-biotin coupling. The secondary antibody was tagged with an alkaline phosphatase (AP) enzyme label. Enzymatic conversion of fluorescein diphosphate (FDP) to fluorescein by AP was measured in real time. Assay demonstrated the applicability of the developed method to detect 78 spores (or 2.6×10^3 spores per ml) in water supply. The assay time from the moment spores were introduced into the system was 30 min, and real-time fluorescence detection was complete in under 1 minutes (Farrell *et al.*, 2005).

Spore identification using instrumental methods

The commonly employed methods of bacteria identification are: microscopy, infrared (IR) spectroscopy, flow cytometry, measurement of physical parameters, detection of cellular compounds such as ATP (bioluminescence), DNA, proteins, carbohydrates, and lipid derivatives (through biochemical methods). These methods might also be applied for spore detection.

Infrared (IR) spectroscopy was used for bacterial identification (Rossi and Warner, 1985). Bacteria are smeared onto an IR cell. IR absorbance spectra are acquired using conventional instrumentation. The main limitation of IR spectroscopy, however, is that it involves an evaluation of the chemical composition of bacteria which is particularly similar at the molecular level. Because of the inherent limitations of this technique, reports of its application to bacterial detection became less frequent from 1960 onwards.

The primary physico-chemical methods of spore identification are those which involve the detection of some naturally occurring components of the bacteria as well as spores. For instance, the Microbial Identification System (Newark, DE) uses gas chromatography to produce a fatty acid profile to detect and identify microorganisms (Swaminathan and Feng, 1994). *B. anthracis* and *B. cereus* are closely related pathogenic organisms that are difficult to differentiate phenotypically or genotypically. By using gas chromatography–mass spectrometry, vegetative cells and spores of the two species were shown to exhibit distinct carbohydrate profiles (Fox *et al.*, 1993). Profiles of vegetative *B. anthracis* typically contained high levels of galactose but did not contain galactosamine, whereas *B. cereus* contained galactosamine and generally low levels of galactose. Spore cultures exhibited unique carbohydrate profiles compared with those of vegetative cultures. *B. anthracis* spore profiles contained rhamnose alone, whereas *B. cereus* spore profiles contained fucose in addition to rhamnose. Furthermore, two spore-specific O-methylated pentoses were discovered. Both *B. anthracis* and *B. cereus* spores contained 3-O-methyl rhamnose, whereas *B. cereus* spores also contained 2-O-methyl rhamnose. Taken together, gas chromatography and mass spectrometry for carbohydrate profiling prove to be powerful tools for differentiating the two closely related species (Fox *et al.*, 1993; Swaminathan and Feng, 1994).

Other methods may be considered such as flow cytometry, immunofluorescence and fluorescence microscopy. The light scattering of the cells gives information on their size, shape and structure, cell mass and bacterial growth (Pinder *et al.*, 1990; Salzman *et al.*, 1990; Lloyd, 1993). Flow cytometry is a highly effective means for rapid analysis of individual cells at rates of up to 1000 cells per second (Melamed *et al.*, 1990; McClelland *et al.*, 1994). By labelling the cells with specific fluorochromes or fluorescent conjugates that bind with high specificity to one particular cellular constituent, it is possible to measure a wide variety of cell constituents, such as proteins, carbohydrates, DNA, RNA and enzymes. Flow cytometry is conveniently used as a bacterial counter in clinical, environmental, and industrial microbiology (Steen *et al.*, 1990; Boye *et al.*, 1991). The advantage of flow cytometry lies in its ability to make rapid, quantitative measurements of multiple parameters of each cell within a large number of cells; this makes it possible to define the properties of the overall population and of component subpopulations. However, flow cytometry has been used almost exclusively for measurements of mammalian or at least eukaryotic cells, while they have remained a rarity in microbiology, including bacteriology. The lack of progress in microbiological applications of flow cytometry was the result of several problems encountered when analysing bacterial spores. The relatively small size of spores and the low number of DNA molecules present to be stained (typically three orders of magnitude less than a mammalian cell) require instruments with higher degree of sensitivity and sophistication. The main experimental difficulty in analysing spores using flow cytometry is that many of their biological characteristics (including size, shape and DNA content) vary depending upon the growth conditions, or the source from which they were obtained (Allman *et al.*, 1993). In addition, flow cytometry of *Bacillus* species has shown, that unlike other species, spores of *B. anthracis* exhibited a red component of fluorescence, which arises from a specific factor present either in or on the spore that is excitable by emission of the fluorescein conjugated to the antibody used in analysis (Stopa, 2000). Finally, the capital cost involved in flow cytometry analyses is still high. Other concerns are related to cross-reactivity with closely related species. A quantitative immunofluorescence assay based on fibre optic microscopy (Phillips *et al.*, 1983) was used to measure the reaction of formalized spores of *B. anthracis* and *B. cereus* isolates with fluorescein conjugates prepared by hyperimmunization with *B. anthracis* Vollum spores. The spores of 11 of the 20 *B. cereus* strains were indistinguishable from *B. anthracis* spores isolates tested (Phillips *et al.*, 1983). However, absorption of the conjugate with spores of *B. cereus* NCTC 8035 and *B. cereus* NCTC 10320 greatly reduced the cross-reaction with *B. cereus* (Phillips *et al.*, 1983).

Since the discovery of microorganism typing by mass spectrometry of whole bacteria (Gutteridge, 1987), numerous attempts were made to automate the typing process by either using matrix-assisted laser desorption/ionization time-of-flight mass spectrometry (MALDI TOF) (Bright *et al.*, 2002; William *et al.*, 2003; Keysa *et al.*, 2004) or pyrolysis mass spectrometry (PyMS) (Meuzelaar *et al.*, 1973, 1974; Freeman *et al.*, 1990; Fenselau *et al.*, 2002). Generating reproducible mass spectra from bacterial samples in a timely fashion at atmospheric pressure remained problematic for many years (Goodacre and Kell, 1996; Fenselau *et al.*, 2002). The time-of-flight mass spectrometer coupled to Direct Analysis in Real Time (AccuTOF DART) is suitable for qualitatively analysing volatile organic compounds, but unsuitable for analysing microbiological samples (i.e. bacteria, spores, whole cells) due to the low volatility of many biomolecules contained in them. To overcome this shortcoming, the AccuTOF DART was re-engineered and used for detection of vegetative bacteria and spores (Alusta *et al.*, 2009) using direct impact corona ionization (DICI) (Wang *et al.*, 2010). Mass spectra were acquired using the AccuTOF DART fed with He at 350°C. Spectral fingerprints of vegetative bacteria and spores were obtained with a high degree of reproducibility (Alusta *et al.*, 2010). The proposed methodology enables rapid, specific, sensitive and reproducible identification of microbiological analytes (genotyping and serotyping) via spectral pattern recognition. The reported method can be used to rapidly identify food-borne pathogens to the genus, subspecies, and serotype level, in fields such as pathogen determination in clinical settings, QA/QC (of drugs, food or cosmetic ingredients), continuous monitoring of air-, water, biowarfare agents (BWA) and the like (Alusta *et al.*, 2010).

Biosensor for spore detection

Currently, most microbiological tests are centralized in large stationary laboratories since complex instrumentation and highly qualified technical personnel are required. In recent years however, intensive research has been undertaken to decentralize such tests to be performed virtually anywhere, under any field conditions (Griffiths *et al.*, 1993; Owen, 1994). Hence, the development of a portable, rapid and sensitive biosensor technology with immediate 'on-the-spot' interpretation of results is well suited for this purpose. Areas for which biosensors show a particular promise are clinical diagnostics, food analysis, bioprocesses and environmental monitoring. The importance of biosensors originated from their high specificity and sensitivity, allowing the detection of a broad spectrum of analytes in complex sample matrices such as blood, serum, urine or food, with a minimum of sample pretreatment (Turner *et al.*, 1986, 1987; Schmid *et al.*, 1989; Hall 1990; Luong *et al.*, 1991; Edelman *et al.*, 1992; Alvarez-Icaza *et al.*, 1993; Deshpande *et al.*, 1994; Rogers *et al.*, 1995; Morgan *et al.*, 1996; Blum, 1997; Kress-Rogers 1997).

Biosensors generally involve a biological recognition component such as receptors, nucleic acids, or antibodies in intimate contact with an appropriate transducer. Depending on the method of signal transduction, biosensors may be divided into four basic groups: optical, mass, electrochemical, and thermal sensors (Goepel 1991;

Sethi 1994; Goepel *et al.*, 1995). In addition, biosensors can be classified into two broad categories: sensors for direct detection of the target analyte and sensors with indirect (labelled) detection. Direct detection biosensors are designed in such a way that the biospecific reaction is directly determined in real time by measuring the physical changes induced by the complex formation. Indirect detection biosensors are those in which a preliminary biochemical reaction takes place and the products of that reaction are then detected by a sensor.

Spores are immunogenic due to the presence of proteins and polysaccharides in their outer coats. This permits the development of immunoassay techniques for bacterial spore detection. In fluorescent immunoassays (FIA), fluorochrome molecules are used to label immunoglobulins. The fluorochrome absorbs short-wavelength light and then emits light at a higher wavelength, which can be detected using fluorescent microscopy. Using an antibody to the protective protein co-expressed with anthrax toxins (Wijesuriya *et al.*, 1994), it has been shown that a device based on this approach could be used as a clinical diagnostic tool for determining the presence of anthrax. A specific monoclonal antibody to the cell surface antigens was used as the capture antibody. As few as 3×10^3 cells/ml of *B. anthrax* spores were detected.

An immunomagnetic assay system (IMAS) has been developed to detect virulent bacteria in biological samples (Yu *et al.*, 1995; Vernozy-Rozand *et al.*, 1997). The IMAS includes a magnetic separator and an electro-chemiluminescent detector. The magnetic separator can capture the antigen. In addition, IMAS was coupled to a flow cytometer and to a continuous fluorimeter. This approach, like other chemiluminescence techniques, offers high signal-to-background ratios and is comparable in sensitivity to radioisotopic methods but has the advantage over other chemiluminescence techniques of being initiated by a voltage potential and thus providing better-controlled luminescence (Yu *et al.*, 1996). In the fluorescence sandwich immunoassay, biotinylated antibody plus streptavidin-coated magnetic beads and fluorescein-conjugated antibodies were used for measurement purposes. The total assay time was 1 hour. The sensitivity of IMAS for *B. anthracis* spore detection is approximately 100 cells/ml in phosphate-buffered saline and 1000 cells/ml in biological samples. In addition, IMAS can also be coupled to a flow cytometer or any analytical instruments for target agent monitoring. Results of this study strongly suggest that IMAS methodology is useful for rapid and sensitive detection.

A simple membrane-strip-based biosensor assay has been combined with a nucleic acid sequence-based amplification (NASBA) reaction for the rapid detection of *B. anthracis* spores (Baeumner *et al.*, 2004). This biosensor is based on a mRNA sequence originating from the protective antigen (*pag*) toxin of *B. anthracis.* To analyse a sample containing as few as ten spores, it was mixed with growth medium and incubated for 30 minutes for spore germination and mRNA production. Soon after, vegetative cells were lysed, RNA was extracted and selectively amplified using NASBA. Specificity analysis revealed no cross-reactivity with other organisms: *B. cereus, B. megaterium, B. subtilis, B. thuringiensis, Lactococcus lactis, Lactobacillus plantarum* and *Clostridium butyricum.* While the biosensor assay requires only 15 minutes assay time, the overall process takes 4 hours to detect as few as 10 viable *B. anthracis* spores, but is significantly shortened if more spores are present (Baeumner *et al.*, 2004).

Piezoelectric-excited millimetre-sized cantilever (PEMC) sensors were designed for the detection of pathogen *B. anthracis* (anti-BA) spores or bovine serum albumin (anti-BSA). PEMC sensors were investigated under both stagnant and flow conditions (Campbell and Mutharasan, 2006). Binding antibodies was quicker under flow conditions. Kinetic analysis was carried out to quantify the effect of flow rate on antibody immobilization and the two types of detection experiments (Campbell and Mutharasan, 2006).

The rapid, sensitive detection of *B. anthracis* spore using the Integrating Waveguide Biosensor followed by rapid release and germination of immunocaptured spores was reported (Hang *et al.*, 2008). Detection was achieved by incubating (10^3) spores simultaneously with capture and detection antibodies ('liquid-phase' assay) prior to capture on capillary tubes/waveguides. Subsequent incubation with BHI broth directly in capillary tubes allowed for rapid germination, outgrowth, and release of spores, resulting in vegetative cells for PCR analysis to differentiate virulent strains (Hang *et al.*, 2008).

A potentiometer traditionally monitors the ion exchange of small ions with a working electrode which holds a selective membrane. A potentiometric sensor was developed based on the surface imprinting technique coupled with a nanoscale transducer: an indium tin oxide (ITO)-coated glass plate (Zhou *et al.*, 2003, 2005; Levon *et al.*, 2004). The potentiometric sensor was used for a selective determination of dipicolinic acid (2,6-pyridinedicarboxylic acid, DPA). This sensor could recognize selectively DPA in concentrations ranging from 1.5×10^{-6} to 0.0194 M within 25 seconds (Zhou *et al.*, 2005). A real-time, portable peptide containing potentiometric biosensor was used for specific recognition of *B. subtilis* and *B. anthracis* Sterne spores (Levon *et al.*, 2004). The sensor can distinguish dead from viable spores and their strains (Zhou *et al.*, 2003, 2005; Levon *et al.*, 2004).

Multiple phage-based magnetoelastic (ME) biosensors were designed and used for the detection of different biological pathogens (Huang *et al.*, 2009). The biosensor was designed by immobilizing a phage and 1 mg/ml BSA (blocking agent) onto the magnetoelastic resonator surface. The detection system included a reference sensor as a control, an E2 phage-coated sensor specific to *S. typhimurium*, and a JRB7 phage-coated sensor specific to *B. anthracis* spores. The sensors were free standing during the test, being held in place by a magnetic field. Upon sequential exposure to single pathogenic solutions, only the biosensor coated with the corresponding specific phage responded. As the cells/spores were captured by the specific phage-coated sensor, the mass of the sensor increased, resulting in a decrease in the sensor resonance frequency. Scanning electron microscopy was used to visually verify the interaction of each biosensor with cells/spores (Huang *et al.*, 2009).

New approaches for spore detection/recognition and inhibition

In previous sections, we described several laboratory and field detection systems, which can be used to identify bacterial spores. However, these systems are relatively slow, sophisticated, expensive, and cumbersome. These systems typically require spore germination and outgrowth of vegetative cells. In general, one can conclude that no current technique fits the ideal – fast, accurate, simple, and cheap – to be used for routine surveillance on a battlefield, in a hospital or in food industry facilities. Since the perfect system does not exist, people continue their studies to find distinct properties of bacterial spores, which might be promising for spore detection and identification. Here, we review six potential approaches that might lead to novel sensors for detection and identification of spores.

Spore interactions with modified particles

Various ligands (peptides, carbohydrates, aptamers, antibodies, lectins, and their conjugates with polymers) have been screened to find a particular, or a group of ligands that are most specific to *Bacillus* spores (Ji *et al.*, 2003; Tarasenko *et al.*, 2003). Among various ligand candidates, peptides seemed to be the most appropriate for the spore detection, since the amino acid sequences specific to the corresponding receptors on the surface or inside the spore can be obtained by well-established methods. In this context, a liposome has been examined (Kazakov *et al.*, 2004) as a vehicle of specific peptides. This liposome may be potentially used as a biomarker for bacterial spore detection. In particular, the surface of liposomes was modified by the sequence Asn-His-Phe-Leu-Pro (NHFLP). The modified liposomes selectively interacted with *B. subtilis* 1A700 spores. Depending on liposome sizes, up to 1000 liposomes may interact with the surface of each spore. As a result of specific interactions between polypeptides of the liposome surface and spore membrane receptors, a pattern of complexation between modified liposomes and spores has been revealed (Fig. 14.4).

The discovered phospholipid shells (2) around every spore (1) are of great potential for developing new methods not only for spore identification, but also for spore decontamination. A variety of nanoparticles can be used as vehicles for different ligands (e.g. polymers, nanospheres, nanotubes, micelles, liposomes, nanogels, nanocapsules). Nevertheless, the most promising entities for prospective development of spore sensors are lipobeads, hydrogel-liposome assemblies, described as drug delivery systems (Kazakov and Levon, 2006).

Lipobeads can be considered as a new type of multifunctional biomarkers for the detection and identification of bacterial spores. An appropriate assembly of a lipid bilayer on a spherical hydrogel surface combines functionality of a modified surface of liposomes with the mechanical stability and a loading capacity of a polymer network. A bi-compartmental structure of lipobeads may, therefore, serve as (i) a functional site for binding various ligands depending on the desired specificity to target spores, and (ii) as a container for loading diverse agents such as fluorescent dyes, quantum dots or other molecules to register a specific interaction.

Several other attempts were made to identify and recognize *Bacillus* spores. Modified liposomes (Kazakov *et al.*, 2004) and glycoconjugate-based ligands were among these methods (Tarasenko *et al.*, 2004; Rasol *et al.*, 2009, 2010). Several specific and selective glycoconjugate-based ligands were identified towards members of *B. cereus* and *B. anthracis* spores (Fig. 14.5) (Rasol *et al.*, 2009, 2010).

Gas chromatography, mass spectrometry, and fluorophore-assisted carbohydrate electrophoresis (FACE) prove to be powerful tools to identify carbohydrate-based receptors expressed on spores of closely related species (Fox *et al.*, 1993; Swaminathan and Feng, 1994; Tarasenko *et al.*, 2004). In addition, these methods help to differentiate spores based on carbohydrate profile specific for each type of spore (Fox *et al.*, 1993; Swaminathan and Feng, 1994; Tarasenko *et al.*, 2004).

Besides recognition (Tarasenko *et al.*, 2004; Rasol *et al.*, 2009, 2010), glycoconjugate-based ligands were also capable to activate resistancy of murine (Fig. 14.6) or human peripheral mononuclear cell (Fig. 14.7) after being exposed to spores (Castleberry *et al.*, 2008; Tarasenko *et al.*, 2008, 2010; Lahiani and Tarasenko, 2010).

Glycoconjugate-based ligands can also neutralize toxins released by bacteria, and prevent cell death (Castleberry *et al.*, 2008; Tarasenko *et al.*, 2008, 2010; Lahiani and Tarasenko, 2010).

Glycoconjugate-based ligands permit defensive decontamination of spores on ceramic tiles (Lone *et al.*, 2007; Tarasenko *et al.*, 2008) and disinfection of bacterial spores on stainless steel surfaces encountered in medical settings (Fig. 14.8) (Pavan *et al.*, 2010).

The different ligands (modified liposomes,

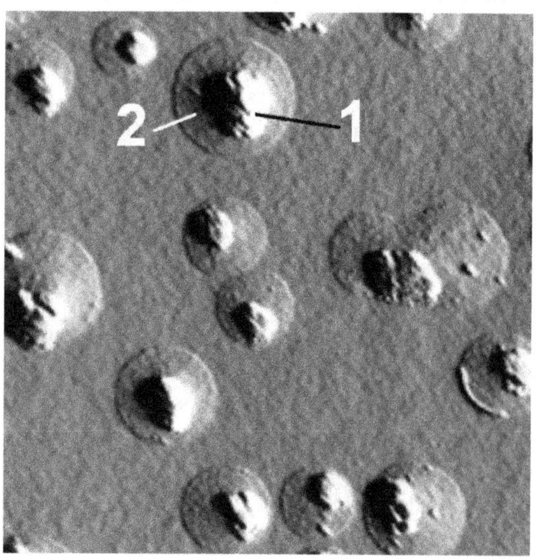

Figure 14.4 Tapping mode AFM imaging of the interactions of functional liposomes with *B. subtilis* spores at a liposome–to-spore ratio of 500: 1. 1 – spore, 2 – phospholipid shell around spore, the frame is $3.5 \times 3.5\,\mu m^2$.

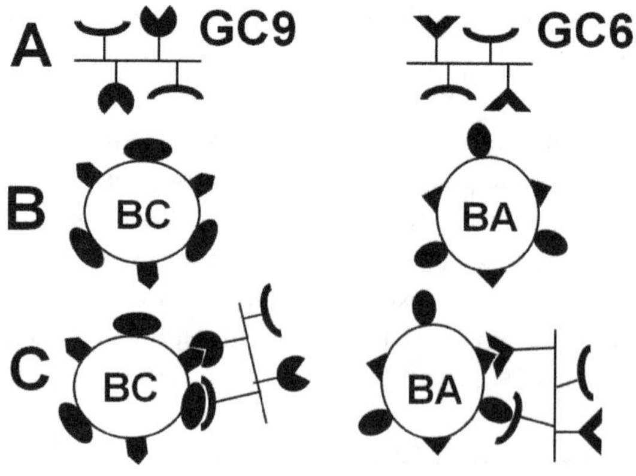

Figure 14.5 Recognition of *B. cereus* and *B. anthracis* spores by glycoconjugate-based ligands (GC9 and GC 6).

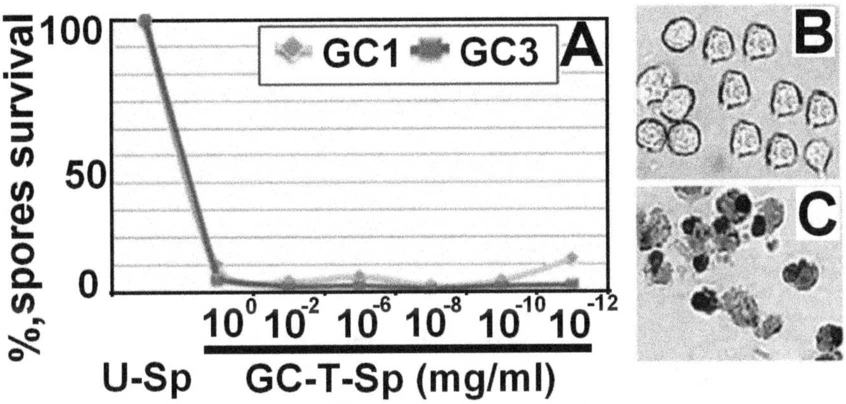

Figure 14.6 Glycoconjugate-based ligands (or GCs) (GC-T-Sp) promote *B. cereus* spores (Sp) killing (%) by murine macrophages (A), $P<0.005$ while protecting macrophage viability (B) from spore- and toxin-induced cell damage (C), $P<10^{-4}$. Cell death by untreated spores (U-Sp) was confirmed by trypan blue (C).

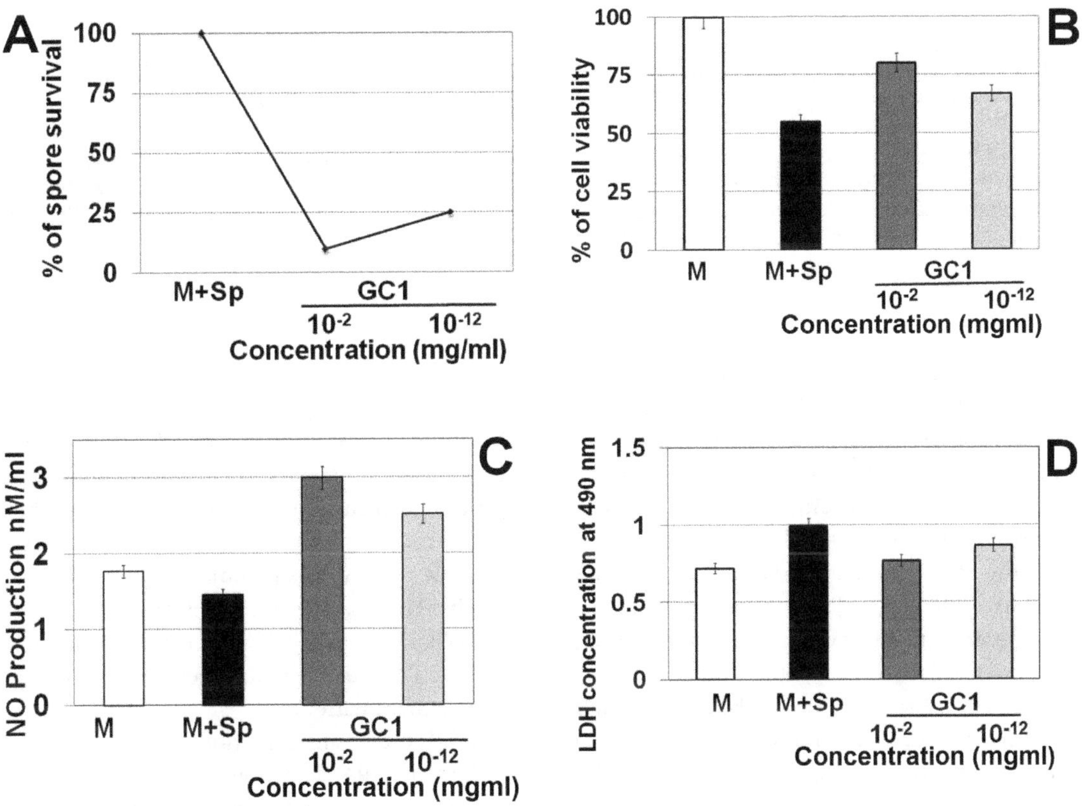

Figure 14.7 Glycoconjugate-based ligands (or GC1) (M+Sp+GC1) promote *B. cereus* spores (Sp) killing (%) by human peripheral mononuclear cells (A), while protecting macrophage viability (B, D) and activating macrophage (C) from cell damage (C). Cell death by untreated spores was confirmed by trypan blue assay.

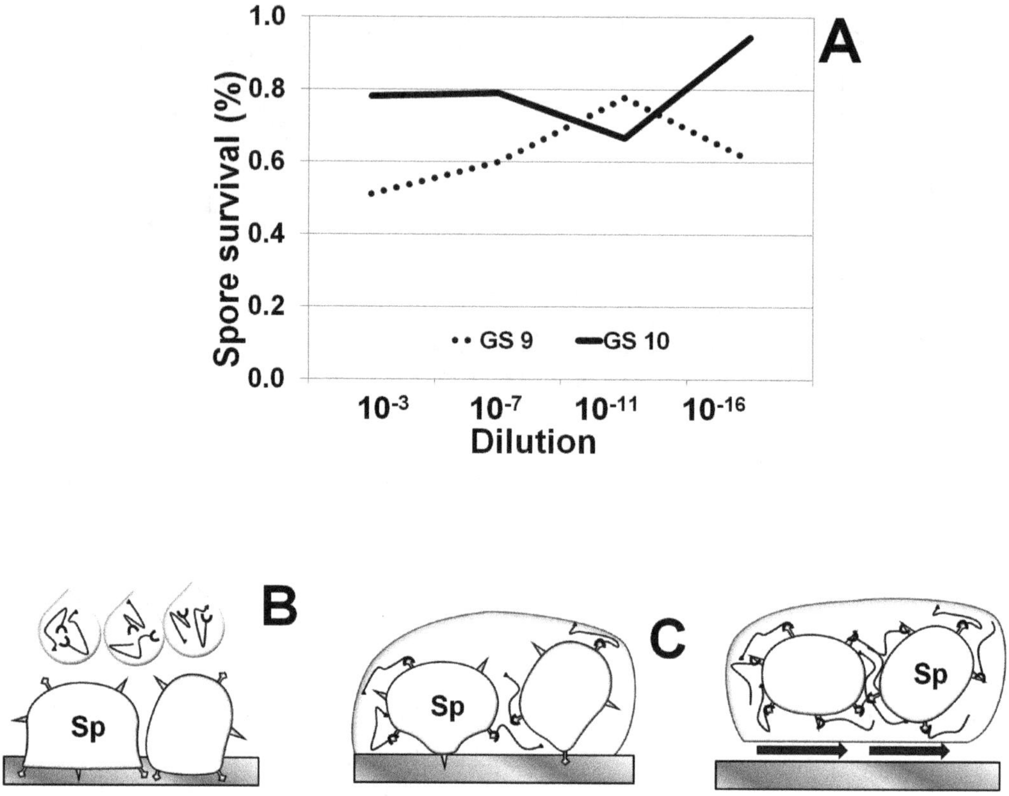

Figure 14.8 *B. cereus* spore survival rate upon disinfection using glycoconjugate-based solutions (GS9 and GS10) in the 0.0–1.0% range (A). Glycoconjugate suspended in an aqueous solution is added to a stainless steel surface that is contaminated by *B. cereus* spores (B). The Glyc groups on the GS monopolize the binding sites of the spore (C) and facilitate disinfection of the stainless steel surfaces (D).

glycoconjugate-based ligands, and others) will allow different approaches to study structural features of spores (Kazakov *et al.*, 2004; Tarasenko *et al.*, 2004), for spore identification (Tarasenko *et al.*, 2004; Kazakov and Levon, 2006), drug delivery (Kazakov and Levon, 2006; Kazakov *et al.*, 2008), for spore decontamination (Lone *et al.*, 2007; Tarasenko *et al.*, 2008; Pavan *et al.*, 2010) and for the development of new immune therapeutics/vaccines (Castleberry *et al.*, 2008; Tarasenko *et al.*, 2008, 2010; Lahiani and Tarasenko 2010) to fight off harmful spores, bacteria, or their toxins. The utility of these glycoconjugates lays in their ability to block germination of spores (Tarasenko *et al.*, 2004), neutralize toxins, and activate of immune cells (Tarasenko *et al.*, 2008, 2010; Lahiani and Tarasenko, 2010).

Detection of secondary products of the spore life cycle

Dipicolinic (pyridine-2,6-dicarboxylic) acid (DPA) is a unique constituent of bacterial spores that has been suggested to be implicated in their dormancy, heat resistance, and germination. It is normally present in equimolar amounts with Ca or other metals. The release of calcium dipicolinate (Ca-DPA) parallels with other germination changes. DPA allows rapid screening of potential spore contamination.

Besides the aforementioned potentiometric method of DPA detection (Zhou *et al.*, 2005), DPA can also be detected spectroscopically. Ocean Optics (www.OceanOptics.com) in collaboration with Los Alamos National Laboratory has developed an integrated system (EDS2000) to detect

DPA and quantify fluorescence from concentrated forms of bacterial spores present in suspect samples. The EDS2000 tests for the presence of DPA, which is extracted from the endospores. Although this technique is not specific, it is a useful tool to detect high concentrations of endospores in locations suspected to harbour spores.

Furthermore, rapid detection of DPA has been accomplished using DPA-sensitized lanthanide luminescence under UV excitation (Cable *et al.*, 2009). Lanthanide-macrocycle binary complexes exhibit highest stability and effective discrimination against competitive ion binding bacterial spores. Specifically, the effectiveness of Sm, Eu, Tb, and Dy complexes with the macrocycle 1,4,7,10-tetraazacyclododecane-1,7-diacetate (DO2A) to the corresponding lanthanide aqua ions were studied. $Ln(DO2A)^+$ binary complexes bind dipicolinic acid (DPA), a major constituent of bacterial spores, with greater affinity and demonstrate significant improvement in bacterial spore detection. It was shown that a Tb complex exhibits the greatest dipicolinate binding affinity [100-fold greater than Tb^{3+} alone, and 10-fold greater than other $Ln(DO2A)^+$ complexes] and highest quantum yield. Moreover, the inclusion of DO2A extends the pH range over which Tb-DPA coordination is stable, reduces the interference of Ca^{2+} nearly 5-fold, and mitigates phosphate interference a 1000-fold compared with free Tb alone. In addition, detection of *B. atrophaeus* bacterial spores was improved by the use of $Tb(DO2A)^+$, yielding a 3-fold increase in the signal-to-noise ratio over Tb^{3+}. This technique can be easily miniaturized and automated for rapid detection of spores or possibly traces of primitive life forms in extreme environments.

Potentiometry: recognition by electrode potential

The outermost layers of bacterial spores carry charged or chargeable groups on their surface, thus creating an electric double layer upon contact with the aqueous phase. When spores come in contact with the surface of an electrode, measurable variations in electrical signal (potential) can be measured by potentiometry. If a spore recognition element (peptide, receptors, nucleic acids, or antibodies) is immobilized on the surface of an electrode, specific bacterial spores can be recognized by the difference in electrical signal. These potentiometric principles for reliable detection and identification of bacterial *Bacillus* spores have been tested (Ji *et al.*, 2003; Levon *et al.*, 2005). It was shown that even low concentrations of spores (< 1000 spore/ml) can cause noteworthy changes in pH and ionic strength of aqueous spore suspensions. Different electrodes were examined to demonstrate that different types of spores can be potentiometrically discerned. Moreover, if the electrodes are modified with certain ligands, selectivity and sensitivity of the potentiometric measurements of spores may be enhanced.

One advantage of the potentiometric system for detecting spores is the use of a technologically affordable electronic device to transform a spore–electrode interaction into a measurable electrical signal that can be reliably recorded, processed and stored in a database for further analysis. The proposed method can employ an integrated system of specifically modified electrodes to discern between different species of *Bacillus* spores. The proposed designed sensor allows the development of a miniature detection system capable of operating 24/7 in any weather.

Potentiometry: recognition by ion exchange kinetics

Spores are known to behave like weak cation-exchangers. The equilibrium of proton binding within *Bacillus subtilis* spores, as well as the kinetics of pH-equilibration in the external aqueous medium substantiate this behaviour (Kazakov *et al.*, 2008). Similar to ionic hydrogels, bacterial spores can be characterized as ionic reservoirs. Time-resolved micro-potentiometry – a method to probe kinetics of proton uptake by dormant spores – has shown that the proton uptake may be attributed to the protective multilayered structure of spores (Fig. 14.9).

Since the morphology, ionic content, and inner structure of dormant spores differ, there is a growing consensus that proton capacity and kinetics of ion exchange may vary from species to species. Thus, the difference in both proton capacity and ion exchange kinetics can be used to discern species.

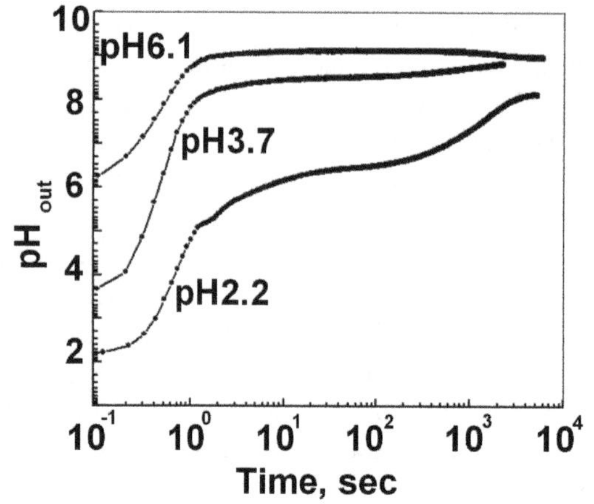

Figure 14.9 The time courses of external pH_{out} measured in *B. subtilis* spore suspensions: prepared by injecting 0.5 ml of the stock suspension (1.2×10^9 sp/ml) into 0.5 ml of aqueous HCl solutions. The pH is indicated for each curve.

Spore ability to swell and shrink: kinetics

As with synthetic ion exchangers ion sensitive spores can swell and shrink depending on (i) the network charge, (ii) counterions associated with the network charge and (iii) the difference of mobile ion concentrations – inside and outside the spores. Experimental observations strongly suggest that the ability to swell and shrink can vary between different species of the *Bacillus* genus, since their compartments benefit from different chemical composition (Kazakov *et al.*, 2008; Kazakov, 2010). Thus, the fluctuations in spore dimensions over time can also be exploited for detection and identification of bacterial spores. The development of a spectroscopic apparatus to examine the swelling/shrinking of soft materials, such as hydrogels, cells, organelles, and spores, in response to different stimuli has been recently reported (Kazakov, 2010). It employs an optical fibre with an open core connected to a spectrophotometer to record the kinetics of structural changes within soft materials deposited on the fibre core. If spores are deposited on the surface during propagation of electromagnetic radiation inside the optical fibre, a portion of the electric field penetrates out of the fibre core and can interact with the spore outermost layers. If physical dimensions of deposited spores change due to variations in external conditions (temperature, humidity, pH, salt content), the integral intensity passing through the fibre core and transmission spectrum of the fibre will vary. However, the spectrum of absorbed/scattered light within collapsed/swollen spores has yet to be determined, since its measurements remain challenging, especially at the nanometre scale. The proposed experimental scheme can provide new information on both the spectral composition of scattered and absorbed light within the structural layers in the immediate vicinity of the fibre core surface. Furthermore, spectrum changes over time at different external conditions were examined. Since this technique is capable of discerning spores based on size and swelling kinetics, it will be integrated into a rapid detection and identification that could even be carried out by an untrained lab technician.

Single spore manipulation

Single spore manipulation will allow one to exclude 'false positive' results in detection and identification of spores. Microscopy used in conjunction with spectroscopy, however, allows monitoring structural and compositional properties of single spores, and analysing changes in various molecular components – DPA, nucleic acids, and proteins – during germination. The methodology that combines external phase contrast microscopy, micro-Raman spectroscopy, and optical tweezers has been reported to monitor a variety of changes during the germination of single spores in both nutrient (L-alanine) and non-nutrient (Ca-DPA) germinants with a temporal resolution of ~2 seconds (Kong *et al.*, 2010). Optical tweezers permit single spore manipulation. Phase contrast microscopy assesses changes in refractivity of individual spores, while Raman spectroscopy provides information on changes in spore-specific molecules. Although this method requires some special training, the combined technique is nonetheless capable of revealing differences in CaDPA releases, and refractivity changes during germination of different spore species. This method would constitute a reliable detection/identification test, since the manipulation of single spores excludes 'false positives'.

Concluding remarks

Examination of published literature has shown that the development of biosensors has earned an increased importance in recent years. Only a few biosensors for bacterial detection are already commercially available or are approaching commercialization. The main reasons for this lag are both technology- and market-related. It is a challenge to create biosensors with necessary properties for reliable and effective use. A biosensor must have the specificity to rapidly distinguish the target spore, the adaptability and sensitivity to detect spores of different species, to detect spores in a multiorganism matrix, all this without pre-enrichment. At the same time, the ideal biosensor should be relatively simple and of inexpensive configurations. A further obstacle is the tendency to exclusively focus on a scientific based technology while excluding other equally important aspects. Research usually proceeds without adhering to defined specifications. To this day, there is no biosensor system available that prides itself on bacterial spore specificity, one of the crucial requirements of today's market. Obviously, enhancing specificity of biosensor systems is an intricate task. This is the main reason for the relatively slow penetration of biosensors into the market. Technical problems facing biosensor development include: the interaction of matrix compounds, methods of sensor calibration, the requirement for reliable and low maintenance functioning over extended periods of time, sterilization (particularly for clinical applications), reproducible fabrication of numerous sensors, the ability to manufacture biosensors at competitive costs, disposable formats, and a clearly identified market. Biosensors should be easy to regenerate after each measurement while generating reproducible results. Spore biosensors should reduce human involvement (to avoid contamination) and hence, automation must be an inherent attribute of any biosensor. Although biosensors available today are improvements over original formulations, they still represent first generation systems. There are several markets, which could potentially support stand-alone biosensors. Highly sensitive and accurate biosensor systems would find wide range applications in medical diagnostics, food quality control, environmental monitoring, defence and other similar industries. The medical diagnostics field offers real opportunities for the exploitation of biosensors for bacterial detection. In fact, the opportunities for biosensors to enter the clinical diagnostic market are wide open since few products have been commercialized. The most viable openings in the food industry will arise where biosensors can rapidly detect total microbial contamination. The largest area of application for the environment lies in the development of biosensors to monitor bacteria in drinking and wastewater, rivers, reservoirs and water supplies.

Ackowledgements

The authors thank A. Rasol, M. Lahiani, and C. Pavan for designing Figures 14.5, 14.7, and 14.8. The authors also thank A. Rasol for a formatting of the references section.

References

Allman, R., Manchee, R., and Lloyd, D. (1993). Flow cytometric analysis of heterogeneous bacterial populations. In Flow Cytometry in Microbiology, Lloyd, D., ed. (Springer-Verlag London, Limited, Germany), pp. 27–47.

Alusta, P., Dorey, C., Parker, R., Bazutu, D., Tarasenko, O., Wilkes, J., and Darsey, J. (2010). Direct Impact Corona Ionization Mass Spectrometry for Bacteria Identification. 5th BioNanoTox and Applications Research Conference. Little Rock, Arkansas. 33 pp. (oral).

Alvarez-Icaza, M., and Bilitewski, U. (1993). Mass production of biosensors. Anal. Chem. *65*, 525A-533A.

Aronson, A.I., and Fitz-James, P. (1968). Biosynthesis of bacterial spore coats. J. Mol. Biol. 33, 199–212.

Aronson, A.I., and Fitz-James, P. (1976). Structure and morphogenesis of the bacterial spore coat. Bacteriol. Rev. *40*, 360–402.

Aronson, A.I., and Pandey, N.K. (1978). Comparative structural and functional aspects of spores coat. In Spores VII, Chambliss, G., and Vary, J.C., eds. (American Society for Microbiology, Washington, D.C.), pp. 54–61.

Ash, C., and Collins, M.D. (1992). Comparative analysis of 23S ribosomal RNA gene sequences of *Bacillus anthracis* and emetic *Bacillus cereus* determined by PCR-direct sequencing. FEMS Microbiol. Lett.73, 75–80.

Ash, C., Farrow, J.A.E., Dorsch, M., Stackebrandt, E., and Collins, M.D. (1991). Comparative analysis of *Bacillus anthracis, Bacillus cereus* 2 and related species on the basis of reverse transcriptase sequencing of the 16S rRNA. Int. J. Syst. Bacteriol. *41*, 343–346.

Baeumner, A.J., Leonard, B., McElwee, J., and Montagna, R.A. (2004). A rapid biosensor for viable *B. anthracis* spores. Anal Bioanal. Chem. *380*, 15–23.

Barnaby, W. (1997). The Plague Markers: The Secret World of Biological Warfare (Vision Paperbacks, London).

Beaman, T.C., Pankratz, H.S., and Gerhardt, P. (1971). Paracrystalline sheets reaggregated from solubilized exosporium of *Bacillus cereus*. J. Bacteriol. *107*, 320–324.

Beaman, T.C., Pankratz, H.S., and Gerhardt, P. (1972). Ultrastructure of the exosporium and underlying inclusions in spores of *Bacillus megaterium* strains. J. Bacteriol. *109*, 1198–1209.

Beatty, M.E., Ashford, D.A., Griffin, P.M., Tauxe, R.V., and Sobel, J. (2003). Gastrointestinal anthrax. Intern. Med. *163*, 2527–2531.

Becker, H., Schaller, G., von Wiese, W., and Terplan, G. (1994). *Bacillus cereus* in infant foods and dried milk products. Int. J. Food Microbiol. *23*, 1–15.

Bhattacharya, M., Plantz, B.A., Swanson-Kobler, J.D., and Nickerson, K.W. (1993). Nonenzymatic glycosylation of lepidopteran-active *Bacillus thuringiensis* protein crystals. Appl. Environ. Microbiol. *2*, 229–233.

Blum, L.J. (1997). Bio- and Chemiluminescent Sensors (World Scientific, Singapore; River Edge, N.J.), pp. 198.

Bottone, E.J. (2010). *Bacillus cereus*, a volatile human pathogen. Clin. Microbiol. Rev. *23*, 382–398.

Boye, E., and Lobner-Olesen, A. (1991). Bacterial growth control studied by flow cytometry. Res. Microbiol. *142*, 131–135.

Bright, J.J., Claydon, M.A., Soufian, M., and Gordon, D.B. (2002). Rapid typing of bacteria using matrix-assisted laser desorption ionization time-of-flight mass spectrometry and pattern recognition software. J. Microbiol. Methods *48*, 127–138.

Buchanan, T.M., Feeley J.C., Hayes, P.S., and Brachman, P.S. (1971). Anthrax indirect microhemagglutination test. J. Immunol. *107*, 1631–1636.

Burrows, W.D., and Renner, S.E. (1999). Environ. Health Persp. *107*, 975–984.

Cable, M.L., Kirby, J.P., Levine, D.J., Manary, M.J., Gray, H.B., and Ponce, A. (2009). Detection of bacterial spores with lanthanide-macrocycle binary complexes. J. Am. Chem. Soc. *131*, 9562–9570.

Campbell, G.A., and Mutharasan, R. (2006). Detection of *Bacillus anthracis* spores and a model protein using PEMC sensors in a flow cell at 1 mL/min. Biosens. Bioelectron. *22*, 78–85.

Carlin, F., Girardin, H., Peck, M.W., Stringer, S.C., Barker, G.C., Martinez, A., Fernandez, A., Fernandez, P., Waites, W.M, Movahedi, S., *et al.* (2000). Research on factors allowing a risk assessment of spore-forming pathogenic bacteria in cooked chilled foods containing vegetables: a FAIR collaborative project. International J. Food Microbiol. *60*, 117–135.

Carr, K., Janes, B., and Hanna, P. (2010). Role of gerP Operon in Germination and Outgrowth of *Bacillus anthracis* Spores. PLoS One *5*, 9128.

Castleberry, J., Scott, A. Alusta, P., Soderberg, L., and Tarasenko, O. (2008). Nitric oxide involved in macrophage activation and neutralization of *Bacillus anthracis* toxins during glycoconjugate treatment. Nitric Oxide *19*, 1.

Charlton, S., Moir, A.J., Baillie, L., and Moir, A. (1999). Characterization of the exosporium of *Bacillus cereus*. J. Appl. Microbiol. *87*, 241–245.

Cherif, A., Borin, S., Rizzi, A., Ouzari, H., Boudabous, A., and Daffonchio, D. (2002). Characterization of a repetitive element polymorphism polymerasechain reaction chromosomal marker that discriminates *Bacillus anthracis* from related species. J. Appl. Microbiol. *93*, 456–462.

Compton, J.A.F. (1987). Military Chemical and Biological Agents (Telford Press, Caldwell, NJ), pp. 458.

Coyne, S.R., Craw, P.D., Teska, J.D., Ezzell, J.W., Shoemaker, D.R., Henchal, E.A., and Knauert, F.K. (2001). Purification of DNA for PCR from *Bacillus anthracis* spores in soil samples. In Program and Abstracts Book of 4th International Conference of Anthrax, Annapolis, MD, Friedlander, M.A., Baillie, L.W.J., Ivins, B.E., Keim, P., Leppla, S.H., Mock, M., Welkos, S.L., eds. (USAMRIID, Fort Detrick, MD), pp. 1–36.

Daffonchio, D., Borin, S., Frova, G., Gallo, R., Mori, E., Fani, R., and Sorlini C. (1999). A randomly amplified polymorphic DNA marker specific for the *Bacillus cereus* group is diagnostic for *Bacillus anthracis*. Appl. Environ. Microbiol. *65*, 1298–1303.

Dando, M. (1994). Biological Warfare in the 21st Century (Brassey's, London), pp. 258.

Day, J.F., Thornburg, R., Thorpe, W.S.R., and Baynes, J.W. (1979). Nonenzymatic glucosylation of rat albumin: studies *in vitro* and *in vivo*. J. Biol. Chem. *254*, 9394–9400.

De Medici, D., Anniballi, F., Wyatt, G.M., Lindström, M., Messelhäußer, U., Aldus, C.F., Delibato E., Korkeala, H., Peck, M.W., and Fenicia L. (2009). Multiplex PCR for detection of botulinum neurotoxin-producing Clostridia in clinical, food, and environmental samples. Appl. Environ. Microbiol. *75*, 6457–6461.

De Vos, V. (1990). The ecology of anthrax in Kruger National Park, South Africa. Salisbury Med. Bull. Suppl *68*, 19–23.

Deshpande, S.S., and Rocco, R.M. (1994). Biosensors and their potential use in food quality-control. Food Technol. *48*, 146–150.

Dolhoffer, R., and Wieland, O.H. (1979). Glycosylation of serum albumin: elevated glycosyl-albumin in diabetic patients. FEBS Lett. *103*, 282–286.

Doyle, M.P., Beuchat, L.P., and Montville, T.J. (1997). Food Microbiology. Fundamentals and Frontiers (American Society of Microbiology Press, Washington D.C.), pp.768.

Doyle, R.J., Nedjat-Haiem, F., and Sing, J.S. (1984). Hydrophobic characteristic of bacillus spores. Curr. Microbiol. *10*, 329- 332.

Dragon, D.C., and Rennie, R.P. (1995). The ecology of anthrax spores: tough but not invincible. Can. Vet. J. *36*, 295–301.

Driks, A. (1999). *Bacillus subtilis* spore coat. Microbiol. Mol. Biol. R. *63*,1–20.

Drobniewski, F.A. (1993). *Bacillus* and related species. Clin. Microbiol. Rev. *6*, 324–338.

Edelman, P.G., and Wang, J. (1992). Biosensors and Chemical Sensors: Optimizing Performance through Polymeric Materials (ACS, Washington D.C.).

Farrell, S., Halsall, H.B., and Heineman, W.R. (2005). Immunoassay for *B. globigii* spores as a model for detecting *B. anthracis* spores in finished water. Analyst *130*, 489–497.

Fenselau, C., and Demirev, P.A. (2001). Characterization of intact microorganisms by MALDI mass spectrometry. Mass Spectrom. Rev. *4*, 157–171.

Fox, A., Black, G.E., Fox, K, and Rostovtseva, S. (1993). Determination of carbohydrate profiles of *Bacillus anthracis* and *Bacillus cereus* including identification of O-methyl methylpentoses by using gas chromatography–mass spectrometry. J. Clin. Microbiol. *31*, 887–894.

Freeman, R., Goodfellow, M., Gould, F.K., Hudson, S.J., and Lightfoot, N.F. (1990). Pyrolysis-mass spectrometry (Py-MS) for the rapid epidemiological typing of clinically significant bacterial pathogens. J. Med. Microbiol. *32*, 283–286.

Frith, G.K., Uppington, H., Redmond, C., Fulop Hibbs, L.S., and Dewey, J. (2001). The development and use of scFv antibody fragments to characterize the vegetative phase S-layer protein EA1 of various *Bacillus* species. In Program and Abstracts Book of 4th International Conference of Anthrax (Annapolis, MD). Friedlander, M.A., Baillie, L.W.J., Ivins, B.E., Keim, P., Leppla, S.H., Mock, M., Welkos, S.L., eds. (USAMRIID, Fort Detrick, MD), pp. 1–36.

Gerhardt, P., and Marquis, R.E. (1989). Spore thermoresistance mechanisms. In Regulation of Prokaryotic Development, Smith, I., Slepecky, R., and Setlow, P., eds. (American Society for Microbiology, Washington, DC), pp. 17–63.

Gerhardt, P., and Ribi, E. (1964). Ultrastructure of the exosporium enveloping spores of *Bacillus cereus*. J. Bacteriol. *88*, 1774–1789.

Goepel, W. (1991). Chemical sensing, molecular electronics, and nanotechnology: interface technologies down to the molecular scale. Sensors Actuators B 4, pp. 7–21.

Goepel, W., and Heiduschka, P. (1995). Interface analysis in biosensor design. Biosens Bioelectron. *10*, 853–883.

Goodacre, R., and Kell, D.B. (1996). Pyrolysis mass spectrometry and its applications in biotechnology. Current. Opin. Biotechnol. *1*, 20–28.

Gordon, R.E., Haynes, W.C., and Pang, C.H.N. (1973). The genus *Bacillus*. U.S. Department of Agriculture Agricultural Handbook no. 427. U.S. (Government Printing Office, Washington, DC).

Griffiths, D., and Hall, G. (1993). Biosensors – What real progress is being made? Trends Biotechnol. *11*, 122–130.

Gutteridge, C.S., and Schweppes, C.P.L.C. (1987). Characterization of microorganisms by pyrolysis mass spectrometry, Meth. Microbiol. 19. (Academic Press Limited.)

Hachizuka, Y., Kojima, K., and Sato, T. (1966). The fine filaments on the outside of the exosporium of *Bacillus anthracis* spores J. Bacteriol. *91*, 2382–2384.

Hall, E.A.H. (1990). Biosensors (Open University Press, Milton Keynes), pp. 351.

Hang, J., Sundaram, A.K., Zhu, P., Shelton, D.R., Karns, J.S., Martin, P.A., Li, S., Amstutz, P., and Tang, C.M. (2008). Development of a rapid and sensitive immunoassay for detection and subsequent recovery of *bacillus anthracis* spores in environmental samples. J. Microbiol. Methods. 73, 242–246.

Hashizuka, Y., and Kozuka, S. (1984). A new test of differentiation of *Bacillus cereus* and *Bacillus anthrax* based on the existence of spores appendages. Microbiol. Immun. 25, 1201–1207.

Helrich, K. (1990). Official Methods of Analysis of the Association of Official Analytical Chemists, Microbiological Methods, 2, 15 ed. (AOAC, Arlington, V.A.), pp. 425–497.

Herman, L.M.F., De Block, J.H.G.E., and Waes, G.M.A.V.J. (1995). A direct PCR detection method for *Clostridium tyrobutyricum* spores in up to 100 milliliters of raw milk. Appl. Environ. Microb. *61*, 4141–4146.

Hiragi, Y. (1972). Physical, chemical and morphological studies of spore coat of *Bacillus subtilis*. J. Gen. Microbiol. 72, 87–99.

Hobson, N.S., Tothill, I., and Turner, A.P.F. (1996). Microbial detection. Biosens. Bioelectron. *11*, 455–477.

Holmes, J.R., Plunkett, T., Pate, P., Roper, W.L., and Alexander, W.J. (1981). Emetic food poisoning caused by *Bacillus cereus*. Arch. Intern. Med. *141*, 766–767.

Holt, S.C., and Leadbetter, E.R. (1969). Comparative ultrastructre of selected aerobic spore-forming bacteria: a freeze-etching study. Bacteriol. Rev. 33, 346–378.

Huang, S., Yang, H., Lakshmanan, R.S., Johnson, M.L., Wan, J., Chen, I.-H., Wikle III, H.C., Petrenko, V.A., Barbaree, J.M., and Chin, B.A. (2009). Sequential detection of *Salmonella typhimurium* and *Bacillus anthracis* spores using magnetoelastic biosensors. Biosens. Bioelectron. 24, 1730–1736.

Hultgren, S.J., Abraham, S., Caparon, M., Falk, P., St Geme, J.W., 3rd, and Normark, S. (1993). Pilus and nonpilus bacterial adhesins: assembly and function in cell recognition. Cell 73, 887–901.

Iacono-Connors, L.C., Novak, J., Rossi, C., Mangiafico, J., and Ksiazek, T. (1994). Enzyme-linked immunosorbent assay using a recombinant *Baculovirus*-expressed *Bacillus anthracis* protective antigen (PA): measurement of human anti-PA antibodies. Clin. Diagn. Lab. Immunol. *1*, 78–82.

Ivins, B.E., Welkos, S.L., Little, S.F., Crumrine, M.H., and Nelson, G.O. (1992). Immunization against anthrax with *Bacillus anthracis* protective antigen combined with adjuvants. Infect. Immun. *60*, 662–668.

Jackson, P.J., Hill, K.K., Pardington, P.E., and Ticknor, L.O. (2001). The relationship of *B. anthracis* to other subgroup 1 *Bacillus* species. In Program and Abstracts Book of 4th International Conference of Anthrax, Annapolis, MD, Friedlander, M.A., Baillie, L.W.J.,

Ivins, B.E., Keim, P., Leppla, S.H., Mock, M., Welkos, S.L., eds. (USAMRIID, Fort Detrick, MD), pp. 1–36.

Jenkinson, H.F., Sawyer, W.D., and Mandelstam, J. (1981). Synthesis and order of assembly of spore coat proteins in *Bacillus subtilis*. J. Gen. Microbiol. *123*, 1–16.

Jensen, G.B., Hansen, B.M., Eilenberg, J., and Mahillon, J. (2003). The hidden lifestyles of *Bacillus cereus* and relatives. Environ. Microbiol. *5*, 631–640.

Johnson-Winegar, A. (1984). Comparison of enzyme-linked immunosorbent and indirect hemagglutination assays for determining anthrax antibodies. J. Clin. Microbiol. *20*, 357–361.

Kaneko, T., Nozaki, R., and Aizawa, K. (1978). Deoxyribonucleic acid relatedness between *Bacillus anthracis* and *Bacillus cereus* and *Bacillus thuringiensis*. Microbiol. Immunol. *22*, 639–641.

Kaspar, C.W., and Tartera, C. (1990). Methods for detecting microbial pathogens in food and water. Methods Microbiol. *22*, 497–530.

Kazakov, S. (2010). Evanescent wave spectroscopy for studying swelling/de-swelling kinetics of soft matter. Soft Material *6*, 191–196.

Kazakov, S., and Levon, K. (2006). Liposome-nanogel structures for future pharmaceutical applications. Curr. Pharm. Desig. *12*, 4713–4728.

Kazakov, S., Kaholek, M., Ji, T., Turnbough, C.L., Jr, and Levon, K. (2004). Interaction of modified liposomes with *Bacillus* spores. Chem. Commun. *4*, 430–431.

Kazakov, S., Bonvouloir, E., and Gazaryan, I. (2008). physicochemical characterization of natural ionic microreservoirs: *Bacillus subtilis* dormant spores. J. Phys. Chem. B. *112*, 2233–2244.

Keim, P., Kalif, A., Schupp, J., Hill, K., Travis, S.E., Richmond, K., Adair, D.M., Hugh-Jones, M., Kuske, C.R., and Jackson, P. (1997). Molecular evolution and diversity in *Bacillus anthracis* as detected by amplified fragment length polymorphism markers. J. Bacteriol. *179*, 818–822.

Keim, P., Price, L.B., Klevytska, A.M., Smith, K.L., Schupp, J.M., Okinaka, R., Jackson, P.J., and Hugh-Jones, M.E. (2000). Multiple-locus variable-number tandem repeat analysis reveals genetic relationships within *Bacillus anthracis*. J. Bacteriol. *182*, 2928–2936.

Keysa, C.J., Dare, D.J., Sutton, H., Wells, G., Lunt, M., McKenna, T., McDowall, M., Shah, H.N. (2004). Compilation of a MALDI-TOF mass spectral database for the rapid screening and characterization of bacteria implicated in human infectious diseases. Infect. Genet. Evol. *4*, 221–242.

Kong, L., Zhang, P., Setlow, P., and Li, Y.-Q. (2010). Characterization of bacterial spore germination using integrated phase contrast microscopy, raman spectroscopy, and optical tweezers. Anal. Chem. *82*, 3840–3838.

Kornberg, A., Spudich, J.A., Nelson, D.L., and Deutscher, M.P. (1968). Origin of proteins in sporulation. Ann. Rev. Biochem. *37*, 51–78.

Koshikawa, Yamazaki, M., Yoshimi, M., Ogawa, S., Yamada, A., Watabe, K., and Torii, M. (1989). Surface hydrophobicity of spores of *Bacillus* spp. J. Gen. Microbiol. *135*, 2717- 22.

Kozuka, S., and Tochikubo, K. (1985). Properties and origin of filamentous appendages on spores of *Bacillus cereus*. Microbiol. Immunol. *29*, 21–37.

Kramer, M.J., and Roth, I.L. (1968). Ultrastructural differences in the exosporium of the Sterne and Vollum strains of *Bacillus antracis*. Can. J. Microbiol. *14*, 1297–1299.

Kress-Rogers, E. (1997). Handbook of Biosensors and Electronic Noses: Medicine, Food, and the Environment (CRC Press, Boca Raton).

Lahiani, M., and Tarasenko, O. (2010). Glycoconjugates effects: do gender and ethnicity influence exposure of pathogen by peripheral mononuclear cells. In Proceedings of the 5th BioNanoTox and Applications International Research Conference, Little Rock, Arkansas, Tarasenko, O., ed. (American Institute of Physics, Melville, NY), *1229*, 29–34.

Levon, K., Yu, B., and Zhou, Y. (2004). Bacterial biosensors. US Patent application 10/888530.

Lindeque, P.M., and Turnbull, P.C. (1994). Ecology and epidemiology of anthrax in the Etosha National Park, Namibia. Onderstepoort J. Vet. Res. *61*, 71–83.

Little, S.F., Leppla, S.H., and Friedlander, A.M. (1990). Production and characterization of monoclonal antibodies against the lethal factor component of *Bacillus anthracis* lethal toxin. Infect. Immun. *58*, 1606–1613.

Lloyd, D. (1993). Flow Cytometry in Microbiology (Springer-Verlag London Limited, Germany).

López-Enríquez, L., Rodríguez-Lázaro, D., and Hernández, M. (2007). Quantitative detection of *Clostridium tyrobutyricum* in milk by real-time PCR. Appl. Environ. Microbiol. *73*, 3747–3751.

Lone, S., Alusta, P., and Tarasenko, O. (2007). Microbiology and microbial degradation. In Glycoconjugates as a Potential Decontaminant Of Bacterial Spores. Starrett, S.K., Hong, J., Wilcock, R.J., Li, Q., and Carson, A., eds. (American Science Press, Houston, TX), pp. 458–464.

Long, G.W., and O'Brien, T. (1999). Antibody-based systems for the detection of *Bacillus anthracis* in environmental samples. J. Appl. Microbiol. *87*, 214.

Longchamp, P., and Leighton, T. (1999). Molecular recognition specificity of *Bacillus anthracis* spore antibodies. J. Appl. Microbiol. *87*, 246–249.

Luong, H.T., Groom, C.A., and Male, K.B. (1991). The potential role of biosensors in the food and drink industries. Biosens. bioelectron. *6*, 547–554.

MacDonald, W.C., Felkner, I.C., Turetsky, A., and Matney, T.S. (1963). Similarities in base composition of deoxyribonucleotes from several strains of *Bacillus anthracis* and *B. cereus*. J. Bacteriol. *85*, 1071–1073.

Madigan, M.T., Martinko, J.M., and Parker, J. (1997). Brock Biology of Microorganisms (A Viacom Company, Upper Saddle River, NJ), pp. 986.

Makino, S.I., Cheun, H.I., Watarai, M., Uchida, I., and Takeshi, K. (2001). Detection of anthrax spores from the air by real-time PCR. Lett. Appl. Microbiol., *33*, 237–240.

Marshall, E. (1997). Bracing for biological nightmare. Science. *275*, 745.

Marz, L.L., Beaman, T.C., and Gerhard, P. (1970). Chemical composition of exosporium from spores of *B. cereus*. J. Bacteriol. *101*, 196–201.

McClelland, R.G., and Pinder, A.C. (1994). Detection of low levels of specific *Salmonella* species by fluorescent antibodies and flow cytometry. J. Appl. Bacteriol. *77*, 440–447.

Melamed, M.R., Lindmo, T., and Mendelson, M.L. (1990). Flow cytometry and sorting (John Wiley, New York), pp. 105–153.

Meng, J.H. (1996). Polymerase chain-reaction for detecting *E. coli* 0157:H7. Int. J. Food Microbiol. *32*, 103–113.

Meuzelaar, H.L.C., and Kistemaker, P.G. (1973). Technique for fast and reproducible fingerprinting of bacteria by pyrolysis mass spectrometry. Anal. Chem. *45*, 587–590.

Meuzelaar, H.L.C., Kistemaker, P.G., and Posthumus, M.A. (1974). Recent advances in pyrolysis mass spectrometry of complex biological materials. Biol. Mass. Spectrometry. *1*, 312–319.

Moberly, B.J., Shafa, F., and Gerhardt, P. (1966). Structural details of anthrax spores during stages of transformation to vegetative cells. J. Bacteriol. *92*, 220–228.

Morgan, C.L., Newman, D.J., and Price, C.P. (1996). Immunosensors: technology and opportunities in laboratory medicine. Clin. Chem. *42*, 193–209.

Owen, V.M. (1994). Market requirements for advanced biosensors in healthcare. Biosens. Bioelectron. *9*, XXIX–XXXV.

Paidhungat, M., and Setlow, P. (2001). Spore Germination and Outgrowth. In *Bacillus subtilis* and its relatives: from genes to cells, Sonenshein, A.L., Hoch, J.A, and Losick, R., eds. (American Society for Microbiology, Washington, D.C.), pp. 537–548.

Patra, G., Sylvestre, P., Ramisse, V., Thérasse, J., and Guesdon, J.L. (1996). Isolation of a specific chromosomic DNA sequence of *Bacillus anthracis* and its possible use in diagnosis. FEMS Immunol. Med. Microbiol. *15*, 223–231.

Patra, G., Vaissaire, J., Weber-Levy, M., Le Doujet, C., and Mock, M. (1998). Molecular characterization of *Bacillus* strains involved in outbreaks of anthrax in France in 1997. J. Clin. Microbiol. *36*, 3412–3414.

Pavan, C., and Tarasenko, O. (2010). Disinfection of preexisting contamination of *Bacillus cereus* on stainless steel when using glycoconjugate solution. Proceedings of the 5th BioNanoTox and Applications International Research Conference, Little Rock, Arkansas, Tarasenko, O. ed. American Institute of Physics (Melville, NY), pp. 55 (poster).

Phillips, A.P., Martin, K.L., and Broster, M.G. (1983). Differentiation between spores of *Bacillus anthracis* and *Bacillus cereus* by a quantitative immunofluorescence technique. J. Clin. Microbiol. *17*, 41–47.

Phillips, A.P., Campbell, A.M., and Quinn, R. (1988). Monoclonal antibodies against spore antigens of *Bacillus anthracis*. FEMS Microbiol. Immunol. *47*, 169–178.

Pinder, A.C., Purdy, P.W., Poulter, S.A., and Clark, D.C. (1990). Validation of flow cytometry for rapid enumeration of bacterial concentrations in pure cultures. J. Appl. Bacteriol. *69*, 92–100.

Pirs, T., Ocepek, M., and Rupnik, M. (2008). Isolation of *Clostridium difficile* from food animals in Slovenia. J. Med. Microbiol. *57*, 790–792.

Previte, J.J. (1972). Radiometric detection of some food-borne bacteria. Appl. Microbiol. *24*, 535–539.

Price, L.B. (1999). Genetic diversity in the protective antigen gene of *Bacillus anthracis*. J. Bacteriol. *181*, 2358–2362.

Quinlan, J.J., and Foegeding, P.M. (1997). Monoclonal antibodies for use in detection of *Bacillus* and *Clostridium* spores. J. Appl. Environ. Microbiol. *63*, 482- 487.

Ramisse, V., Patra, G., Garrigue, H., Guesdon, J.L., and Mock, M. (1996). Identification and characterization of *Bacillus anthracis* by multiplex PCR analysis of sequences on plasmids pXO1 and pXO2 and chromosomal DNA. FEMS Microbiol. Lett. *145*, 9–16.

Rasol, A., Eassa, S., and Tarasenko, O. (2009). Binding affinity of glycoconjugates to *Bacillus* spores and toxins. In Proceedings of the 4th BioNanoTox and Applications International Research Conference, Little Rock, Arkansas, Tarasenko, O., ed. (American Institute of Physics, Melville, NY). *1229*, 67-72.

Rasol, A., Eassa, S., and Tarasenko, O. (2010). Binding affinity of glycoconjugates to *Bacillus anthracis* spores and their respective toxins. PMSE *102*, 130–131.

Redmond, C., Baillie, L.W.J., Hibbs, S., Moir, A.J.G., and Moir, A. (2004). Identification of proteins in the exosporium of *Bacillus anthracis*. Microbiol. *150*, 355–363.

Rodriguez-Palacios, A., Staempfli, H.R., Duffield, T., Weese, J.S. (2007). *Clostridium difficile* in retail ground meat, Canada. Emerg. Infect. Dis. *13*, 485–487.

Rodriguez-Palacios, A., Reid-Smith, R.J., Staempfli, H.R., Daignault, D., Janecko, N., Avery, B.P., Martin, H., Thomspon, A.D., McDonald, L.C., Limbago, B., *et al.* (2009). Possible seasonality of *Clostridium difficile* in retail meat, Canada. Emerg Infect Dis. *15*, 802–805.

Rogers, K.R., Mulchandani, A., and Zhou, W. (1995). Biosensor and Chemical Sensor Technology: Process Monitoring and Control (ACS, Washington, DC), pp. 187.

Rossi, T.M., and Warner, M. (1985). Bacterial identification using fluorescense spectroscopy. In Instrumental Methods for Rapid Microbiological Analysis, Nelson, W.H., ed. (VCH Publishers, Deefield Beach, FL), pp. 1–50.

Salzman, G., Singham, S., Johnston, R., and Bohren, C. (1990). Light scattering and cytometry. In Flow Cytometry and Sorting, Melamed, M.R., Lindmo, T., and Mendelsohn, M.L., eds. (John Wiley, New York), pp.105–153.

Scherer, P.S., and Somerville, H.J. (1977). Membrane fractions from the outer layers of spores of *Bacillus thuringiensis* with toxicity to lepidopterous larvae. Eur. J. Biochem. 72, 479–490.

Schmid, R.D., and Scheller, F. (1989). Biosensors: application in medicine, environmental protection, and process control (VCH, New York).

Schmitt, N., Bowmer, E.J., and Willoughby, B.A. (1976). Food poisoning outbreak attributed to *Bacillus cereus*. Can. J. Public Health *67*, 418–422.

Schupp, J.M., Klevytska, A.M., Zinser, G., Price, L.B., and Keim P. (2000). vrrB, a hypervariable open reading frame in *Bacillus anthracis*. J. Bacteriol. *182*, 3989–3997.

Seki, T., Chung, C.-K., Mikami, H., and Oshima, Y. (1978). Deoxyribonucleic acid homology of the genus *Bacillus*. Int. J. Syst. Bacteriol. *28*, 182–189.

Sethi, R.S. (1994). Transducer aspects of biosensors. Biosens. Bioelectron. *9*, 243–264.

Setlow, P. (1994). Mechanisms which contribute to the long-term survival of spores of *Bacillus* species. J. Appl. Bacteriol. *176*, 49S-60S.

Setlow, P. (1995). Mechanisms for the prevention of damage to DNA in spores of *Bacillus* species. Annu. Rev. Microbiol. *49*, 29–54.

Shangkuan, Y.H., Yang, J.-F., Lin, H.-C. and Shaio, M.-F. (2000). Comparison of PCR-RFLP, ribotyping and ERIC-PCR for typing *Bacillus anthracis* and *Bacillus cereus* strains. J. Appl. Microbiol. *89*, 452–462.

Shiota, M., Saitou, K., Mizumoto, H., Matsusaka, M., Agata, N., Nakayama, M., Kage, M., Tatsumi, S., Okamoto, A., Yamaguchi, S., *et al.* (2010). Rapid Detoxification of Cereulide in *Bacillus cereus* Food Poisoning. Pediatrics *125*, 951–955.

Smith, T.J., Blackman, S.A., and Foster, S.J. (2000). Autolysins of *Bacillus subtilis*: multiple enzymes with multiple functions. Microbiol. *146*, 249–262.

Songer, J.G., Trinh, H.T., Killgore, G.E., Thompson, A.D., McDonald, L.C., and Limbago, B.M. (2009). *Clostridium difficile* in retail meat products, U.S.A., 2007. Emerg. Infect. Dis. *15*, 819–821.

Sousa, J.C., Silva, M.T., and Balassa, G. (1976). An exosporium-like outer layer in *Bacillus subtilis* spores. Nature. *263*, 53–54.

Sousa, J.C., Silva, M.T., and Balassa, G. (1978). Ultrastructure and development of an exosporium-like outer spore envelope in *Bacillus subtilis*. Ann. Microbiol. *129*, 339–362.

Sperveslage, J., Stackebrandt, E., Lembke, F.W., and Koch, C. (1996). Detection of bacterial-contamination, including *Bacillus* spores, in dry growth media and in milk by identification of their 16S RDNA by polymerase chain-reaction. J. Microbiol. Methods. *26*, 219–224.

Steen, H.B., Skaratad, K., and Boye, E. (1990). DNA measurement of bacteria. In Methods in Cell Biology, Darzynkiewicz, Z., and Crissman, H., eds. (Academic Press, San Diego), pp. 519–526.

Stopa, P.J. (2000). The flow cytometry of *Bacillus anthracis* spores revisited. Cytometry *41*, 237–244.

Stratis, D., Griffin, G.D., Mobley, J., and Vo-Dinh, T. (2007). In Intensified Biochip System Using Chemiluminescence for the Detection of *Bacillus globiggii* Spores. Anal Bioanal Chem. *391*, 1655–1660.

Swaminathan, B., and Feng, P. (1994). Rapid detection of food-borne pathogenic bacteria. Annu. Rev. Microbiol. *48*, 401–426.

Sylvestre, P., Tosi-Couture, E., and Mock, M. (2001). Analysis of *Bacillus anthracis* spore surface antigens with monoclonal antibodies. In Program and Abstracts Book of 4th International Conference of Anthrax, Annapolis, MD, Friedlander, M.A., Baillie, L.W.J., Ivins, B.E., Keim, P., Leppla, S.H., Mock, M., Welkos, S.L., eds. (USAMRIID, Fort Detrick, MD), pp. 1–36.

Takumi, K., Kinouchi, T., and Kawata, T. (1979). Isolation and partial characterization of exosporium from spores of highly sporogenic mutant of *Clostridium botulinum* type A. Microbiol. Immunol. *28*, 443–454.

Tarasenko, O., Islam, S., Paquiot, D., and Levon, K. (2004). Glycoconjugates for the recognition of *Bacillus* spores. Carbohydr. Res. *339*, 2859–2870.

Tarasenko, O., Nourbakhsh, S., Kuo, S.P., Bakhtina, A., Alusta, P., Kudasheva, D., Cowman, M., and Levon, K. (2006). Scanning electron and atomic force microscopy to study plasma torch effects on *B. cereus* Spores. IEEE Transactions on Plasma Science. *34*, 1281–1289.

Tarasenko, O., Jones, A., Castleberry, J., Akpore, D., Hester, K., Park Kim, M., McManus, D., Soderberg, L., and Alusta, P. (2008a). Study of *Bacillus anthracis* PA-EF-LF, PA-EF, and PA-LF toxin complexes neutralization during phagocytosis using glycoconjugates. PMSE 98867–98868.

Tarasenko, O., Burton, E., Soderberg, L., and Alusta, P. (2008b). Polymeric glycoconjugates protect and activate macrophages to promote killing of *Bacillus cereus* spores during phagocytosis. Glycoconj. J. *25*, 473–480.

Tarasenko, O., Lone, S., and Alusta, P. (2008c). Defensive and simultaneous actions of glycoconjugates during spore decontamination. Carbohydr. Res. *343*, 2243–2250.

Tarasenko, O., Soderberg, L., Hester, K., Park Kim, M., McManus, D., and Alusta, P. (2008d). Glycoconjugates enhanced the intracellular killing of *Bacillus* spores, increasing macrophage viability and activation. Arch. Microbiol. *189*, 579–587.

Tarasenko, O, Scott, A., Soderberg, S., Ponnappan, U., and Alusta, P. (2010). Killing of *Bacillus* spores is mediated by nitric oxide and nitric oxide synthase during glycoconjugate – enhanced phagocytosis. Glycoconj. J. *27*, 13–25.

Turner, A.P.F., Karube, I., and Wilson, G.S. (1987). Biosensors: Fundamentals and applications (Oxford University Press, Oxford), pp. 770.

Taylor, A.J., and Gilbert, R.J. (1975). *Bacillus cereus* food poisoning: a provisional serotyping scheme. J. Med. Microbiol. *8*, 543–550.

Tietjen, M., and Fung, D.Y.C. (1995). *Salmonella* and food safety. Crit. Rev. Microb. *21*, 53–83.

Turnbull, P.C.B., and Kramer, J.M. (1991). *Bacillus*. In Manual of clinical microbiology, Balows, A., Hausler, W.J., Herrmann, K.L., Jr., Isenberg, H.D., and Shadomy, H.J., eds. (American Society for Microbiology, Washington, DC), pp. 296–303.

Turnbull, P.C.B., Broster, M.G., Carman, J.A., Manchee, R.J., and Melling, J. (1986). Development of antibodies to protective antigen and lethal factor components of anthrax toxin in humans and guinea pigs and their relevance to protective immunity. Infect Immun. *52*, 356–363.

Turnbull, P.C.B., Hutson, R.A., Ward, M.J., Jones, M.N., Quinn, C.P., Finnie, N.J., Duggleby, C.J., Kramer, J.M., and Melling, J. (1992). *Bacillus anthracis* but not always anthrax. J. Appl. Bact. 72, 21–28.

Turner, A.P.F., Allen, M., Schneider, B.H., Swain, A., and Taylor, F. (1989). An inexpensive method for ultra-rapid detection of microbial contamination in industrial fluids. Int. Bio. 25, 137–145.

Van Ness, G.B. (1971). Ecology of anthrax. Science. *172*, 1303–1307.

Vernozy-Rozand, C., Mazuy, C., Ray-Gueniot, S., Boutrand-Loeï, S., Meyrand, A., and Richard, Y. (1997). Detection of *Escherichia coli* O157 in French food samples using an immunomagnetic separation method and the VIDAS *E. coli* O157. Lett. Appl. Microbiol. *25*, 442–444.

Wang, H., Sun, W., Zhang, J., Yang, X., Lin, T. and Ding, L. (2010). Desorption corona beam ionization source for mass spectrometry. Royal Soc. Chem. *135*, 688–695.

Warth, A.D., Ohey, D.F., and Murrell, W.G. (1963). The composition and structure of bacterial spores. J. Cell Biol. *16*, 578–592.

Watson, A., and Keir, D. (1994). Information on which to base assessments of risk from environments contaminated with anthrax spores. Epidemiol. Infect. *113*, 479–490.

Weese, J.S., Avery, B.R., Rousseau, J., and Reid-Smith, R.J. (2009). Detection and enumeration of *Clostridium difficile* spores in retail beef and pork. Appl. Environ. Microbiol. 5009–5011.

Weese, J.S., Avery, B., Rousseau, J. and Reid-Smith, R. (2009). Detection and enumeration of *Clostridium difficile* in retail meat, abstr. P2037. 19th Eur. Conf. Clin. Microbiol. Infect. Dis. (Helsinki, Finland).

Wijesuriya, D.C., Anderson, G.P., and Ligler, F.S. (1994). A rapid and sensitive immunoassay for bacterial cells. In Proceedings of the 1993 ERDEC Scientific Conference on Chemical Defense Research, Berg, D.A., Williams, J.D., Reeves, J., Reeves, P.J., eds. U.S. Army Chemical and Biological Defense Agency (Aberdeen Proving Ground, MD), pp.16–19.

Williams, T.L., Andrzejewski, D., Lay, J.O., and Musser, S.M. (2003). Experimental factors affecting the quality and reproducibility of MALDI TOF mass spectra obtained from whole bacteria cells. J. Am. Soc. Mass. Spectrom. *14*, 342–351.

Yu, H., and Bruno, J.G. (1996). Immunomagnetic-electrochemiluminescent detection of *E. coli* O157 and *Salmonella typhimurium* in foods and environmental water samples. Appl. Environ. Microbiol. *62*, 587–592.

Yu, H., and Stopa, P.J. (1995). Application of an immunomagnetic assay system for detection of virulent bacteria in biological samples. In Environmental Immunochemical Methods, Van Emon, J.M., Gerlach, C.L., and Johnson, J.C., eds. (ACS, Washington, DC).

Zhou, Y., Yu, B., and Levon, K. (2003). Potentiometric sensing of chiral amino acids. Chem. Mater. *15*, 2774–2779.

Zhou, Y., Yu, B., and Levon, K. (2005). Potentiometric sensor for dipicolinic acid. Biosens. Bioelectron. *20*, 1851–1855.

Zilinskas, R.A. (1997). Iraq's biological weapons: the past as future? JAMA *278*, 418–424.

Expression of Recombinant Proteins using *Bacillus subtilis* Spores

15

Luis Carlos de Souza Ferreira and Wolfgang Schumann

Abstract

Surface display of peptides and proteins on recombinant phages and bacterial cells has developed into a powerful technique to identify proteins with desired properties with multiple applications. This technique is of utmost importance as a tool for fundamental and applied research in the fields of microbiology, biotechnology and vaccination. Another promising bioparticle are spores produced by *Bacillus subtilis* cells that allow fusion of recombinant proteins to one of several outer spore coat proteins. There are several reports were antigens and proteins of biotechnological interest have been successfully displayed on the outside of spores. A major advantage of spores over bacterial cells is their intracellular biogenesis that does not require specific secretion sequences of the cellular apparatus. Therefore, heterologous proteins do not have to cross membranes and correct folding of these proteins can be achieved with the aid of cytoplasmic folder chaperones.

Introduction

Surface display is based on expressing recombinant (heterologous) (poly) peptides fused to sorting signals of carrier proteins that direct their incorporation on the surface of bioparticles. This technology has a variety of applications including the development of life vaccines, generation of biosensors and biocatalysts and screening of peptides libraries (Stahl and Uhlén, 1997). Large combinatorial libraries of encoded proteins or peptides (up to 10^{11} members), specific proteins or random protein fragments can be similarly displayed (Daugherty, 2007). The displayed proteins or peptides with the desired property can be selected from the library and decoded by sequencing of the appropriate gene. Bioparticles can be phages (Paschke, 2006; Sidhu and Koide, 2007; Bratkovic, 2010) or eukaryotic viruses (Michelfelder *et al.*, 2007; Makela and Oker-Blom, 2008), bacterial (Desvaux *et al.*, 2006; Kwon *et al.*, 2007) or eukaryotic cells (Pepper *et al.*, 2008; Ho and Pastan, 2009). Recently, endospores formed by *Bacillus subtilis* and related species have been used for surface display (Isticato *et al.*, 2001; Ricca and Cutting, 2003; Kim and Schumann, 2009; Knecht *et al.*, 2011). Methods to display desired (poly) peptides on the surface of bioparticles involve the construction of gene fusions that code for the carrier protein anchored on the outside and heterologous protein forming a chimeric protein.

Formation of endospores

Endospores (spores) are extremely robust life forms formed by about 60% of the *B. subtilis* cells upon entering the stationary growth phase. These spores are metabolically inactive, can survive for millions of years in the absence of nutrients (Cano and Borucki, 1995) and can resist extreme temperature, UV irradiation and exposure to toxic chemicals (Henriques and Moran, 2007). The formation of spores is largely an adaptive response to starvation but the cell integrates a huge number of external and internal signals before entering the sporulation programme. Sporulation involves asymmetric cell division where the septum is formed close to one cell pole (arbitrarily chosen) rather than at midcell, and during which a copy of the genome is partioned into each sister cell (Errington, 2003; Piggot and Hilbert, 2004). The smaller cell designated the prespore, develops into the mature endospore and the larger cell, the mother cell, contributes to the differentiation process and undergoes autolysis to release the endospore into the surrounding medium. Soon after the asymmetric division, the mother cell membrane engulfs the prespore, creating a double-membrane system, the inner and the outer spore membrane (Errington, 2003; Piggot and Hilbert, 2004). Following its formation, the prespore undergoes a dramatic morphological transition resulting in its dehydration and mineralization. The spore surface (spore coat) is formed by more than 70 different proteins which are organized into an inner and an outer layer (Henriques and Moran, 2000; Kim *et al.*, 2006).

Spores of *Clostridium*, of the three *Bacillus* species *B. cereus*, *B. anthracis* and *B. thuringiensis*, all three pathogens of animals and insects, are surrounded by an additional outer layer called exosporium (Henriques and Moran, 2007). It is composed of proteins, lipids, and carbohydrates and consists of a paracrystalline layer and an external hair-like nap.

Targeting heterologous proteins to the spore surface

Spores have the following advantages over other bioparticles for surface display.

1. *B. subtilis* has been given the GRAS (generally regarded as safe) status by the American Food and Drug Administration. Microorganisms with the GRAS status can be used in food production.
2. Since spores are formed in the cytoplasm, the passenger protein has not to cross any membrane.
3. Cytoplasmic chaperones can aid the passenger protein to fold correctly.
4. There is no codon bias. Because of these advantages, spores of *B. subtilis* and other Bacilli have been used for surface display. In principle, two different classes of proteins have been immobilized on the spore surface, proteins with enzymatic functions and antigens.

Three different proteins mainly associated with the outer coat (Isticato *et al.*, 2004; Kim *et al.*, 2005) have been selected for their surface location or relative abundance as carrier proteins: CotB (66 kDa), CotC (12 kDa) and CotG (24 kDa) and one inner layer protein OxdD (43.5 kDa). CotB, CotC, and CotG, possibly the most represented proteins within the spore coat depend on CotE for assembly. All three proteins also undergo extensive multimerization during their assembly at the surface (Zilhao *et al.*, 2004). CotB forms multimers and one of which, an abundant form of about 64 kDa is presumed to result from its direct interaction with CotG (Zilhao *et al.*, 2004). CotC can form multimeric species with itself and the nearly identical CotU protein at the surface (Isticato *et al.*, 2008, 2010). The CotG gene encodes a protein of 24 kDa, but the protein isolated for microsequencing by SDS-PAGE migrates as a 36 kDa protein (Henriques *et al.*, 1998) and undergoes extensive multimerization following spore release from the mother cell (Zilhao *et al.*, 1999, 2004). Normally, the 3′ end of CotB containing repetitive sequences and coding for the C-terminal 105 amino acid residues, is not included in the amplified fragment to avoid potential stability problems in the gene fusion (Isticato *et al.*, 2001). This C-terminal end of CotB codes for three 27-amino-acid repeat which might cause a genetic stability problem of the fusion proteins. In spores of *cotG* insertional mutants, the normal structural organization of the outer coat is severely disturbed (Henriques *et al.*, 1998), because CotQ and CotU are partially misassembled (Kim *et al.*, 2006). OxdD is a minor component of the inner-coat layers, which is highly similar to OxdC, which is specifically produced during growth of *B. subtilis* under acidic conditions (Tanner *et al.*, 2001). Both OxdD and OxdC exhibit oxalate decarboxylase activity (Tanner *et al.*, 2001; Costa *et al.*, 2004).

Anchoring of functional proteins on the outside of endospores

Until now, most applications of *B. subtilis* spores are focused on delivery of surface-expressed antigens (see below). The very first enzyme which has been anchored on the spore surface was the β-galactosidase of *E. coli* encoded by the *lacZ* gene, the most utilized reporter protein in bioscience and biotechnology. β-Galactosidase is a rather large protein of 116 kDa being active as a homotetramer. The coding region of the enzyme was fused to *cotG* and *cotE* serving as anchoring motifs; the spores expressed β-galactosidase activity (Ban *et al.*, 2003; Kwon *et al.*, 2007). It was shown that the enzyme was stable in the presence of non-polar hydrophobic organic solvents, and the enzyme was further stabilized by chemically cross-linking it with glutaraldehyde (Kwon *et al.*, 2007). Furthermore, the spore surface immobilized CotG-β-galactosidase was active in a transgalactosylation reaction. When lactose was inoculated with octanol, octyl-β-D-galactopyranoside was synthesized. In a recent report, the *nanA* gene derived from *E. coli* was translationally fused to *cotG* (Xu *et al.*, 200). The NanA enzyme condenses ManNAc with pyruvate to produce *N*-acetyl-D-neuraminic acid (Neu5Ac). Under optimal conditions, Neu5Ac was produced at a high concentration (54.7 g/l) and a high yield (90.2%). Therefore, spore-based display represents a suitable and inexpensive alternative method for the synthesis of Neu5Ac.

The second protein displayed on *B. subtilis* spores was streptavidin (Kim *et al.*, 2005). This 60-kDa protein produced by *Streptomyces avidinii*, binds to biotin very tightly with a K_d of 10^{-15}, and active streptavidin is composed of four monomers with four active sites for biotin binding in the active tetrameric streptavidin. Expression of streptavidin on the external surface of spores fused to CotG was verified by the fluorescence-activated cell sorter (FACS) technique. Purified spores were first treated with anti-streptavidin antibodies and then by FITC-labelled secondary antibodies. They exhibited improved fluorescence intensity compared with control spores (Kim *et al.*, 2005). To prove the activity of streptavidin, purified spores were treated by FITC-biotin and analysed by confocal microscopy. A distinctive fluorescence could be observed.

The green fluorescent protein (GFP) of the jellyfish *Aequoria victoria* emits fluorescence when excited by UV or blue light (Chalfie *et al.*, 1994). Therefore, it is being used as a reporter for gene expression and regulation and protein localization within the cell. Variants of GFP have been isolated such as GFP_{UV} exhibiting an improvement in excitation (Crameri *et al.*, 1996). GFP_{UV} was fused to the C-terminal end of CotG, and expression of the fusion protein on *B. subtilis* spores was confirmed by FACS and by confocal microscopic analysis (Kim *et al.*, 2007a). Biosorption is a method that uses the ability of biological material to bind and accumulate pollutants such as heavy metals. Microorganisms have been engineered with immobilized proteins or peptides on their outside for biosorption of heavy metals (Lee *et al.*, 2003). Fusion of synthetic polyhistidines to the *E. coli* outer membrane protein LamB resulted in cells efficiently binding Cd^{2+} (Sousa *et al.*, 1996). Synthetic DNA coding for 18 histidine residues were cloned in-frame to the 3′ end of the truncated *cotB*. Spores expressing the recombinant protein were efficient in adsorbing nickel ions (Hinc *et al.*, 2010a). To evaluate the expression and display of the A subunit of the *Helicobacter acinonychis* urease (UreA), its coding region was fused to *cotB*, *cotC* and *cotG*. Extracted proteins from spores were analysed by Western-blot and immuno fluorescence microscopy. While about 1×10^3 recombinant molecules were displayed per spore when fused to CotB, $7–15\times10^3$ fusion proteins were anchored per spore using CotC or CotG. In the case of CotC-UreA, the fusion proteins were not displayed on the surface of the spores, and the CotG–UreA fusion proteins appeared to be partially processed (Hinc *et al.*, 2010b). While most fusion proteins exhibited a size of about 55 kDa, a fraction of these molecules is probably processed to yield a protein of about 30 kDa being smaller than the UreA alone.

Biosensors are biological tools which are used to sense the physicochemical composition of the surroundings, e. g., in detecting pollutants in normal settings. The application of biosensors in various fields has been discussed in several reviews (Harms *et al.*, 2006; Sorensen *et al.*, 2006; Yagi, 2007). Besides whole cells, spores have been developed as a storage method for biosensing systems. The very first report involved spores of *B. subtilis* and *B. megaterium* for luminescent sensing systems for arsenic and zinc, respectively.

Specifically, genes that confer resistance to arsenic together with the *lacZ* reporter gene were inserted into a plasmid which was subsequently recombined in the *B. subtilis* chromosome (Date *et al.*, 2007). In a second experiment, the *smtB* gene coding for a zinc-binding protein and its regulatory region fused to *egfp* (enhanced green fluorescent protein) were integrated into a plasmid. The SmtB protein is a repressor that binds zinc causing dissociation from its regulatory region thereby allowing synthesis of EGFP. The recombinant plasmid was then transformed into *B. megaterium*. In both cases, cells were sporulated, and the spore-based sending systems turned to be stable up to at least 24 months (Date *et al.*, 2007). Using both sensing systems, the same authors could detect arsenic concentrations as low as 1×10^{-7} M in freshwater and serum samples and 1×10^{-6} M zinc in serum samples (Date *et al.*, 2010b). To create a portable system where spore germination, incubation with samples and signal detection are all integrated, the authors developed a miniaturized device (Date *et al.*, 2010a). The sensing spores are incorporated into a portable centrifugal microfluidic compact disk platform enabling the rapid and sensitive detection of the analytes. In a similar approach, spores were used as colorimetric sensors for zinc and the antibiotic bacitracin (Fantino *et al.*, 2009). Here, the *czcD* gene induced by the presence of Zn^{2+} and a bacitracin-inducible gene were fused to *lacZ*. When recombinant spores called 'sposensors' spotted on filter paper were either incubated with zinc or bacitracin, the *lacZ* gene was turned on and allowed detection of zinc concentrations down to 10 μM and the minimal concentration of bacitracin was 0.035 μM.

Using the 43.5-kDa monomeric inner coat protein OxdD, the *B. subtilis* phytase was fused to the C-terminal end of this coat protein and to CotG. Phytase is a commonly used feed enzyme for monogastric animals to improve the nutritional value. Surface display with both carrier proteins was shown by immuno fluorescence microscopy which disappeared after treatment with trypsin (Potot *et al.*, 2010). These spores expressed phytase activity. When the enzymatic activities were measured a 2-fold decrease was measured with the OxdD fusion protein as compared with the CotG hybrid. Based on these results, the authors suggested that although the location of OxdD may offer protection of the enzyme from the environment, it could also hinder enzyme substrate accessibility. In another experiment, the *E. coli* β-glucuronidase encoded by *uidA* gene was fused to OxdD. Most probably, this enzyme functions as a homotetramer. About 40% of the spores expressed OxdD-UidA and exhibited β-glucuronidase activity (Potot *et al.*, 2010). To increase the number of passenger molecules per spore, the hybrid CotB-AmyQ protein (the *amyQ* gene codes an α-amylase) was transcriptionally fused to the IPTG-inducible Pgrac promoter (Nguyen *et al.*, 2007). Full induction was obtained in the presence of 1 mM IPTG, and recombinant spores expressed γ-amylase activity (Nguyen and Schumann, 2011). When *amyE* was expressed from the *cotB* promoter, 4.6×10^3 molecules were present per spore, and 9.7×10^3 with Pgrac. This experiment proves that the authentic promoter can be replaced by an inducible one.

Spore display has also been used to perform directed evolution of the CotA enzyme. This enzyme codes for a laccase, which catalyses a large range of substrates including diammonium 2,2-azinobis-3-ethylbenzothiazoline-6-sulfonate (ABTS) and 4-hydroxy-3,5-dimethoxbenzaldehyde azine (SGZ). A library of mutated and native *cotA* genes were ligated into a plasmid, transformed into a *B. subtilis cotA* knockout mutant (Gupta and Farinas, 2010). Both control spores and mutant spores obtained from the mutagenesis library were assayed for CotA activity by measuring changes in absorbance of ABTS and SGZ substrates. The authors could identify one mutant CotA, which was 120 times more specific for ABTS than the native laccase. This publication was the first to demonstrate that the display of proteins on the spore surface can be used to create and analyse mutagenesis libraries.

Spore surface display of vaccine antigens

The initial aim of expressing heterologous proteins on the surface of spores was based on the possibility to generate a new class of oral-delivered live vaccine vectors that could combine the probiotic effects of *B. subtilis* spores with the remarkable heat-stability of endospores. Such live vectors would offer several advantages over other live bacterial vectors such as vegetative cells of attenuated Gram-negative or Gram-positive bacterial species. Isticato and colleagues (2001) were the first to demonstrate that the tetanus toxin fragment C polypeptide (TTFC) could be successfully expressed on the surface of *B. subtilis* spores following C-terminal fusion with the CotB spore coat protein following insertion of the recombinant gene into the chromosome of the host bacterial strain (Isticato *et al.*, 2001). Based on this expression approach, the recombinant gene was kept as single copy integration at a non-essential genetic locus of the bacterial genome preserving the native gene CotB-encoding gene. These initial results showed that approximately 1.5×10^3 recombinant proteins can be expressed and exposed on the surface of each spore. The antigenicity of the recombinant antigen was preserved indicating that the protein at the spore surface was recognized and accessible to anti-TTFC antibodies at the spore surface. The subsequent proof that genetically modified *B. subtilis* spores could be used as vaccine vectors came later with mice immunized via mucosal or parenteral routes and detection of induced serum and mucosal TTFC-specific antibody responses (Duc *et al.*, 2003). Mice immunized with the recombinant spores, via oral or nasal administration routes, developed anti-TTFC antibody responses that could be measured both in serum and faeces of animals receiving the recombinant spores but not in those treated with non-recombinant spores. More relevantly, mice orally immunized with the recombinant spores were fully protected to a lethal challenge carried out with purified tetanus toxin. These results demonstrated that *B. subtilis* spores were proficient as live vaccine vectors allowing both transit of the antigen through the gastrointestinal tract and proper targeting of the recombinant antigen to the host immune cells. Nonetheless, efficient mucosal immunization required multiple vaccine doses (nine to six doses for oral and nasal immunizations, respectively) and rather large number of spores per dose (from 1×10^9 to 2×10^{10} spores/dose). This procedure is frequently used for mucosal immunizations with on non-pathogenic Gram-positive bacterial species unable to transiently colonize the mammalian intestinal tract, such as *Lactococcus lactis* and some *Streptococcus* species. Based on this initial report, several other studies have been published showing that *B. subtilis* strains displaying recombinant antigens on the spore-coat surface can act as delivered vectors under experimental conditions and induce a specific immune response and, when available, confer protection to lethal challenges to mice, such as *Bacillus anthracis* (anthrax) (Duc *et al.*, 2007), *Clostridium perfringens* (necrotic enteritis) (Hoang *et al.*, 2008), *Clostridium difficile* (causing nosocomial infections) (Permpoonpattana *et al.*, 2011), *Clonorchis sinensis* (a nematode causing clonorchiasis) (Zhou *et al.*, 2007, 2008), and rotavirus (Lee *et al.*, 2010a).

The same approach has also shown promising results in arthropods hosts including the silkworm (*Bombyx mori*) and shrimp species (*Procambarus clarkii*) (Fu *et al.*, 2008; Li *et al.*, 2011). Indeed, animals fed with recombinant *B. subtilis* spores expressing proteins derived from pathogenic virus causing important economic losses to the silk and shrimp industries showed partial protection to virus infection. These results suggest that application of the vaccine technology based on spore surface display approaches may find broader applications than conventional vaccines for mammalian or avian hosts.

So far display of proteins, including vaccine antigens, on *B. subtilis* spores has been carried out with three major spore coat proteins, CotB, CotC and Cot G (Isticato *et al.*, 2001, 2007; Duc *et al.*, 2003, 2007; Mauriello *et al.*, 2004; Uyen *et al.*, 2007; Hoang *et al.*, 2008; Lee *et al.*, 2010b). The stable expression of the recombinant gene is achieved by integration into the genome due to the high instability of most episomal vectors of *B. subtilis* (Isticato *et al.*, 2001; Duc *et al.*, 2003, 2007; Uyen *et al.*, 2007). Gene fusions with *cotB*, but not with *cotC* or *cotG*, require the maintenance of the native gene in order to avoid decrease in the sporulation efficiency and the structural of the spore coat. Considering that approximately 50 proteins are present in the spore coat of *B. subtilis*, additional candidates for fusion with heterologous antigen such as OxdD, an inner spore coat protein (Kim *et al.*, 2005, 2007b; Hinc *et al.*, 2010b). There are, therefore, potential to generate recombinant spores expressing multiple antigens fused to different compartment of the spore coat.

The amount of antigen molecules expressed on the surface the recombinant spores is an important concern in the development of spore-based vaccine vectors with enhanced immunogenicity. Fusion of heterologous antigens at the C-terminal end of CotB results in approximately 10^3 molecules exposed at the spore surface (Mauriello *et al.*, 2004; Isticato *et al.*, 2007). Fusions with CotC seems to depend on the site of insertion of the heterologous proteins and insertions located at the N-terminal end of CotC result in quantities up to 5-fold higher than fusions at its C-terminal end (Isticato *et al.*, 2007). In a recent study, the expression of a single antigen, the A subunit of *H. acinonychis* urease, has been tested after fusion with three different (CoC, CotB and CotG) outer spore coat proteins (Hinc *et al.*, 2010b). The results further confirmed that all three spore coat proteins can display heterologous proteins on the spore coat. In addition, fusion with CotC resulted in higher expression levels of the target antigen than fusions with CotB but, in contrast with the CotB fusion, part of the expressed protein was not exposed at the spore surface while fusions with CotG remain largely unstable and susceptible to proteolytic cleavage (Hinc *et al.*, 2010b). These results illustrate the versatility of spore surface expression system but also indicate that further studies will be needed before *B.subtilis* spores could be used to efficiently activate immune responses.

Fusion to a spore coat protein may affect the correct folding of a target antigen and alter the specificity of the antibodies that are required to neutralize pathogens like virus and bacteria. So far no direct evaluation of the conformation changes of recombinant antigens, following fusion with spore coat protein, has been carried out but recent evidences indicated that genetic fusions with OxdD, an inner spore coat protein, partially preserve the activity of feed enzymes, such as the endogenous phytase (usually expressed as a cytoplasmatic protein) and an heterologous β-glucuronidase suggesting that correct folding of recombinant proteins, as evaluated by the enzymatic activity, was achieved after fusion with the *B. subtilis* OxdD protein (Potot *et al.*, 2010). Similar evaluations shall be performed with different inner and outer spore coat proteins to allow precise quantitative determination of both expression levels and immunological activity of target antigens following the fusion process.

Expression of heterologous proteins by germinating *B. subtilis* spores

Expression of recombinant proteins following spore germination has also been successfully applied as an approach to induce antigen-specific immune responses in mice immunized with purified of *B. subtilis* spores. Several reports have demonstrated that expression of antigens under control of promoters activated immediately after spore germination can promote significant immune response in mice immunized with spores via mucosal or parenteral administration routes (Paccez *et al.*, 2006, 2007; Uyen *et al.*, 2007; Luiz *et al.*, 2008; Lee *et al.*, 2010a). Such expression approach did not require fusion of the passenger antigen to *B. subtilis* spore proteins but demands strong promoters that allow expression of the target antigen immediately after germination, such as the *gsiB* promoter activated by stress-inducing conditions (Paccez *et al.*, 2007). The experimental demonstration that *B. subtilis* spores germinate during transit through the gastrointestinal tract and in phagolysosome of antigen-presenting cells, such as macrophages, support the notion that *in vivo* antigen expression suffices for the generation of strong immune responses in animals immunized with purified *B. subtilis* spores (Duc *et al.*, 2003, 2004; Paccez *et al.*, 2006, 2007; Tam *et al.*, 2006; Fu *et al.*, 2008, Ceragioli *et al.*, 2009).

Factors affecting spore surface display

The following factors affect the amount of passenger proteins anchored per spore and will be discussed: costs, instability due to proteases, increasing the number of passenger proteins per spore, reusability of spores and immobilization of fusion proteins on all spores.

1 Spore-displayed biocatalysts and antigens can be prepared readily and inexpensively. The high specific activities of spore-displayed enzymes reflect the high active ratio of surface area to volume of spores.
2 *B. subtilis* cells produce several exoproteases

when entering the stationary phase (Grossman, 1995). These will come into contact with the spores when the mother cells lyse. Use of an 8-fold protease-minus strain (Wu *et al.*, 2002) will help to increase stability of the passenger proteins.

3 The number of fusion proteins anchored per spore is dependent on different factors, including the promoter strength and folding of the fusion protein. To deal with the promoter strength, two strategies can be envisaged: Use of strong regulatable promoters and fusion to two and more different *cot* genes. Correct folding of the passenger protein can be enhanced by concomitant overexpression of folder chaperones (De and De, 2004; De *et al.*, 2007) or inserting an aliphatic amino acid linker of about ten residues between the carrier and the passenger protein to minimize any potential steric hindrance between the two proteins that could influence the correct folding of either protein (Potot *et al.*, 2010).

4 Another aspect of the spore immobilization system is its reusability. If spores with anchored enzymes have been used to catalyse the appropriate enzymatic reaction, spores can be separated from the reaction products by standard micro filtration processes. If the enzymes have been denatured or inactivated, spores can be germinated into vegetative cells which can produce spores again with native enzymes.

5 The last point deals with the observation published with some fusion proteins that there are spores with no detectable passenger proteins anchored. It is completely unknown why up to 50% of the spores do not contain passenger proteins or in very low amounts to be demonstrated.

6 An interesting observation is that, for various coat proteins used as carriers, the genetic background of the host cell differently affects the surface expression of the fusion proteins. CotC-based hybrid proteins are expressed to a higher level in the absence of a wild-type allele of *cotC* but in the presence of *cotU*, coding for a CotC homologue, known to interact with CotC (Isticato *et al.*, 2008). On the other hand, CotB chimeras are only expressed in the presence of a wild-type allele of *cotB*, and the CotG–UreA fusion protein is only expressed in the absence of a wild-type allele of CotG (Hinc *et al.*, 2010b).

Surface display of heterologous proteins on the spore surface of other endospore formers

The first two reports deal with spores of *Bacillus thuringiensis*, which have been engineered to produce micropatterns (Park *et al.*, 2004). Micropatterns are spatially well-defined, two-dimensional microstructures, which can serve as sensors or detectors. These micropatterns have been produced by fusing EGFP to the spore surface protein InhA. Next, these spores were treated with anti-GFP antibodies followed by the addition of protein-A-biotin. These complexes were added to glass slides containing immobilized streptavidin and further analysed under the microscope. Using spores from the same species and insecticidal protoxin Cry1Ac as anchor, the GFP and a single-chain antibody gene have been displayed on the spore surface (Du *et al.*, 2005).

Two collagen-like glycoproteins present on the exosporium surface of *B. anthracis* are BclA and BclB (Steichen *et al.*, 2003; Thompson *et al.*, 2007). The N-terminal domains of both proteins are responsible for the targeting and incorporation of these proteins into the exosporium. These domains were used as targeting domains to incorporate the GFPuv and DsRed reporter fluorescent proteins into the exosporium (Thompson and Stewart, 2008).

References

Ban, J.G., Choi, S.G., Jung, H.C., Kim, B.G., and Kim, J.H. (2003). Method for expression of proteins on spore surface. KR20030065534, South-Korea, Patent.

Bratkovic, T. (2010). Progress in phage display: evolution of the technique and its application. Cell Mol. Life Sci. *67*, 749–767.

Cano, R.J., and Borucki, M.K. (1995). Revival and identification of bacterial spores in 25- to 40-million-year-old Dominican amber. Science *268*, 1060–1064.

Ceragioli, M., Cangiano, G., Esin, S., Ghelardi, E., Ricca, E., and Senesi, S. (2009). Phagocytosis, germination and killing of *Bacillus subtilis* spores presenting heterologous antigens in human macrophages. Microbiology *155*, 338–346.

Chalfie, M., Tu, Y., Euskirchen, G., Ward, W.W., and Prasher, D.C. (1994). Green fluorescent protein as a marker for gene expression. Science *263*, 802–805.

Costa, T., Steil, L., Martins, L.O., Volker, U., and Henriques, A.O. (2004). Assembly of an oxalate decarboxylase produced under sigmaK control into the *Bacillus subtilis* spore coat. J. Bacteriol. *186*, 1462–1474.

Crameri, A., Whitehorn, E.A., Tate, E., and Stemmer, W.P.C. (1996). Improved green fluorescent protein by molecular evolution using DNA shuffling. Nat. Biotechnol. *14*, 315–319.

Date, A., Pasini, P., and Daunert, S. (2007). Construction of spores for portable bacterial whole-cell biosensing systems. Anal. Chem. *79*, 9391–9397.

Date, A., Pasini, P., and Daunert, S. (2010a). Integration of spore-based genetically engineered whole-cell sensing systems into portable centrifugal microfluidic platforms. Anal. Bioanal. Chem. *398*, 349–356.

Date, A., Pasini, P., Sangal, A., and Daunert, S. (2010b). Packaging sensing cells in spores for long-term preservation of sensors: a tool for biomedical and environmental analysis. Anal. Chem. *82*, 6098–6103.

Daugherty, P.S. (2007). Protein engineering with bacterial display. Curr. Opin. Struct. Biol. *17*, 474–480.

De, M.A., and De, M.V. (2004). Bacteria co-transformed with recombinant proteins and chaperones cloned in independent plasmids are suitable for expression tuning. J. Biotechnol. *109*, 45–52.

De, M.A., Deuerling, E., Mogk, A., Tomoyasu, T., and Bukau, B. (2007). Chaperone-based procedure to increase yields of soluble recombinant proteins produced in *E. coli*. BMC. Biotechnol. *7*, 32.

Desvaux, M., Dumas, E., Chafsey, I., and Hébraud, M. (2006). Protein cell surface display in Gram-positive bacteria: from single protein to macromolecular protein structure. FEMS Microbiol. Lett. *256*, 1–15.

Du, C., Chan, W.C., McKeithan, T.W., and Nickerson, K.W. (2005). Surface display of recombinant proteins on *Bacillus thuringiensis* spores. Appl. Environ. Microbiol. *71*, 3337–3341.

Duc, L.H., Hong, H.A., and Cutting, S.M. (2003). Germination of the spore in the gastrointestinal tract provides a novel route for heterologous antigen delivery. Vaccine *21*, 4215–4224.

Duc, L.H., Hong, H.A., Uyen, N.Q., and Cutting, S.M. (2004). Intracellular fate and immunogenicity of *B. subtilis* spores. Vaccine *22*, 1873–1885.

Duc, L.H., Hong, H.A., Atkins, H.S., Flick-Smith, H.C., Durrani, Z., Rijpkema, S., Titball, R.W., and Cutting, S.M. (2007). Immunization against anthrax using *Bacillus subtilis* spores expressing the anthrax protective antigen. Vaccine *25*, 346–355.

Errington, J. (2003). Regulation of endospore formation in *Bacillus subtilis*. Nat. Rev. Microbiol. *1*, 117–126.

Fantino, J.R., Barras, F., and Denizot, F. (2009). Sposensor: a whole-bacterial biosensor that uses immobilized *Bacillus subtilis* spores and a one-step incubation/detection process. J. Mol. Microbiol. Biotechnol. *17*, 90–95.

Fu, L.L., Li, W.F., Du, H.H., Dai, W., and Xu, Z.R. (2008). Oral vaccination with envelope protein VP28 against white spot syndrome virus in *Procambarus clarkii* using *Bacillus subtilis* as delivery vehicles. Lett. Appl. Microbiol. *46*, 581–586.

Grossman, A.D. (1995). Genetic networks controlling the initiation of sporulation and the development of genetic competence in *Bacillus subtilis*. Annu. Rev. Genet. *29*, 477–508.

Gupta, N., and Farinas, E.T. (2010). Directed evolution of CotA laccase for increased substrate specificity using *Bacillus subtilis* spores. Protein Eng. Des. Sel. *23*, 679–682.

Harms, H., Wells, M.C., and van der Meer, J.R. (2006). Whole-cell living biosensors – are they ready for environmental application? Appl. Microbiol. Biotechnol. *70*, 273–280.

Henriques, A.O., Melsen, L.R., and Moran, C.P., Jr. (1998). Involvement of superoxide dismutase in spore coat assembly in *Bacillus subtilis*. J. Bacteriol. *180*, 2285–2291.

Henriques, A.O., and Moran, C.P., Jr. (2000). Structure and assembly of the bacterial endospore coat. Methods *20*, 95–110.

Henriques, A.O., and Moran, C.P., Jr. (2007). Structure, assembly, and function of the spore surface layers. Annu. Rev. Microbiol. *61*, 555–588.

Hinc, K., Ghandili, S., Karbalaee, G., Shali, A., Noghabi, K.A., Ricca, E., and Ahmadian, G. (2010a). Efficient binding of nickel ions to recombinant *Bacillus subtilis* spores. Res. Microbiol. *161*, 757–764.

Hinc, K., Isticato, R., Dembek, M., Karczewska, J., Iwanicki, A., Peszynska-Sularz, G., De Felice, M., Obuchowski, M., and Ricca, E. (2010b). Expression and dispaly of UreA of *Helicobacter acinonychis* on the surface of *Bacillus subtilis* spores. Microb. Cell Fact. *9*, 2.

Ho, M., and Pastan, I. (2009). Mammalian cell display for antibody engineering. Methods Mol. Biol. *525*, 337–52, xiv.

Hoang, T.H., Hong, H.A., Clark, G.C., Titball, R.W., and Cutting, S.M. (2008). Recombinant *Bacillus subtilis* expressing the *Clostridium perfringens* alpha toxoid is a candidate orally delivered vaccine against necrotic enteritis. Infect. Immun. *76*, 5257–5265.

Isticato, R., Cangiano, G., Tran, H.T., Ciabattini, A., Medaglini, D., Oggioni, M.R., De, F.M., Pozzi, G., and Ricca, E. (2001). Surface display of recombinant proteins on *Bacillus subtilis* spores. J. Bacteriol. *183*, 6294–6301.

Isticato, R., Esposito, G., Zilhao, R., Nolasco, S., Cangiano, G., De, F.M., Henriques, A.O., and Ricca, E. (2004). Assembly of multiple CotC forms into the *Bacillus subtilis* spore coat. J. Bacteriol. *186*, 1129–1135.

Isticato, R., Di Mase, D.S., Mauriello, E.M., De, F.M., and Ricca, E. (2007). Amino terminal fusion of heterologous proteins to CotC increases display efficiencies in the *Bacillus subtilis* spore system. BioTechniques 42, 151–2, 154, 156.

Isticato, R., Pelosi, A., Zilhao, R., Baccigalupi, L., Henriques, A.O., De, F.M., and Ricca, E. (2008). CotC–CotU heterodimerization during assembly of the *Bacillus subtilis* spore coat. J. Bacteriol. *190*, 1267–1275.

Isticato, R., Pelosi, A., De, F.M., and Ricca, E. (2010). CotE binds to CotC and CotU and mediates their interaction during spore coat formation in *Bacillus subtilis*. J. Bacteriol. *192*, 949–954.

Kim, H., Hahn, M., Grabowski, P., McPherson, D.C., Otte, M.M., Wang, R., Ferguson, C.C., Eichenberger, P., and Driks, A. (2006). The *Bacillus subtilis* spore coat protein interaction network. Mol. Microbiol. *59*, 487–502.

Kim, J., and Schumann, W. (2009). Display of proteins on *Bacillus subtilis* endospores. Cell Mol. Life Sci. *66*, 3127–3136.

Kim, J.H., Lee, C.S., and Kim, B.G. (2005). Spore-displayed streptavidin: a live diagnostic tool in biotechnology. Biochem. Biophys. Res. Commun. *331*, 210–214.

Kim, J.H., Roh, C., Lee, C.W., Kyung, D., Choi, S.K., Jung, H.C., Pan, J.G., and Kim, B.G. (2007a). Bacterial surface display of GFP(uv) on *Bacillus subtilis* spores. J. Microbiol. Biotechnol. *17*, 677–680.

Kim, J.H., Wong, S.L., and Kim, B.G. (2007b). Optimization of staphylokinase production in *Bacillus subtilis* using inducible and constitutive promoters. Biotechnol. Bioprocess. Eng. *6*, 167–172.

Knecht, L.D., Pasini, P., and Daunert, S. (2011). Bacterial spores as platforms for bioanalytical and biomedical applications. Anal. Bioanal. Chem. In press.

Kwon, S.J., Jung, H.C., and Pan, J.G. (2007). Transgalactosylation in a water-solvent biphasic reaction system with beta-galactosidase displayed on the surfaces of *Bacillus subtilis* spores. Appl. Environ. Microbiol. 73, 2251–2256.

Lee, S., Belitsky, B.R., Brinker, J.P., Kerstein, K.O., Brown, D.W., Clements, J.D., Keusch, G.T., Tzipori, S., Sonenshein, A.L., and Herrmann, J.E. (2010a). Development of a *Bacillus subtilis*-Based Rotavirus Vaccine. Clin. Vaccine Immunol. *17*, 1647–1655.

Lee, S., Belitsky, B.R., Brown, D.W., Brinker, J.P., Kerstein, K.O., Herrmann, J.E., Keusch, G.T., Sonenshein, A.L., and Tzipori, S. (2010b). Efficacy, heat stability and safety of intranasally administered *Bacillus subtilis* spore or vegetative cell vaccines expressing tetanus toxin fragment C. Vaccine *28*, 6658–6665.

Lee, S.Y., Choi, J.H., and Xu, Z. (2003). Microbial cell-surface display. Trends Biotechnol. *21*, 45–52.

Li, G., Tang, Q., Chen, H., Yao, Q., Ning, D., and Chen, K. (2011). Display of *Bombyx mori* nucleopolyhedrovirus GP64 on the *Bacillus subtilis* spore coat. Curr. Microbiol. *62*, 1368–1373.

Luiz, W.B., Cavalcante, R.C., Paccez, J.D., Souza, R.D., Sbrogio-Almeida, M.E., Ferreira, R.C., and Ferreira, L.C. (2008). Boosting systemic and secreted antibody responses in mice orally immunized with recombinant *Bacillus subtilis* strains following parenteral priming with a DNA vaccine encoding the enterotoxigenic Escherichia coli (ETEC) CFA/I fimbriae B subunit. Vaccine *26*, 3998–4005.

Makela, A.R., and Oker-Blom, C. (2008). The baculovirus display technology– an evolving instrument for molecular screening and drug delivery. Comb. Chem. High Throughput. Screen. *11*, 86–98.

Mauriello, E.M., Duc, L.H., Isticato, R., Cangiano, G., Hong, H.A., De, F.M., Ricca, E., and Cutting, S.M. (2004). Display of heterologous antigens on the *Bacillus subtilis* spore coat using CotC as a fusion partner. Vaccine 22, 1177–1187.

Michelfelder, S., Lee, M.K., Lima-Hahn, E., Wilmes, T., Kaul, F., Muller, O., Kleinschmidt, J.A., and Trepel, M. (2007). Vectors selected from adeno-associated viral display peptide libraries for leukemia cell-targeted cytotoxic gene therapy. Exp. Hematol. 35, 1766–1776.

Nguyen, D.H., Phan, T.T.P., and Schumann, W. (2007). Expression vectors for the rapid purification of recombinant proteins in *Bacillus subtilis*. Current Microbiol. 55, 89–93.

Nguyen, Q.A., and Schumann, W. (2011). Anchoring recombinant proteins on the *Bacillus subtilis* spore surface using an IPTG-inducible promoter. Protein Expr. Purif. In press.

Paccez, J.D., Luiz, W.B., Sbrogio-Almeida, M.E., Ferreira, R.C., Schumann, W., and Ferreira, L.C. (2006). Stable episomal expression system under control of a stress inducible promoter enhances the immunogenicity of *Bacillus subtilis* as a vector for antigen delivery. Vaccine *24*, 2935–2943.

Paccez, J.D., Nguyen, H.D., Luiz, W.B., Ferreira, R.C., Sbrogio-Almeida, M.E., Schuman, W., and Ferreira, L.C. (2007). Evaluation of different promoter sequences and antigen sorting signals on the immunogenicity of *Bacillus subtilis* vaccine vehicles. Vaccine 25, 4671–4680.

Park, T.J., Lee, K.B., Lee, S.J., Park, J.P., Lee, Z.W., Choi, S.K., Jung, H.C., Pan, J.G., Lee, S.Y., and Choi, I.S. (2004). Micropatterns of spores displaying heterologous proteins. J. Am. Chem. Soc. *126*, 10512–10513.

Paschke, M. (2006). Phage display systems and their applications. Appl. Microbiol. Biotechnol. *70*, 2–11.

Pepper, L.R., Cho, Y.K., Boder, E.T., and Shusta, E.V. (2008). A decade of yeast surface display technology: where are we now? Comb. Chem. High Throughput. Screen. *11*, 127–134.

Permpoonpattana, P., Hong, H.A., Phetcharaburanin, J., Huang, J.M., Cook, J., Fairweather, N.F., and Cutting, S.M. (2011). Immunisation with *Bacillus* spores expressing toxin A peptide repeats protects against infection with toxin A+ B+ strains of *Clostridium difficile*. Infect. Immun. In press.

Piggot, P.J., and Hilbert, D.W. (2004). Sporulation of *Bacillus subtilis*. Curr. Opin. Microbiol. 7, 579–586.

Potot, S., Serra, C.R., Henriques, A.O., and Schyns, G. (2010). Display of recombinant proteins on *Bacillus subtilis* spores, using a coat-associated enzyme as the carrier. Appl. Environ. Microbiol. *76*, 5926–5933.

Ricca, E., and Cutting, S.M. (2003). Emerging applications of bacterial spores in nanobiotechnology. J. Nanobiotechnology *1*, 6.

Sidhu, S.S., and Koide, S. (2007). Phage display for engineering and analyzing protein interaction interfaces. Curr. Opin. Struct. Biol. *17*, 481–487.

Sorensen, S.J., Burmolle, M., and Hansen, L.H. (2006). Making bio-sense of toxicity: new developments in whole-cell biosensors. Curr. Opin. Biotechnol. *17*, 11–16.

Sousa, C., Cebolla, A., and de, L., V. (1996). Enhanced metalloadsorption of bacterial cells displaying poly-His peptides. Nat. Biotechnol. *14*, 1017–1020.

Stahl, S., and Uhlén, M. (1997). Bacterial surface display: trends and progress. Trends Biotechnol. *15*, 185–192.

Steichen, C., Chen, P., Kearney, J.F., and Turnbough, C.L., Jr. (2003). Identification of the immunodominant protein and other proteins of the *Bacillus anthracis* exosporium. J. Bacteriol. *185*, 1903–1910.

Tam, N.K., Uyen, N.Q., Hong, H.A., Duc, L.H., Hoa, T.T., Serra, C.R., Henriques, A.O., and Cutting, S.M. (2006). The intestinal life cycle of *Bacillus subtilis* and close relatives. J. Bacteriol. *188*, 2692–2700.

Tanner, A., Bowater, L., Fairhurst, S.A., and Bornemann, S. (2001). Oxalate decarboxylase requires manganese and dioxygen for activity. Overexpression and characterization of *Bacillus subtilis* YvrK and YoaN. J. Biol. Chem. *276*, 43627–43634.

Thompson, B.M., and Stewart, G.C. (2008). Targeting of the BclA and BclB proteins to the *Bacillus anthracis* spore surface. Mol. Microbiol. *70*, 421–434.

Thompson, B.M., Waller, L.N., Fox, K.F., Fox, A., and Stewart, G.C. (2007). The BclB glycoprotein of *Bacillus anthracis* is involved in exosporium integrity. J. Bacteriol. *189*, 6704–6713.

Uyen, N.Q., Hong, H.A., and Cutting, S.M. (2007). Enhanced immunisation and expression strategies using bacterial spores as heat-stable vaccine delivery vehicles. Vaccine *25*, 356–365.

Wu, S.C., Yeung, J.C., Duan, Y.J., Ye, R.Q., Szarka, S.J., Habibi, H.R., and Wong, S.-L. (2002). Functional production and characterization of a fibrin-specific single-chain antibody fragment from *Bacillus subtilis*: Effects of molecular chaperones and a wall-bound protease on antibody fragment production. Appl. Environ. Microbiol. *68*, 3261–3269.

Xu, X., Gao, C., Zhang, X., Che, B., Ma, C., Qiu, J., Tao, F., and Xu, P. (2011). Production of *N*-acetyl-D-neuramic acid using an efficient spore surface display system. Appl. Environ. Microbiol. 77, 3197–3201.

Yagi, K. (2007). Applications of whole-cell bacterial sensors in biotechnology and environmental science. Appl. Microbiol. Biotechnol. 73, 1251–1258.

Zhou, Z., Hu, X., Huang, Y., Hu, H., Ma, C., Chen, X., Hu, F., Xu, J., Lu, F., Wu, Z., *et al.* (2007). Molecular cloning and identification of a novel *Clonorchis sinensis* gene encoding a tegumental protein. Parasitol. Res. *101*, 737–742.

Zhou, Z., Xia, H., Hu, X., Huang, Y., Li, Y., Li, L., Ma, C., Chen, X., Hu, F., Xu, J., *et al.* (2008). Oral administration of a *Bacillus subtilis* spore-based vaccine expressing *Clonorchis sinensis* tegumental protein 22.3 kDa confers protection against *Clonorchis sinensis*. Vaccine *26*, 1817–1825.

Zilhao, R., Naclerio, G., Henriques, A.O., Baccigalupi, L., Moran, C.P., Jr., and Ricca, E. (1999). Assembly requirements and role of CotH during spore coat formation in *Bacillus subtilis*. J. Bacteriol. *181*, 2631–2633.

Zilhao, R., Serrano, M., Isticato, R., Ricca, E., Moran, C.P., Jr., and Henriques, A.O. (2004). Interactions among CotB, CotG, and CotH during assembly of the *Bacillus subtilis* spore coat. J. Bacteriol. *186*, 1110–1119.

Index

A

abrB 10, 12
acetylmuramic acid 115
Acid-soluble spore proteins (SASP) 75–76, 122
  α/β-type SASPs 76–82
Acoustic 223
Actinomycin D 27
Adenosine 112
Advantages of heterogeneity 211
Aerial hyphae 228
African green monkeys 172
Alanine 111–112
Alanine racemase 5
Aldolase 30
Alkaline phosphatase (AP) 245
Alternative vaccine delivery methods 182
Aluminium adjuvant 180–183
Amino acid 90–91, 188
  alanine 188
  histidine 188
Amplified fragment length polymorphism (AFLP) method 243
Anthrax 1, 108, 169–170, 176, 178, 180, 238, 240, 243
  cutaneous 108
  detection 248
  oropharyngeal 108
  disease 170
  toxins 248
  vaccine 2, 182, 185, 245
Anti-capsule vaccines 184
Anti-PA antibody 245
Anti-spore vaccination strategies 185
Antibody 244, 247–249
  capture and detection 248
Antibiotics 169, 176–177, 180
  production of 10
  resistance to 10, 12, 180
Antigen 244, 248
  capture
    dipstick assay 244
    ELISA 245
Antimicrobial peptide 174
Appendages 239–240
Applied Biosystems gas-phase sequencer 242
Archangium 228
Archebiosis 4
Aretaeus of Cappadocia 2
Asparagine 111
Assay
  chromatin precipitation 10
  electrophoretic mobility shift 10, 11
  transcription profiling 10
Atomic force microscopy (AFM) 57, 63, 218
atxA 222
Automation of assay procedures 237
Azotobacter 216

B

Bacillus 73, 76, 82, 89–94, 98, 100, 102–103, 182, 216
  Bacillus alvei 219
  Bacillus anthracis 74, 107–110, 112–114, 169, 170–180, 182, 184–185, 188–189, 216, 237–240, 242–250
    7702 183
    9602 186
    Ames 173–174, 242
    aerosol challenge 182–186
    detection through immunological technique 244
    disease progression within macrophages 173
    DNA vaccine 185
    exosporium 242
    infection 176
    isolates 176, 179
    location 242
    spore resistance 240
    protective antigen (pag) toxin 248
    spores 171, 174–178, 180, 183, 186, 240, 244, 247–248, 250
    spore detection 248
    spore fluorescence 247
    spore profiles 246
    STI-1 183–184
    strain 169, 171, 176, 177, 179–180, 186
    subgroups 243
    vaccine 185–186
    vegetative cells 246
    Vollum spores 183, 247
    gerX 112–113
    gerA 112

gerY 112
gerAA 112
gerYB 112
gerYC 112
Bacillus atrophaeus 241, 253
spore detection using microscopy method 241
Bacillus brevis 219
Bacillus cereus 74, 77, 111, 171, 172, 217, 237–244, 246–250
appendages 240
diarrheal food-borne illness 239
Emetic food-borne illness 238–239
exosporium 186, 240
long-incubator form 238
pore profiles 246
short-incubation form 238
spore surface 244
spores 247, 250
spores antibodies 244
spores detected using microscopy method 241
Bacillus circulans 219
Bacillus clausii 218
Bacillus coagulans 219
Bacillus genera 107
Bacillus globigii spores 245
Bacillus licheniformis 219
Bacillus macerans 219
Bacillus megaterium 77, 110–111, 113, 218, 238–239, 248
gerUA 110
gerUC 110
gerVB 110
gerWB 110
Bacillus mycoides 219, 238–239
Bacillus of Nicolaier 3
Bacillus pseudomycoides 219
Bacillus pumilus 218
Bacillus species 169
Bacillus spore 180
Bacillus subtilis 9, 10, 15, 25–27, 73–80, 82, 110, 114, 216, 240, 248–249, 250, 253, 263
developmental states of 9
exosporium 240
GerA 110–111
GerB 110–111
GerK 110–111
spore coat 240
spores 249, 253
Bacillus thuringiensis 217, 218, 238–239, 240, 248
Bacillus weihenstephanensis 219
Basal layer 64–66, 68
BclA 219
vaccine 188
β-galactosidase 29
Bicistronic 114
Biocides
spores as 6
Bioinformatics 97
Biological markers 242, 249–250
Biological warfare weapons 237–239
Biomarkers *see* biological markers
Bioparticles 263, 264.
Biosensor 237, 247–249, 255
categories 248
development problems 255
direct detection 248
groups 247
importance 247
indirect detection 248
integrating waveguide 255
potentiometric 249
spore 255
system 255
technology 247
Biosynthesis
exopolysaccharide 12
cytochrome 12
Biowarfare agents (BWA) 247
BL genes 225
Black Bain 2
Book of Georgics 1
Bottle cap model (deposition of proteins in the basal layer) 66
Botulinum neurotoxin (BoNT) 243
Botulinum toxin 169
botulism 3
botulus 3
Bronchoalveolar lavage (BAL) 110
Buoyant density gradient centrifugation 201
Byssophaga 228

C

Ca-DPA 76–77, 81. *See also* DPA
content 202, 203
release 201, 205–207
Ca2+-DPA *see* calcium-dipicolinate
Calcium-dipicolinate (Ca2+-DPA) 121–122, 124, 127–128, 133, 137, 139, 216
Callithrix jacchus 172
Cantilever 223
capBCAD 176
Capture ELISA 245
Carbuncles (see Charbonous diseases)
Casamino acid 34
CFLE 126
CFU/ml 220
Charbonous diseases 2
Chemokine receptors 175
Chemotaxis 11
Chondromyces 228
Chromosome remodeling 10
Clostridia 9, 73, 97, 111, 225
Clostridial 169
SpoA 226
toxins A 226
toxins B 226
Clostridium 76, 89–91, 93, 98, 100, 103, 216
Clostridium acetobutylicum 225
Clostridium beijerinckii 225
Clostridium botulinum 111, 169, 216, 225, 227
type A exosporium 240
Clostridium butyricum 248

Clostridium difficile 111, 169, 216, 225, 227
Clostridium genera 107
Clostridium mangenoti 79
Clostridium perfringens 77, 111, 113, 225
cortex depolymerization 133. *See also* cortex
enterotoxin (CPE) 226
food poisoning 238
germination 133
Clostridium phytofermentans 225
Clostridium specis 169
Clostridium tetani 169, 225
Clostridium thermocellum 225
Clostridium tyrobutyricum
spore detection 243
vegetative cell detection 243
Coat 57–68, 74, 147–155, 157
appendage 64
layer 121
proteins 5, 59–68
water in 148–150
Competence 11
Conidia 216
Core 5, 144–155, 157–158
water content 144–146, 202, 203–204
Cortex 5, 22, 23, 25, 26, 31, 35, 75, 122, 144–145, 147–152, 154–155, 157–158, 215. *See also* Peptidoglycan and Outer layer
breakdown 124
depolymerization in *Clostridium perfringens* 133
formation 9
hydrolysis 128
lytic activity 125, 134
lytic enzymes 91, 92, 100–101, 124, 127
generic peptidoglycan 138
general properties 124
peptidoglycan 132, 138
role of 147–148
structure modification 126
Cortical fragmant lytic enzymes (CFLEs) 125, 133–135
CotA/B/C/D/E/G/Q 264, 265, 266, 267, 268, 269
cross-linking, of coat proteins 60–61
Cry Toxin 225
cspABC genes 133. *See also* CspA, CspB, CspC
mutant 133
csp-sleC genetic locus 133
Clostridium perfringens 133
protease 133, 135
CwlD spore 129
CwlJ 130–131, 136–137
selB 128. *See also* selB
denaturation 131
homologue 135
muramidase activity 131
mutant 128
product 127
protein of *B. subtilis* 134
Cynomolgus macaques 172
Cystobacter 228
Cysts 216, 238
Cyt Toxin 225
Cytometry 220

D

d-alanine 5
d-cycloserine 24
DAP 24,25
Dendritic cells 109–110, 169
in disease progression 175
Desiccation resistance 78
Desulfosporomusa 216
Detection 237
DIC microscopy 200, 205
Dictyostelium disciodeueum 32
Dideoxyglucopyranoide 65. *See also* anthrose
Differential interference contrast microscopy *see* DIC microscopy
Differentiation 239
Dipicolinic acid 5–6, 76–77, 80–81, 252–253, 254. *See also* Ca-DPA
detection 253
Dipteran larvae 63
Direct culture 241
Direct impact corona ionization (DICI) 247
DNA
repair enzymes 78
replication 11
vaccines 185
DNA–DNA hybridization 243
DPA *see* dipicolinic acid
Draining lymph nodes 175
Drug delivery 250, 252
Dry heat resistance 77
Dry powder vaccine 184

E

EA1 220
Edman degradation 242
EDS2000 252–253
Electrical conductivity *see* spore electrical conductivity
Electrically active polyaniline coated magnetic nanoparticles (EAPM) 223
Electro-chemiluminescent detector 248
Electrochemical sensors 247
Electromagnetic radiation 254
Electron microscopy 201
Emetic toxin genes 225
Emile von Ermengem 3
Endogen 24
Endospore 19,22,24,25,26,34, 215, 253. *See also* Spores
bacterial 121
development of in bacilli 9
formation of in bacilli 9
Epimerase 115
Epithelial cells 216
Epitopes 244
Epizootics 1–2
Escherichia coli 79, 245
Eusebius' epidemic 1–2
Exodus 1
Exogen 23, 27
Exogenous lysozyme 128
Exopolysaccharide 30
Exosporial proteins 57, 65–68

Exosporium 5, 57–59, 63–68, 74, 123, 150–151, 154, 216, 218, 239–240. *See also* Outer layer
fragments 244
glycoprotein *Bacillus* 173

F

False positive exclusion 254
Feed forward loop see Forward feedback loop
Fibre core 254
integral intensity 254
transmission 254
Filament, in Streptomyces spores fruiting body 6
Firmicutes 73
Flow cytometry 201, 246
cost 247
Bacillus species 247
Fluorescence sandwich immunoassay 248
Fluorescent immunoassay (FIA) 248
Fluorophore-assisted carbohydrate electrophoresis (FACE) 250
Flux 143, 146, 148–151, 154, 157
of ions 151, 154, 157
of water 148–151
Food contamination 245
Food quality control 255
Food safety 238
Food poisoning 237, 238
definition 238
symptoms 238
Food-spoilage 237, 238
Forespore 9–15, 59
developmental states 9
sigma factor σF 10
sporulation cycle 9
Forward feedback loops (FFL)
coherent 10, 11, 13, 14, 15
incoherent 10

G

Galactosamine 246
Galactose 246
Gastrointestinal anthrax 171, 176
Gene expression 9, 10, 11, 13, 14, 15
Gene regulation 9, 10
Genes 92
gerAA 92
gerAB 92
gerAC 92
GeneXpert Q-PCR DNA detection system 222
Ger receptor 138
gerA receptors 216
Germ cell wall 75
Germination 81–82, 143, 145, 147–148, 151–152, 154–155, 157–159
at the site of infection 175
chemical 124
receptors 74
germination-specific protease (GSP) 133
germination-triggered reaction 121
protease Gpr. 122
Germination-specific lytic enzymes (GSLEs) 113–114, 124–125, 138
Bacillus 126
conservation 134
fluorescence spectroscopy 125
function 134
immune-electron microscopy 125
GerQ 130, 135
cross-linking 130
homologue 135
GFP fusion 59
Global regulators 74
Glucosamine 25
Glucosaminidase 115
Glucose 98
effect on coat exosporium 66
Glycoconjugate-based ligands 250, 252
Glycosylation 57, 65
Glyoxylate 27
Gram-positive species 136
GroEL protein 240
Guinea pig model 172

H

Haemolymph 63
Haemolysin 225
Haliangium 228
Halobacterium halobium 79
Heat activation 203, 205, 206
Heterogeneity
advantages 211
physical and chemical 202–203
physiological 203–204
reasons for 210–211
spore germination 204–206
spore outgrowth 206
Heterologous 263, 264, 266, 267, 268, 269.
HGMF 226
Hippocrates 2
Histidine 112–113
History of anthrax vaccines 177
HPLC 24
Human peripheral mononuclear cell 250–251
Human Vaccine 179
Hydrogel-liposome assemblies *See* drug delivery systems
Hylangium 228

I

Identification 237
Immune
cell activation 252
therapeutics *See* immune vaccines
vaccines 252
Immonofluorescence assay 247
Immuno-electron microscopy 129
Immuno-radiometric (IRMA) assays 220
Immunochromographic assays 244
Immunodiagnostic techniques 244
Immunofluorescence (IF) 220
Immunogold labeling 65
Immunomagnetic assay system (IMAS) 248

Improvement of AVA through use of a different adjuvant 180
Imunofluorescence 246
Inactivated spore vaccine 186
Indirect haemagglutination assay 245
Inflammatory response 174
Inner membrane (of spores) 5, 75–76. *See also* spore inner membrane
Interspace 67
Intradermal 216
Ionic hydrogels 253
Ionizing radiation resistance 80
Ions
  flux of *see* Flux of ions
  roles of 154–157. *See also* Mineral content of spores
  transport of 157–160
Isocitrate dehydrogenase 27
IsoFA 27

J

Johann Heinrich Ferdinand Autenrieth 3
Justinus Kerner 3

K

Killing factors 10
Kin genes *see* sensor kinases
Kinetics 253–254

L

Lactobacillus 73, 182
  gasseri vaccine 183
  acidophilus vaccine 183
  plantarum 248
Lactococcus lactis 248
Lag period 200, 205, 209
Laminar 224
Lanthanide-macrocycle binary complexes 253
Laser Doppler vibrometer (LDV) 223
Laser tweezers Raman spectroscopy (LTRS) 201, 204, 205
Lethal toxin 170
Lipobeads 250
  structure 250
Liposome 249–250
Listeria 73
Luminex Mag-Plex microsphere 245
Lymph nodes 108–110
LysM
  domain of SleL 135
  motif 132
Lysozyme 102, 115, 123, 126, 215
  sensitivity 66–67
Lytic 107–113
  bacilli 121–122
  clostridia 121–122
  enzyme 121–122, 136
  enzyme activation 138
  identity 121
  mechanism 121
  protein 125, 133
  transglycosylase activity 127

M

Macrophages 109–110
  after phagocytosis 174
  release of vegetative bacilli 175
Mannitol egg yolk polymyxin (MYP) 224
MAPK cascade 175
Mass sensors 247, 250
Mechanisms
  of bacterial spore survival 73
  of spore resistance 77
Mellitangium 228
Membrane-strip-based biosensor assay 248
Michaelis-Menten 99
Microbial identification system 246
Microscopy
  atomic force 241–242
  combined with spectroscopy 254
  external phase contrast 254
  fluorescence 246
  optical 241–242
  scanning electron 241–242,249
Microsphere-based flow cytometric assays 245
Mineral content 151–154
MLVA of *Bacillus anthracis* 243
Mobile ion concentrations 254
Molecular markers 237
Monomorphism 243
Mother cell 9–15
  inactive sigma factor σE 10
  lysis of 9, 15
  sporulation cycle 9
Motility 11
Multiplex PCR 242
  spore detection 243
Muramic 114
  acid 24, 25, 75, 122
  δ-lactam 25, 121–122, 125–126, 130
Muramidase 24
Murine 250–251
Muropeptide 24
Mycelia 216
Myxobacterium 19, 82
  smegmatis 78
  tuberculosis 78
Myxobolus
  cerebralis 228
  kofleria 228
  stipitatis 228
Myxococcus 216, 227
  xanthus 19–24,31–35
Myxospore 20–24, 27, 28, 31, 34

N

N-acetyl-muramic acid 24
N-acetylgalactosamine 219
N-acetylglucosaminidase 126–127, 132
Nannocystis 228
Nano small ion mass spectroscopy (nano-SIMS) 201
nap 64–65, 68
NASBA amplification 222
Nattō 6

Neonates 169
Network charge 254
Neutralize toxins 252
Nitric oxide 175
- stress 174
- synthase 174

Non-enzymatic glycosylation 244
Non-homologous-end joining 78
Non-nutrient germinants 254
Nucleic acid sequence-based amplification (NASBA) 248
Nucleotide excision repair 82
Nutrient germinants 254
- receptors (nGRs) 204–205, 207–210

O

O-methylated pentoses 246
Oedema toxin 170
Operon 30,33, 92
- physiology 90
- bicistronic of gerQ 130

Optical
- fibre 254
- sensors 247
- tweezers 254

Oral administration (of vaccines) 182, 183
Oscillospira 216
Osmotic stress 138
Outer membrane 5
Oxidation of tyrosine 60. *See also* dityrosine
Oxidizing agents 79–80

P

pag 222
- promoter 110

Panzootic 2
Paramagnetic Dynal bead-based sandwich immunoassay 245
Passive immunization 185
Pasteur vaccine 177–178
Pathogen 107, 237–238, 240–241, 245
- determination 247
- spores 241
- pathogenesis 107
- pathogenicity 108

PCR assay 241–242
- *B. anthracis* detection 241–242
- C. tyrobutyricum detection 241

Peptidoglycan 121, 124
PepSY
- domain 130
- motif 130

Peptidoglycan 5, 23–25, 113–114
Permease 30
Petpidoglycan 75
Phage-based magnetoelastic (ME) biosensors 249
Phage-based magnetoelastic wireless biosensor 223
Phago-endosomal acidification 174
Phagocytes 108–110
Phagocytosis of spores by macrophages 173
Phagolysosomes 109
Phase contrast microscopy 200, 201, 205
Phenotypic analysis of mutants 127
Phosphatases Rap proteins 12
Phosphorylation system 12
Phylloplane 61
Physical and chemical heterogeneity 202–203
Physiological heterogeneity 203–204
Piezoelectric-excited millimetre-sized cantilever (PEMC) sensors 224, 248
Pigments 80
Polyangium 228
polyclonal 109
Polycyclic isoprenoids 80
Polyglutamic acid 170
Polymerase chain reaction (PCR) 241
Polymyxin egg yolk mannitol bromothymol blue agar (PEMBA) 224
Post-exponential phase 10
Potentiometric method of DPA detection 252
Predivisional cell σH 10
PrkC 91, 98, 100
Protease ClpC 13
Protection from nitric oxide 174
Proteolytic cleavage, in coat protein deposition 60
Proteolytic 21
Pseudomembranous colitis 3, 169
Pseudomonas aeruginosa 31
'Pull-down' assays 61
Pulmonary 216
pXO1 113, 217
pXO2 217
Pyrolysis mass spectrometry (PyMs) 247
Pyxicoccus 228

Q

Quantum dots (Q-dots) 220
Quartz crystal microbalance (QCM) 224

R

Radiometric measurement of 14CO2 244
Radiometric measure detection time
- spores 244
- vegetative cells 244

RAW 264.7 cells 110
Receptor 92–94, 96–100, 102
- germinant 9, 13

Recombinant 263, 265, 266, 267, 268.
- PA-based vaccines in development 180
- protective antigen vaccine 181

Regulation, negative 11
Regulatory Protein 20
Rhamnopyranoside 65
Ribotyping 242
RNAP 28
RNase III 41
RocA protein 240
rPA vaccine 181–182, 184
rRNA gene sequences 239
Ruthenium red 57, 65
- stain 220

S

Salmonella 182
  acedomyceticus 229
  albidoflavus 229
  alboniger 229
  albosporeus 229
  albus 229
  anulatus 229
  coelicolor 228
  diastaticus 229
  griseoplanus 229
  olivaceus 229
  typhimurium 249
SafA 131
Sap S-layer 220
SASP 6, 75–76. *See also* α/β-type SASPs, Acid-soluble spore proteins
  α-β analyses 219
  profile 219
Schaeffer–Fulton stain 226
Schiff base 244
S1eB protein 126–127, 129, 134–135, 138. *See also* cortex lytic enzyme
  enzymatic domain 134
S1eC 133, 135–136, 138
  mutant 133
S1eL 126, 131–132, 135–136
  cwlJ 128
  homologue 135
  protein 128
SleM 133–136
  muramidase activity 134
SDS-PAGE 59–60, 64, 242
selC 91
Sensitivity to lysozyme 59
Sensor kinase 10, 12
Septicaemia 170–171
Septum 9, 10, 12–14
Serine 112–113
Sig B Regulatory 28
Sigma factor 5, 9–15, 28, 31
  activation *see* coherent feed forward loop
  repression *see* incoherent feed forward loop
Silica, in the spore coat 64
Single spore manipulation 254
Small acid-soluble proteins (SASP) 240
SOD genes 174
Sorangium 228
Spectral pattern recognition 247
Spectroscopy micro-Raman 254
Spo 0A-P 12, 13
SpoA-P 10
Spodoptera frugiperda (Sf-9) cell lysates 245
spoIIAA 12, 13
spoIIAB 12, 13
SpoIIB 13
SpoIID protein 134
SpoIIID 14
SpoIIID 226
SpoIIIE 14
Sporangia 239
Sporangiospores 216
Sporangium 133
Spore 237, 239–255, 263, 264, 266. *See also* Sporulation
  architecture 74
  coat 239–240
    coat-permeabilized 125
    coat proteins 9
  contamination 252
  core 75–76, 121. *See also* core
  cortex 13, 121. *See also* cortex
    Bacillus 127
    germination 121
    hydrolysis 136
    lytic enzyme (SCLEs) 122, 125, 133, 135, 138
    modification 135
    structure 122
    synthesis 122
  cultures 246
  decontamination 252
  defensive decontamination 250
  dehydration 240
  detection 237, 249–250, 253–255
    using chemiluminescence techniques 248
    enrichment method 241
    genetic methods 237
    immunological methods 237
    microbiological method 237, 240, 244
    microscopy method 241–242
    radiometric technique 244
  diagram 239
  differentiation 250
  dimensions 254
  disinfectants 250
  dispersion 243
  dormant 121–122, 253
  electrical conductivity 151
  forespore 122
  germination 74, 89–91, 97, 102, 121, 245, 249, 254
    changes in molecular components 254
    effectors 188
    inhibition 252
  identification 249–250, 252, 254
    using infrared spectroscopy 246
    using instrumental method 246
    using physico-chemical methods 246
  inner membrane 145–149, 150–151, 157–158
  inoculums 244
  interaction with electrode 253
  ion sensitive 254
  membrane receptors 249
  mineral content 76. *See also* mineral content
  mineralization 76
  photoproduct (SP) 76, 81–82
    SP lyase 82
  potentiometric measurements 253
  proteins as vaccine candidates 186
  quantitative model 207–210
  recognition element 253
  release of 9
  resistance 240
  sensors 250

shrinking 254
size 202
specificity 255
structure 9, 239, 250
swelling 254
synthesis 122
water content 75
Sporolactobacillus 216
Sporosarcina 216
Sporulation 9, 12, 19–21, 23, 26, 29, 30–32, 34, 35, 113, 143–145, 147, 150, 152–159, 215, 263, 267. *See also* Spores/endospores
cell differentiation in 9
delay in 10
network 74
regulatory network 74
schematic 239
sigmas 11
Sporulenes 80. *See also* Polycyclic isoprenoids
SpoVID 131
sspB promoter 110
Staining 202
Standard mouse bioassay (SMB) 243
Staphylococcus 73
Sterne Vaccine 178
Stigmatella aurantiaca 228
Stigmetella auraantiaca 228
Stigmetella erecta 228
Streptavidin 265, 269
Streptococcus 73
Streptomyces 39, 82, 216, 228
ABC-transporter 42
aerial hyphae 39–46, 49
actinorhodin 39, 40
autolytic enzymes 50
RpfA 50
Swl 50
bld gene 40, 41, 50
BldD 40, 41
cell division 43–45
localization of proteins 44
chaplins 41, 42
domain 42
chromosome condensation 40, 42, 43 48
chromosome segregation 40, 43, 45, 46
CrgA 44
*crg*A gene 44, 45
mutant 45
development 39
DivIC 44, 45
DivIVA 39
DNA condensation 48
Dps-like proteins 48, 49
dps loci 49
dpsA 49
dspB 49
dspC 49
extra-cytoplasmic function 41
ECF *See* Extra-cytoplasmic function
extra-cytoplasmic protein 50
Foci 46
fts gene 43–45, 47, 48
mutant 45, 47
FtsI 50
FtsI 44
FtsK 46- 48
FtsL 44, 45
FtsQ 44, 45
FtsW 44
FtsZ 43–46, 48, 50
GntR family transcription factor 42
HU-like proteins 48
genes 48
HupA 48
HupS 48
integral membrane proteins 50
lantibiotic synthase 42
leucyl tRNA 41
membrane proteins 45
MreB 46, 49, 50
MreC 50
MreD 50
mre gene 49, 50
MreB-like proteins 50
Mbl 50
SCO 6166 50
nucleoprotein complex 45
ParA 45–47
ParAB 47
ParB 45, 46, 48
ParJ 46, 47
par gene 45–48
par system 48
PBP 2 50
peptidoglycan hydrolases 49
hydrolase mutant 49
RpfA 49
SwlA 49
SwlB 49
SwlC 49
polyketide-derived pigment 40, 43
pbp 2 50
prespore compartments 40, 46
prsH 41
*ram*CSAB operon 41
ram genes 41, 42
RasP 45
RodZ 50
SALPs 42, 44, 50
mutant 45
SsgA 42, 44, 45
SsgB 42, 44
SsgE 50
SsgF 50
SCO 3901 50
SapB 41, 42
ScpA 48
ScpB 48
SEDS 43
Sfr 50
sfr 50
SffA 46–48

SmeA 46–48
smeA-sffA operon 47, 48
septal ingrowth 45
septation 40, 42, 44, 46, 47
sigH 41
sigma factor 41, 43
Anti-anti 41
Late sporulation-specific 42, 43
Smc locus 48
SMC proteins 48
spore maturation 40, 48, 49
spore pigmentation 43
sporulation 39–43
classical regulators 40
gene products 42
SsgA-like proteins *See* SALPS
Streptomyces coelicolor 39, 40
genome 41
sigma factors 41
structural maintenance of chromosomes *See* SMC
transcriptional regulators 43
undecylprodigiosin 39, 40
vegetative growth 39
vegetative mycelium 39
whi gene 40, 42–44, 50
early *whi* mutation 42, 48
late *whi* mutation 42
Whi proteins 42, 43
mutants 42, 43
WhiA 42
WhiB 42
WhiD
Z-rings 43–46
Sulphur-rich coat proteins 80
Superdormant spores 200, 206, 209

T

Tetanus 2–3
Tetanus antitoxin 3
Tetratricopeptide 31
Tgl transglutaminase 130
Therapeutic product 215
Thermal sensors 247
Thermoactinomyces 216
Thin multilayered coat (CT) 57
Time-of-flight mass spectrometer coupled to Direct Analysis in Real Time (AccuTOFDART) 247
Time-resolved micro-potentiometry 253
Timing of commitment to germinate 206–207, 209–210
Toxaemia 108, 171–172
Toxin neutralization 250
Toxins 238, 245, 248, 250, 252
Tracheobronchial lymph glands 173
Transcription activation. *See* Transcription factor Spo 0A
Transcription factor ComA 12
Transcription factor, Spo
0A 9, 10, 11
0B 10
0F 10
Transcription regulation. *See* SpoA-P
Transcriptional control. *See* Sporulation
Transmission electron microscopy (TEM) 218
Tripolypeptide complex 14
Tryptose–sulphite–cycloserine (TSC) 226
Turgon pressure 129
Two-step PCR method 243
Tyndallization 4

U

Undercoat 57–58, 64
Urea/Urease 265, 268, 269
UV photoproduct 80. *See also* Spore photoproduct
UV radiation resistance 80

V

Vaccine 169–171, 179–182
Vaccines for Animals 178
Vaccines for use in humans 179
Variable-number tandem repeat (VNTR) 243
Vegetative
Vegetative cells 20, 26, 29, 239, 242, 246, 248
detection 247
Virgil 1

W

Water 143–152, 154, 157–158
content (in spore core) 144–146
dynamics of (in spore coat) 148–150
mobility of 147
radiolysis 80
Wet-heat resistance 77, 203
Wet-heat killing 77
Woolsorter's disease 171

Y

Yaah *see* SleL
YdhD 135
YkvT protein 134
YpeB 128, 130, 135
YvbX 135

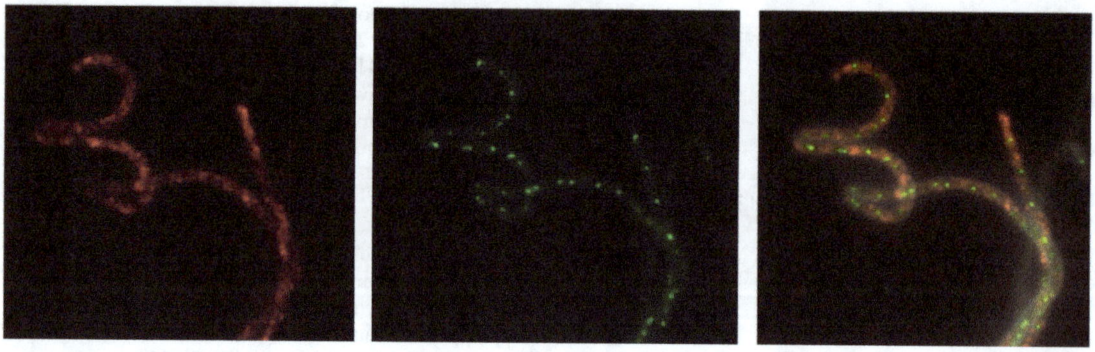

Plate 4.3 (A) Localization of ParA and ParB within the developing aerial hyphae. The ParA helix (left) is visualized using immunofluorescence; ParB (centre) is translationally fused to GFP, and in the merged image (right), the cell wall is stained with wheatgerm agglutin (WGA) Alexa Fluor 350 and false coloured grey (images courtesy of Dr Dagmara Jakimowicz).

www.ingramcontent.com/pod-product-compliance
Lightning Source LLC
LaVergne TN
LVHW062225200926
848484LV00006B/1476

* 9 7 8 1 9 0 8 2 3 0 0 0 3 *